Unsupervised and Semi-Supervised Learning

Series Editor

M. Emre Celebi, Department of Computer Science and Engineering, Conway, AR, USA

Springer's Unsupervised and Semi-Supervised Learning book series covers the latest theoretical and practical developments in unsupervised and semi-supervised learning. Titles – including monographs, contributed works, professional books, and textbooks – tackle various issues surrounding the proliferation of massive amounts of unlabeled data in many application domains and how unsupervised learning algorithms can automatically discover interesting and useful patterns in such data. The books discuss how these algorithms have found numerous applications including pattern recognition, market basket analysis, web mining, social network analysis, information retrieval, recommender systems, market research, intrusion detection, and fraud detection. Books also discuss semi-supervised algorithms, which can make use of both labeled and unlabeled data and can be useful in application domains where unlabeled data is abundant, yet it is possible to obtain a small amount of labeled data.

Topics of interest in include:

- Unsupervised/Semi-Supervised Discretization
- Unsupervised/Semi-Supervised Feature Extraction
- Unsupervised/Semi-Supervised Feature Selection
- Association Rule Learning
- Semi-Supervised Classification
- Semi-Supervised Regression
- Unsupervised/Semi-Supervised Clustering
- Unsupervised/Semi-Supervised Anomaly/Novelty/Outlier Detection
- Evaluation of Unsupervised/Semi-Supervised Learning Algorithms
- Applications of Unsupervised/Semi-Supervised Learning

While the series focuses on unsupervised and semi-supervised learning, outstanding contributions in the field of supervised learning will also be considered. The intended audience includes students, researchers, and practitioners.

** Indexing: The books of this series indexed in zbMATH **

Michal Kawulok • Jolanta Kawulok •
Bogdan Smolka • M. Emre Celebi
Editors

Super-Resolution for Remote Sensing

Springer

Editors
Michal Kawulok
Department of Algorithmics and Software
Silesian University of Technology
Gliwice, Poland

Jolanta Kawulok
Department of Algorithmics and Software
Silesian University of Technology
Gliwice, Poland

Bogdan Smolka
Department of Data Science and Engineering
Silesian University of Technology
Gliwice, Poland

M. Emre Celebi
Department of Computer Science and Engineering
University of Central Arkansas
Conway, AR, USA

ISSN 2522-848X ISSN 2522-8498 (electronic)
Unsupervised and Semi-Supervised Learning
ISBN 978-3-031-68105-9 ISBN 978-3-031-68106-6 (eBook)
https://doi.org/10.1007/978-3-031-68106-6

This Springer imprint is published by the registered company Springer Nature Switzerland AG
The registered company address is: Gewerbestrasse 11, 6330 Cham, Switzerland

Preface

Even though the computer vision field has recently taken great leaps, the capabilities of many systems are limited due to insufficient spatial resolution of the available input images. This is particularly relevant for remote sensing (including satellite images) when it is often very costly or even impossible to acquire data of sufficient resolution. This obstacle has motivated many researchers to explore the topic of super-resolution (SR)—a common term for a variety of techniques whose goal is to reconstruct a high-resolution (HR) image from a low-resolution (LR) observation. The emerging techniques underpinned with deep learning allow for obtaining very promising results in terms of both generating visually plausible outcomes and concerning the capabilities of reconstructing the actual HR information. New research directions include in particular real-world SR aimed at enhancing the original (rather than simulated) images, as well as new evaluation procedures that would allow for assessing the quality of super-resolved images and understanding their value in different contexts.

The goal of this volume is to provide a summary of the recent advances in SR aimed at enhancing remotely sensed images. The volume opens with the chapter entitled "Introduction to Super-Resolution for Remotely Sensed Hyperspectral Images," in which Kowaleczko et al. introduce the landscape of SR techniques applied to remotely sensed hyperspectral images. The discussed categories include single-image and multi-image techniques employed in a band-wise manner, as well as the solutions that exploit panchromatic or multispectral images that serve as a source of HR information during reconstruction. Each of these categories is reviewed in detail, and the analysis is complemented with presenting the loss functions exploited to train the SR deep networks and the evaluation strategies.

In the chapter "Real-World Unsupervised Remote Sensing Image Super-Resolution: Addressing Challenges, Solution and Future Prospects," Mishra and Hadar tackle the problem of real-world SR performed from single images. It provides an overview of the existing techniques considering different approaches to train SR algorithms, including self-supervised, few-shot, and zero-shot learning. In addition to that, the chapter contains the results of an extensive experimental study that compares a number of SR techniques applied to enhance real-world satellite and aerial images.

In the chapter entitled "Advancements in Deep Learning-Based Super-Resolution for Remote Sensing: A Comprehensive Review and Future Directions," Tabesh et al. present an extensive survey of generic SR techniques, including single-image and multi-image approaches with particular attention given to the learning-based methods. The authors clearly demonstrate how the existing generic techniques have been applied in the field of remote sensing and they conclude with identifying several research directions that are critical for advancing the state of the art in SR for remotely sensed images.

Tarasiewicz and Kawulok in their chapter entitled "Multi-Image Super-Resolution Using Graph Neural Networks" present the variety of SR techniques employed for enhancing remotely sensed images, with particular attention given to single-image and multi-image approaches. The authors demonstrate the benefits of representing multiple images of the same scene with a graph that is subsequently processed using a graph neural network to obtain the super-resolved image. Also, the chapter includes the outcome of a thorough experimental study performed over the simulated data and a real-world PROBA-V benchmark dataset.

In the chapter "Effectiveness Analysis of Example-Based Machine Learning and Deep Learning Methods for Super-Resolution Hyperspectral Images," Polat and Simsek compare several approaches to super-resolving hyperspectral images, including multi-image and interpolation-based methods, the example-based approaches that exploit sparse representation and dictionary learning, as well as those underpinned with deep learning. Also, this study presents these techniques from the perspective of the processing time which is important for real-world applications.

The chapter "Synergy of Images: Multi-Image Fusion Empowering Super-Resolution in Remote Sensing" by Lu et al. presents the field of multi-image fusion based on pansharpening that allows for enhancing multispectral and hyperspectral images. The authors outline the taxonomy of existing approaches and they provide a thorough review of techniques in each category, considering the recent achievements. Furthermore, the chapter describes the commonly used metrics that serve for evaluating the SR outcome and it presents the results of an extensive experimental study.

The chapter "Unsupervised Pansharpening Using ConvNets" by Ciotola and Scarpa revolves around the problem of unsupervised training for pansharpening multi-spectral remotely sensed images. The authors introduce a new unsupervised training framework and they evaluate the spectral and spatial fidelity of the pansharpened images. The chapter tackles an important problem of existing supervised frameworks requiring HR ground-truth images, which are very often hard to obtain, especially for remote sensing scenarios.

Pereira-Sánchez et al. in the chapter entitled "A Comprehensive Overview of Satellite Image Fusion: From Classical Model-Based to Cutting-Edge Deep Learning Approaches" focus on satellite image fusion, showing the broad landscape of the existing techniques, ranging from classic model-based approaches to the data-driven methods underpinned with deep learning. For the latter, the chapter contains an overview of training and validation phases, including the loss functions exploited during training. Also, the chapter includes the discussion on how to assess the quality of fused products and it reviews the existing satellite image datasets.

The chapter entitled "Super-Resolution for Spectral Image" by Li and Wang concludes the volume. The authors focus on the problem of preserving spectral consistency when enhancing spatial resolution of images with multiple spectral channels. They present and discuss a variety of convolutional neural network architectures that have been employed for this purpose. Also, the chapter introduces two new techniques and it includes an experimental study that compares the SR methods with the particular attention given to the problem of preserving spectral information.

We hope that this volume has succeeded in demonstrating the recent progress in developing new SR techniques and applying them to remotely sensed data and that the discussed challenges will motivate and help the researchers to further advance this important field.

Gliwice, Poland
June 2024

Michal Kawulok
Jolanta Kawulok
Bogdan Smolka
M. Emre Celebi

Acknowledgements

The work on accomplishing this volume was partially funded by the National Science Centre, Poland, under Research Grant 2019/35/B/ST6/03006.

Dr. Celebi's work was supported by the National Science Foundation (Award No. OIA-1946391). Any opinions, findings, and conclusions or recommendations expressed in this material are those of the authors and do not necessarily reflect the views of the National Science Foundation.

Contents

Contributors

Masoud Babadi Ataabadi Department of Geography and Planning, Queen's University, Kingston, ON, Canada

Guangsheng Chen College of Computer and Control Engineering, Northeast Forestry University, Harbin, China

Dongmei Chen Department of Geography and Planning, Queen's University, Kingston, ON, Canada

Matteo Ciotola Department of Electrical Engineering and Information Technology, University of Naples Federico II, Naples, Italy

Joan Duran Department of Mathematics and Computer Science and Institute of Applied Computing and Community Code (IAC3), Universitat de les Illes Balears (UIB), Palma (Islas Baleares), Spain

Ofer Hadar School of Electrical and Computer Engineering, Ben Gurion University of the Negev, Beersheva, Israel

Lirong Han Department of Technology of Computers and Communications, University of Extremadura, Cáceres, Spain

Juan M. Haut Department of Technology of Computers and Communications, University of Extremadura, Cáceres, Spain

Saba Hosseini Tabesh Department of Geography and Planning, Queen's University, Kingston, ON, Canada

Weipeng Jing College of Computer and Control Engineering, Northeast Forestry University, Harbin, China

Michal Kawulok Department of Algorithmics and Software, Silesian University of Technology, Gliwice, Poland
KP Labs, Gliwice, Poland

Daniel Kostrzewa Department of Applied Informatics, Silesian University of Technology, Gliwice, Poland
KP Labs, Gliwice, Poland

Pawel Kowaleczko Institute of Computer Science, Warsaw University of Technology, Warsaw, Poland
KP Labs, Gliwice, Poland

Qiang Li School of Artificial Intelligence, Optics and Electronics (iOPEN), Northwestern Polytechnical University, Xi'an, Shaanxi, China

Hailiang Lu College of Computer and Control Engineering, Northeast Forestry University, Harbin, China

Divya Mishra School of Electrical and Computer Engineering, Ben Gurion University of the Negev, Beersheva, Israel

Julia Navarro Department of Mathematics and Computer Science and Institute of Applied Computing and Community Code (IAC3), Universitat de les Illes Balears (UIB), Palma (Islas Baleares), Spain

Mercedes E. Paoletti Department of Technology of Computers and Communications, University of Extremadura, Cáceres, Spain

Ivan Pereira-Sánchez Department of Mathematics and Computer Science and Institute of Applied Computing and Community Code (IAC3), Universitat de les Illes Balears (UIB), Palma (Islas Baleares), Spain

Ediz Polat School of Electrical Engineering and Computer Science, University of North Dakota, Grand Forks, ND, USA

Eloi Sans Department of Mathematics and Computer Science and Institute of Applied Computing and Community Code (IAC3), Universitat de les Illes Balears (UIB), Palma (Islas Baleares), Spain

Giuseppe Scarpa Department of Engineering, University of Naples Parthenope, Naples, Italy

Murat Simsek Artificial Intelligence Engineering Department, Ostim Technical University, Ankara, Turkiye

Tomasz Tarasiewicz Department of Algorithmics and Software, Silesian University of Technology, Gliwice, Poland

Qi Wang School of Artificial Intelligence, Optics and Electronics (iOPEN), Northwestern Polytechnical University, Xi'an, Shaanxi, China

Maciej Ziaja KP Labs, Gliwice, Poland
Department of Algorithmics and Software, Silesian University of Technology, Gliwice, Poland

Introduction to Super-Resolution for Remotely Sensed Hyperspectral Images

Pawel Kowaleczko, Maciej Ziaja, Daniel Kostrzewa, and Michal Kawulok

1 Introduction

Super-resolution (SR) algorithms aim to enhance the spatial resolution of digital images, although in some cases, SR is tailored to enhance spectral resolution as well [34]. This objective addresses a significant challenge in the field of computer vision (CV), attracting considerable attention from researchers. In the early stages of SR development, classic approaches for single-image super-resolution (SISR) relied on nonuniform interpolations, frequency domain analysis, deterministic and stochastic regularization, and projection onto convex sets, among others [7, 10, 71, 89]. For tasks such as pansharpening and hyperspectral/multispectral image fusion, classic methods included component substitution (CS), multi-resolution

P. Kowaleczko
Institute of Computer Science, Warsaw University of Technology, Warsaw, Poland

KP Labs, Gliwice, Poland
e-mail: pawel.kowaleczko.dokt@pw.edu.pl

M. Ziaja
KP Labs, Gliwice, Poland

Department of Algorithmics and Software, Silesian University of Technology, Gliwice, Poland
e-mail: mziaja@kplabs.pl

D. Kostrzewa
Department of Applied Informatics, Silesian University of Technology, Gliwice, Poland

KP Labs, Gliwice, Poland
e-mail: daniel.kostrzewa@polsl.pl

M. Kawulok (✉)
Department of Algorithmics and Software, Silesian University of Technology, Gliwice, Poland

KP Labs, Gliwice, Poland
e-mail: michal.kawulok@polsl.pl

M. Kawulok et al. (eds.), *Super-Resolution for Remote Sensing*, Unsupervised and Semi-Supervised Learning, https://doi.org/10.1007/978-3-031-68106-6_1

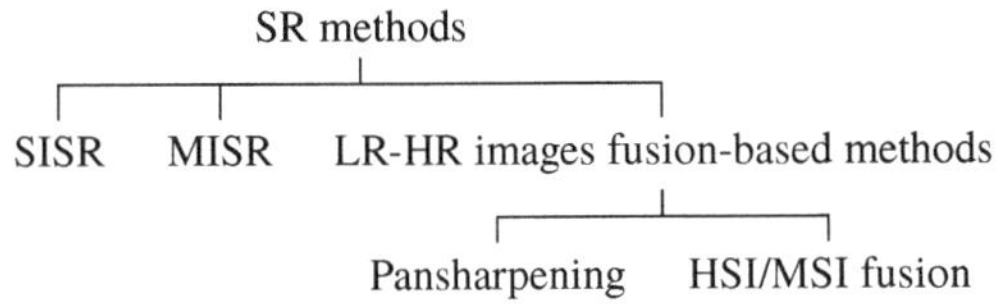

Fig. 1 A general taxonomy of SR methods for hyperspectral imagery: SISR [2, 12, 13, 63, 95, 108], MISR [68, 111], pansharpening [15, 95], and hyperspectral/multispectral image fusion (HSI/MSI fusion) [95, 109]

analysis (MRA), variational optimization (VO), spectral unmixing, and Bayesian models [15, 61, 109].

With the rapid advancement of computational hardware and substantial increases in computational efficiency observed over the past decade, numerous new solutions have emerged, primarily leveraging deep neural networks (DNNs), which have demonstrated remarkable performance gains over previous techniques [54]. These DNN-based solutions either draw inspiration from classic models or introduce entirely novel concepts.

In contrast to methods designed for natural grayscale or color images, those tailored for hyperspectral data often demand specialized approaches. This necessity stems from the unique characteristics of hyperspectral data, which may involve an increased number of spectral bands or the availability of multiple images representing the same objects at varying spatial and spectral resolutions. Broadly, three main SR approaches can be distinguished based on the nature of the available data: SISR, multi-image super-resolution (MISR), and low-resolution (LR) and high-resolution (HR) image fusion-based methods, as illustrated in Fig. 1.

The primary objective of this chapter is to introduce the reader to DNN-based SR techniques and evaluation strategies, focusing specifically on their application to hyperspectral satellite imagery. Firstly, the concept of SR based on DNNs is introduced, highlighting the most important achievements in this field. Then, the chapter presents the most common loss functions used for SR DNNs, along with SR hyperspectral imagery datasets and evaluation strategies. In the latter part of the chapter, each group of methods presented in the taxonomy in Fig. 1 has its dedicated section, where the main contributions and ideas behind those methods are briefly described. Finally, the chapter is summarized, and the authors' ideas for future research are outlined.

2 Super-Resolution Reconstruction

SR poses an inherently ill-posed problem, wherein applying various degradation processes $\mathbb{F}$ can yield many different LR images I_{LR} from a single HR image and vice versa. Let us denote $f \in \mathbb{F}$ as a specific degradation process, where $\mathbb{F}$ represents the set of all degradation functions. In real-world scenarios, when the

HR image is unavailable, the closest approximation of HR is interpreted as the light rays falling on the camera matrix, and f becomes dependent on factors such as the camera matrix native resolution, electronic noise, lighting conditions, and more. Additionally, LR images can be simulated from the original images, and the HR images serve as a reference. In this case, f is characterized by parameters such as the downsampling factor, blur kernel, random noise distribution, etc. [97, 114, 116].

Hence, the relationship between I_{HR} and I_{LR} is represented as:

$$I_{LR} = f(I_{HR}). \tag{1}$$

The SR system is designed to model the inverse function f^{-1} to the degradation function f, or rather, to a wide group of degradation functions $\mathbb{F}$:

$$I_{HR} = f^{-1}(I_{LR}). \tag{2}$$

Over the last decade, most state-of-the-art SR systems have been based on DNNs [21, 101].

2.1 *Super-Resolution Approaches*

Usage of convolutional neural networks (CNNs) for SR dates back to 2015, when Dong et al. [21] proposed SRCNN—an architecture composed of only three convolutional layers where each was designated for one specific task, namely, feature patch extraction, nonlinear mapping, and reconstruction. SRCNN, like many other SR models cited later in this chapter, requires a pre-upsampling step—the LR image has to be upsampled to the target resolution before being fed to the SR network. This is achieved by applying one of the interpolation methods, in the case of SRCNN, the bicubic interpolation. Therefore, the network is not responsible for the enhancement of the resolution of the LR image itself, but rather for the enhancement of the upsampled image quality (removing blur, noise, etc.).

In 2016, Kim et al. introduced another significant work in the SR field [47]. In their study, the authors utilized the VGG architecture [81] for the SR task, resulting in a network dubbed VDSR. Characterized by a higher number of parameters, VDSR demonstrated better performance compared to SRCNN. Additionally, the authors proposed a novel approach to image reconstruction. Unlike SRCNN, which generates individual pixel values, VDSR predicts the residual image. The reconstructed HR image is obtained by adding the residual image to the bicubically upsampled LR image.

Another common solution for SR is the post-upsampling method, which reduces the number of training parameters in comparison with previous methods. This method involves the inclusion of upsampling layers within the SR network itself, typically located at the end of the architecture to limit the number of training parameters even further. In the work of Dong et al. [22], deconvolution/transposed convolution layers were employed for upsampling, while Shi et al. [79] used sub-pixel convolution layers for the same purpose.

ResNet, released in 2015 [35], stood as another groundbreaking DNN architecture, widely employed across various applications, including SR [53]. In efforts to enhance ResNet's performance for SR tasks, numerous adaptations have been proposed. For instance, multistage processing [28], cascading mechanism [1], and the removal of batch normalization layers [58] have been suggested.

Recursive networks [48, 82], built on the concept of parameter sharing among convolutional layers, have emerged as a solution to reduce both the number of trainable parameters and computational complexity.

Laplacian pyramids have proven to be a useful tool, particularly in SR tasks requiring high upscaling factors. Building upon this concept, Lai et al. proposed a gradual upscaling of the reconstructed image in their work [52]. Different scales of LR images are also used in [39], although, in this case, they are processed in separate, parallel network branches.

Another breakthrough in the deep learning domain was proposed in 2017 attention mechanism [90]. At the core of this mechanism lies the idea of applying trainable weights to the data propagated through the DNN, allowing different parts of the training data to have varying impacts on the output. The transformer architecture, with attention as its backbone, has become the foundation for many solutions achieving state-of-the-art results in various natural language processing (NLP), CV, and digital signal processing (DSP) tasks. One of the most recognizable and impressive applications of the transformer is OpenAI's chatbot—generative pretrained transformer (GPT). The attention mechanism was also successfully adapted for SR tasks [118].

In recent years, a significant increase in the performance of generative models, such as generative adversarial networks (GANs) [32, 44, 120] or diffusion models [75], has been observed. This has garnered attention not only from researchers but also, perhaps primarily, from the general audience. The main objective of these networks is to generate images of high perceptual quality rather than focusing solely on pixel-wise similarity to reference images. Notably, the SR task also necessitates the generation of information, making GANs a viable choice [53, 76, 98]. However, due to their specific design based on the zero-sum two-player game, GANs are prone to a phenomenon called hallucination, where some parts of the generated images are plausible but not necessarily close to the ground truth. This can be disadvantageous if high accuracy is required. Additionally, GANs are well known for their training instability, as they are sensitive to issues such as vanishing gradients or mode collapse.

2.2 Super-Resolution Loss Functions

The most commonly applied loss functions for training SR networks are L1 or L2 metrics, computed as:

$$L1 = \frac{1}{n}\sum_{i=1}^{n}|\hat{y}_i - y_i|, \tag{3}$$

and

$$L2 = \frac{1}{n}\sum_{i=1}^{n}(\hat{y}_i - y_i)^2, \tag{4}$$

where n is the total number of pixels, and $\hat{y}_i$ and y_i are the reconstructed and reference pixels values, respectively. These losses are based on a pixel-wise difference between the generated and target images and are desired if the generated pixel values must be as close to ground-truth values as possible. If the high perceptual quality of reconstructed images is required, more sophisticated losses, such as the adversarial [32], perceptual [42], and cycle consistency [120], should be used. In most current SR setups a combination of those losses is utilized to obtain the highest reconstruction quality.

2.2.1 Per-Pixel Loss Functions

Per-pixel loss functions, such as L1 and L2, are designed to minimize the differences between corresponding pixels of reconstructed SR and target HR images (Fig. 2). L2 is directly linked with peak signal-to-noise ratio (PSNR), which is one of the most common metrics used to assess the quality of SR reconstruction. L1 minimizes the absolute difference between the images, whereas L2 minimizes the squares of those differences. L2 is more commonly used, but it is prone to data with a high number of outlier values. These outliers can result in a few large error values, which may significantly contribute to the total loss value and prevent the network from finding a global minimum. In such cases, the L1 loss is often considered a better option.

Fig. 2 The SR network training setup with L1 or L2 losses. The per-pixel losses are based on differences between the values of the reconstructed SR image and the reference HR target image

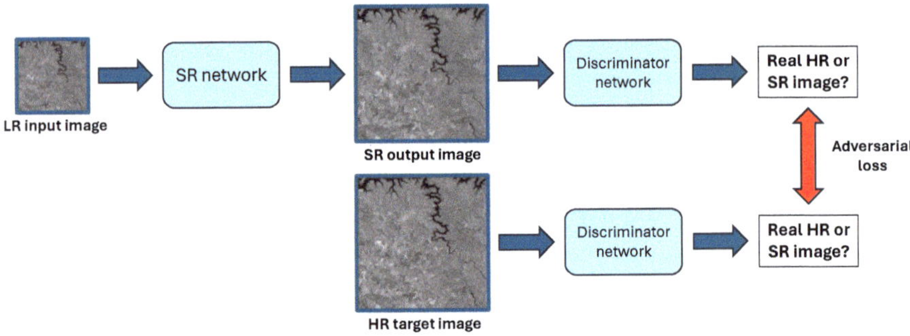

Fig. 3 The SR network training setup with adversarial loss. In this setup, SR network is treated as GAN generator network. This adversarial training process encourages the SR network to produce high-quality SR/GAN images from input LR images that closely resemble the target HR images

2.2.2 Adversarial Loss

The concept of adversarial loss originates from the original GAN paper by Goodfellow et al. [32]. The GAN setup comprises two major components: the generator network G and the discriminator network D. These two networks engage in a two-player zero-sum minmax game, where they are optimized to opposing objectives. The role of the generator is to produce samples that mimic the distribution of the training data, while the discriminator is trained to distinguish between generated samples and real ones. The opposing targets improve the performance of both networks leading to better quality of the generated samples (Fig. 3).

Let us denote z as the input to the generator (e.g., random noise or LR image), x as a sample from the training dataset (e.g., HR image), $G(z)$ as a sample generated by the generator (e.g., SR image), and $D(x)$ and $D(G(z))$ as the outputs of the discriminator for real and generated samples, respectively. These outputs may be interpreted as the probabilities of the samples being real or generated. Then, the task of the discriminator is to maximize the following term:

$$\max \ (\log D(x) + \log(1 - D(G(z)))) , \tag{5}$$

whereas the generator minimizes the opposite:

$$\min \ \log(1 - D(G(z))). \tag{6}$$

2.2.3 Perceptual Loss

Perceptual loss, akin to adversarial loss, prioritizes enhancing the overall perceptual quality of reconstructed images over minimizing differences between individual pixels. The concept revolves around leveraging high-level features extracted by a

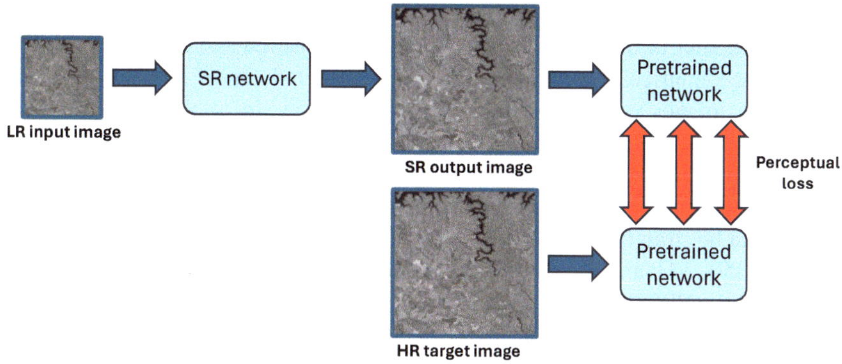

Fig. 4 The SR network training setup with perceptual loss. Both reconstructed SR and target HR images are fed to the pretrained network. The outputs of the intermediate layers of this network for the SR and HR images are compared with each other using L1 loss

pretrained DNN. In this approach, the compared images are fed into the network, and the outputs of intermediate layers are compared using L1 or L2 loss functions (Fig. 4). Typically, the network is based on the VGG architecture [81], pretrained for the classification task on ImageNet [16] dataset. This allows the perceptual loss to capture semantic similarities between images, leading to improved perceptual quality in the reconstructed images.

3 Evaluation Strategies and Hyperspectral Imagery SR Datasets

In this section, the most prominent and popular hyperspectral imagery datasets are presented along with the evaluation strategies and challenges associated with them.

3.1 Hyperspectral Imagery SR Datasets

Hyperspectral imagery datasets are relatively scarce due to the requirement of technologically advanced sensors, especially when compared to the availability of natural images. Moreover, real-world hyperspectral datasets suitable for straightforward use in SR tasks are even rarer, as both the LR and HR images should preferably be captured by the same device.

One such dataset known to the authors of this chapter is the PROBA-V dataset [24, 64], captured by the satellite of the same name, which is owned by the European Space Agency (ESA) and was launched in 2013. The PROBA-V SR dataset was published in 2018 by ESA as part of the PROBA-V SR Challenge [25], with the main purpose of developing new MISR methods for satellite imagery (these are outlined later in the chapter). This dataset contains red and near-infrared (NIR) bands of geometrically and radiometrically corrected top-of-atmosphere (TOA) reflectance images captured within a 30-day period. In total, 74 regions were selected, with 1160 training and 290 testing scenes. Each scene contains at least 9 LR images (19 on average) at a 300-m ground sampling distance (GSD) and a single HR image at a 100-m GSD. Although the dataset was designed for MISR tasks, it can be straightforwardly used for SISR.

MuS2 [50, 51] and WorldStrat [14] are other real-world datasets; however, contrary to the PROBA-V, LR and HR images were captured by different satellites. For both datasets, the LR (with GSD ranging from 60 m to 10 m) images were captured by Sentinel-2 satellites. The MuS2 dataset contains 14–15 LR images for each scene, captured at different revisits, whereas the WorldStrat contains 16 LR images per scene. For MuS2, the HR images come from the WorldView-2 satellite [27]. They include 8 spectral bands along with a single panchromatic image, all at 3.3-m GSD. On the other hand, the WorldStrat HR images come from the Airbus SPOT 6/7 dataset, which contains 5 spectral bands at 6-m GSD and one panchromatic image at 1.5-m GSD. Both MuS2 and WorldStrat datasets can be used for training and benchmarking newly developed MISR and SISR networks.

As mentioned earlier, real-world hyperspectral imagery SR datasets are rare, leading to a more popular approach of simulating LR data from the original images and treating the original images as HR references. This approach is outlined in Eq. 1, where f represents the degradation function. For spatial degradation, this function typically involves downsampling, blurring, and additive noise [92]. If spectral degradation is necessary, spectral response functions are utilized. These operations can be applied to any hyperspectral or multispectral datasets, with the most popular ones including Sentinel-2 [26, 31], WorldView-1,2,3 [27], ROSIS Pavia Center and University [33], Indian Pines [33], Botswana [33], Chikusei [110], CAVE (non-satellite) [107], and Harvard (non-satellite) [11].

3.2 *Evaluation Strategies*

The evaluation of SR models can be conducted in a twofold manner: quantitatively and qualitatively. Qualitative assessment typically involves a mean opinion score (MOS) survey, where participants provide subjective feedback on the reconstruction quality and its similarity to the HR image if it is available. In [51], it was demonstrated that this subjective evaluation aligns well with quality assessment using the learned perceptual image patch similarity (LPIPS) evaluation metric [115].

Quantitative assessment of the reconstruction is based on both popular image similarity metrics and those specifically designed for hyperspectral imagery SR. The most frequently reported metrics include PSNR [37], structural similarity index (SSIM) [37, 100], spectral angle mapper (SAM) [112], erreur relative globale adimensionnelle de synthèse (ERGAS) index [73, 91], and LPIPS [115].

PSNR is perhaps the most common metric used to compare two signals/images. It is a per-pixel metric, relying solely on differences between values of corresponding pixels, making it susceptible to spatial shifts and overall differences in brightness. Conversely, SSIM is based on differences between windowed statistical features of images, which resolves many issues present with PSNR.

SAM, as the name suggests, evaluates the preservation of spectral properties of images. Each hyperspectral pixel is treated as a vector, and SAM is the mean arccos between the corresponding pixels.

ERGAS index, another per-pixel metric based on root mean square error (RMSE), was specifically designed for SR tasks as it accounts for the upsampling factor.

The idea behind LPIPS is to compare deep features extracted from images by pretrained DNNs. Although the perceived quality of reconstruction aligns with the LPIPS score, it is susceptible to the effect of hallucination [8], which is undesired for remote sensing applications.

Other metrics often found in works on SR include cross-correlation (CC), image quality index (Q) [99], quality index with no reference (QNR), spatial distortion index (D_s), and spectral distortion index (D_λ) [3]. The last three are used for unsupervised learning, where a reference image is not required.

The purpose of SR is to enhance spatial resolution while preserving the spectral properties of images, which are crucial for many remote sensing applications such as precision farming, environmental monitoring, and air pollution estimation. The higher spatial resolution should ideally lead to improved performance in these real-world applications of hyperspectral images. Therefore, reconstruction quality can also be indirectly assessed using evaluation metrics designed for these real-world use cases [45, 95, 121].

For simulated datasets, the evaluation process is relatively straightforward—the LR images are explicitly derived from the HR images, and the SR algorithm should ideally reverse a specific degradation process. Thus, any differences between the reconstructed SR and reference HR images should primarily stem from imperfections in the SR method.

However, for real-life datasets, the situation is more complex. Discrepancies between SR and HR images can arise not only from limitations in the SR method but also from differences in LR and HR sensor characteristics (if they are different), various acquisition times and conditions, etc. This challenge was addressed in the paper that introduced the MuS2 dataset [51], where different masks were proposed to mitigate the influence of these undesired factors on the final SR outcome score.

4 Single-Image Super-Resolution for Satellite Imagery

SISR [2, 12, 13, 63, 95, 108] aims to enhance the spatial resolution of images, whether grayscale, color (RGB), or hyperspectral, using only a single LR image as input. Figure 5 illustrates a typical SISR processing pipeline. The SR term is often used interchangeably with SISR, as the majority of SR DNNs are SISR networks. While most are initially designed for grayscale or natural color images, they can generally be adapted for hyperspectral imagery, albeit with suboptimal performance compared to models specifically tailored for hyperspectral data.

In the work by Li et al. [56], a novel setup is proposed where two SR networks are trained simultaneously—one for RGB images and one for hyperspectral images. The RGB SR network serves as an auxiliary task, regularizing the training of the main HSI SR network. Both models share the same encoders but have separate decoders. The use of the RGB SR network is motivated by the abundance of grayscale or RGB training data compared to hyperspectral data, which may introduce additional information. Moreover, the authors introduce spectral mix-up, an augmentation technique and semi-supervised training aimed at enhancing SR network performance.

The solution proposed by Tang et al. [83] involves adding the network's output (the residual image) to the bicubically upsampled LR image. However, the residual image is obtained recursively. The input LR HSI is divided into groups of images containing adjacent spectral bands, and each image is recursively processed to obtain the final residual.

Hu et al. [38] propose the intrafusion network, an architecture that separately addresses spatial and spectral information. The network takes as an input the LR spectral band along with the consecutive band, generates a spectral difference map from these bands, and feeds it to the triple-branch network. The reconstructed image is returned by the intrafusion module, which merges the output of the triple-branch network.

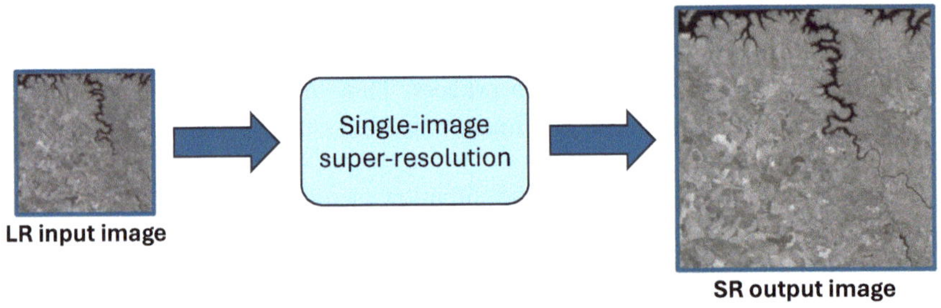

Fig. 5 SISR pipeline. The input to the model is a single image (can be panchromatic, multispectral, or hyperspectral) and the output is the SR image with the same number of spectral bands and increased spatial resolution

Deep image prior (DIP) [86] is a technique based on the assumption that the structure of a CNN is sufficient for certain information generation tasks, such as SR, without the need for large training datasets. Sidorov et al. [80] utilize DIP for HSI SR, where the network's objective is to produce corrupted HR HSI using only random noise as input. Interrupting the training at a specific moment reportedly yields reconstructed images of quality close to the non-disturbed image. In the case of SR DIP, the corrupted image is the LR image upsampled using one of the interpolation methods.

SR networks are tasked not only with increasing the spatial resolution of images but also with retaining their spectral properties. While this is relatively straightforward for DNNs based on 2D convolutions when dealing with grayscale or RGB images, it becomes more complex for hyperspectral imagery. To address this challenge, 3D convolutions are often employed, allowing the network to process multiple spectral bands simultaneously and thus learn correlations between them. However, a major drawback of 3D convolution is the typically large number of parameters it entails.

In their work, Li et al. [57] utilize both 2D and 3D convolutions to handle hyperspectral imagery. 3D convolutions extract both spatial and spectral features of the hyperspectral image, while 2D convolutions help balance the parameter count. To enhance computational efficiency, the authors employ separable convolutions. This involves replacing a single convolution with a kernel of size $c \times h \times w$ (spanning spectral c and two spatial h, w dimensions) with two convolutions using kernels $c \times 1 \times 1$ and $1 \times h \times w$, reducing the parameter count from chw to $c + hw$.

Bi-3DQRNN [29] adopts a U-Net-like architecture comprising units called Bi-3DQRU, where the LR image is processed along the spectral dimension in both directions using 3D convolutions, facilitating learning correlations between bands.

In the approach presented by Wang et al. [93] bands are processed sequentially. Two inputs are fed into separate branches: a single LR band is directed to a 2D convolutional feature extractor, while adjacent bands are fed into the 3D separable convolutions branch. After processing the first branch, additional input is introduced in the form of features extracted from previous bands. These extracted features are then combined to produce the final SR output.

Attention mechanism, as introduced by Vaswani et al. [90], holds promise for HSI SISR, as the relevance of information contained in spectral bands may vary across the image. Weighting the influence of spectral bands on the SR outcome via attention mechanism may enhance performance.

In the work by Li et al. [55], a combination of GANs and band attention is proposed. Liu et al. [59] group highly correlated spectral bands and process them with the same kernel to improve computational efficiency. The covariance statistics of features are utilized by the attention mechanism to extract spectral priors. Similarly, Jiang et al. [40] employ group convolution along with band attention. Neighboring bands are grouped and fed into separate 2D convolution branches. The outputs of those branches are then upsampled and combined to yield the reconstructed image.

Wang et al. [96] also leverage spectral band attention in their method called super-resolution via recurrent feedback embedding and spatial–spectral consistency regularization (RFSR). Spectral bands are grouped and treated as a sequence, recursively fed along with outputs from previous blocks to the residual attention blocks. The reconstructed image is obtained from a regularization block containing separable 3D convolutions placed at the end of the RFSR architecture.

These approaches demonstrate the potential of attention mechanism to improve HSI SISR performance by selectively attending to relevant spectral information during the SR process.

5 LR-HR Image Fusion-Based Methods

The previous section introduced SISR methods, which are focused on enhancing spatial resolution while preserving spectral properties. However, the capabilities of sensors for capturing hyperspectral images are constrained. The necessity for a high number of narrow spectral bands makes them susceptible to noise interference. In particular, the narrow spectral band results in a weaker signal compared to wider bands, while noise remains at the same level. Consequently, to mitigate this influence, the physical size of the photosensitive elements in sensor matrices is increased, leading to a reduction in the spatial resolution of captured images.

Notably, many satellites can capture multiple images simultaneously, each with varying spatial and spectral resolutions, thereby presenting new opportunities for SR, specifically through fusion-based methods. These methods aim to enhance both spectral and spatial resolutions of images. Fusion networks typically require at least two inputs: a low spatial but high spectral resolution image LR HSI and a high spatial but low spectral resolution image HR MSI. These input images are processed by the fusion network to produce an output image with high spatial and spectral resolution HR HSI. One particular instance of HSI/MSI fusion is pansharpening, where the multispectral image is substituted with the panchromatic image.

5.1 Multispectral and Hyperspectral Image Fusion

HSI/MSI fusion [95, 109] can be approached from two perspectives: increasing the spatial resolution of HSI or enhancing the number and transferring the style of HSI spectral bands to the MSI. These perspectives can be formulated using observation models [104, 105]:

$$Y = XR, \tag{7}$$

$$Z = CX, \tag{8}$$

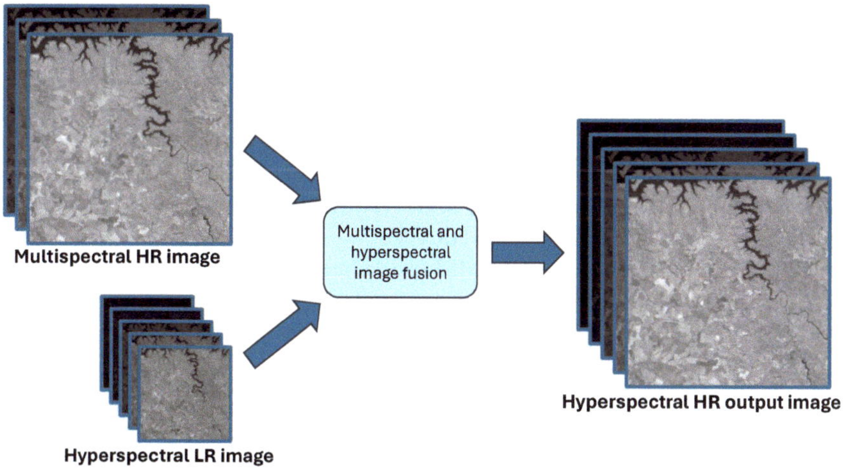

Fig. 6 HSI/MSI image fusion pipeline. There are two inputs to the model—HR MSI and LR HSI. The model's product is the image, which has the same number of spectral bands as the input LR hyperspectral image and the same spatial resolution as the input HR multispectral image

where X, Y, and Z represent the HR HSI, HR MSI, and LR HSI, respectively. R and C denote the spectral response of the sensor responsible for spectrally downsampling the HR HSI and a linear operator representing the spatial downsampling operation and degradation. Therefore, the objective of HSI/MSI fusion is to estimate X by evaluating R and C (Fig. 6).

In the section discussing SISR methods, many approaches utilize 3D convolutions due to their capability to capture correlations between spectral bands. This is no different in the case of HSI/MSI fusion. Palsson et al. [70] employ 3D convolutions, where they first reduce the number of spectral bands of the HSI using principal component analysis (PCA) to increase computational efficiency. This reduced spectrally HSI is then fused with downsampled spatially MSI. Similarly, Ma et al. [62] leverage 3D convolutions to exploit spatio-spectral correlations, combined with attention mechanism (in the form of transformers) to capture spatial priors, within deep unfolded neural networks [23, 105] for HSI reconstruction.

SSR-NET [117] is an interpretable network comprising three consecutive blocks. In the initial block, termed cross-domain message inserting, LR HSI and HR MSI are preliminarily fused using bilinear upsampling and convolutional layers. The resulting preliminary HR HSI is then processed by the spatial reconstruction network and spectral reconstruction network to obtain reconstructed HR HSI.

Deep unfolded neural networks are also employed to solve the problem posed by the transformed observation models (Eqs. 7 and 8) [104, 105]. A similar methodology was proposed in [23], where both observation models and deep unfolded neural networks were utilized.

Dian et al. [20] propose the CNN-Fus network, which leverages singular value decomposition (SVD) to learn the spectral subspace of the LR HSI and eliminate redundant information. Subsequently, a pretrained CNN denoiser is applied to the alternating direction method of multipliers (ADMM) algorithm to estimate coefficients, which, along with the spectral subspace, are used for HR HSI reconstruction.

In [19], HSI/MSI fusion is divided into three consecutive steps. Firstly, the HR HSI is initialized by solving the proposed Sylvester equation, which includes the LR HSI and HR MSI, along with interpretable elements. Then, the residual image is learned by a DNN and added to the initialized HR HSI. Finally, another Sylvester equation is solved to obtain the final HR HSI.

Wang et al. [94] decompose HSI/MSI fusion into learning observation and fusion models, akin to the GAN setup. The fusion model learns the mapping from LR HSI and HR MSI to HR HSI, while the observation model learns the inverse mapping from estimated HR HSI to LR HSI and HR MSI.

Most SR methods are supervised, meaning that the target image is available during training. However, real-world hyperspectral datasets for SR, where LR HSI, HR MSI, and HR HSI images are all available, are rather scarce. As a result, the SR HSI/MSI fusion datasets are often simulated. To simulate the HR MSI, the neighboring bands of the HR HSI are typically aggregated using a sensor-specific spectral response function or by simply averaging the bands. LR HSI images are most often obtained by applying downsampling and degradation operations following Wald's protocol [92]. During training, the original HR HSI is treated as the target image.

This approach raises some issues, primarily concerning the generalization abilities of the trained SR network. The network learns to reverse very specific degradation algorithms with a low number of assumptions and parameters. However, in real life, the relationship between LR HSI, HR MSI, and HR HSI is much more complicated. Some works tackle SR with unsupervised or self-supervised approaches, where the target HR HSI is not required. For instance, in [30] Gao et al. initialize HR HSI using the adaptive Gram–Schmidt transformation (GSA). This preliminary HR HSI is then fed to a CNN, and the output of this CNN is compared, using spectral and spatial loss functions proposed by the authors, with the upsampled LR HSI and preliminary HR HSI, respectively. In such a setup, training data with targets is not mandatory.

5.2 Pansharpening

Pansharpening [15, 95], depicted in Fig. 7, is a special case of MSI and HSI fusion where the HR panchromatic single-channel image (PAN) replaces the MSI. Many methods used for MSI and HSI fusion, after slight adjustments, can also be applied to pansharpening. Pansharpening methods can generally be divided into four groups: three conventional and one based on deep learning. The conventional methods include CS, MRA, and VO [15, 65], as illustrated in Fig. 8. However, in this chapter, only the DNN-based methods are discussed, as their performance is typically much better than that of the conventional methods.

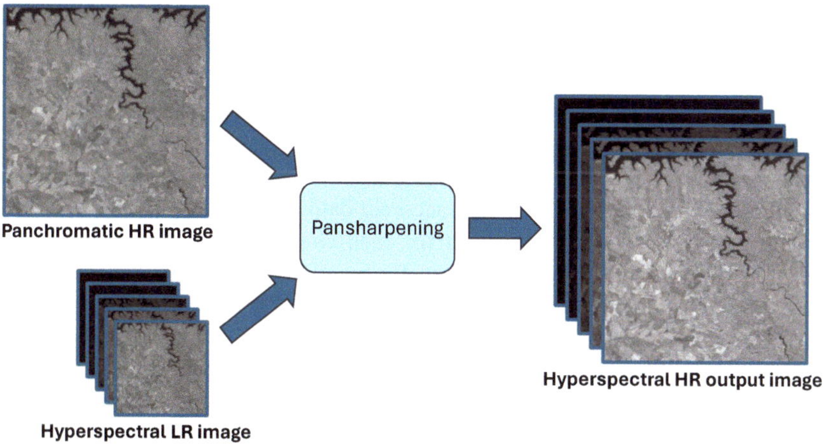

Fig. 7 Pansharpening pipeline, which is a special case of HSI/MSI fusion. Similarly, there are two inputs to the model—HR PAN image and LR HSI. The model's product is the image, which has the same number of spectral bands as the input LR HSI and the same spatial resolution as the input PAN image

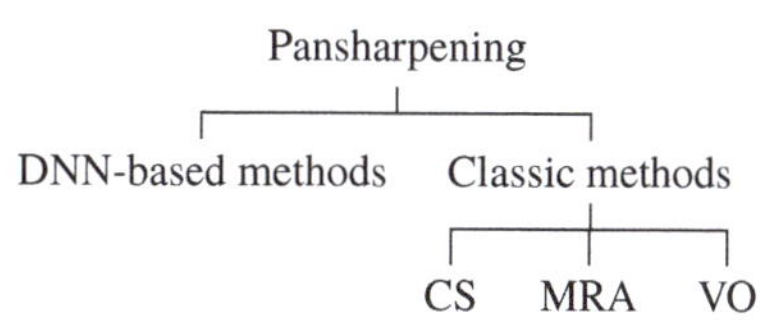

Fig. 8 Taxonomy of pansharpening methods, with enlisted classic methods—CS, MRA, and VO [15, 65]

HyperTransformer, proposed by Bandara et al. [5], utilizes the attention mechanism in a transformer architecture to fuse the panchromatic image with LR HSI. Two separate CNNs are employed to extract features from the upsampled LR HSI, sequentially downsampled and upsampled panchromatic image, and source panchromatic image, which are then used as queries, keys, and values for the attention mechanism, respectively. In addition to the L1 loss, the authors incorporate perceptual loss, which focuses on reconstructing higher-level features to yield visually pleasing results.

In another work by Bandara et al. [6], the DIP method is used to estimate the preliminary HR HSI by optimizing spatial and spectral loss functions. DIP is trained with additional constraints—the PAN image is estimated from the preliminary HR HSI by the learnable spectral response function and compared with the original PAN image. This preliminary HR HSI is then fused with PAN in the HyperKite network, which is trained to estimate the residual map for the preliminary HR HSI image. The residual map is subsequently added to the estimated HR HSI obtained by DIP to enhance the quality of the HR HSI.

Wu et al. [103] propose the dynamic cross-feature fusion network (DCFNet), a triple-branch CNN architecture. In this model, a pre-fused panchromatic image and an upsampled four-times LR hyperspectral image are processed in one branch. Simultaneously, the original LR HSI and an upscaled two-time LR HSI are processed in two additional branches. Intermediate features are exchanged between the branches to ensure feature processing at different spatial scales. A similar approach is employed by Jin et al. [41], who utilize Laplacian pyramid decomposition (LPD) in their Laplacian pyramid pansharpening network (LPPN) to achieve processing at different spatial scales. To ensure accurate data simulation, the authors replace the constant Gaussian kernel present for each channel in the Laplacian pyramid with the specific sensor modulation transfer function (MTF). Recursive layers that share parameters within the sub-networks of the LPPN are utilized to increase computational efficiency.

Xu et al. [106] propose an approach based on observation models, which are combined with DIP. The model consists of two key parts: the MS block and the PAN block. In each MS block, the residual for LR HSI obtained from the current estimation of HR HSI is learned to match the ground truth LR HSI. Subsequently, the next HR HSI estimation is produced by adding this residual. Similarly, in the PAN block, the estimated panchromatic image is compared with the real panchromatic image. The pairs of blocks are stacked and learned sequentially, with the input of the PAN block being the output of the MS block, and vice versa.

In [113], Zhang et al. propose two networks: TNet and PNet. TNet is a DNN that learns to reconstruct the gradient image of the PAN image in both horizontal and vertical directions from the HR HSI gradients. It is trained independently from PNet. PNet, on the other hand, serves as the actual pansharpening network. It learns residuals that are later added to the upsampled LR MSI. During training, PNet is guided by the pretrained TNet, with the guidance being realized by incorporating TNet as part of the loss function.

In [17] by Deng et al., an approach inspired by conventional models, namely, CS and MRA, is proposed. The authors introduce two interpretable DNNs: MRA-Net and CS-Net. However, these networks require certain prior knowledge assumptions, similar to the methods they are based on. Building upon these two solutions, the authors propose Fusion-Net, which operates on the residuals of the upsampled HR HSI and panchromatic image (of which the number of channels is matched with HR HSI by simple duplication). This ensures the injection of the spectral information into the pansharpening network.

Similarly to HSI/MSI fusion, most pansharpening methods are designed for supervised training, where the original images serve as reference HSIs, and the training/testing data is simulated using Wald's protocol (for LR HSI) or spectral response function/band averaging (for panchromatic images). However, there are a few unsupervised pansharpening methods worth noting.

PerepPan [119] by Zhou et al. is based on a GAN architecture. Initially, it generates the HR HSI using LR HSI and the panchromatic image and then reconstructs LR HSI and the panchromatic images from this generated HR HSI. The authors incorporate L1 loss, GAN loss, and perceptual loss to optimize the network. The best results are achieved when the HR HSI generator is initialized not with random weights, but with weights pretrained on RGB images for SR tasks.

In [72] by Qu et al., an unsupervised pansharpening (UP) approach is combined with a self-attention mechanism in UP-SAM network. The attention mechanism facilitates the extraction of LR HSI details at a sub-pixel level. This is motivated by the fact that a single pixel often covers multiple materials, each with distinct spectral signatures. The authors decompose each LR HSI pixel into a linear combination of spectral signatures, which are then used to extract high-frequency details from the panchromatic image. This decomposition is also known as spectral unmixing. These details are then used in the detail injection step to obtain the reconstructed HR HSI.

Inspired by DIP, Uezato et al. propose the guided deep decoder (GDD) method in [85]. This unsupervised approach involves extracting features from the panchromatic image using a U-Net-like architecture at various levels and scales. These features are then injected into the DIP decoder, with the LR HSI as input. The DIP decoder outputs the reconstructed HR HSI.

6 Multi-Image Super-Resolution

Multi-image fusion that processes a series of images captured at different times allows for higher reconstruction accuracy than SISR [68, 111], but it poses significant challenges including image co-registration and managing temporal variability among images. The state-of-the-art techniques in MISR are aimed at processing single-band images, and they can be applied for multispectral or hyperspectral imagery in a band-wise manner (i.e., each band is processed independently) (Fig. 9).

The DeepSUM network (the winner of the PROBA-V SR Challenge [64]) was the first end-to-end deep learning solution for MISR [66], later enhanced with nonlocal operations [67]. It is composed of three parts that (*i*) extract deep features from each LR image, (*ii*) co-register the feature maps with sub-pixel precision, and (*iii*) perform their final fusion, whose outcome is added to the image obtained as an average of bicubically upsampled LR inputs. DeepSUM assumes a fixed number of LR inputs and requires long training, being the result of fusing the upsampled LR images. These downsides were addressed in other MISR solutions, including

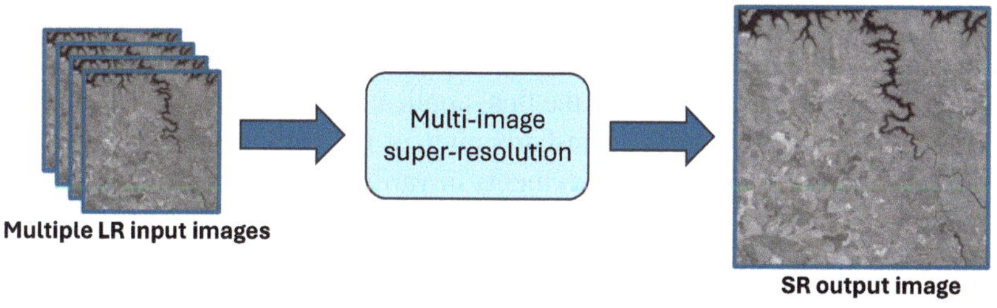

Fig. 9 MISR pipeline. The input to the model is a series of images (most often representing a single spectral band) captured at different times and the output is the SR image with enhanced spatial resolution

HighRes-net which combines the latent LR representations in a recursive manner to obtain the global representation which is upsampled to obtain the SR image [18]. Also, the attention mechanism was found useful for selecting the most valuable features extracted from LR inputs in the residual attention multi-image SR (RAMS) network [77], and a recurrent network with gated recurrent units (MISR-GRU) was proposed in [74]. In [4] the authors focused on simplifying the training with the use of transformers and reported competitive results for the PROBA-V dataset. Another approach, proposed by Tarasiewicz et al. [84], is to represent an input set of LR images as a graph which is processed with a graph neural network that produces the SR image. Recently, Valsesia and Magli introduced the PIUnet network that enforces permutation invariance within a set of LR inputs—it leads to significant improvements in super-resolving PROBA-V images and Sentinel-2 B03 band with simulated LR [87]. Multi-temporal fusion may also be performed to combine images captured by different satellites, including Sentinel-2 and Landsat-8 [102].

CNN-based multi-temporal fusion is also involved in burst-image [9] and video SR [43], thoroughly discussed in an excellent review by Liu et al. [60]. Video SR can be effectively tackled with 3D CNNs [49] and some recent techniques combine SISR and MISR models to better deal with a video stream [36, 78]. Such techniques have been applied to satellite imagery, including enhancing satellite image bursts [69]. These approaches commonly benefit from the high and fixed temporal frequency of the input frames, and they are often tailored to deal with moving objects, by estimating the motion fields. This makes them different from the multi-image fusion which processes an unordered set of images with unknown timestamps rather than their time series [18].

Some of the existing MISR techniques have been employed to enhance multispectral and hyperspectral images; however, they were applied in a band-wise manner. Each band is super-resolved independently by fusing multiple instances of the same spectral band extracted from images captured at different times. In particular, such an approach was found effective for Sentinel-2 imagery [46, 88].

7 Conclusions and Future Research

Over the last few years, DNN-based solutions have dominated the field of SR for hyperspectral images, outperforming classic solutions in almost every aspect. Depending on the main purpose of the final application of reconstructed images and the available data, SISR, MISR, pansharpening, or HSI/MSI fusion methods can be utilized. SR products enable enhancements in various remote sensing applications, such as precision farming and environmental monitoring.

However, most SR methods are developed using simulated LR data, which provides an overoptimistic estimation and does not generalize well over real-world datasets. Generalization across different simulated datasets is also challenging due to variability in sensor characteristics, acquisition conditions, noise distributions, differences in data post-processing, and other factors.

Therefore, the authors of this chapter believe that future research should focus on developing more general SR methods capable of successful use with a wide variety of datasets. This may involve the exploration of new DNN methods/architectures or more sophisticated data simulation procedures. Additionally, the authors hold hope for the wider availability of new hyperspectral imagery datasets in the future, which could further facilitate advancements in the field.

Acknowledgments This work was funded by the European Space Agency (the ROMEO project: "Real-world hyperspectral image super-resolution").

References

1. Ahn, N., Kang, B., Sohn, K.A.: Fast, accurate, and lightweight super-resolution with cascading residual network. In: Proceedings of the European Conference on Computer Vision (ECCV), pp. 256–272. Springer, Berlin (2018). https://doi.org/10.1007/978-3-030-01249-6_16
2. Al-Mekhlafi, H., Liu, S.: Single image super-resolution: a comprehensive review and recent insight. Front. Comput. Sci. **18**, 181702 (2023). https://doi.org/10.1007/s11704-023-2588-9
3. Alparone, L., Aiazzi, B., Baronti, S., Garzelli, A., Nencini, F., Selva, M.: Multispectral and panchromatic data fusion assessment without reference. ASPRS J. Photogramm. Eng. Remote Sensing **74**, 193–200 (2008). https://doi.org/10.14358/PERS.74.2.193
4. An, T., Zhang, X., Huo, C., Xue, B., Wang, L., Pan, C.: TR-MISR: multiimage super-resolution based on feature fusion with transformers. IEEE J. Sel. Top. Appl. Earth Observ. Remote Sensing **15**, 1373–1388 (2022). https://doi.org/10.1109/JSTARS.2022.3143532
5. Bandara, W.G.C., Patel, V.M.: HyperTransformer: a textural and spectral feature fusion transformer for pansharpening. In: Proceedings of the IEEE/CVF Conference on Computer Vision and Pattern Recognition (CVPR), pp. 1767–1777 (2022). https://doi.org/10.1109/CVPR52688.2022.00181
6. Bandara, W.G.C., Valanarasu, J.M.J., Patel, V.M.: Hyperspectral pansharpening based on improved deep image prior and residual reconstruction. IEEE Trans. Geosci. Remote Sensing **60**, 1–16 (2022). https://doi.org/10.1109/TGRS.2021.3139292
7. Bashir, S.M.A., Wang, Y., Khan, M., Niu, Y.: A comprehensive review of deep learning-based single image super-resolution (2022). https://doi.org/10.48550/arXiv.2102.09351
8. Bhadra, S., Kelkar, V., Brooks, F., Anastasio, M.: On hallucinations in tomographic image reconstruction. IEEE Trans. Med. Imaging **PP**, (2021). https://doi.org/10.1109/TMI.2021.3077857
9. Bhat, G., Danelljan, M., Van Gool, L., Timofte, R.: Deep burst super-resolution. In: Proceedings of the IEEE/CVF Conference on Computer Vision and Pattern Recognition (CVPR), pp. 9209–9218 (2021). https://doi.org/10.1109/CVPR46437.2021.00909
10. Borman, S., Stevenson, R.: Super-resolution from image sequences—a review. In: Proceedings of the IEEE Midwest Symposium on Circuits and Systems (MWSCAS), pp. 374–378 (1998). https://doi.org/10.1109/MWSCAS.1998.759509
11. Chakrabarti, A., Zickler, T.: Statistics of real-world hyperspectral images. In: Proceedings of the IEEE/CVF Conference on Computer Vision and Pattern Recognition (CVPR), pp. 193–200 (2011). https://doi.org/10.1109/CVPR.2011.5995660
12. Chauhan, K., Patel, S.N., Kumhar, M., Bhatia, J., Tanwar, S., Davidson, I.E., Mazibuko, T.F., Sharma, R.: Deep learning-based single-image super-resolution: a comprehensive review. IEEE Access **11**, 21811–21830 (2023). https://doi.org/10.1109/ACCESS.2023.3251396

13. Chen, H., He, X., Qing, L., Wu, Y., Ren, C., Sheriff, R.E., Zhu, C.: Real-world single image super-resolution: a brief review. Information Fusion **79**, 124–145 (2022). https://doi.org/10.1016/j.inffus.2021.09.005
14. Cornebise, J., Orsolic, I., Kalaitzis, F.: Open high-resolution satellite imagery: the WorldStrat dataset – with application to super-resolution. In: Proceedings of the Conference on Neural Information Processing Systems (NeurIPS) (2022). https://openreview.net/forum?id=DEigo9L8xZA
15. Dadrass Javan, F., Samadzadegan, F., Mehravar, S., Toosi, A., Khatami, R., Stein, A.: A review of image fusion techniques for pan-sharpening of high-resolution satellite imagery. ISPRS J. Photogramm. Remote Sensing **171**, 101–117 (2021). https://doi.org/10.1016/j.isprsjprs.2020.11.001
16. Deng, J., Dong, W., Socher, R., Li, L.J., Li, K., Fei-Fei, L.: ImageNet: a large-scale hierarchical image database. In: Proceedings of the IEEE/CVF Conference on Computer Vision and Pattern Recognition (CVPR), pp. 248–255 (2009). https://doi.org/10.1109/CVPR.2009.5206848
17. Deng, L.J., Vivone, G., Jin, C., Chanussot, J.: Detail injection-based deep convolutional neural networks for pansharpening. IEEE Trans. Geosci. Remote Sensing **59**(8), 6995–7010 (2021). https://doi.org/10.1109/TGRS.2020.3031366
18. Deudon, M., Kalaitzis, A., Goytom, I., Arefin, M.R., Lin, Z., Sankaran, K., Michalski, V., Kahou, S.E., Cornebise, J., Bengio, Y.: HighRes-net: Recursive fusion for multi-frame super-resolution of satellite imagery (2020). https://doi.org/10.48550/arXiv.2002.06460
19. Dian, R., Li, S., Guo, A., Fang, L.: Deep hyperspectral image sharpening. IEEE Trans. Neural Netw. Learn. Syst. **29**(11), 5345–5355 (2018). https://doi.org/10.1109/TNNLS.2018.2798162
20. Dian, R., Li, S., Kang, X.: Regularizing hyperspectral and multispectral image fusion by CNN denoiser. IEEE Trans. Neural Netw. Learn. Syst. **32**(3), 1124–1135 (2021). https://doi.org/10.1109/TNNLS.2020.2980398
21. Dong, C., Loy, C.C., He, K., Tang, X.: Learning a deep convolutional network for image super-resolution. In: Proceedings of the European Conference on Computer Vision (ECCV), pp. 184–199. Springer, Berlin (2014). https://doi.org/10.1007/978-3-319-10593-2_13
22. Dong, C., Loy, C.C., Tang, X.: Accelerating the super-resolution convolutional neural network. In: Proceedings of the European Conference on Computer Vision (ECCV), pp. 391–407. Springer, Berlin (2016). https://doi.org/10.1007/978-3-319-46475-6_25
23. Dong, W., Zhou, C., Wu, F., Wu, J., ShiGuangming, Li, X.: Model-guided deep hyperspectral image super-resolution. IEEE Trans. Image Process. **30**, 5754–5768 (2021). https://doi.org/10.1109/TIP.2021.3078058
24. European Space Agency: Earth Online (official web resources about PROBA-V mission and data). https://earth.esa.int/eogateway/missions/proba-v
25. European Space Agency: PROBA-V super resolution (2018). https://kelvins.esa.int/proba-v-super-resolution/
26. European Space Agency: Sentinel-2 mission guide (2023). https://sentinels.copernicus.eu/web/sentinel/missions/sentinel-2
27. European Space Agency: WorldView-2 (2023). https://earth.esa.int/eogateway/missions/worldview-2
28. Fan, Y., Shi, H., Yu, J., Liu, D., Han, W., Yu, H., Wang, Z., Wang, X., Huang, T.S.: Balanced two-stage residual networks for image super-resolution. In: Proceedings of the IEEE/CVF Conference on Computer Vision and Pattern Recognition (CVPR), pp. 1157–1164 (2017). https://doi.org/10.1109/CVPRW.2017.154
29. Fu, Y., Liang, Z., You, S.: Bidirectional 3D quasi-recurrent neural network for hyperspectral image super-resolution. IEEE J. Sel. Top. Appl. Earth Observ. Remote Sensing **14**, 2674–2688 (2021). https://doi.org/10.1109/JSTARS.2021.3057936
30. Gao, J., Li, J., Jiang, M.: Hyperspectral and multispectral image fusion by deep neural network in a self-supervised manner. Remote Sensing **13**(16) (2021). https://doi.org/10.3390/rs13163226

31. Gatti, A., Bertolini, A.: Sentinel-2 products specification document (2015). https://sentinel.esa.int/documents/247904/349490/s2_msi_product_specification.pdf
32. Goodfellow, I., Pouget-Abadie, J., Mirza, M., Xu, B., Warde-Farley, D., Ozair, S., Courville, A., Bengio, Y.: Generative adversarial nets. In: Proceedings of the Advances in Neural Information Processing Systems (NeurIPS), pp. 2672–2680 (2014)
33. Graña, M., Veganzons, M., Ayerdi, B.: Hyperspectral remote sensing scenes (2021). https://www.ehu.eus/ccwintco/index.php/Hyperspectral_Remote_Sensing_Scenes#Pavia_Centre_and_University
34. He, J., Yuan, Q., Li, J., Xiao, Y., Liu, D., Shen, H., Zhang, L.: Spectral super-resolution meets deep learning: Achievements and challenges. Inform. Fusion **97**, 101812 (2023). https://doi.org/10.1016/j.inffus.2023.101812. https://www.sciencedirect.com/science/article/pii/S1566253523001215
35. He, K., Zhang, X., Ren, S., Sun, J.: Deep residual learning for image recognition. In: Proceedings of the IEEE Conference on Computer Vision and Pattern Recognition (CVPR), pp. 770–778 (2016). https://doi.org/10.1109/CVPR.2016.90
36. He, Z., Li, J., Liu, L., He, D., Xiao, M.: Multiframe video satellite image super-resolution via attention-based residual learning. IEEE Trans. Geosci. Remote Sensing **60**, 1–15 (2022). https://doi.org/10.1109/TGRS.2021.3072381
37. Horé, A., Ziou, D.: Image quality metrics: PSNR vs. SSIM. In: Proceedings of the International Conference on Pattern Recognition (ICPR), pp. 2366–2369 (2010). https://doi.org/10.1109/ICPR.2010.579
38. Hu, J., Jia, X., Li, Y., He, G., Zhao, M.: Hyperspectral image super-resolution via intrafusion network. IEEE Trans. Geosci. Remote Sensing **58**(10), 7459–7471 (2020). https://doi.org/10.1109/TGRS.2020.2982940
39. Hui, Z., Wang, X., Gao, X.: Fast and accurate single image super-resolution via information distillation network. In: Proceedings of the IEEE/CVF Conference on Computer Vision and Pattern Recognition (CVPR), pp. 723–731 (2018). https://doi.org/10.1109/CVPR.2018.00082
40. Jiang, J., Sun, H., Liu, X., Ma, J.: Learning spatial-spectral prior for super-resolution of hyperspectral imagery. IEEE Trans. Comput. Imaging **6**, 1082–1096 (2020). https://doi.org/10.1109/tci.2020.2996075
41. Jin, C., Deng, L.J., Huang, T.Z., Vivone, G.: Laplacian pyramid networks: a new approach for multispectral pansharpening. Inform. Fusion **78**, 158–170 (2022). https://doi.org/10.1016/j.inffus.2021.09.002
42. Johnson, J., Alahi, A., Fei-Fei, L.: Perceptual losses for real-time style transfer and super-resolution. In: Proceedings of the European Conference on Computer Vision (ECCV), pp. 694–711. Springer, Berlin (2016). https://doi.org/10.1007/978-3-319-46475-6_43
43. Kappeler, A., Yoo, S., Dai, Q., Katsaggelos, A.K.: Video super-resolution with convolutional neural networks. IEEE Trans. Comput. Imaging **2**(2), 109–122 (2016). https://doi.org/10.1109/TCI.2016.2532323
44. Karras, T., Laine, S., Aila, T.: A style-based generator architecture for generative adversarial networks. In: Proceedings of the IEEE/CVF Conference on Computer Vision and Pattern Recognition (CVPR) (2019). https://doi.org/10.1109/TPAMI.2020.2970919
45. Kawulok, M., Kowaleczko, P., Ziaja, M., Nalepa, J., Kostrzewa, D., Latini, D., De Santis, D., Salvucci, G., Petracca, I., La Pegna, V., Bartalis, Z., Del Frate, F.: Understanding the value of hyperspectral image super-resolution from PRISMA data. In: Proceedings of the IEEE International Geoscience and Remote Sensing Symposium (IGARSS), pp. 1489–1492 (2023). https://doi.org/10.1109/IGARSS52108.2023.10283013
46. Kawulok, M., Tarasiewicz, T., Nalepa, J., Tyrna, D., Kostrzewa, D.: Deep learning for multiple-image super-resolution of Sentinel-2 data. In: Proceedings of the IEEE International Geoscience and Remote Sensing Symposium (IGARSS), pp. 3885–3888 (2021). https://doi.org/10.1109/IGARSS47720.2021.9553243
47. Kim, J., Lee, J.K., Lee, K.M.: Accurate image super-resolution using very deep convolutional networks. In: Proceedings of the IEEE/CVF Conference on Computer Vision and Pattern Recognition (CVPR), pp. 1646–1654 (2016). https://doi.org/10.1109/CVPR.2016.182

48. Kim, J., Lee, J.K., Lee, K.M.: Deeply-recursive convolutional network for image super-resolution. In: Proceedings of the IEEE/CVF Conference on Computer Vision and Pattern Recognition (CVPR), pp. 1637–1645 (2016). https://doi.org/10.1109/CVPR.2016.181
49. Kim, S.Y., Lim, J., Na, T., Kim, M.: Video super-resolution based on 3D-CNNs with consideration of scene change. In: Proceedings of the IEEE International Conference on Image Processing (ICIP), pp. 2831–2835 (2019). https://doi.org/10.1109/ICIP.2019.8803297
50. Kowaleczko, P., Tarasiewicz, T., Ziaja, M., Kostrzewa, D., Nalepa, J., Rokita, P., Kawulok, M.: Data for: MuS2: a benchmark for Sentinel-2 multi-image super-resolution (2022). https://doi.org/10.7910/DVN/1JMRAT
51. Kowaleczko, P., Tarasiewicz, T., Ziaja, M., Kostrzewa, D., Nalepa, J., Rokita, P., Kawulok, M.: A real-world benchmark for sentinel-2 multi-image super-resolution. Sci. Data **10**(1), 644 (2023). https://doi.org/10.1038/s41597-023-02538-9
52. Lai, W.S., Huang, J.B., Ahuja, N., Yang, M.H.: Deep Laplacian pyramid networks for fast and accurate super-resolution. In: Proceedings of the IEEE/CVF Conference on Computer Vision and Pattern Recognition (CVPR), pp. 5835–5843 (2017). https://doi.org/10.1109/CVPR.2017.618
53. Ledig, C., Theis, L., Huszar, F., Caballero, J., Cunningham, A., Acosta, A., Aitken, A., Tejani, A., Totz, J., Wang, Z., Shi, W.: Photo-realistic single image super-resolution using a generative adversarial network. In: Proceedings of the IEEE/CVF Conference on Computer Vision and Pattern Recognition (CVPR), pp. 105–114. IEEE Computer Society (2017). https://doi.org/10.1109/CVPR.2017.19
54. Lepcha, D.C., Goyal, B., Dogra, A., Goyal, V.: Image super-resolution: a comprehensive review, recent trends, challenges and applications. Inform. Fusion **91**, 230–260 (2023). https://doi.org/10.1016/j.inffus.2022.10.007
55. Li, J., Cui, R., Li, B., Song, R., Li, Y., Dai, Y., Du, Q.: Hyperspectral image super-resolution by band attention through adversarial learning. IEEE Trans. Geosci. Remote Sensing **58**(6), 4304–4318 (2020). https://doi.org/10.1109/TGRS.2019.2962713
56. Li, K., Dai, D., Konukoglu, E., Van Gool, L.: Hyperspectral image super-resolution with spectral mixup and heterogeneous datasets (2021). https://doi.org/10.48550/ARXIV.2101.07589
57. Li, Q., Wang, Q., Li, X.: Mixed 2D/3D convolutional network for hyperspectral image super-resolution. Remote Sensing **12**(10), 1660 (2020). https://doi.org/10.3390/rs12101660
58. Lim, B., Son, S., Kim, H., Nah, S., Lee, K.M.: Enhanced deep residual networks for single image super-resolution. In: Proceedings of the IEEE/CVF Conference on Computer Vision and Pattern Recognition (CVPR), pp. 1132–1140 (2017). https://doi.org/10.1109/CVPRW.2017.151
59. Liu, D., Li, J., Yuan, Q.: A spectral grouping and attention-driven residual dense network for hyperspectral image super-resolution. IEEE Trans. Geosci. Remote Sensing **59**(9), 7711–7725 (2021). https://doi.org/10.1109/TGRS.2021.3049875
60. Liu, H., Ruan, Z., Zhao, P., Dong, C., Shang, F., Liu, Y., Yang, L., Timofte, R.: Video super-resolution based on deep learning: a comprehensive survey. Artif. Intell. Rev. 1–55 (2022). https://doi.org/10.1007/s10462-022-10147-y
61. Loncan, L., de Almeida, L.B., Bioucas-Dias, J.M., Briottet, X., Chanussot, J., Dobigeon, N., Fabre, S., Liao, W., Licciardi, G.A., Simões, M., Tourneret, J.Y., Veganzones, M.A., Vivone, G., Wei, Q., Yokoya, N.: Hyperspectral pansharpening: a review. IEEE Geosci. Remote Sensing Mag. **3**(3), 27–46 (2015). https://doi.org/10.1109/MGRS.2015.2440094
62. Ma, Q., Jiang, J., Liu, X., Ma, J.: Learning a 3D-CNN and transformer prior for hyperspectral image super-resolution. Inform. Fusion **100**, 101907 (2023). https://doi.org/10.1016/j.inffus.2023.101907
63. Maral, B.C.: Single image super-resolution methods: a survey (2022). https://doi.org/10.48550/arXiv.2202.11763
64. Märtens, M., Izzo, D., Krzic, A., Cox, D.: Super-resolution of PROBA-V images using convolutional neural networks. Astrodynamics **3**(4), 387–402 (2019). https://doi.org/10.1007/s42064-019-0059-8

65. Meng, X., Xiong, Y., Shao, F., Shen, H., Sun, W., Yang, G., Yuan, Q., Fu, R., Zhang, H.: A large-scale benchmark data set for evaluating pansharpening performance: Overview and implementation. IEEE Geosci. Remote Sensing Mag. **9**(1), 18–52 (2021). https://doi.org/10.1109/MGRS.2020.2976696
66. Molini, A.B., Valsesia, D., Fracastoro, G., Magli, E.: DeepSUM: deep neural network for super-resolution of unregistered multitemporal images. IEEE Trans. Geosci. Remote Sensing **58**(5), 3644–3656 (2020). https://doi.org/10.1109/TGRS.2019.2959248
67. Molini, A.B., Valsesia, D., Fracastoro, G., Magli, E.: DeepSUM++: Non-local deep neural network for super-resolution of unregistered multitemporal images. In: Proceedings of the IEEE International Geoscience and Remote Sensing Symposium (IGARSS), pp. 609–612 (2020). https://doi.org/10.1109/IGARSS39084.2020.9324418
68. Nasrollahi, K., Moeslund, T.B.: Super-resolution: a comprehensive survey. Mach. Vis. Appl. **25**(6), 1423–1468 (2014). https://doi.org/10.1007/s00138-014-0623-4
69. Nguyen, N.L., Anger, J., Davy, A., Arias, P., Facciolo, G.: PROBA-V-REF: Repurposing the PROBA-V challenge for reference-aware super resolution. In: Proceedings of the IEEE International Geoscience and Remote Sensing Symposium (IGARSS), pp. 3881–3884. IEEE, Piscataway (2021). https://doi.org/10.1109/IGARSS47720.2021.9554479
70. Palsson, F., Sveinsson, J.R., Ulfarsson, M.O.: Multispectral and hyperspectral image fusion using a 3D-convolutional neural network. IEEE Geosci. Remote Sensing Lett. **14**(5), 639–643 (2017). https://doi.org/10.1109/LGRS.2017.2668299
71. Park, S.C., Park, M.K., Kang, M.G.: Super-resolution image reconstruction: a technical overview. IEEE Signal Process. Mag. **20**(3), 21–36 (2003). https://doi.org/10.1109/MSP.2003.1203207
72. Qu, Y., Baghbaderani, R.K., Qi, H., Kwan, C.: Unsupervised pansharpening based on self-attention mechanism. IEEE Trans. Geosci. Remote Sensing **59**(4), 3192–3208 (2021). https://doi.org/10.1109/TGRS.2020.3009207
73. Renza, D., Martinez, E., Arquero, A.: A new approach to change detection in multispectral images by means of ERGAS index. IEEE Geosci. Remote Sensing Lett. **10**(1), 76–80 (2013). https://doi.org/10.1109/LGRS.2012.2193372
74. Rifat Arefin, M., Michalski, V., St-Charles, P.L., Kalaitzis, A., Kim, S., Kahou, S.E., Bengio, Y.: Multi-image super-resolution for remote sensing using deep recurrent networks. In: Proceedings of the IEEE/CVF Conference on Computer Vision and Pattern Recognition (CVPR), pp. 206–207 (2020). https://doi.org/10.1109/CVPRW50498.2020.00111
75. Rombach, R., Blattmann, A., Lorenz, D., Esser, P., Ommer, B.: High-resolution image synthesis with latent diffusion models. In: Proceedings of the IEEE/CVF Conference on Computer Vision and Pattern Recognition (CVPR), pp. 10684–10695 (2022). https://doi.org/10.1109/CVPR52688.2022.01042
76. Sajjadi, M.S.M., Schölkopf, B., Hirsch, M.: EnhanceNet: single image super-resolution through automated texture synthesis. In: Proceedings of the IEEE International Conference on Computer Vision (ICCV), pp. 4501–4510 (2017). https://doi.org/10.1109/ICCV.2017.481
77. Salvetti, F., Mazzia, V., Khaliq, A., Chiaberge, M.: Multi-image super resolution of remotely sensed images using residual attention deep neural networks. Remote Sensing **12**(14), 2207 (2020). https://doi.org/10.3390/rs12142207
78. Shen, H., Qiu, Z., Yue, L., Zhang, L.: Deep-learning-based super-resolution of video satellite imagery by the coupling of multiframe and single-frame models. IEEE Trans. Geosci. Remote Sensing **60**, 1–14 (2022). https://doi.org/10.1109/TGRS.2021.3121303
79. Shi, W., Caballero, J., Huszár, F., Totz, J., Aitken, A.P., Bishop, R., Rueckert, D., Wang, Z.: Real-time single image and video super-resolution using an efficient sub-pixel convolutional neural network. In: Proceedings of the IEEE/CVF Conference on Computer Vision and Pattern Recognition (CVPR), pp. 1874–1883 (2016). https://doi.org/10.1109/CVPR.2016.207
80. Sidorov, O., Hardeberg, J.Y.: Deep hyperspectral prior: single-image denoising, inpainting, super-resolution. In: Proceedings of the IEEE/CVF International Conference on Computer Vision Workshop (ICCVW), pp. 3844–3851 (2019). https://doi.org/10.1109/ICCVW.2019.00477

81. Simonyan, K., Zisserman, A.: Very deep convolutional networks for large-scale image recognition (2014). CoRR abs/1409.1556. https://doi.org/10.48550/arXiv.1409.1556
82. Tai, Y., Yang, J., Liu, X.: Image super-resolution via deep recursive residual network. In: Proceedings of the IEEE/CVF Conference on Computer Vision and Pattern Recognition (CVPR), pp. 2790–2798 (2017). https://doi.org/10.1109/CVPR.2017.298
83. Tang, Z., Xu, Q., Wu, P., Shi, Z., Pan, B.: Feedback refined local-global network for super-resolution of hyperspectral imagery. Remote Sensing **14**(8), 1944 (2022). https://doi.org/10.3390/rs14081944
84. Tarasiewicz, T., Nalepa, J., Kawulok, M.: A graph neural network for multiple-image super-resolution. In: Proceedings of the IEEE International Conference on Image Processing (ICIP), pp. 1824–1828 (2021). https://doi.org/10.1109/ICIP42928.2021.9506070
85. Uezato, T., Hong, D., Yokoya, N., He, W.: Guided deep decoder: Unsupervised image pair fusion. In: Proceedings of the European Conference on Computer Vision (ECCV), pp. 87–102. Springer, Berlin (2020). https://doi.org/10.1007/978-3-030-58539-6_6
86. Ulyanov, D., Vedaldi, A., Lempitsky, V.: Deep image prior. In: Proceedings of the IEEE/CVF Conference on Computer Vision and Pattern Recognition (CVPR), pp. 9446–9454 (2018)
87. Valsesia, D., Magli, E.: Permutation invariance and uncertainty in multitemporal image super-resolution. IEEE Trans. Geosci. Remote Sensing **60**, 1–12 (2022). https://doi.org/10.1109/TGRS.2021.3130673
88. Valsesia, D., Magli, E.: Super-resolved multi-temporal segmentation with deep permutation-invariant networks. In: Proceedings of the IEEE International Geoscience and Remote Sensing Symposium (IGARSS), pp. 995–998 (2022). https://doi.org/10.1109/IGARSS46834.2022.9884811
89. van Ouwerkerk, J.: Image super-resolution survey. Image Vis. Comput. **24**(10), 1039–1052 (2006). https://doi.org/10.1016/j.imavis.2006.02.026
90. Vaswani, A., Shazeer, N., Parmar, N., Uszkoreit, J., Jones, L., Gomez, A.N., Kaiser, Ł., Polosukhin, I.: Attention is all you need. In: Proceedings of the Advances in Neural Information Processing Systems, pp. 5998–6008 (2017). https://www.bibsonomy.org/bibtex/2a08c93d224dfcfb83550246c3d6a178f/stefan.ernst
91. Wald, L.: Quality of high resolution synthesised images: Is there a simple criterion? In: Proceedings of the Third conference Fusion of Earth Data: Merging Point Measurements, Raster Maps and Remotely Sensed Images, pp. 99–103. SEE/URISCA (2000). https://hal.archives-ouvertes.fr/hal-00395027
92. Wald, L., Ranchin, T., Mangolini, M.: Fusion of satellite images of different spatial resolutions: assessing the quality of resulting images. Photogramm. Eng. Remote Sensing **63**, 691–699 (1997)
93. Wang, Q., Li, Q., Li, X.: Hyperspectral image superresolution using spectrum and feature context. IEEE Trans. Ind. Electron. **68**(11), 11276–11285 (2021). https://doi.org/10.1109/TIE.2020.3038096
94. Wang, W., Fu, X., Zeng, W., Sun, L., ZhanRonghui, Huang, Y., Ding, X.: Enhanced deep blind hyperspectral image fusion. IEEE Trans. Neural Netw. Learn. Syst. 1–11 (2021). https://doi.org/10.1109/TNNLS.2021.3105543
95. Wang, X., Hu, Q., Cheng, Y., Ma, J.: Hyperspectral image super-resolution meets deep learning: a survey and perspective. IEEE/CAA J. Autom. Sin. **10**(8), 1668–1691 (2023). https://doi.org/10.1109/JAS.2023.123681
96. Wang, X., Ma, J., Jiang, J.: Hyperspectral image super-resolution via recurrent feedback embedding and spatial–spectral consistency regularization. IEEE Trans. Geosci. Remote Sensing **60**, 1–13 (2022). https://doi.org/10.1109/TGRS.2021.3064450
97. Wang, X., Xie, L., Dong, C., Shan, Y.: Real-ESRGAN: training real-world blind super-resolution with pure synthetic data. In: Proceedings of the IEEE/CVF International Conference on Computer Vision Workshops (ICCVW), pp. 1905–1914 (2021). https://doi.org/10.1109/ICCVW54120.2021.00217
98. Wang, X., Yu, K., Wu, S., Gu, J., Liu, Y., Dong, C., Qiao, Y., Loy, C.C.: ESRGAN: enhanced super-resolution generative adversarial networks. In: Proceedings of the European Conference

on Computer Vision (ECCV), pp. 63–79. Springer, Berlin (2019). https://doi.org/10.1007/978-3-030-11021-5_5
99. Wang, Z., Bovik, A.: A universal image quality index. IEEE Signal Process. Lett. **9**(3), 81–84 (2002). https://doi.org/10.1109/97.995823
100. Wang, Z., Bovik, A., Sheikh, H., Simoncelli, E.: Image quality assessment: from error visibility to structural similarity. IEEE Trans. Image Process. **13**(4), 600–612 (2004). https://doi.org/10.1109/TIP.2003.819861
101. Wang, Z., Chen, J., Hoi, S.C.H.: Deep learning for image super-resolution: a survey. IEEE Trans. Pattern Anal. Mach. Intell. **43**(10), 3365–3387 (2021). https://doi.org/10.1109/TPAMI.2020.2982166
102. Wu, J., Lin, L., Li, T., Cheng, Q., Zhang, C., Shen, H.: Fusing Landsat 8 and Sentinel-2 data for 10-m dense time-series imagery using a degradation-term constrained deep network. Int. J. Appl. Earth Observ. Geoinform. **108**, 102738 (2022). https://doi.org/10.1016/j.jag.2022.102738
103. Wu, X., Huang, T.Z., Deng, L.J., Zhang, T.J.: Dynamic cross feature fusion for remote sensing pansharpening. In: Proceedings of the IEEE/CVF International Conference on Computer Vision (ICCV), pp. 14667–14676 (2021). https://doi.org/10.1109/ICCV48922.2021.01442
104. Xie, Q., Zhou, M., Zhao, Q., Meng, D., ZuoWangmeng, Xu, Z.: Multispectral and hyperspectral image fusion by MS/HS Fusion Net. In: Proceedings of the IEEE/CVF Conference on Computer Vision and Pattern Recognition (CVPR), pp. 1585–1594 (2019). https://doi.org/10.1109/CVPR.2019.00168
105. Xie, Q., Zhou, M., Zhao, Q., Xu, Z., Meng, D.: MHF-Net: an interpretable deep network for multispectral and hyperspectral image fusion. IEEE Trans. Pattern Anal. Mach. Intell. **44**(3), 1457–1473 (2020)
106. Xu, S., Zhang, J., Zhao, Z., Sun, K., Liu, J., Zhang, C.: Deep gradient projection networks for pan-sharpening. In: Proceedings of the IEEE/CVF Conference on Computer Vision and Pattern Recognition (CVPR), pp. 1366–1375 (2021). https://doi.org/10.1109/CVPR46437.2021.00142
107. Yasuma, F., Mitsunaga, T., Iso, D., Nayar, S.: Generalized assorted pixel camera: postcapture control of resolution, dynamic range, and spectrum. IEEE Trans. Image Process. **19**, 2241–53 (2010). https://doi.org/10.1109/TIP.2010.2046811
108. Ye, S., Zhao, S., Hu, Y., Xie, C.: Single-image super-resolution challenges: a brief review. Electronics **12**(13), 2975 (2023). https://doi.org/10.3390/electronics12132975
109. Yokoya, N., Grohnfeldt, C., Chanussot, J.: Hyperspectral and multispectral data fusion: a comparative review of the recent literature. IEEE Geosci. Remote Sensing Mag. **5**(2), 29–56 (2017). https://doi.org/10.1109/MGRS.2016.2637824
110. Yokoya, N., Iwasaki, A.: Airborne hyperspectral data over Chikusei. Tech. Rep. SAL-2016-05-27, Space Application Laboratory, University of Tokyo, Japan (2016)
111. Yue, L., Shen, H., Li, J., Yuan, Q., Zhang, H., Zhang, L.: Image super-resolution: the techniques, applications, and future. Signal Process. **128**, 389–408 (2016). https://doi.org/10.1016/j.sigpro.2016.05.002
112. Yuhas, R.H., Goetz, A.F.H., Boardman, J.W.: Discrimination among semi-arid landscape endmembers using the spectral angle mapper (SAM) algorithm. In: Proceedings of the Third Annual JPL Airborne Geoscience Workshop. Volume 1: AVIRIS Workshop (1992)
113. Zhang, H., Ma, J.: GTP-PNet: a residual learning network based on gradient transformation prior for pansharpening. ISPRS J. Photogramm. Remote Sensing **172**, 223–239 (2021). https://doi.org/10.1016/j.isprsjprs.2020.12.014
114. Zhang, K., Liang, J., Van Gool, L., Timofte, R.: Designing a practical degradation model for deep blind image super-resolution. In: Proceedings of the IEEE/CVF International Conference on Computer Vision (ICCV), pp. 4771–4780 (2021). https://doi.org/10.1109/ICCV48922.2021.00475
115. Zhang, R., Isola, P., Efros, A.A., Shechtman, E., Wang, O.: The unreasonable effectiveness of deep features as a perceptual metric. In: Proceedings of the IEEE/CVF Conference on

Computer Vision and Pattern Recognition (CVPR), pp. 586–595 (2018). https://doi.org/10.1109/CVPR.2018.00068
116. Zhang, W., Shi, G., Liu, Y., Dong, C., Wu, X.: A closer look at blind super-resolution: Degradation models, baselines, and performance upper bounds. In: Proceedings of the IEEE/CVF Conference on Computer Vision and Pattern Recognition Workshops (CVPRW), pp. 526–535 (2022). https://doi.org/10.1109/CVPRW56347.2022.00068
117. Zhang, X., Huang, W., Wang, Q., Li, X.: SSR-NET: spatial–spectral reconstruction network for hyperspectral and multispectral image fusion. IEEE Trans. Geosci. Remote Sensing **59**(7), 5953–5965 (2021). https://doi.org/10.1109/TGRS.2020.3018732
118. Zhang, Y., Li, K., Li, K., Wang, L., Zhong, B., Fu, Y.: Image super-resolution using very deep residual channel attention network. In: Proceedings of the European Conference on Computer Vision (ECCV), pp. 294–310. Springer, Berlin (2018). https://doi.org/10.1007/978-3-030-01234-2_18
119. Zhou, C., Zhang, J., Liu, J., Zhang, C., Fei, R., Xu, S.: PercepPan: towards unsupervised pan-sharpening based on perceptual loss. Remote Sensing **12**(14), 2318 (2020). https://doi.org/10.3390/rs12142318
120. Zhu, J.Y., Park, T., Isola, P., Efros, A.A.: Unpaired image-to-image translation using cycle-consistent adversarial networks. In: Proceedings of the IEEE International Conference on Computer Vision (ICCV), pp. 2242–2251 (2017). https://doi.org/10.1109/ICCV.2017.244
121. Ziaja, M., Kowaleczko, P., Nalepa, J., Kostrzewa, D., Latini, D., De Santis, D., Salvucci, G., Petracca, I., Pegna, V.L., Del Frate, F., Kawulok, M.: Hyperspectral image pansharpening: the PRISMA case study. In: Proceedings of the IEEE International Geoscience and Remote Sensing Symposium (IGARSS), pp. 1633–1636 (2023). https://doi.org/10.1109/IGARSS52108.2023.10282612

Real-World Unsupervised Remote Sensing Image Super-Resolution: Addressing Challenges, Solution, and Future Prospects

Divya Mishra and Ofer Hadar

1 Introduction to Image Super-Resolution

Image super-resolution (ISR) refers to enhancing the resolution and level of detail in a digital image while zoomed in or upscaling. This technique involves generating a high-resolution (HR) version of an image from a lower-resolution (LR) source. Super-resolution aims to recover finer details and textures that are not readily differentiable in the original low-resolution image. It finds applications in various fields, including remote sensing [35, 36], medical imaging [15, 41], surveillance [2], and photography [14], where improved image clarity and sharpness are essential for analysis, diagnosis, or visual appeal. Usually, deep learning-based ISR models are designed using ground-truth HR pairs for the same scene as training samples to extract high-level feature details for reconstruction.

Reconstruction-based SR algorithms deal with SR problems inversely [27]. It first creates a fake LR image by blurring followed by downsampling, mostly bicubic interpolation as presented in Eq. 1:

$$I_{LR} = (I_{HR} * k)_{\downarrow s} \tag{1}$$

where k is the standard Gaussian kernel, I_{HR} is the original high-resolution image, I_{LR} is the artificially created low-resolution image, and s is the desired downscaling factor.

In applications that require real-time processing, arranging HR images for exact LR scenes is almost tricky and sometimes impossible for a few scenes that cannot be

D. Mishra (✉) · O. Hadar
School of Electrical and Computer Engineering, Ben Gurion University of the Negev, Beersheva, Israel

M. Kawulok et al. (eds.), *Super-Resolution for Remote Sensing*, Unsupervised and Semi-Supervised Learning, https://doi.org/10.1007/978-3-031-68106-6_2

recreated. In such scenarios, unsupervised image super-resolution is essential where obtaining paired training data is impractical or impossible. It allows enhancement of image quality and detail without relying on ground-truth high-resolution images, making it a valuable tool in various fields and applications. A broad classification of unsupervised ISR methods according to the availability of datasets is shown in Fig. 1.

However, suppose the training dataset already has some improved spatial resolution and at least a few noticeable visual features. In that case, this notion of making fake LR or paired samples is a useful one as usually done in self-supervised learning-based approaches. However, such methods do not perform well with extremely low-resolution real-world images captured under arbitrary degradations. In this chapter, a study is examined for two such real-world datasets: CubeSat satellite imagery (from BGUSat [34]) and plant root images (from RootCam [32] that are underground images of roots). To understand better what we mean by such real-world images, a glimpse of the BGUSat dataset and its associated patches at different locations is presented and shown in Fig. 2. One could observe the image quality is pretty low, the edges are greatly affected by blurring and blocky pattern

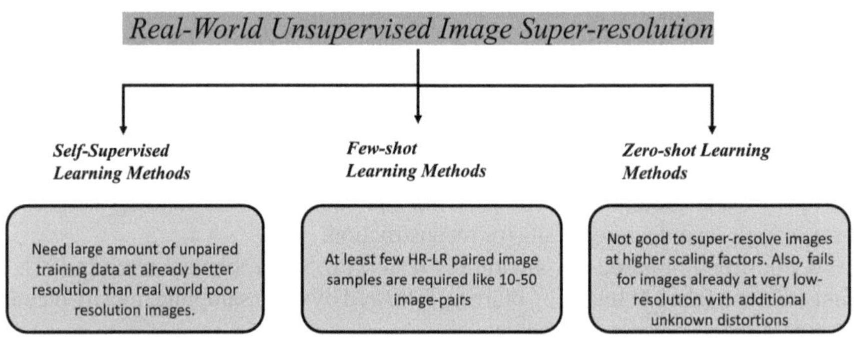

Fig. 1 Broad classification of unsupervised image super-resolution methods based on deep-learning: (1) self-supervised methods [4, 6, 18, 19, 40, 54], (2) few-shot learning methods [25, 28, 53], and (3) zero-shot learning based methods [3, 7, 8, 13, 36, 44, 57]

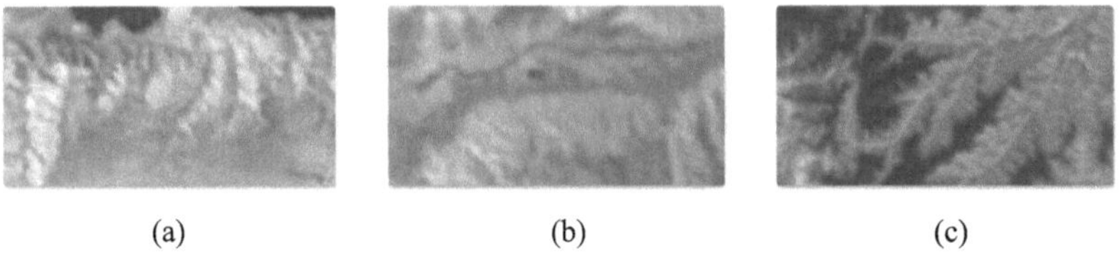

Fig. 2 An illustration of 600 m/pixel spatial resolution: an example extremely low-resolution real-world remote sensing image from BGUSat. (**a**) Edges are not greatly visible to distinguish. (**b**) Texture and edges both are not clearly visible. (**c**) Fine details and edges between the valley and ground area are not visible

effects, and the texture can also be described as very rough, which is understandable considering the 600 m per pixel spatial resolution.

1.1 Motivation

Enhancing the super-resolution of real-world remote sensing images stems from the critical need to overcome the inherent limitations of the compact designed image-capturing devices while maximizing their potential in various applications. With the small form factor and constrained resources, CubeSats often struggle to capture high-resolution images [43] comparable to more extensive and sophisticated satellite systems. This limitation impacts their effectiveness in applications such as remote sensing, scientific research, environmental monitoring, disaster management, and even commercial endeavors. The primary motivations for super-resolution techniques, particularly for real-world remote sensing images, are:

- Improved spatial detail [45]: Super-resolution methods generate images with greater spatial information when zoomed in by different upscale than individual low-resolution captures can achieve. This improved resolution can enable better identification and analysis of objects, features, and changes on the Earth's surface.
- Enhanced remote sensing [16, 30]: Higher spatial resolution is essential for accurate and detailed remote sensing applications. Super-resolution provides a clearer view of land cover, vegetation health, urban development, and natural resources, enabling more informed decision-making in areas like agriculture, forestry, and land-use planning.
- Scientific research [9]: Super-resolution can aid scientific research by providing more explicit imagery for studying geological formations, climatic changes, ocean dynamics, and environmental phenomena. The finer details captured through enhanced resolution can lead to deeper insights and an improved understanding of these processes.
- Disaster management [42]: During emergencies such as natural disasters or environmental hazards, high-resolution imagery is crucial for assessing damages, planning relief efforts, and monitoring recovery progress. Super-resolution can enhance the utility of CubeSats in such critical scenarios.
- Commercial applications [1]: In commercial sectors like cartography, urban planning, and infrastructure monitoring, high-resolution imagery is essential. By enabling CubeSats to capture clearer images, super-resolution can expand the range of commercial services offered by these satellites.
- Budgetary considerations: CubeSats are favored for their cost-effectiveness, making them an attractive option for various missions. Enhancing their imaging capabilities through super-resolution allows organizations to achieve high-resolution results without investing in significantly larger and more expensive satellites.

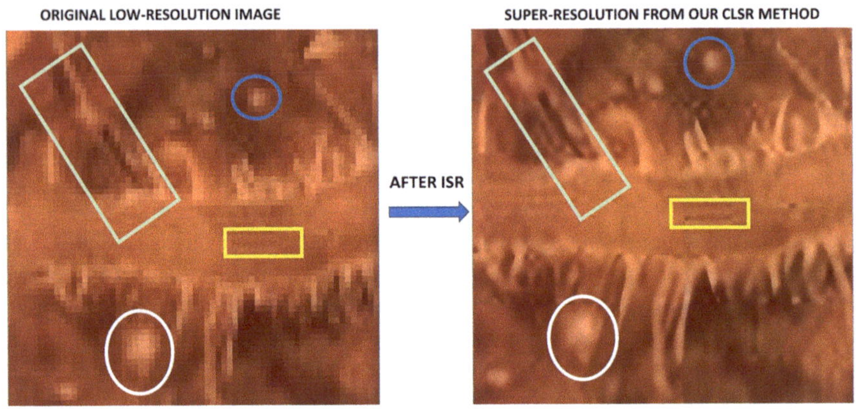

Fig. 3 Visualization of real-world plant root image (bell pepper) before and after ISR (CLSR [35])

- Technological advancements [35, 36]: Advancements in image processing, machine learning, and computational techniques have made it possible to effectively apply super-resolution algorithms to low-resolution input images, offering a practical solution for enhancing CubeSat imagery.

1.2 Contributions

Our contributions to this study are summarized as:

- Broad taxonomy of unsupervised image super-resolution is classified into three categories that include self-supervised learning methods, few-shot learning-based methods, and zero-shot learning-based methods.
- A qualitative and quantitative comparison using image quality measures of the ISR algorithms has been carried out across multiple real-world datasets.
- Using real-world datasets with arbitrary degradations, the study demonstrated some of the most common challenges and motivation to super-resolve them.
- Lastly, some of the potential future research directions have been discussed in detail.

2 Related Work

2.1 Supervised Learning Methods

Supervised learning methods like SRCNN [12], SRGAN [24], ESRGAN [51], etc. employ the relationship from paired HR-LR data. The existing supervised learning-based algorithms [20] that use downsampling techniques such as bicubic interpolation [21] are artificial degradations that come close to matching real-world degradations. These approaches aim to iteratively minimize the difference between the reconstructed (or predicted) and original ground-truth images until the loss function saturates. Also, these supervised learning-based approaches need paired ground-truth data to extract features and learn the relationship, which is expensive and sometimes complicated to arrange in the case of remote-sensing images. Also, models are complex in terms of architectural design. These challenges are a big hurdle for real-world images without paired HR counterpart images. These approaches are based on the learned nonlinear relationship to reconstruct missing information. They will require a massive number of paired (HR-LR) datasets for training, which could not only require a lot of computational time and resources but will also be complicated to train due to many trainable parameters. Additionally, these methodologies usually fail to apply to various types of image datasets with varying distributions and without paired ground-truth images. To tackle these challenges, unsupervised methods for ISR are required.

2.2 Unsupervised Learning Methods

In 2018, the first unsupervised CNN-based network was designed for the task of ISR, called zero-shot super-resolution (ZSSR) [44], which aims to perform super-resolution from a single low-resolution image without any training. It utilizes a self-example-based method, where the algorithm learns from the image itself to generate a higher-resolution version. ZSSR leverages the self-similarity present in natural images. The method has limitations when used for no-ideal datasets and, hence, is not feasible for real-world image ISR. Further, a deeper feed-forward network DBPN [17] was introduced that focuses on iterative up-projection of features from low to high resolution. It uses a deep back-projection network to progressively refine the HR estimation by integrating information from multiple intermediate upscaled feature maps. This method is computationally intensive but can achieve good performance. Another self-supervised learning method, Noise2Noise super-resolution (N2N-SR) [46], was developed for fluorescence microscopy images to improve the denoising performance of existing relatively low-performing methods in real scene noise removal. N2N-SR leverages the idea that training on noisy data can lead to robust models. By introducing noise during training and learning to map noisy images to clean high-resolution images, N2N-SR aims to handle real-world noisy inputs without explicit noise modeling. Another novel concept is deep image prior (DIP) [47], where the structure of the deep neural network itself serves as a prior for image reconstruction. During training, the network is not provided with pairs of low- and high-resolution images. Instead, it starts with a random initialization and optimizes the weights to reconstruct the input image. Recently, a fusion-based framework [33, 37] was introduced that focused mainly on extremely low-resolution real-world remote sensing datasets like BGUSat.

The approach fuses different features extracted from a single LR image by an optimum weight. These fusion weights are decided by iterative optimization of two NR-IQM scores, BRISQUE [38] and NIQE [39]. Based on this framework, a network called Self-FuseNet [36] is designed to add a customized UNet for image reconstruction and noise artifact removal for the generated fused image. The same BGUSat data is further enhanced using more advanced contrastive learning methods. Contrastive learning [5] is a self-supervised comparison of two samples to identify characteristics and traits that distinguish one data class from another, improving performance on visual tasks. Existing methods are based on regular sample construction approaches (e.g., considering the dissimilar samples of the same dataset as negative samples and data-augmented samples of a similar scene as positive samples) instead of exploring a task-friendly one. We proposed CLSR [35]: a naive sample construction method where multiple positive samples of the same scene are produced using pre-trained image super-resolution deep neural networks, and any open-source image dataset can be taken into consideration as a set of negative samples whose image distribution matches that of the low-resolution test dataset. Further, this network is slightly modified to NST-CL [34], for more promising results specially on higher upscaling factor-like 8. The framework utilizes the output of CLSR [35] as an "artificial style image," which is utilized as a style image, and the original low-resolution image as a "content image" for neural style transfer learning for image super-resolution in an unsupervised manner. The idea is to use high-resolution textures and features as a style and transfer on an original low-resolution content image as base elements. The proposed framework has three benefits: (1) The framework is capable of super-resolving different modalities of data like single-band remote sensing images, multi-spectral band images, RGB remote sensing images, and real-world natural images. The proposed CLSR [35] method is also used to test plant root images [31] and help to extract various improved root trait measurements essential for overall plant growth and precision agriculture.

These methods represent a range of approaches to unsupervised image super-resolution, each with its strengths and limitations. The choice of method depends on factors such as computational resources, the quality of results required after ISR, and the specific characteristics of the input images.

3 Challenges for Real-World Image Datasets

Challenges in this domain stem from the inherent complexities of remote sensing data, including varying weather conditions, sensor limitations, and geometric distortions. The absence of paired high-resolution and low-resolution images for training increases the difficulty of designing effective super-resolution models. Adapting algorithms to diverse landscapes and imaging conditions poses a substantial challenge. Some of the key challenges include:

- Limited hardware: For example, CubeSats and underground image acquisition systems with restricted physical dimensions and limited payload capacity often use small lenses and sensors. These hardware limitations directly impact the quality and amount of information captured in each image, making it challenging to extract sufficient details for super-resolution.
- Noise and artifacts [49]: Remote sensing images are susceptible to noise and artifacts due to factors such as sensor limitations, vibrations, and thermal variations.
- Data availability: Acquiring high-resolution remote sensing images for training neural networks is expensive and challenging, especially for dynamic or rapidly changing scenes, as well as for missions with limited resource opportunities.
- Motion and distortion: CubeSats often experience movement, rotation, and orbital variations, which can introduce geometric distortions in images. Super-resolution algorithms must account for and rectify these distortions to ensure accurate reconstruction.
- Processing constraints: Systems having limited onboard computational resources for image processing and implementing computationally intensive super-resolution algorithms in real-time or near-real-time scenarios can strain these resources and impact the efficiency of the satellite's operation.
- Validation and evaluation [10, 52]: Validating the effectiveness of super-resolution methods for remote sensing can be challenging due to the lack of ground-truth high-resolution data. Accurately assessing the improvement in spatial resolution and overall image quality is vital but can be difficult without reliable reference data.
- Algorithm adaptation: Many super-resolution algorithms are designed for more extensive and more traditional satellite systems, assuming specific sensor properties and image acquisition characteristics. Adapting these algorithms to other hardware and constraints requires careful consideration and adjustments.
- Trade-offs and resources: There is a trade-off between achieving higher spatial resolution and efficiently utilizing the limited resources of CubeSats. Decisions need to be made regarding allocating power, bandwidth, and memory for super-resolution processing versus other mission priorities.
- Real-time constraints [11]: Some applications require real-time or near-real-time image analysis. Implementing super-resolution techniques within tight time constraints while maintaining the desired image quality can be demanding. Also, training deep learning models for super-resolution on large datasets can be computationally expensive and require substantial computational resources.
- Algorithm selection [55]: Choosing the appropriate super-resolution algorithm for particular remote sensing images is critical. Different algorithms have varying strengths and weaknesses, and selecting the right one that aligns with distortions and constraints in different datasets is crucial.

Addressing these challenges requires a combination of advanced algorithm development, sensor improvements, smart mission planning, and careful optimization to ensure that super-resolution techniques can effectively enhance remote sensing imagery while considering the unique constraints of these platforms, too.

4 Potential Solutions

Unsupervised real-world remote sensing image super-resolution presents a unique challenge due to the lack of ground-truth high-resolution images for training. There could be some potential solutions that researchers and practitioners might explore:

- **Unsupervised representation learning:** Leveraging unsupervised learning algorithms such as autoencoders or variational autoencoders, one can learn a compact representation of remote sensing images. By reconstructing the high-resolution images from the learned representation, super-resolution can be achieved without explicit supervision.
- **Weakly supervised learning:** Instead of having precise high-resolution ground-truth labels, weakly supervised learning techniques operate with coarse annotations or partial information. Methods like image-level labels or bounding boxes can guide the learning process without pixel-level supervision.
- **Transfer learning from similar domains:** Transfer learning leverages pre-trained models on related tasks or datasets to bootstrap the training process on a new dataset with limited annotations. Models pre-trained on natural image datasets like ImageNet can be fine-tuned on remote sensing data to learn features relevant to super-resolution. Our proposed network NST-CL [34] is an example of using this idea for unsupervised image super-resolution.
- **Domain adaptation:** Remote sensing images may exhibit domain shifts due to differences in sensor characteristics, atmospheric conditions, or geographic locations. Domain adaptation techniques aim to align feature distributions between the source domain (e.g., synthetic data or labeled remote sensing images) and the target domain (real-world remote sensing images) to improve generalization performance.
- **Multimodal fusion:** Combining information from multiple sources such as multispectral, hyperspectral, LiDAR, or radar data can provide complementary cues for super-resolution. Multimodal fusion techniques integrate these diverse data sources to enhance the spatial resolution of remote-sensing images. Our proposed network Self-FuseNet [36] utilizes this idea for unsupervised image super-resolution.
- **Active learning:** Active learning strategies select the most informative samples for annotation to maximize the learning efficiency. By iteratively selecting samples that are most uncertain or would provide the most improvement in model performance, active learning can reduce the annotation burden for training super-resolution models.
- **Semi-supervised learning:** Semi-supervised learning utilizes a combination of labeled and unlabeled data for training. By leveraging both the small set

of labeled high-resolution images and the abundant unlabeled low-resolution images, semi-supervised learning methods can improve super-resolution performance. Our proposed network CLSR [35] utilizes this idea for unsupervised image super-resolution.

These approaches can be explored individually or in combination to effectively address the challenges of unsupervised real-world remote sensing image super-resolution.

5 Experimental Settings

5.1 Datasets and Metrics

5.1.1 Dataset

Four different datasets have been used for demonstration purposes. Out of them, BGUSat and RootCam are real-world datasets, while the other two are standard remote sensing datasets whose corresponding high-resolution is available.

- **BGUSat: A 600 m/pixel spatial resolution CubeSat Satellite data.** BGUsat is Israel's first single SWIR band nanosatellite designed for educational purposes, in collaboration with Ben-Gurion University of the Negev (BGU), Israel Aerospace Industries (IAI), and the Israeli space agency; the dataset comprises nearly 2000 images of various scenes over different regions. The images have an extremely low spatial resolution of 600 m/pixel and have arbitrary degradations. Few ISR methods have been proposed for particularly BGUSat dataset [33–36], which have the potential to enhance the spatial resolution of such real-world degraded imagery significantly, unlocking new possibilities for different applications in future.
- **RootCam: A plant root image data.** RootCam [31] is a fully automated tool designed for monitoring the root images to be utilized for further image processing and computer vision tasks for the sake of overall plant growth and precision agriculture [32]. RootCam is a device intended to track the roots of a plant into the ground. Researchers, farmers, and gardeners who wish to understand more about the growth and development of their plants' root systems can utilize it. One can examine the data on your own or with their premium service.
- **UC-Merced dataset [56]:** It is a 0.3-m high spatial resolution remote sensing dataset published primarily for image classification tasks. However, it is also used for the ISR task by downsampling the ground truth followed by Gaussian blur to generate the corresponding LR.
- **PROBA-V dataset [29]:** A multispectral satellite image dataset from 74 hand-selected regions around the globe at different points in time. It is available in two bands, RED and NIR, in different spatial resolutions, 100 m and 300 m.

5.1.2 Image Quality Assessment Metrics

The quantitative evaluation is done by no-reference image quality assessment metrics (NR-IQAM) like DOM [22], BRISQUE [38], NIQE [39], and PIQUE [48].

- **BRISQUE:** BRISQUE (blind/referenceless image spatial quality evaluator) [38] is designed to evaluate the perceived quality of digital images without relying on a reference (i.e., without comparing the image to an original or high-quality version). BRISQUE aims to predict the perceptual quality of an image by analyzing its spatial domain statistics. It uses a set of handcrafted features derived from the spatial domain statistics of an image. These features capture information about the distribution of pixel intensities, gradients, and other characteristics. The feature set is then used to train a support vector machine (SVM) model. The SVM is trained on a dataset of images with corresponding quality scores, learning the relationship between the handcrafted features and perceived image quality. BRISQUE finds applications in various fields, including image and video processing, where it can be used to automatically assess the quality of images without the need for human-generated reference scores. While BRISQUE is an applicable no-reference quality metric, its performance can vary depending on the type of distortions present in the images, and it may not capture all aspects of perceptual quality. For BRISQUE, a lower score denotes higher perceptual quality.
- **NIQE:** The natural image quality evaluator, or NIQE [39], is a completely blind image quality analyzer that relies only on measurable differences from statistical regularities found in natural images. It does not undergo any exposure to distorted images or training on human-rated images. However, all existing state-of-the-art general-purpose no-reference (NR) IQA algorithms require training examples and matching human opinion ratings that provide information about expected distortions. Based on a straightforward and effective space domain natural scene statistic (NSS) approach, it consists of building a quality-aware collection of statistical characteristics. These characteristics come from an unaltered, realistic image dataset. The findings of the experiments indicate that the new index performs comparable with the best-performing NR IQA models, which need to be trained on substantial datasets of subjective assessments of distorted images. A smaller score indicates better perceptual quality for NIQE.
- **PIQUE:** Like NIQE, perception image quality evaluator (PIQUE) [48], also derives image quality evaluations. The gray-level co-occurrence matrix (GLCM) and its function characterize an image's texture by determining how frequently pairs of pixels with particular values occur in an image. After GLCM is created, statistical measures are extracted from this matrix to provide information about the texture of the image, such as contrast, energy, etc. At the same time, the LBP features encode the local information that is derived from GLCM. These descriptor results are fed into SVM to train the machine; the machine's training success determines how accurate the outcome is. For PIQE, a lower score denotes higher perceptual quality.

- **DOM:** Difference of medians (DOM) [22] is a metric to quantify the sharpness of an image. The metrics are primarily intended to measure the sharpness of document page images performing well on natural scenes. It is specifically made to predict sharpness in instances when camera motion (or handshake) causes images to become blurry. For DOM, a higher score denotes higher perceptual quality that signifies more sharpness in the predicted image.

5.2 *Evaluation Results and Analysis*

To reconstruct a super-resolution image, we used the following methods for comparison as the baseline: ZSSR [44], SRCNN [12], SRGAN [24], ESGRAN [51], LapSRN [23], EDSR [26], MIP [50], Self-FuseNet [36], CLSR [35], and NST-CL [34] of which three methods—**NST-CL [34], CLSR [35], and Self-FuseNet [36]**—for unsupervised remote sensing ISR are developed by us. All the experiments are conducted for the scaling factor of 2, 3, 4, and 8 for different modalities of the test datasets.

Four benchmark datasets are used for testing, including BGUSat [34] (600 m/pixel spatial resolution), UC-Merced [56] (0.3 m spatial resolution), root-hair data [32] (real-world plant root dataset), and PROBA-V [29] (100 m/pixel spatial resolution) multispectral dataset.

5.2.1 Qualitative Results

The following observations could be seen for different modalities of the dataset:

- For BGUSat [34], two test images of the Israel Dead Sea region and China border region are reproduced by various ISR methods as shown in Figs. 4 and 5, respectively.
- RootCam dataset [32]: We used two different plant images for testing the root hair structure, bell pepper as shown in Fig. 6 and Arabidopsis in Fig. 7.

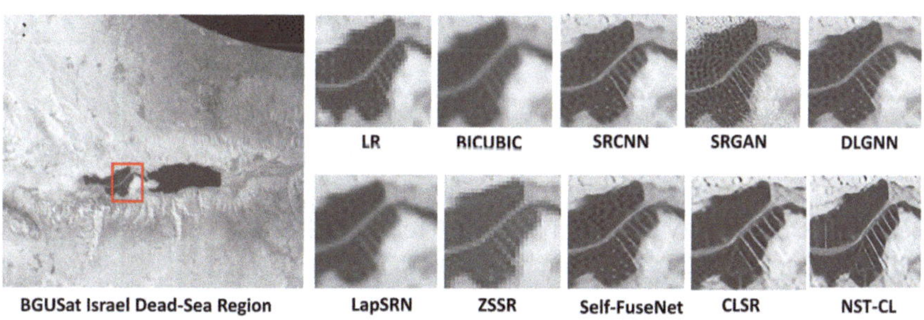

Fig. 4 Qualitative comparison of different ISR methods vs. ours (**NST-CL [34], CLSR [35], and Self-FuseNet [36]**) for upscaling factor of ×2 on BGU-Sat Israel Dead Sea image

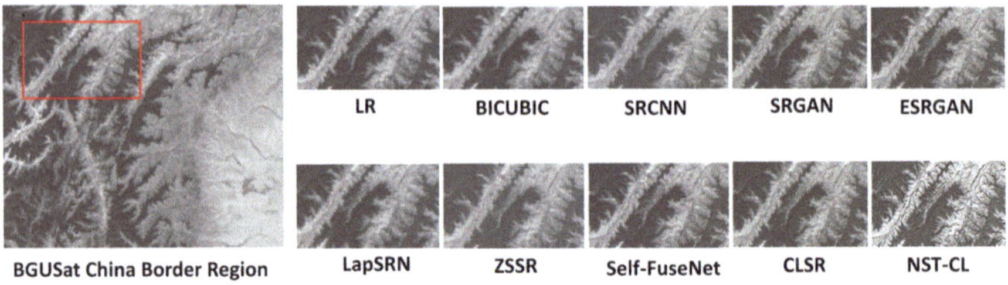

Fig. 5 Qualitative comparison of different ISR methods vs. ours (**NST-CL [34], CLSR [35], and Self-FuseNet [36]**) for upscaling factor of x2 on BGU-Sat China border image

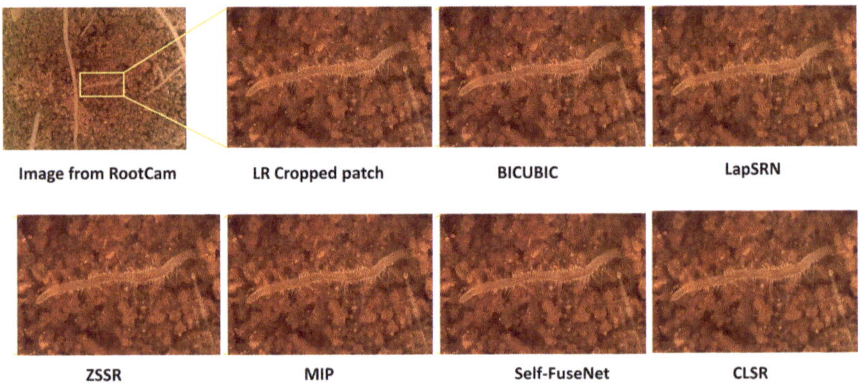

Fig. 6 Qualitative comparison of different ISR methods for the upscaling factor of ×4 on RootCam data: bell pepper plant root image

- UC-MERCED [56]: It is a high resolution of 0.3-m remote sensing dataset, whose ground truth is also available. We used five different text images, each from a different class: airplane in Fig. 8, building in Fig. 9, harbor in Fig. 10, runway in Fig. 11, and tennis court in Fig. 12. Supervised methods were also used for two test images, harbor in Fig. 13 and airplane in Fig. 14.
- PROBA-V [29]: A multispectral 100-m resolution dataset as shown in Fig. 15 and Fig. 16.

5.2.2 Quantitative Results

For image quality assessment, four different NR-IQM measures are used, including DOM [22], BRISQUE [38], NIQE [39], and PIQE [48]. The SR results by different methods are objectively evaluated in Table 1 for the BGUSat Israel Dead Sea region and Table 2 for the BGUSat China border region.

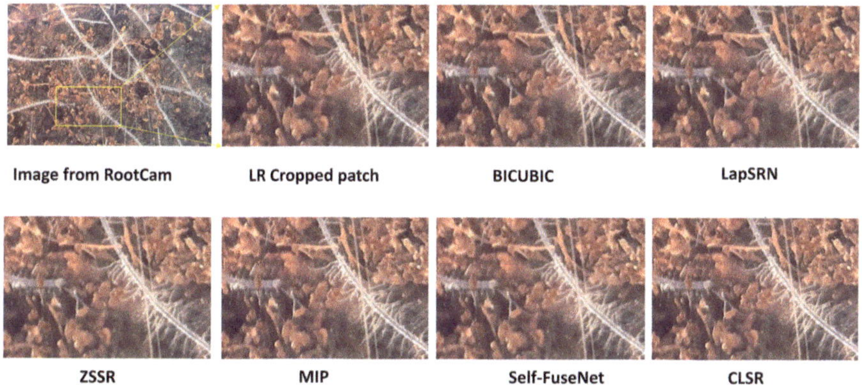

Fig. 7 Qualitative comparison of different ISR methods for the upscaling factor of ×4 on RootCam data: Arabidopsis plant root image

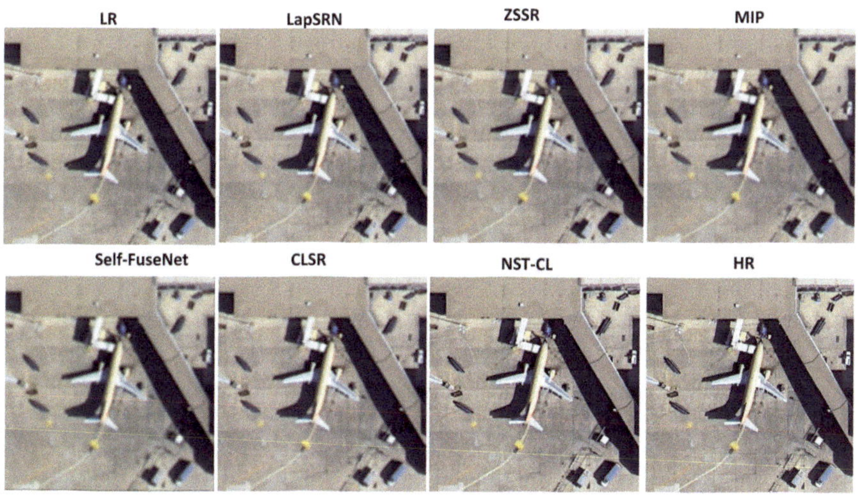

Fig. 8 Visual comparison of different ISR methods for the upscaling factor of ×4 on UC-MERCED data from class airplane

5.3 *Discussion*

The comparison findings on visual quality, objective quality, and complexity allow us to draw the following conclusions: (i) Compared unsupervised ISR methods are more recent and definitely better reconstruction methods of obtaining super-resolved images than traditional interpolation (such as "bicubic"). (ii) Overall, there is a clear difference in the visual effects, particularly texture and edge areas, between the super-resolved image and the equivalent LR version. For instance, over-smoothing or oversharpening artifacts are seen in specific ISR results like LapSRN [23] and

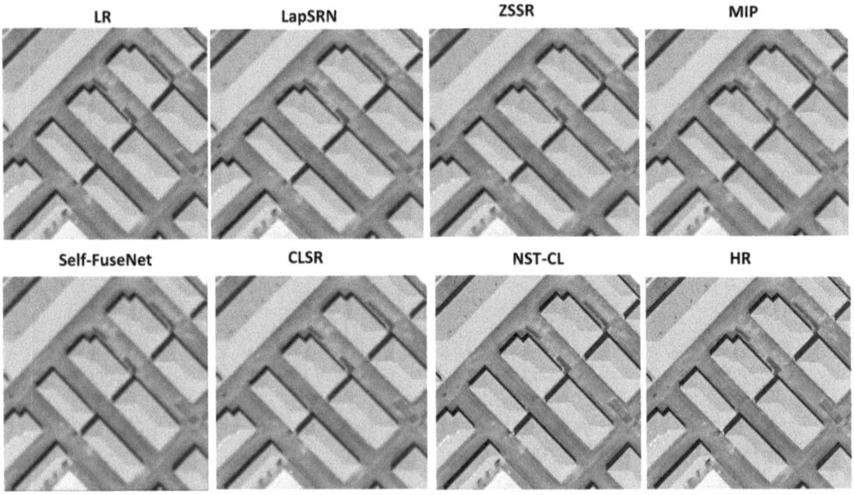

Fig. 9 Visual comparison of different ISR methods for the upscaling factor of ×4 on UC-MERCED data from class building

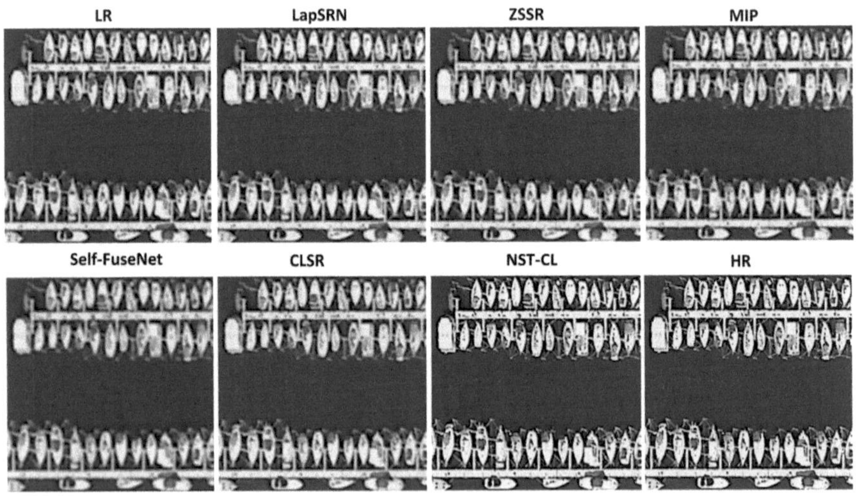

Fig. 10 Visual comparison of different ISR methods for the upscaling factor of ×4 on UC-MERCED data from class harbor

Self-FuseNet [36]. (iii) The results of CLSR [35] and NST-CL [34] are more convincing and significantly improve the sharp features of the LR image. (iv) The findings of subjective and objective assessments differ in a few ways. Furthermore, there is a possibility that the outcomes of several objective evaluation measures may not always be consistent with one another. (v) In general, unsupervised learning-based SR approaches for real-world remote sensing data have been successfully evolved to produce upsampled images.

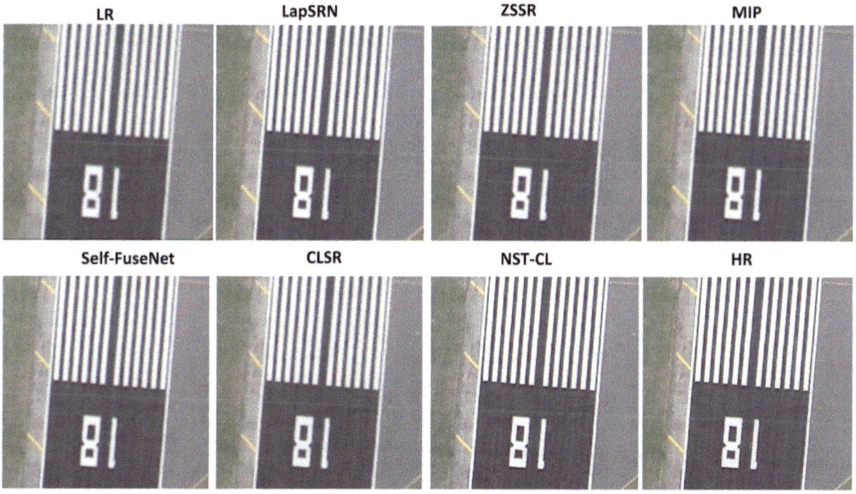

Fig. 11 Visual comparison of different ISR methods for the upscaling factor of ×4 on UC-MERCED data from class runway

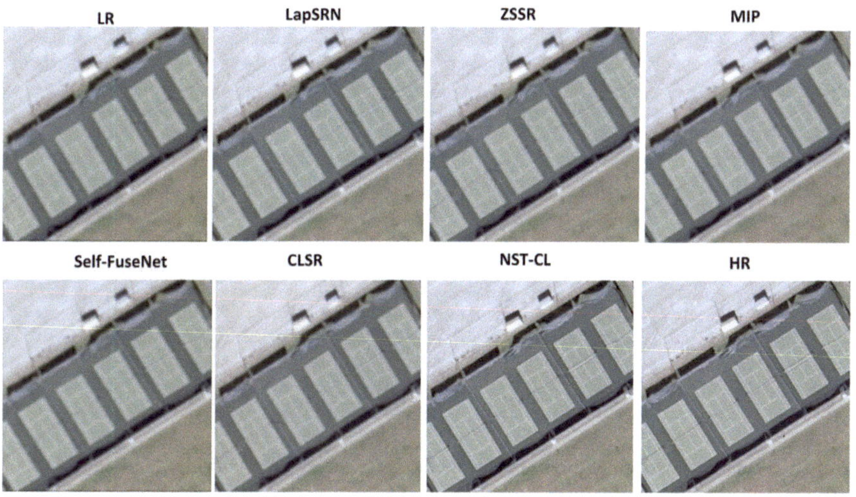

Fig. 12 Visual comparison of different ISR methods for the upscaling factor of ×4 on UC-MERCED data from class tennis court

6 Conclusion and Future Applications

The growing need for high-resolution imaging has encouraged the investigation of super-resolution techniques to improve the spatial features of remote sensing images. In remote sensing applications, where data is collected from a distance, the growing need for super-resolution techniques is driven by the desire for higher

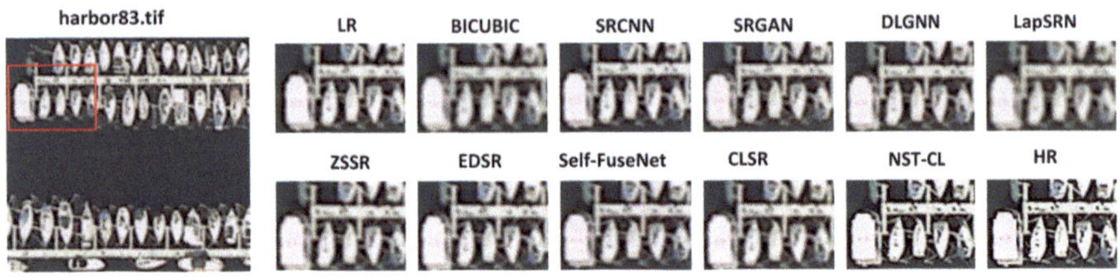

Fig. 13 Visual comparison of different ISR methods for the upscaling factor of ×2 on UC-MERCED data from class harbor

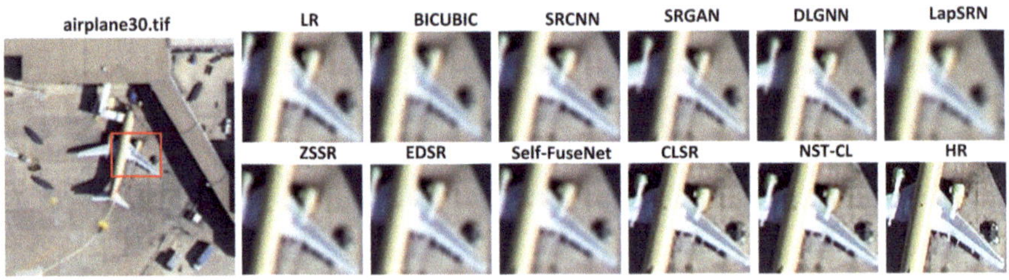

Fig. 14 Visual comparison of different ISR methods for the upscaling factor of ×2 on UC-MERCED data from class airplane

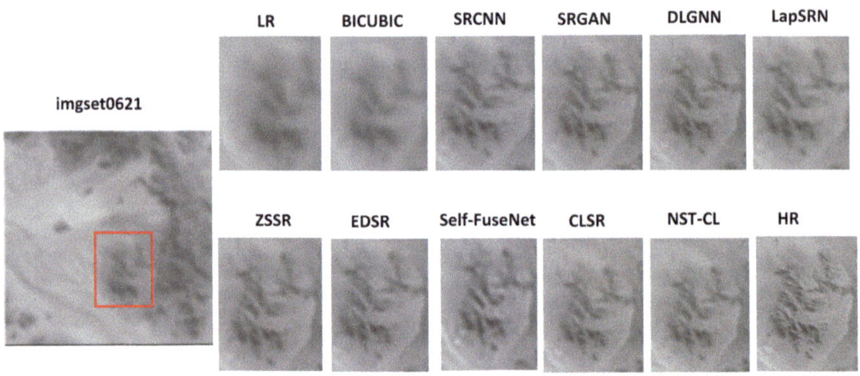

Fig. 15 Visual comparison of different ISR methods for the upscaling factor of ×4 on PROBA-V NIR band image

spatial resolution, increased accuracy, and improved interpretability of the acquired imagery.

Specific areas within remote sensing where the demand for super-resolution is evident include (1) precision agriculture, where super-resolution in satellite or drone imagery helps in obtaining higher-resolution maps of agricultural fields and enables the farmers to monitor crop health, detect diseases, and optimize resource usage with greater precision; (2) environmental monitoring which is crucial for tracking changes in land cover, studying ecosystems, and monitoring

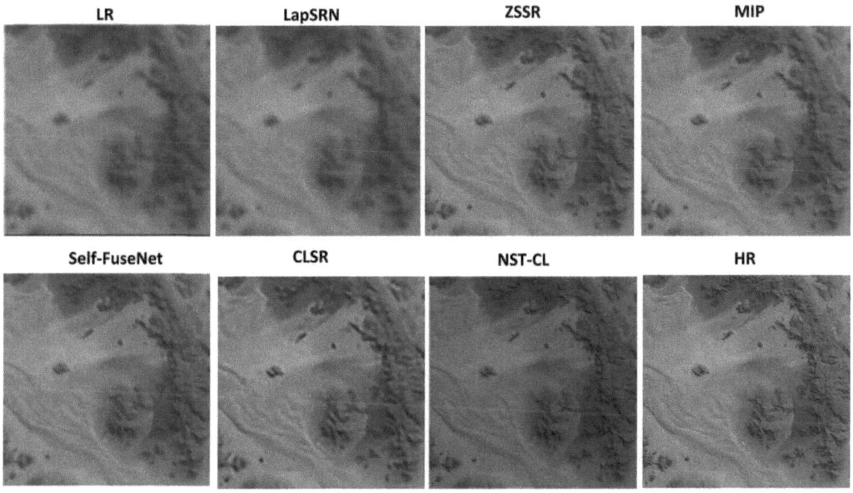

Fig. 16 Visual comparison of different ISR methods for the upscaling factor of ×4 on PROBA-V RED band image

the impact of climate change; (3) urban planning where super-resolution aids in obtaining clearer and more detailed images for urban planning purposes—this includes monitoring infrastructure, assessing land use, and managing city growth—(4) disaster management, where the effectiveness of enhanced satellite imagery for disaster response and management allows for better identification of affected areas, assessment of damage, and planning of rescue operations; (5) forestry management for assessment of tree health, identifying species, and monitoring deforestation in infrastructure monitoring where super-resolved images are valuable for monitoring critical infrastructure such as bridges, roads, and pipelines that further enables the identification of potential issues and aids in preventive maintenance; and (6) water resource management for detecting pollution sources and managing water resources more effectively and in wildlife conservation that allows for more accurate tracking of animal movements, population assessments, and habitat analysis.

As technology advances and the demand for more precise and detailed information from remote sensing data increases, super-resolution techniques will continue to play a crucial role in enhancing the capabilities of these applications. Extracting more information from remotely sensed data contributes to better decision-making and a deeper understanding of the Earth's surface and its changes over time. Domain adaptation strategies may enhance the model's ability to be generalized across various geographic locations and imaging scenarios. Investigating multimodal data fusion, such as combining information from several sensors, may improve the precision and resilience of super-resolution models even more. Furthermore, developing interpretability methods and explainable AI can help us comprehend how complicated super-resolution networks make decisions. This research highlights the importance of real-world unsupervised remote sensing image super-resolution and its consequences in several fields.

Table 1 Quantitative evaluation based on NR-IQM metrics at different scaling factors for Israel Dead Sea image

Metrics	Factor	Bicubic [21]	SRCNN [12]	SRGAN [24]	ESGRAN [51]	LapSRN [23]	ZSSR [44]	EDSR [26]	MIP [50]	NST-CL [34]
↓ BRISQUE	2	85.6251	63.4562	68.6563	58.4568	59.3646	81.4603	62.6754	60.2445	**57.7302**
	3	89.3646	73.3321	67.4352	**34.4579**	66.2995	75.6403	73.4092	72.5466	65.2307
	4	60.2995	55.6785	56.7842	55.3281	**31.4603**	59.8727	55.6755	57.6931	48.9223
	8	62.4635	64.5678	57.4564	54.5333	50.4322	62.3457	60.2345	57.8723	**49.5610**
↓ NIQE	2	35.2022	23.6543	21.6524	20.3245	37.5195	28.6001	24.2672	22.2808	**18.7461**
	3	27.0630	24.5678	21.3465	19.1625	28.5463	28.0028	26.6543	25.8653	**17.2069**
	4	15.7967	16.2344	15.3248	**13.6425**	15.0766	15.2996	14.7654	14.9772	15.0058
	8	16.7544	16.4532	18.6265	15.5265	17.4236	16.4328	19.6822	16.7863	**15.1973**
↓ PIQUE	2	48.7736	45.3245	43.2265	**39.5647**	46.6041	42.4608	46.4078	48.4070	40.0078
	3	58.5377	44.4562	43.4625	**42.4566**	54.3625	58.4825	55.7865	55.7865	52.2875
	4	72.2268	55.7852	56.6545	54.3365	65.9236	65.8729	63.7865	63.7865	**53.5932**
	8	73.4245	62.4288	57.1293	**46.6125**	62.4322	65.4354	65.7824	55.456	68.4562
↑ DOM	2	0.8981	0.7645	0.8565	0.9243	0.8994	0.6567	0.8125	0.8125	**1.1722**
	3	0.7418	0.7234	0.9665	0.9052	0.7654	0.5276	0.7266	0.7266	**1.1104**
	4	0.5898	0.6542	0.7255	0.7965	0.7415	0.4132	0.5422	0.5422	**0.9212**
	8	0.3536	0.4242	0.5132	0.5025	0.3934	0.3423	0.4428	0.4624	**0.5450**

Bold indicates first best score and underlined score indicates second best score

Table 2 Quantitative evaluation based on NR-IQM metrics at different scaling factors for China border image

Metrics	factor	Bicubic [21]	SRCNN [12]	SRGAN [24]	ESGRAN [51]	LapSRN [23]	ZSSR [44]	EDSR [26]	MIP [50]	NST-CL [34]
↓ BRISQUE	2	51.3264	70.4587	52.8963	<u>50.4276</u>	71.0053	74.3124	67.8643	61.0053	**48.2878**
	3	70.0239	52.4672	57.7628	<u>50.3688</u>	60.5473	62.0652	57.8324	60.5473	**47.9947**
	4	68.8509	62.5363	48.4797	**45.7523**	67.6931	<u>47.8643</u>	59.7854	57.6931	51.9007
	8	72.3456	67.5524	63.4568	<u>62.4567</u>	78.3467	64.4592	64.2744	53.4566	**58.3525**
↓ NIQE	2	42.6981	30.3278	26.5426	<u>25.1765</u>	46.2808	40.8822	28.8432	28.2808	**19.3043**
	3	24.8298	20.2465	<u>19.2525</u>	19.8245	27.8653	22.5658	20.5564	20.5658	**15.1974**
	4	19.2829	19.4425	21.6565	19.6784	35.9772	19.2898	<u>18.7654</u>	19.9772	**14.7884**
	8	25.4352	23.6665	22.4365	**20.2784**	39.8823	27.8432	25.6782	24.3454	<u>16.9626</u>
↓ PIQUE	2	49.6784	45.8291	40.3256	<u>25.2688</u>	50.4179	30.6788	45.4921	**22.6224**	30.4078
	3	59.6526	47.8293	45.2389	43.5793	58.8757	37.2265	54.7593	<u>32.5422</u>	**31.7865**
	4	69.4145	54.3822	47.8842	<u>47.4335</u>	62.0158	69.4455	62.7986	62.9758	**43.2825**
	8	65.3244	67.2381	<u>52.4497</u>	53.2765	67.8322	72.4578	72.7844	54.2367	**50.8934**
↑ DOM	2	0.6214	0.7672	0.8478	<u>0.9582</u>	0.6253	0.5678	0.7021	0.8253	**1.1738**
	3	0.6668	0.7467	0.8125	<u>0.8245</u>	0.6684	0.4897	0.6154	0.7684	**1.1072**
	4	0.5191	0.7782	0.8288	<u>0.8632</u>	0.6812	0.4254	0.5543	0.6812	**0.9406**
	8	0.3428	0.4212	0.4892	<u>0.5243</u>	0.3578	0.3667	0.4423	0.4534	**0.5972**

Bold indicates first best score and underlined score indicates second best score

References

1. Camps, A.: Nanosatellites and applications to commercial and scientific missions. In: Satellites Missions and Technologies for Geosciences, pp. 145–169 (2020)
2. Caner, G., Tekalp, A.M., Heinzelman, W.: Super resolution recovery for multi-camera surveillance imaging. In: 2003 International Conference on Multimedia and Expo. ICME'03. Proceedings (Cat. No. 03TH8698), vol. 1, pp. I–109. IEEE, Piscataway (2003)
3. Cha, Z., Xu, D., Tang, Y., Jiang, Z.: Meta-learning for zero-shot remote sensing image super-resolution. Mathematics **11**(7), 1653 (2023)
4. Chen, H., He, X., Yang, H., Wu, Y., Qing, L., Sheriff, R.E.: Self-supervised cycle-consistent learning for scale-arbitrary real-world single image super-resolution. Expert Syst. Appl. **212**, 118657 (2023)
5. Chen, T., Kornblith, S., Norouzi, M., Hinton, G.: A simple framework for contrastive learning of visual representations. In: International Conference on Machine Learning, pp. 1597–1607. PMLR (2020)
6. Chen, Y., Jiang, C., Pan, Y.: Single-photon image super-resolution via self-supervised learning. In: ICASSP 2023-2023 IEEE International Conference on Acoustics, Speech and Signal Processing (ICASSP), pp. 1–5. IEEE, Piscataway (2023)
7. Cheng, X., Fu, Z., Yang, J.: Zero-shot image super-resolution with depth guided internal degradation learning. In: Computer Vision–ECCV 2020: 16th European Conference, Glasgow, UK, August 23–28, 2020, Proceedings, Part XVII 16, pp. 265–280. Springer, Berlin (2020)
8. Cheng, Z., Xiong, Z., Chen, C., Liu, D., Zha, Z.J.: Light field super-resolution with zero-shot learning. In: Proceedings of the IEEE/CVF Conference on Computer Vision and Pattern Recognition, pp. 10010–10019 (2021)
9. Chin, A., Coelho, R., Nugent, R., Munakata, R., Puig-Suari, J.: Cubesat: the pico-satellite standard for research and education. In: AIAA Space 2008 Conference & Exposition, p. 7734 (2008)
10. Chiu, T.Y., Zhao, Y., Gurari, D.: Assessing image quality issues for real-world problems. In: Proceedings of the IEEE/CVF Conference on Computer Vision and Pattern Recognition, pp. 3646–3656 (2020)
11. Das, S., Dai, R., Koperski, M., Minciullo, L., Garattoni, L., Bremond, F., Francesca, G.: Toyota smarthome: Real-world activities of daily living. In: Proceedings of the IEEE/CVF International Conference on Computer Vision, pp. 833–842 (2019)
12. Dong, C., Loy, C.C., He, K., Tang, X.: Image super-resolution using deep convolutional networks. IEEE Trans. Pattern Anal. Mach. Intell. **38**(2), 295–307 (2016). https://doi.org/10.1109/TPAMI.2015.2439281
13. Emad, M., Peemen, M., Corporaal, H.: Dualsr: zero-shot dual learning for real-world super-resolution. In: Proceedings of the IEEE/CVF Winter Conference on Applications of Computer Vision, pp. 1630–1639 (2021)
14. Fatima, N.: Ai in photography: scrutinizing implementation of super-resolution techniques in photo-editors. In: 2020 35th International Conference on Image and Vision Computing New Zealand (IVCNZ), pp. 1–6. IEEE, Piscataway (2020)
15. Greenspan, H.: Super-resolution in medical imaging. Comput. J. **52**(1), 43–63 (2009)
16. Guilherme Teixeira Crusiol, L., Sun, L., Chen, R., Sun, Z., Zhang, D., Chen, Z., Wuyun, D., Rafael Nanni, M., Lima Nepomuceno, A., Boucas Farias, J.R.: Assessing the potential of using high spatial resolution daily NDVI-time-series from planet cubesat images for crop monitoring. Int. J. Remote Sensing **42**(18), 7114–7142 (2021)
17. Haris, M., Shakhnarovich, G., Ukita, N.: Deep back-projection networks for super-resolution (2018). CoRR abs/1803.02735. http://arxiv.org/abs/1803.02735
18. He, J., Yuan, Q., Li, J., Xiao, Y., Zhang, L.: A self-supervised remote sensing image fusion framework with dual-stage self-learning and spectral super-resolution injection. ISPRS J. Photogramm. Remote Sensing **204**, 131–144 (2023)

19. Hong, D., Yao, J., Li, C., Meng, D., Yokoya, N., Chanussot, J.: Decoupled-and-coupled networks: self-supervised hyperspectral image super-resolution with subpixel fusion. IEEE Trans. Geosci. Remote Sensing (2023)
20. Hu, J.: A review of super-resolution reconstruction based on supervised learning. In: 2019 International Conference on Artificial Intelligence and Advanced Manufacturing (AIAM), pp. 768–773. IEEE, Piscataway (2019)
21. Keys, R.: Cubic convolution interpolation for digital image processing. IEEE Trans. Acoust. Speech Signal Process. **29**(6), 1153–1160 (1981)
22. Kumar, J., Chen, F., Doermann, D.: Sharpness estimation for document and scene images. In: Proceedings of the 21st International Conference on Pattern Recognition (ICPR2012), pp. 3292–3295. IEEE, Piscataway (2012)
23. Lai, W.S., Huang, J.B., Ahuja, N., Yang, M.H.: Fast and accurate image super-resolution with deep laplacian pyramid networks. IEEE Trans. Pattern Anal. Mach. Intell. **41**(11), 2599–2613 (2019). https://doi.org/10.1109/TPAMI.2018.2865304
24. Ledig, C., Theis, L., Huszar, F., Caballero, J., Cunningham, A., Acosta, A., Aitken, A., Tejani, A., Totz, J., Wang, Z., Shi, W.: Photo-realistic single image super-resolution using a generative adversarial network. In: 2017 IEEE Conference on Computer Vision and Pattern Recognition (CVPR), vol. 2017, pp. 105–114. IEEE, Piscataway (2017). https://doi.org/10.1109/CVPR.2017.19
25. Li, X., Jin, X., Fu, J., Yu, X., Tong, B., Chen, Z.: Few-shot real image super-resolution via distortion-relation guided transfer learning (2021). arXiv preprint arXiv:2111.13078
26. Lim, B., Son, S., Kim, H., Nah, S., Lee, K.M.: Enhanced deep residual networks for single image super-resolution. In: 2017 IEEE Conference on Computer Vision and Pattern Recognition Workshops (CVPRW), vol. 2017, pp. 1132–1140. IEEE, Piscataway (2017). https://doi.org/10.1109/CVPRW.2017.151
27. Lin, Z., Shum, H.Y.: On the fundamental limits of reconstruction-based super-resolution algorithms. In: Proceedings of the 2001 IEEE Computer Society Conference on Computer Vision and Pattern Recognition. CVPR 2001, vol. 1, pp. I–I. IEEE, Piscataway (2001)
28. Ma, Y., Xiong, K., Hou, X., Zhang, W., Chen, X., Li, L., Yang, S.: Cascade neural approximating for few-shot super-resolution photoacoustic angiography. Appl. Phys. Lett. **121**(10), 103701 (2022)
29. Märtens, M., Izzo, D., Krzic, A., Cox, D.: Super-resolution of PROBA-V images using convolutional neural networks. Astrodynamics **3**(4), 387–402 (2019)
30. Mhangara, P.: The emerging role of cubesats for Earth observation applications in South Africa. Photogramm. Eng. Remote Sens **86**, 333–340 (2020)
31. Mishra, D., Chemweno, S., Hadar, O., Ben-Tovim, O., Lazarovitch, N., Ephrath, J.E.: Unsupervised image super-resolution for root hair enhancement and improved root traits measurements. IEEE Trans. AgriFood Electron. **2**(1), 81–90 (2024). https://doi.org/10.1109/TAFE.2024.3359660
32. Mishra, D., Chemweno, S., Hadar, O., Lazarovitch, N., Ephrath, J.E.: Deep-learning-based image super-resolution for enhanced root hair visualization and root traits analysis. In: Neale, C.M.U., Maltese, A. (eds.) Remote Sensing for Agriculture, Ecosystems, and Hydrology XXV, vol. 12727, p. 1272711. International Society for Optics and Photonics, SPIE (2023)
33. Mishra, D., Dror, I., Hadar, O., Choukroun, D., Maman, S., Blumberg, D.G.: A fusion-based framework for unsupervised single image super-resolution. In: Dolev, S., Gudes, E., Paillier, P. (eds.) Cyber Security, Cryptology, and Machine Learning, pp. 85–95. Springer, Cham (2023)
34. Mishra, D., Hadar, O.: Accelerating neural style-transfer using contrastive learning for unsupervised satellite image super-resolution. IEEE Trans. Geosci. Remote Sensing (2023). https://doi.org/10.1109/TGRS.2023.3314283
35. Mishra, D., Hadar, O.: Clsr: Contrastive learning for semi-supervised remote sensing image super-resolution. IEEE Geosci. Remote Sensing Lett. **20**, 1–5 (2023). https://doi.org/10.1109/LGRS.2023.3294595
36. Mishra, D., Hadar, O.: Self-fusenet: data free unsupervised remote sensing image super-resolution. IEEE J. Sel. Top. Appl. Earth Observ. Remote Sensing **16**, 1710–1727 (2023)

37. Mishra, D., Hadar, O., Dror, I., Maman, S., Zagrizak, L., Aspir, L., Choukroun, D., Blumberg, D., Geltser, Y., Nisany, O., Shyriayev, A.: General architecture for unsupervised single-image super-resolution of single band nano-satellitebgusat via image-specific feature extraction. Proceedings of the International Astronautical Congress, IAC 2022 (2022). Funding Information: This work is supported by a grant (Grant No. 3-17380) from the Ministry of Science and Technology, Israel. Publisher Copyright: Copyright © 2022 by the International Astronautical Federation (IAF). All rights reserved.; 73rd International Astronautical Congress, IAC 2022 ; Conference date: 18-09-2022 Through 22-09-2022
38. Mittal, A., Moorthy, A.K., Bovik, A.C.: No-reference image quality assessment in the spatial domain. IEEE Trans. Image Process. **21**(12), 4695–4708 (2012). https://doi.org/10.1109/TIP.2012.2214050
39. Mittal, A., Soundararajan, R., Bovik, A.C.: Making a "Completely blind" image quality analyzer. IEEE Signal Process. Lett. **20**(3), 209–212 (2013). https://doi.org/10.1109/LSP.2012.2227726
40. Nguyen, N.L., Anger, J., Davy, A., Arias, P., Facciolo, G.: Self-supervised multi-image super-resolution for push-frame satellite images. In: Proceedings of the IEEE/CVF Conference on Computer Vision and Pattern Recognition, pp. 1121–1131 (2021)
41. Robinson, M.D., Chiu, S.J., Toth, C.A., Izatt, J.A., Lo, J.Y., Farsiu, S.: New applications of super-resolution in medical imaging. In: Super-Resolution Imaging, pp. 383–412. CRC Press, Boca Raton (2017)
42. Santilli, G., Vendittozzi, C., Cappelletti, C., Battistini, S., Gessini, P.: Cubesat constellations for disaster management in remote areas. Acta Astronaut. **145**, 11–17 (2018)
43. Schwartz, N., Brzozowski, W., Milanova, M., Morris, K., Todd, S., Ali, Z., Sauvage, J.F., Ward, A., Lunney, D., MacLeod, D.: High-resolution deployable cubesat prototype. In: Space Telescopes and Instrumentation 2020: Optical, Infrared, and Millimeter Wave, vol. 11443, pp. 565–577. SPIE (2020)
44. Shocher, A., Cohen, N., Irani, M.: Zero-shot super-resolution using deep internal learning. In: Proceedings of the IEEE Computer Society Conference on Computer Vision and Pattern Recognition, pp. 3118–3126 (2018). https://doi.org/10.1109/CVPR.2018.00329
45. Symolon, W., Dagli, C.: A survey of super-resolution techniques for a potential cubesat imagery system architecture. In: Recent Trends and Advances in Model Based Systems Engineering, pp. 91–99 (2022)
46. Tian, X., Wu, Q., Wei, H., Zhang, Y.: Noise2sr: learning to denoise from super-resolved single noisy fluorescence image. In: International Conference on Medical Image Computing and Computer-Assisted Intervention, pp. 334–343. Springer, Berlin (2022)
47. Ulyanov, D., Vedaldi, A., Lempitsky, V.S.: Deep image prior (2017). CoRR abs/1711.10925. http://arxiv.org/abs/1711.10925
48. Venkatanath N, Praneeth D, Maruthi Chandrasekhar Bh, Channappayya, S.S., Medasani, S.S.: Blind image quality evaluation using perception based features. In: 2015 Twenty First National Conference on Communications (NCC), pp. 1–6. IEEE, Piscataway (2015). https://doi.org/10.1109/NCC.2015.7084843
49. Wang, J., Shao, Z., Huang, X., Lu, T., Zhang, R., Li, Y.: From artifact removal to super-resolution. IEEE Trans. Geosci. Remote Sensing **60**, 1–15 (2022)
50. Wang, J., Shao, Z., Lu, T., Huang, X., Zhang, R., Wang, Y.: Unsupervised remoting sensing super-resolution via migration image prior. In: 2021 IEEE International Conference on Multimedia and Expo (ICME), pp. 1–6. IEEE, Piscataway (2021)
51. Wang, X., Yu, K., Wu, S., Gu, J., Liu, Y., Dong, C., Qiao, Y., Change Loy, C.: Esrgan: Enhanced super-resolution generative adversarial networks. In: Proceedings of the European Conference on Computer Vision (ECCV) Workshops (2018)
52. Wang, Z.: Objective image quality assessment: facing the real-world challenges. Electron. Imaging **2016**(13), 1–6 (2016)
53. Wu, J., Hovakimyan, N., Hobbs, J.: Genco: an auxiliary generator from contrastive learning for enhanced few-shot learning in remote sensing (2023). arXiv preprint arXiv:2307.14612

54. Xiao, Y., Yuan, Q., Jiang, K., He, J., Wang, Y., Zhang, L.: From degrade to upgrade: Learning a self-supervised degradation guided adaptive network for blind remote sensing image super-resolution. Inform. Fusion **96**, 297–311 (2023)
55. Yang, D., Li, Z., Xia, Y., Chen, Z.: Remote sensing image super-resolution: challenges and approaches. In: 2015 IEEE International Conference on Digital Signal Processing (DSP), pp. 196–200. IEEE, Piscataway (2015)
56. Yang, Y., Newsam, S.: Bag-of-visual-words and spatial extensions for land-use classification. In: Proceedings of the 18th SIGSPATIAL International Conference on Advances in Geographic Information Systems, pp. 270–279 (2010)
57. Ye, Z., Lyu, F., Li, L., Fu, Q., Ren, J., Hu, F.: Sr-gan: semantic rectifying generative adversarial network for zero-shot learning. In: 2019 IEEE International Conference on Multimedia and Expo (ICME), pp. 85–90. IEEE, Piscataway (2019)

Advancements in Deep Learning-Based Super-resolution for Remote Sensing: A Comprehensive Review and Future Directions

Saba Hosseini Tabesh, Masoud Babadi Ataabadi, and Dongmei Chen

1 Introduction

The limitations in the spatial and spectral resolution of Earth observation (EO) satellites are primarily attributed to constraints in optic and sensor technology, as well as the high costs involved in satellite launch and maintenance, especially for older satellites [1]. These constraints create significant challenges in critical fields such as change monitoring. The insufficient resolution hinders the ability to accurately discern fine details, identify objects, and extract meaningful information from the imagery [2–5]. For long-term building change detection research, for example, the only available option for building extraction during the late twentieth century and early twenty-first century is relatively low-resolution (LR) remotely sensed images [6–8].

To tackle the challenges faced in remote sensing, the utilization of image super-resolution (ISR) techniques has emerged as a crucial approach to enhance the quality, interpretability, and accuracy of remote sensing data [9]. ISR techniques, as described by [10], aim to enhance the spatial resolution from low-resolution (LR) images to high-resolution (HR) images beyond the initial resolution provided by the sensor. These techniques have been specifically developed to improve the spatial and spectral quality of remotely sensed images [11]. Out of the multiple applications of this approach, one is to aid in the analysis of satellite images. There are several applications where these can be used, for example, military surveillance, monitoring vegetation and animals, and monitoring areas prone to natural calamities [12]. Super-resolution (SR) was initially introduced by Tsai and Huang [13] for the restoration of bandlimited signals in multi-frame images. Research on SR is part of

S. Hosseini Tabesh · M. Babadi Ataabadi · D. Chen (✉)
Department of Geography and Planning, Queen's University, Kingston, ON, Canada
e-mail: 23ldp2@queensu.ca; 22mba1@queensu.ca; chendm@queensu.ca

M. Kawulok et al. (eds.), *Super-Resolution for Remote Sensing*, Unsupervised and Semi-Supervised Learning, https://doi.org/10.1007/978-3-031-68106-6_3

the image and signal processing study field, which includes two different branches. One branch aims to improve the interpretability of an optical imaging system, and the other focuses on reconstruction to obtain a higher-resolution image from low-resolution images. Most studies in remote sensing image RSISR belong to the latter branch.

Over the years, researchers have introduced many SR methods in this field. They can be categorized based on different criteria including underlying principles, data inputs, computational complexity, and application domains. For example, Lepcha et al. [14] categorized the SR techniques into four main groups based on the approaches used including interpolation-based, reconstruction-based, learning-based, and transformer-based algorithms. Based on the number of inputs, two kinds of SR can be distinguished in remote sensing: single-image super-resolution (SISR) and multi-image super-resolution (MISR). In the SISR method, only a single LR image of the objective scene is used to generate the HR output. In the MISR, several LR images of the same scene are used to create a higher-resolution result.

In recent years, there has been significant progress in deep learning (DL) techniques applied to various computer vision tasks in the field of remote sensing. Researchers such as P. Wang et al. [15] have made notable contributions to DL-based remote sensing image super-resolution (RSISR) methods. They have focused on improving scene classification, object detection, image segmentation, and image SR using DL techniques. The use of DL methods has also led to substantial improvements in the quality of satellite images. While there have been review papers on DL-based image super-resolution methods for natural images in computer vision [16–18], the remote sensing field has seen similar progress, particularly in single-image super-resolution (SISR) algorithms. However, there is a lack of comprehensive studies that cover both SISR and multi-image super-resolution (MISR) methods in the latest progress. Salvetti et al. [19] have made some progress in MISR, but a comprehensive review that encompasses both areas is rare. Tsagkatakis et al. [11] provide broad perspectives on different enhancement tasks but offer limited insights into DL-based SISR. Although some review articles have explored the application of DL methods to classification, segmentation, object detection, and pan-sharpening problems in remote sensing, there is currently no comprehensive review specifically focused on DL-based SISR and MISR algorithms. Therefore, the objective of our study is to conduct a comprehensive review of DL-based SISR and MISR methods, with a particular emphasis on DL models.

2 SISR Methods

In the field of remote sensing and photogrammetry, single-image super-resolution (SISR) has emerged as an essential tool for enhancing the resolution and quality of LR images [20]. SISR utilizes a variety of approaches to elevate the resolution of LR images [21], recover missing high-frequency details, and significantly improve the visual appearance of images in remote sensing and photogrammetry applications [20]. They enable the generation of HR versions of images without the need for additional information from multiple images or sensor data. These representations are then used to reconstruct high-quality images from low-quality ones by understanding the underlying features of the scene. The versatility of SISR has been widely recognized and applied in various domains, including medical imaging, surveillance, and digital photography [20].

In the context of single-image super-resolution (SISR), there are two main approaches: non-blind and blind. Blind SISR methods do not require auxiliary multispectral (MS) information or impose prior assumptions, with the simplest approach being an interpolation [12]. Some SISR methods utilize prior knowledge, often employing an example-based strategy, using internal similarities within the same image [22–25] or learning mapping functions from pairs of external LR and HR exemplars [22, 26, 27].

The HYSR framework proposed by Wu et al. [28] used a novel approach to address the issue of prior knowledge. It consists of two stages: MISR and SISR. MISR methods involve combining multiple LR images, each with its own unique misalignment, to capture more fine details when reconstructing the HR image [29, 30]. In the SISR stage, deep learning algorithms analyze large amounts of training data to uncover nonlinear relationships, and the co-occurrence prior, based on statistical pixel relationships within a single image, guides the algorithm to enhance the HR image further. CNN-based methods, such as the MRNN proposed by Lu et al. [31] and the modified super-resolution generative adversarial network (SRGAN), have been successful in addressing the prior knowledge of the SISR problem in remote sensing images. However, DL-based SISR for RSIs still faces challenges, as the current RSI datasets are not designed for real-world situations, and the LR–LR mapping deviates from the desired LR–HR mapping, adversely affecting the practical performance of DL models in super-resolving RSIs.

In the past, classical SR methods such as statistical methods, prediction-based methods, patch-based methods, edge-based methods, and sparse representation methods were used to achieve SR [32]. We call them the traditional method in this study. Most recent SISR techniques rely on neural networks (deep convolutional neural network (CNN), attention mechanism methods for image processing, and generative adversarial network (GAN)-based methods) and traditional methods [33] to learn complex representations of different versions of a region observed at different resolutions [33]. In this study, we categorize them into four groups, namely, traditional methods, convolutional neural networks (CNN)-based methods, generative adversarial network (GAN)-based methods, and attention mechanism methods.

2.1 Traditional Methods

The first category of SISR techniques is traditional methods that can be classified into interpolation-based, edge-based, and statistical methods. In the field of image processing, interpolation is commonly used as a traditional method in SR to estimate the intensity values of new pixels based on the LR image input [34]. It is an elementary kind of SR technique and is commonly utilized for the simplicity of zooming images.

2.1.1 Interpolation-Based

By estimating the pixel intensities on an up-sampled grid, interpolation techniques produce an HR image from its LR version. In this technique, permanent function kernels, local variance coefficients [21], and adaptive structure kernels [35, 36] are commonly utilized. There are several well-known interpolation-based methods such as nearest neighbor, bicubic interpolation, bilinear interpolation, and Lanczos interpolation method. The area of consideration in each method of interpolation is different (Fig. 1) [34].

Nearest neighbor interpolation is a computationally efficient technique for image upscaling, where each pixel in the LR image is assigned the value of the nearest pixel in the HR output image. The pixel value from the LR image is then replicated or copied to that position in the HR image [37]. While it has advantages of simplicity and low computational overhead, it suffers from limitations such as producing jagged edges and aliasing artifacts [38]. Recent studies have proposed novel approaches to enhance nearest neighbor interpolation by considering the

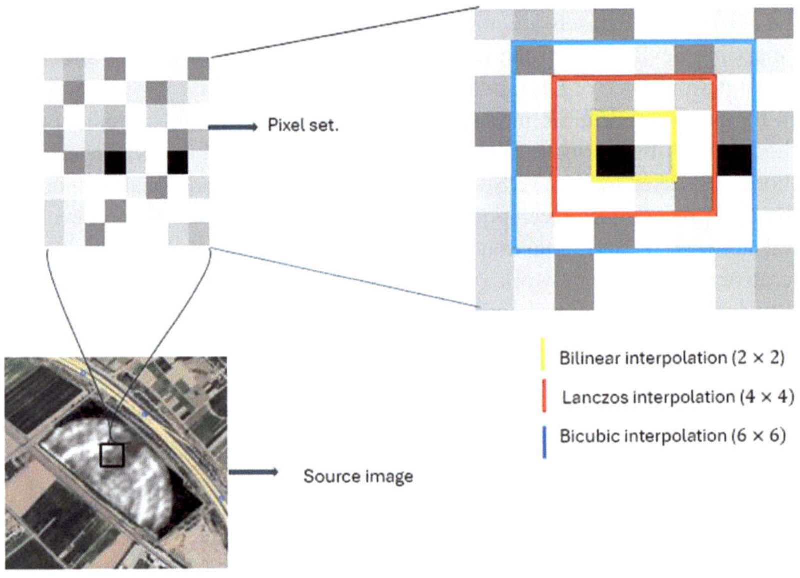

Fig. 1 This image demonstrates the varying pixel neighborhoods used by three common image interpolation methods: Lanczos, bicubic, and bilinear. Lanczos considers the largest area, mostly (6*6) and (8*8); bicubic, a moderate-sized area (4*4); and bilinear, the smallest (2*2)

nearest pixel value, utilizing logical transform [39], and incorporating multi-surface fitting [40]. These advancements aim to improve image quality, preserve details, and address the limitations of traditional nearest-neighbor interpolation.

Bilinear interpolation involves estimating pixel values at positions between neighboring data points by taking a weighted average based on their proximity. One of the seminal works in this area is the paper by Crow [41], which introduced the concept of bilinear interpolation and its application in shading calculations for computer-generated images and laid the foundation for subsequent research and applications of bilinear interpolation in computer graphics. In addition to computer graphics, bilinear interpolation finds broad application in image-processing tasks. A comprehensive overview of these applications can be found in the book *Digital Image Processing* by Gonzalez [37], in which the authors discuss the use of bilinear interpolation in image warping, texture mapping, and image registration, among other image-processing tasks.

Bicubic interpolation (BI) has been widely utilized in SR research due to its simplicity and computational efficiency. Originally proposed by Hou and Andrews [42], this concept has undergone extensive study, with subsequent variations suggested by Yang et al. [43] and Hurdale and Khandekar [44], aimed at enhancing its performance. BI has been commonly used as a baseline method for comparison with more complex SR algorithms. Studies such as by Park et al. [45] demonstrated the effectiveness of bicubic interpolation in producing visually appealing results in SR reconstruction. However, BI has limitations, including vulnerability to edge artifacts, the presence of chessboard and ringing effects, and its limited capabilities in complex scenarios. Researchers have explored combining BI with other techniques, such as directional interpolation methods [46], support vector regression (SVR) [47], and iterative approaches [46], to enhance the quality of super-resolved images. The integration of bicubic interpolation with SVR has shown promising results in enhancing visual quality and perceptual sharpness. Despite its limitations, bicubic interpolation remains a valuable tool in SR research, and it has also been combined with deep neural networks to achieve state-of-the-art performance.

Lanczos interpolation is another widely used technique in image processing for resampling or resizing images. It aims to provide higher-quality interpolation compared to simple methods such as bilinear or bicubic interpolation. The Lanczos interpolation was initially proposed by Gonzalez [37] which introduced the concept of Lanczos interpolation and its application in mitigating aliasing artifacts during image resampling. Research studies such as [48] discuss its application in image resizing and highlight its ability to preserve image details and reduce artifacts [49]. While Lanczos interpolation offers advantages in detail preservation and artifact reduction, it also has limitations such as increased computational complexity and the potential for overshooting and ringing artifacts along high-contrast edges or details in the image. Therefore, before considering Lanczos interpolation, it is important to weigh its benefits against the computational and memory costs, as well as the potential for introducing artifacts.

Interpolation methods suffer from blurriness and artifacts due to aliasing effects, which are especially evident with high scaling factors, and that renders them

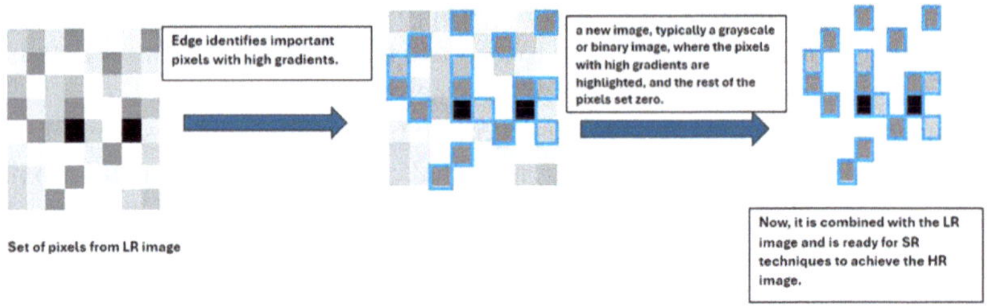

Fig. 2 The process of edge-based methods for pixels in the image

ineffective for HSIs. Nowadays, convolutional neural networks (CNNs) and deep learning methods are considered the most effective approaches for SISR [33].

2.1.2 Edge-Based Methods

In the realm of remote sensing, conventional techniques such as the Canny edge detector, Sobel operator, and Robert's operator have been commonly utilized to detect and extract edge information, as shown in Fig. 2, from satellite images [50]. These methods have been widely employed to identify objects, boundaries, and structures by leveraging edge-based feature extraction techniques such as edge histograms, edge orientations, and edge connectivity [21, 51].

To enhance the quality of satellite images while preserving their edge details, remote sensing researchers have extensively investigated edge-aware image enhancement techniques. Approaches such as bilateral filters, guided filters, and weighted least squares filters have been extensively studied and applied to improve the visual appearance of satellite imagery [52]. Furthermore, DL methods, particularly convolutional neural networks (CNNs) such as U-Net, FCN, and Deep Lab, have demonstrated remarkable progress in analyzing satellite images based on their edge information [53, 54]. In the domain of remote sensing, there has been a growing interest in fusing edge information with spectral and contextual data. Researchers have explored integration techniques such as multi-sensor data fusion, graph-based fusion, and deep fusion networks to combine edge information with other data sources. This integration enables a more comprehensive analysis and interpretation of satellite imagery, leading to an improved understanding of the observed phenomena [55, 56].

2.1.3 Statistical Modeling

Statistical modeling methods in SR imaging refer to a set of techniques employed to enhance the resolution and detail of low-resolution images. These methods utilize statistical approaches to tackle challenges associated with low-resolution data and improve the quality of image reconstruction [33]. Their objective is to restore degraded images without using additional data. However, this problem is known as ill-posed. To address this, several image priors have been proposed to capture the domain knowledge of natural images. These priors include Gaussian priors [57], Markov random field (MRF) priors [58], sparsity priors [59], and low-rank priors [60]. However, due to the complex structure of real-world images, many of these priors struggle to accurately model image properties. As a result, modeling image structure remains an ongoing and challenging task [61].

Patch-based methods, including sparse representation-based SR [43] and the nonlocal means (NLM) algorithm [62], exploit the self-similarity and redundancy present in natural images to achieve superior reconstruction quality. Another statistical modeling approach, based on Markov random field (MRF), captures dependencies between neighboring pixels in an image, offering an effective means for SR [63]. Bayesian compressive sensing (BCS) methods, which combine compressive sensing and Bayesian estimation, leverage the sparsity of natural images to estimate high-resolution images [64].

These statistical modeling techniques address various challenges associated with low-resolution data, such as outliers and limited data fidelity. They serve as valuable tools for enhancing the quality and accuracy of the SR process, with applications extending to diverse fields, including satellite imagery. For a comprehensive exploration of statistical modeling methods in the context of SR, the book *Super-Resolution Imaging* by Peyman Milanfar provides an extensive overview, covering patch-based approaches, Bayesian frameworks, sparse coding, and DL methods. By utilizing statistical modeling techniques, it becomes possible to enhance low-resolution data, resulting in high-resolution images with improved details and visual quality.

2.2 *Convolutional Neural Network (CNN)-Based Methods*

In SISR, algorithms aim to enhance the resolution of a single input image, as depicted in Fig. 3, despite the challenges posed by rapidly changing scenes in Earth observation missions. While traditional interpolation methods such as bicubic interpolation provide straightforward solutions, recent advancements in machine learning, particularly in computer vision, have led to the development of deep learning techniques using convolutional neural networks (CNNs). These CNN-based methods can learn from large datasets to enhance image data in a more information-theoretical manner. Unlike traditional interpolation methods that rely on local neighborhood information, CNN-based approaches have broader applicability and can be used for tasks such as visual inspection and preprocessing steps in remote sensing applications. However, it is important to note that CNNs trained on generic images may not perform well on remote sensing images due to their unique properties, including multispectral or hyperspectral data [65].

In a study proposed by Liebel and Körner [65], they applied a CNN-based approach to super-resolution of remote sensing (RS) data, specifically Sentinel-

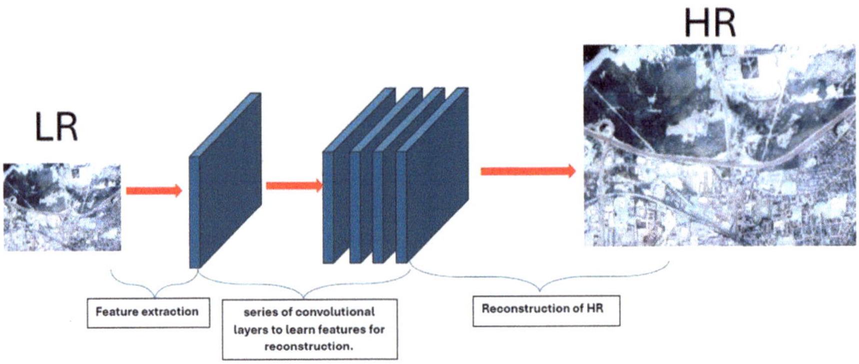

Fig. 3 Process of CNN-based methods for the reconstruction of HR image from LR image

2 images. They found that fine-tuning the SRCNN with the dataset led to the successful application of CNN methods to RS data. However, their focus was limited to the third band of multispectral images. Tuna et al. [66] compared SRCNN and VDSR models on satellite images and found that VDSR performed better on both panchromatic (PAN) and multispectral (MS) images. Liu et al. [67] reported that directly applying VDSR to Sentinel-2A images did not yield satisfactory results, so they proposed a remote sensing deep residual-learning (RS-DRL) network that outperformed VDSR on Sentinel-2A images.

Other researchers have also proposed different models for RSISR. Lei et al. [68] designed a local–global combined network (LGCNet) that learned both local and global features of RS images. Xu et al. [69] proposed a deep memory connected network (DMCN) that combined image detail with environmental information. Jiang et al. [70] introduced a deep distillation recursive network (DDRN) with ultradense residual blocks and a distillation mechanism. Deeba et al. [71] proposed a super-resolution wide remote sensing residual network (WRSR) that improved training loss performance and super-resolved results.

Recently, Shao et al. proposed a coupled sparse autoencoder (CSAE) for RSISR that combined sparse learning and autoencoder techniques. Fu et al. implemented a convolutional neural network-based SR method. Lu et al. [31] used a multiscale residual neural network (MRNN) for accurate SR reconstruction. Wang et al. [72] introduced an adaptive multiscale feature fusion network (AMFFN) for RSISR. Keshk and Yin utilized a multiscale feature mapping structure in their enhancement deep convolutional neural network algorithm (EDCNN). Deeba et al. presented a transferred wide residual SISR deep neural network model (WRSR) that reduced memory costs. Ren et al. [73] proposed an enhanced residual convolutional neural network (ERCNN) with a dual-luminance scheme (DLS) for RSISR. J. Wang et al. [74] introduced a feedback ghost residual dense network (FGRDN) that reduced computation burden and parameter count. Chang and Luo [75] proposed a bidirectional convolutional LSTM neural network (BiConvLSTM) for RSISR. Huan et al. [76] proposed a pyramidal multiscale residual network (PMSRN) to address the loss of LR image information. Huan et al. developed a progressive residual depth neural network (PRDNN) for RSISR.

Some methods explored training models on different frequency bands or incorporating wavelet transforms. Wang et al. [77] developed a CNN-based method that used wavelets to create SR images from aerial platform data. Ma et al. [78] combined wavelet transforms and recursive Res-Net for HR image reconstruction. Wagner et al. [79] adapted VDSR to handle different spatial resolutions of individual Sentinel-2 bands. Müller et al. [80] utilized the SFIM pan-sharpening method for RSISR of multispectral images.

Despite these advancements, there is still a need for further research on utilizing the abundant information in multispectral images and addressing the challenge of super-resolving real satellite images [81] proposed a realistic training data generation model for commercial satellite imagery products to improve the quality of super-resolved images. Some studies focused on better methods than SISR-based methods in SISR. Sui et al. proposed a dual transformer residual network (DTRN) by introducing a transformer branch and a CNN-based residual branch for global and local feature extraction and found that DTRN outperformed existing SISR methods for remote sensing images.

In recent years, deep convolutional neural network (CNN) methods have demonstrated notable advancements in comparison to traditional SR techniques [82–86] as well. Dong et al. [84] introduced SRCNN, a three-layer CNN for image SR, while subsequent works such as VDSR and DRCN by Kim et al. increased the network depth to 20, resulting in significant improvements over SRCNN [20, 85, 87]. This effective residual learning strategy was adopted in various other CNN-based SR methods [88–92]. Lim et al. [89] further expanded the network width with enhanced deep super-resolution (EDSR) and network depth with multiscale deep super-resolution system (MDSR), using simplified residual blocks, and achieved remarkable performance enhancements in SR [89].

Deep convolutional neural networks (CNNs) have revolutionized image classification with a string of significant advancements [93–96]. These deep networks seamlessly combine low-, mid-, and high-level features along with classifiers in a multilayered architecture, allowing for end-to-end learning. The depth of the network, determined by the number of stacked layers, enhances the richness and complexity of the learned features [95]. This literature review explores the applications and advancements of four prominent deep CNN architectures: residual networks (ResNets), dense networks (DenseNets), U-Net architecture, and MobileNet-based approaches but totally U-net acquires outstanding accuracy in segmentation, and ResNets have impressive performance in appropriate object detection and image classification. However, the network aims at handling natural image processing. From the abovementioned, the RS image is usually distinct from the natural image and explores suitable network structure for an RS image classification.

2.2.1 Residual Networks (ResNets)

Residual networks (ResNets) have sparked a fascination in the field of remote sensing, offering exciting possibilities for advancing image analysis and classification tasks. Researchers have explored the application of ResNet and its variants in diverse remote sensing applications, uncovering promising results and novel insights. Cheng et al. [97] took an extensive survey approach, providing a captivating overview of scene classification in remote sensing. They delved into the state-of-the-art techniques and benchmarks, highlighting the remarkable impact of ResNet and other DL models in this domain. In their groundbreaking work, Chen et al. [98] introduced the captivating DeepLab architecture for semantic image segmentation in remote sensing. By harnessing the power of ResNet as the backbone network and incorporating innovative techniques such as atrous convolution and fully connected conditional random fields (CRFs), they achieved mesmerizingly accurate and detailed segmentation results.

In recent research, a compact and efficient distance attention residual network (DARN) was proposed by Wang [99] to strike a balance between model accuracy and complexity. DARN outperformed existing lightweight remote sensing single-image super-resolution (SISR) methods in terms of parameter count, computation requirements, and inference speed. Another notable approach, the multiscale residual dense network (MRDN), has proven to be highly effective in reconstructing remote sensing images (RSIs) by preserving multiscale features throughout the transmission process, as demonstrated by Kong et al. [100]. MRDN not only surpasses existing methods on various remote sensing datasets but also shows exceptional performance in natural image SR, real-world image SR, and small object recognition tasks, highlighting its versatility and effectiveness. Additionally, Patnaik et al. [101] introduced the convolutional block attention residual network with joint adversarial mechanisms (CRNJAMs) to capture finer details in remote sensing images, effectively reducing noise in the generated super-resolved images.

These advancements highlight the importance of depth in representation for image SR. However, it is worth noting that merely stacking residual blocks to construct deeper networks does not necessarily guarantee better improvements. The potential contributions of deeper networks to image SR and the appropriate construction of highly deep trainable networks are areas that require further exploration.

2.2.2 Dense Networks (DenseNets)

Dense networks, a groundbreaking innovation presented by Huang et al. [4], introduced dense connections, paving the way for remarkable advancements in image-processing tasks. DenseNets establish direct connections between all preceding layers, resulting in a dense interconnection pattern that fosters feature reuse and facilitates the flow of gradients. One compelling application of DenseNets lies in SISR. Xiaolong Wang et al. [102] harnessed the power of DenseNet-based architectures to tackle the challenge of SR. By leveraging the dense connections within the network, they achieved a remarkable improvement in capturing high-level features and facilitating seamless information flow. As a result, their proposed DenseNet-based models exhibited enhanced performance and produced super-resolved images of superior quality. The inherent strength of dense connections in DenseNets has propelled the capabilities of CNN architectures, particularly in the realm of image super-resolution. The pioneering work by Huang et al. [103] and the subsequent advancements by K. Wang et al. [104] have deepened our understanding of the potential of dense connections and their transformative impact on the field of image processing.

In recent advancements, researchers have introduced a gating mechanism called the gated content-oriented residual dense network (GCoRDN) for hyperspectral image super-resolution (HSI SR). This mechanism serves as a post-processing step to further enhance the performance of SR. By leveraging the GCoRDN, the study achieved notable improvements in detail preservation and the generation of sharper high-resolution (HR) images [105]. Moreover, another significant contribution in this domain is the feedback multiscale residual dense network (FMDN) proposed by Lin et al. [106]. FMDN incorporates an error feedback mechanism along with a dense residual mechanism to effectively fuse multiscale features. Experimental results demonstrate that FMDN outperforms the conventional method of simply concatenating features, showcasing its superior performance.

These advancements highlight the continuous efforts to enhance SR techniques and provide more accurate and visually appealing results in hyperspectral image processing.

2.2.3 U-Net Architecture

U-Net architecture was first introduced by Ranneberger [107] primarily for biomedical image segmentation. This versatile architecture offers a unique framework consisting of a contracting path and an expansive path, allowing it to capture both high-level contextual information and intricate details. Its distinctive design, encompassing contracting and expansive paths, empowers the network to effectively handle various image-processing tasks, including remote sensing image SR. It has also demonstrated its efficacy in addressing image super-resolution (SR) challenges in remote sensing.

The potential of the U-Net architecture for remote sensing image SR was further explored [108]. They proposed a novel U-Net-based SR network customized for remote sensing applications. By incorporating skip connections within the U-Net structure, they successfully enhanced the reconstruction of high-frequency details, leading to considerable improvements in the quality of super-resolved images. The ability of the U-Net architecture to capture both local and global information played a pivotal role in achieving superior results in image SR tasks.

The U-Net architecture has demonstrated immense potential in various image-processing tasks, particularly in computer vision super-resolution (SR) techniques.

However, its utilization in remote sensing super-resolution (RSISR) has been relatively limited thus far. Given the remarkable capabilities of the U-Net architecture, it is highly recommended to explore and employ this method more extensively in future RSISR research.

2.2.4 MobileNet-based Approaches

MobileNet-based approaches have emerged as powerful solutions in the field of DL for computer vision tasks. These approaches leverage the MobileNet architecture, which is specifically designed to achieve high performance while being computationally efficient, making it well-suited for resource-constrained environments such as mobile devices.

Sandler et al. [109] introduced the MobileNetV2 architecture, an upgraded version of MobileNet, which further improved the efficiency and accuracy of mobile-based models. This architecture incorporates depth-wise separable convolutions, which reduce the computational cost by factorizing conventional convolutions into separate depth-wise and pointwise convolutions. By doing so, MobileNetV2 significantly reduces the number of parameters and computational operations required while maintaining competitive performance. Sadeghi-Tehran et al. [110] have extracted low-dimensional features from the fully connected layers by fine-tuning the pre-trained MobileNets. They utilize a deep convolutional neural network (CNN) model as a feature extractor to derive deep feature representations from the imaging data, and their results demonstrate that the proposed scheme is effective and considerably faster than conventional indexing structures. Yandouzi et al. [111] built a DL model specifically to detect wildfires using transfer learning techniques from the best-trained DL computer vision architectures available nowadays, such as MobileNet, MobileNetV2, and NAS Net Mobile. Their approach attained a detection rate of more than 99.9% over multiple metrics, proving that it could be used in real-world forest fire detection applications. So, it can be claimed that this method has the potential to be used more in the remote sensing field.

2.3 Generative Adversarial Network (GAN)-Based Methods

Generative adversarial networks (GANs) have emerged as a powerful framework for various image-processing tasks including image super-resolution. The GAN framework has been recently considered for the remote sensing SISR [15]. GANs are a popular area of research in artificial intelligence [112]. GANs consist of two main components: a generator (G) and a discriminator (D) (Fig. 4). The training process of GANs involves using adversarial learning techniques to reach a point called Nash equilibrium. This equilibrium represents a balance where the generated samples from the GAN cannot be distinguished from real samples. The discriminator is not able to accurately tell apart the real samples from the generated

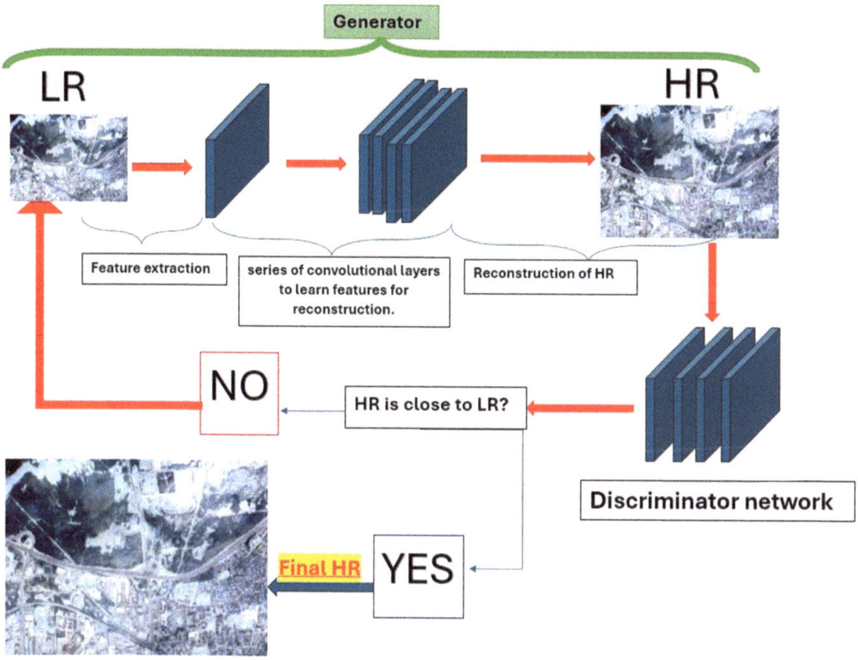

Fig. 4 GAN-based methods generate the closest HR image to the LR image

ones. In other words, the GAN's objective is to generate samples that are so realistic that the discriminator cannot tell the difference between them and real data [113]. This literature review explores the applications and advancements of GAN-based methods, specifically focusing on super-resolution GAN (SRGAN), enhanced SRGAN (ESRGAN), and CycleGAN-based approaches.

2.3.1 Super-resolution GAN (SRGAN)

Super-resolution generative adversarial network (SRGAN) has emerged as a significant advancement in remote sensing for enhancing the spatial resolution of images. Introduced by Ledig et al. [88], SRGAN utilizes a generative adversarial network (GAN) framework to generate HR representations from LR inputs.

The key concept behind SRGAN is the adversarial training process between a generator network and a discriminator network. The generator network aims to produce HR images from LR inputs, while the discriminator network is trained to distinguish between HR images and those generated by the generator. This adversarial interplay pushes the generator network to generate increasingly realistic and visually pleasing super-resolved images. Some variants of SRGAN were studied. A GAN-based method transferred generative adversarial network TGAN and removed the batch normalization layers to reduce the computational burden and improve the accuracy [114]. Another one is edge-enhanced GAN (EEGAN) by Jiang et al. [115]. This architecture consists of an ultradense subnetwork (UDSN) and an edge-enhancement subnetwork (EESN). It showed superior performance on various datasets compared to other super-resolution approaches. Wang et al. [72] proposed ultra-dense GAN (udGAN) that reformed the internal layout of the residual block into a two-dimensional matrix topology.

To enhance the visualization and improve the image quality, Wang et al. [113] extensively explored the three essential elements of SRGAN: network design, adversarial loss, and perceptual loss. They enhanced it to create ESRGAN (enhanced super-resolution generative adversarial network).

2.3.2 ESRGAN (Enhanced SRGAN)

ESRGAN, proposed by Xintao Wang et al. [116], is an enhanced version of SRGAN that further improves the quality of super-resolved images. ESRGAN incorporates a novel architecture with a feature extraction network and a generator network. The feature extraction network captures high-level features from both LR and super-resolved images, and the generator network utilizes these features to generate enhanced images. Moreover, ESRGAN introduces a new perceptual loss function based on the feature representations extracted from a pre-trained deep CNN, such as the visual geometry group (VGG) network.

ESRGAN has shown significant improvements in preserving image details and textures, resulting in visually appealing super-resolved images. Lanaras et al. [117] proposed an improved version of ESRGAN trained with WorldView–Sentinel image pairs for super-resolved multispectral Sentinel-2 output. Pineda et al. [118] developed a GAN-based SR model to improve the spatial resolution of Sentinel-2 images using PeruSat1 images as the training dataset. Gong et al. [119] introduced enlighten GAN, which used enlightened blocks, self-supervised hierarchical perceptual loss, and other strategies to enhance performance. Y.-Z. Chen et al. [120] improved EEGAN with diagonal connections and a 2D topology for better information conversion and gradient optimization. To enhance the realism of textures in generated images of ESRGAN, the authors in [121] presented an improved version of the ESRGAN model, called ESRGAN+. Their solution introduced a residual block between each pair of layers in the RRDB architecture and applied random Gaussian noise to enhance the results.

2.3.3 CycleGAN-based Approaches

CycleGAN, proposed by Zhu et al. [122], is a popular GAN-based framework for image translation tasks. CycleGAN enables unsupervised image-to-image translation by leveraging cycle consistency loss. This framework has been extended and applied to image SR tasks, where LR images are translated into HR images. CycleGAN-based approaches for SR offer the advantage of not requiring paired

LR and HR training data, making them more flexible and versatile. By training on unpaired datasets, CycleGAN-based approaches have demonstrated the ability to learn meaningful transformations and generate visually plausible HR images.

The traditional GAN is a one-way generated network, and CycleGAN is an A to B one-way GAN plus a B to A one-way GAN. Two GANs share two generators, and then each has a discriminator, so there are two discriminators and two generators in total. The image output by the first generator can be used as the input image of the second generator, and the output image of the second generator should match the original image. Conversely, the output image of the second generator can be used as the input image of the first generator, and the result should match the input image of the second generator [123]. Inspired by CycleGAN, Yuan et al. [124] proposed CinCGAN, using two CycleGANs to reconstruct the image [125]. To train the HR to LR degradation model using unpaired LR-HR images, the degradation model generates LR images, which, along with the HR images, are then used to train the SR network. Zhang et al. [126] used a CycleGAN to drag the LR-visible natural images to the remote sensing domain; then, they applied an SR network to upscale these simulated remote sensing domain LR images. They achieved superior results with state-of-the-art unpaired and remote sensing SR methods on several challenging remote sensing image datasets. Ghelichkhani et al. [127] used a trained CycleGAN architecture to change the land cover from vegetation to barren land and vice versa demonstrated the validity of the methods using Sentinel-2 images and highlighted their potential applications and limitations.

2.4 Attention Mechanism-Based Methods for Image Processing

Attention mechanisms have gained significant attention in recent years as a powerful tool in DL models for image-processing tasks. Many current SR techniques focus on increasing the model's depth to improve performance (Fig. 5). However, deeper models require significant computational resources and memory. Although CNN-based methods have shown remarkable results in various tasks, they treat all channels equally and cannot effectively utilize high-frequency information in low-resolution images. To address these issues, attention mechanisms have been proposed to capture deeper interdependencies between channels. Attention-guided algorithms, including channel attention, spatial attention, and pixel attention, have been developed for extracting informative features in tasks such as image super-resolution [9].

Researchers such as [9, 128–130] have made notable contributions in proposing different attention-based methods. This literature review discusses the applications and advancements of attention mechanisms, specifically focusing on self-attention mechanisms, channel attention mechanisms, and spatial attention mechanisms.

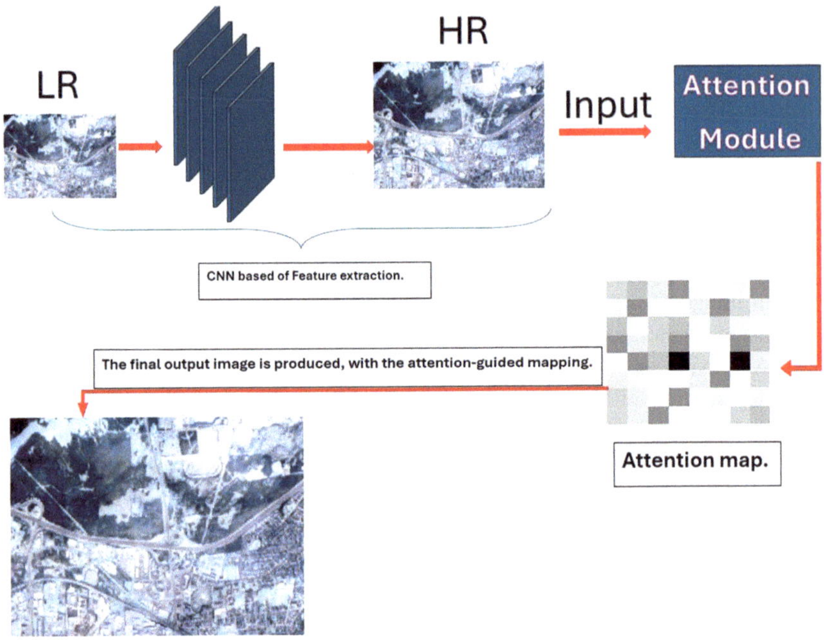

Fig. 5 Process of attention mechanism based on SR

2.4.1 Self-Attention Mechanisms

Self-attention mechanisms have emerged as a powerful tool for remote sensing applications, enabling models to dynamically focus on relevant spatial regions and capture long-range dependencies within images.

The seminal work on transformer models by Vaswani [131] introduced self-attention as a means of modeling relationships between different parts of a sequence, achieving state-of-the-art performance in natural language processing tasks. This breakthrough was soon extended to computer vision by [116], who proposed the nonlocal neural network, which employs self-attention to capture long-range dependencies in images and achieved impressive results in video recognition. Motivated by the unique challenges of remote sensing data, such as the prevalence of mixed pixels where a single pixel covers multiple materials, researchers have further adapted self-attention mechanisms for specialized remote sensing tasks. Qu et al. [132] proposed a self-attention-based method that operates at the sub-pixel level to better represent these complex spatial relationships within satellite imagery.

Beyond spatial modeling, self-attention has also shown promise in enhancing spectral feature representations for remote sensing applications. A self-attention mechanism is developed to improve the focus on important spectral channels, which they combined with remote sensing indices to mitigate edge information loss in mangrove semantic segmentation. Moreover, the integration of self-attention with generative modeling techniques has led to advancements in remote sensing image synthesis and enhancement. Xu et al. [133] introduced TE-SAGAN, which combines self-attention with weight normalization in a generative adversarial network (GAN) architecture, effectively removing artifacts and enhancing the texture of remote sensing imagery.

These examples showcase the versatility and effectiveness of self-attention mechanisms in modeling complex spatial and spectral relationships within remote sensing data, leading to improved performance across a variety of image-processing and analysis tasks. As the field of remote sensing continues to evolve, self-attention is poised to play an increasingly crucial role in unlocking the full potential of these advanced techniques.

2.4.2 Channel Attention Mechanisms

Attention mechanisms have been widely adopted in remote sensing applications to improve the performance of various tasks, such as object detection and feature representation. The focus is on modeling dependencies across different channels of a feature map. These mechanisms learn channel-wise attention weights to emphasize informative channels and suppress less relevant ones. Hu et al. [134] introduced the squeeze-and-excitation (SE) block, which employs channel attention mechanisms to improve the representational power of CNNs. The SE block adaptively recalibrates channel-wise feature responses, leading to improved accuracy in image classification tasks. Li et al. [135] introduced the channel attention mechanism into multiple receptive field fusion modules to suppress irrelevant background information in remote sensing images. Wang [136] attached the channel attention mechanism to the backbone network to enhance the capability of extracting ship features from complex backgrounds in remote sensing. Hu et al. and Qin et al. [137, 138] incorporated both a spatial attention mechanism and a channel attention mechanism to highlight the ships in remote sensing images. Chen et al. [139] designed a coordinate attention module that effectively combines spatial attention and channel attention to enhance the ability of ship feature representation.

2.4.3 Cascading CNN Architectures

Cascading has emerged as a powerful solution to improve the performance of super-resolution (SR) algorithms, particularly in remote sensing applications. Cascaded CNN architectures involve stacking multiple convolutional neural networks (CNNs) to create a hierarchical model that can progressively extract more detailed and high-level features from the input data.

For example, Zhang et al. [140] addressed a novel cascaded CNN framework with two subnetworks using different convolutional kernel sizes to achieve multiscale feature mapping for SISR. Wei et al. [141] constructed a cascading dense network to fully utilize the hierarchical features from all the convolutional layers for accurate SISR. Lan et al. [142] proposed a cascading residual network (CRN) and local wider residual blocks (LWRBs) for SISR, as well as an enhanced residual network (ERN) employing a dual global pathway structure and multiscale block (MSB).

Most recently, Zhang et al. [143] presented a cascaded visual attention network (CVANet) for SISR, which simulates the visual attention mechanism for improving the quality of LR satellite images. The cascaded architecture is a key innovation in these remote sensing SISR approaches, as it allows the models to progressively build up more sophisticated and multiscale feature representations compared to traditional single-stage CNN designs. This cascading strategy has proven to be highly effective in improving the quality and fidelity of super-resolved remote sensing imagery.

3 MISR Methods

Multi-image super-resolution (MISR) is an image-processing technique that aims to enhance the resolution and quality of an image by utilizing multiple input low-resolution (LR) images of the same scene in remote sensing [43, 88, 144, 145]. This is in contrast to SISR, which has been more heavily researched using deep learning.

MISR methods need to address additional challenges compared to SISR, such as aligning the multiple input images (co-registration) and effectively combining the information from them (information fusion) [30]. These extra steps make MISR more complex to implement and optimize.

The earliest MISR work used frequency-domain techniques to combine multiple images with sub-pixel displacements [13]. Later, spatial-domain MISR techniques were developed to better incorporate prior information about the high-resolution (HR) image [146]. MISR methods involve degrading the original image through blurring, warping, noise, and decimation and then reversing this degradation by solving an ill-posed optimization problem [146].

MISR is well-suited for remote sensing applications, as it can fuse multiple LR satellite images to overcome issues such as atmospheric turbulence and sensor noise [30]. However, MISR tasks face two key challenges: the long time span between image acquisitions can lead to uncontrollable environmental factors, and the order of the images may even be unknown [147]. Most existing MISR methods [19, 148, 149] are sensitive to the input sequence order, while some [19, 150] improve robustness by randomly shuffling the inputs. Additionally, due to varying clearness (pixel quality), some methods [19, 150–152] discard unclear images, resulting in information loss. Introducing attention mechanisms, such as in transformer models [131], can better utilize multiple images by capturing long-term features and adapting to variable-length sequences, regardless of order.

The Kelvins website, hosted by the European Space Agency (ESA), showcases various advanced concept competitions, including the PROBA-V Super Resolution challenge that focused on multi-image super-resolution (MISR) in remote sensing [153, 154]. This challenge, proposed by the ESA in 2018–2019, aimed to advance

MISR, where a sequence of low-resolution images is used to reconstruct a high-resolution image. Notable approaches that emerged were DeepSUM, a CNN architecture [148], and HighResNet, with an encoder, recursive module, and decoder [153]. Subsequent research, such as [155], built upon these by incorporating consistency losses to improve reconstruction quality. The field has gained more attention recently, with new datasets [156, 157] and strategies using attention [19] or transformers [158].

MISR approaches can be categorized into three main types: reconstruction-based methods, transformer methods, and learning-based methods. Transformer and learning-based methods leverage a training set of HR images to learn a transfer function or mapping to enhance the resolution of the input multi-image [131, 159, 160]. Reconstruction-based methods focus on reconstructing an HR image from LR inputs using priors or regularization [30]. Overall, MISR is an active area of research, with scholars working on improving sub-pixel-level reconstruction algorithms [28].

3.1 Reconstruction-Based Methods

In recent years, the application of image (SR) reconstruction in various fields has received increasing attention, and many models based on deep neural networks have achieved excellent results [82, 161]. SR refers to the technique of reconstructing HR images from observed LR images using relevant image-processing and algorithm models [161]. HR images contain more texture features and detailed information, making them valuable in monitoring equipment, satellite imagery, and medical imaging [161].

The reconstruction-based maximum posterior (MAP) method offers flexibility, allowing the addition of specific constraints to address specific problems [161]. For example, the Markov random field model has been integrated into the MAP framework to achieve SR reconstruction of sequence images [161]. Additionally, an improved method based on maximum posterior-Markov random field (MAP-MRF) has been proposed to enhance the spatial resolution of hyperspectral images [161]. This method utilizes nonlinear programming technology to solve the joint energy minimization problem based on MAP-MRF, resulting in HR hyperspectral images close to the original HR images [161].

Iterative back projection techniques for SR reconstruction have faced challenges such as the sawtooth effect and noise in the edge of reconstructed images [161]. To address this, researchers have proposed innovative approaches, including an image SR reconstruction method based on a high-frequency enhancement curve and iterative back projection [161]. Furthermore, DL-based methods have been introduced, utilizing feature extraction blocks, information extraction blocks, and reconstruction modules to effectively restore details and improve visual quality [82, 104, 161]. These approaches have shown promising results in benchmark datasets [82, 161].

In the field of ground-penetrating radar (GPR) signal analysis, an SR signal reconstruction method based on wavelet theory has been proposed, leveraging wavelet multi-resolution analysis to overcome limitations caused by insufficient signal sampling [82, 161, 162]. This method has successfully reconstructed HR signals, demonstrating its potential for wider applicability in engineering applications [82].

3.2 Transformer-Based Method

Transformer-based models have achieved promising results across many computer vision and natural language processing (NLP) tasks. The transformer architecture, introduced by Vaswani [131], has revolutionized NLP by effectively capturing long-range dependencies. Building on this, the work of [163] introduced vision transformers (ViTs) for image recognition. ViTs interpret an image as a sequence of patches and process it using a conventional transformer encoder.

The success of ViTs in generic visual data has sparked interest in the remote sensing community. Compared to RNNs and CNNs, the transformer model offers key advantages, such as fully parallel computation for improved efficiency and a global receptive field for better capturing global information. These benefits have led to significant progress in applying transformer models to image super-resolution research, particularly for multi-image super-resolution (MISR) tasks in remote sensing l [164–167].

An et al. [158] were the first to apply a transformer to the remote sensing image MISR task, proposing TR-MISR, and through extensive experiments, they demonstrated the effectiveness of the transformer model in multi-frame fusion and in improving the model's robustness to noise. Luo et al. [164] reduced alignment errors by combining optical flow estimation and deformable convolution and proposed burst super-resolution transformer (BSRT), utilizing blocks and groups based on the Swin transformer as the backbone structure to more effectively extract global contextual information. Dudhane et al. [168] introduced Burstormer, a transformer-based hierarchical model, for burst image restoration and enhancement tasks. This model improved feature alignment and fusion by introducing enhanced deformable alignment modules and reference-free feature enrichment modules, utilizing multiscale local and nonlocal features. Song et al. [169] proposed a novel end-to-end CoT-MISR network. The CoT-MISR network compensates for local and global information by leveraging the strengths of both convolution and transformer techniques.

3.3 Learning-Based Methods

Recently, learning-based algorithms have gained popularity and are outperforming classical methods in various fields. In the context of (SR) tasks, which aim to enhance the resolution and quality of images, learning-based approaches utilizing machine learning techniques have shown superior performance. Spectral SR has proven to be a prevalent framework for current remote sensing image fusion (RSIF) applications such as MS super-resolution and hyperspectral super-resolution [170]. The SR methods used by pan-sharpening can be broken down into two classes: those that rely on learning for reconstruction and those that rely on compressive sensing. In recent years, learning techniques have spread to nearly all applications in remote sensing imaging, including MS pan-sharpening [171–173].

There are three major types of learning-based SR techniques: neighbor embedding (NE), sparse coding, and DL approaches [174–176]. NE-based algorithms operate on the assumption that small patches of LR and HR images exhibit similar local geometry and form low-dimensional nonlinear manifolds. Locally linear embedding (LLE) is an example of an NE-based approach that generates HR patches using LLE techniques. LLE assumes that, given appropriate samples, every sample and its neighbors lie on or near a locally linear patch [174]. Another NE-based technique is nonnegative neighbor embedding, which utilizes the least squares (LS) approximation of nonnegative LR patches to determine nonnegative weights [175]. Dual geometric neighbor embedding (DGNE) explores the multiview properties and local spatial neighbors of patches to find an SR spatial multiple embedding function [176].

Sparse coding-based approaches leverage the idea that LR and HR characteristics share similar sparse components. One approach involves reconstructing the HR image patch as a sparse linear combination of learned HR dictionary atoms. However, the optimization function used in these methods, which involves the regularity of the L1 norm, can lead to computationally intensive calculations for each LR input feature. To address this issue, orthogonal matching pursuit (OMP) and principal component analysis (PCA) have been utilized to reduce dimensionality and enhance sparse representation, respectively [27].

Other techniques, such as manifold regularization with sparse support regression [177], graph embedding super-resolution (GESR) [14], and a piecewise linear regression strategy [178], aim to improve image resolution and visual quality by preserving the intrinsic geometrical structure of the HR image patch manifold, extracting multiscale information, and learning the mapping relation between LR and HR spaces.

In addition to the above techniques, there are zoom-based SR methods [179], support vector regression (SVR) with sparse image representation [180] statistical approaches using one-to-many correspondences [181], and learned dictionaries [178, 182]. Rapid SR algorithms based on regressors and sparse representations anchored to dictionary atoms have also been explored [183].

One interesting learning-based study was done by Mishra and Hadar [184]. They used a self-supervised deep learning approach called contrastive learning, which identifies similar and distinct types of images and learns general features of the dataset. The proposed framework shows outstanding visual and quantitative performance compared to existing unsupervised and supervised deep learning methods and even performs close to the supervised learning-based method ESRGAN, despite being learned in a completely unsupervised setting. The technique also generalizes well to other datasets such as multispectral band images, real-world colored remote sensing images, and real-world natural images.

DL algorithms, which are a subset of the broader machine learning family, have been extensively studied for SR tasks. These algorithms directly learn the relationship between input and output from data and have demonstrated promising results in recovering essential features for SR tasks.

3.3.1 Deep Learning (DL)

DL is a branch of machine learning algorithms that aims at learning the hierarchical representations of data [174]. DL-based approaches provide an end-to-end solution for learning complex mapping functions and are rapidly and successfully applied to SR tasks. In contrast to traditional task-specific learning algorithms that select useful handcrafted features with expert domain knowledge, DL algorithms aim to learn informative hierarchical representations automatically and then leverage them to achieve the final purpose (Fig. 6), where the whole learning process can be seen as an entirety [174, 175].

DL methods in SR can be broadly categorized into two types: SISR and MISR [185]. While both approaches have been explored, it has been observed that learning-based methods, including DL, generally exhibit superior performance in the context of MISR [30]. Therefore, in this particular study, DL is considered a learning-based method within the framework of MISR.

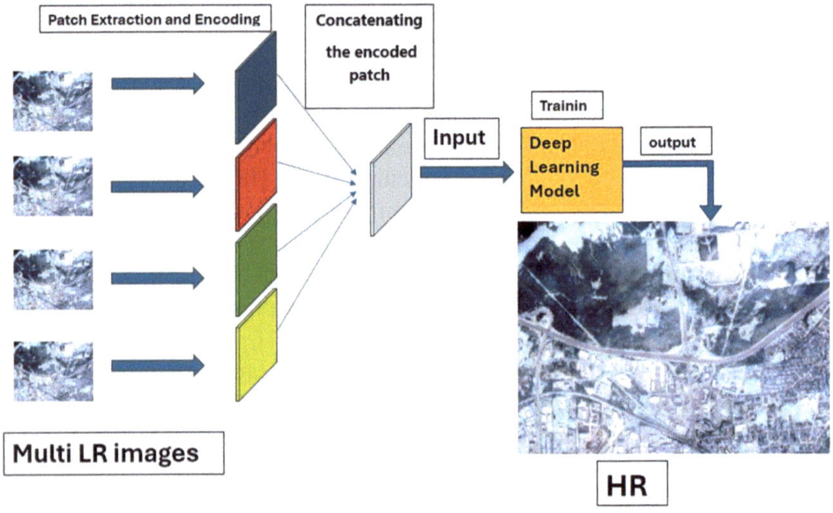

Fig. 6 Presenting how deep learning makes HR from multi-LR images

When applying DL algorithms for SR to satellite images, several challenges need to be addressed. Multispectral and hyperspectral data have a higher dimensionality ranging from four to dozens of bands. Atmospheric conditions, haze, clouds, and cloud shadows add additional variation to the measured values. Land cover characteristics vary globally to a high degree, and a model trained on images of temperate forest areas in Europe might fail when applied to images of tropical forests in Southeast Asia. Large amounts of data are encoded in each observation because of the large field of view and the distance between the sensor and the scene [176]. Moreover, many satellite sensors record image bands in different resolutions. For example, Sentinel-2 images have a resolution of 10, 20, or 60 m depending on the band. The availability of a panchromatic band, which captures light in a broader spectral range, is also common. This panchromatic band achieves high spatial resolution but provides limited information about reflectance at specific wavelengths. Example datasets include Landsat-8, Satellite Pour l'Observation de la Terre (SPOT), and Pléiades, which have different bands with varying resolutions [179].

The differences in geometric resolution can be leveraged in the SR image enhancement process, allowing DL algorithms to effectively enhance the quality of satellite images and provide valuable insights for various applications [14, 176].

3.3.2 Deep Learning-Based Fusion Methods

Due to the wide variety of earth-observing satellites, remote sensing image fusion (RSIF) has become a prominent subfield in image fusion. Satellite images are either PAN images or MS images [170]. Deep learning (DL)-based fusion methods offer several advantages, such as learning complex alignment transformations and effectively combining aligned images, resulting in high-resolution outputs with improved details and clarity [172, 186].

The reconstruction-based methods for image fusion aim to calculate the reflectance of the center fusion pixel through a weighting function that considers the spectral, temporal, and spatial information in similar pixels. Gao et al. [187] proposed the spatiotemporal adaptive reflection fusion model (STARFM), Hilker et al. [188] developed the spatiotemporal fusion algorithm for mapping reflectance change (STAARCH), Roy et al. [189] proposed a semi-physical method considering the bidirectional reflectance effect, Zhu et al. [190] developed the enhanced STARFM (ESTARFM) model, Wang and Atkinson [191] introduced the Fit-FC algorithm, and Kwan et al. [192] proposed a simple two-step method. Jia et al. [193] proposed a hybrid deep learning-based STF method (HDLSFM) that combines a nonlinear deep learning-based radiometric normalization, a deep learning-based super-resolution, and a linear-based fusion to address radiation differences, land cover, and phenological changes. Shi et al. [194] presented the reliable and adaptive spatiotemporal data fusion (RASDF) method, which analyzes the effects of sensor differences, introduces a reliability index to improve robustness, and utilizes adaptive global and local unmixing models to enhance the retrieval of strong temporal changes. Ma et al. [195] designed a new encoder–decoder network that learns the noise distribution and employs iterative noise removal and outperforming seven state-of-the-art algorithms on Landsat-7 and Landsat-5 datasets in terms of root mean square error and preserving clearer details during abrupt phenological changes.

3.3.3 Deep Learning-Based Depth-Guided Super-resolution

Depth estimation using deep models has seen significant advancements due to the powerful representation-learning capabilities of deep neural networks. Various deep architectures, such as CNNs and encoder–decoder networks, have been employed for depth estimation tasks [196]. These models are trained on large-scale datasets with depth annotations to learn the mapping between input images and corresponding depth maps. By leveraging the learned representations, deep models can estimate depth maps accurately, even in challenging scenarios with occlusions or textureless regions.

Depth-guided SR architectures utilize the estimated depth maps to guide the SR process. These architectures typically consist of two main components: a depth-guided fusion module and a SR network. The depth-guided fusion module integrates the depth information with the LR input image to produce a depth-guided feature map. This fusion process ensures that the SR network focuses on enhancing the details and structures consistent with the estimated depth. The SR network, usually based on deep convolutional or residual networks, then upscales the depth-guided feature map to produce the HR output image [197].

Several studies have explored depth-guided SR architectures using DL. For instance, Ma [198] proposed a depth-guided residual network that incorporates depth information into the SR process. Li [199] introduced a depth-attention network that adaptively adjusts the attention weights based on the depth map. Guo [197] proposed a progressive fusion network that progressively refines the depth-guided features for better SR performance.

However, there are still ongoing research efforts to improve depth-guided SR methods. One challenge is the availability of accurate depth maps for training. Obtaining high-quality depth annotations can be challenging and expensive. Therefore, exploring self-supervised or unsupervised depth estimation methods can alleviate the reliance on ground truth depth maps during training [200]. Another research direction is the integration of additional cues, such as surface normal or reflectance information, to further enhance the depth-guided SR performance [201].

In conclusion, these methods leverage deep models for depth estimation and utilize the estimated depth maps to guide the SR process. Despite the progress made, addressing challenges related to depth estimation training data and exploring additional cues for guidance remains important for advancing depth-guided super-resolution.

3.3.4 Algorithm Unrolling in Deep Learning

Multispectral image fusion and (SR) microscopy are important computational imaging problems that have benefited from the technique of algorithm unrolling. Unrolling involves representing each iteration of an iterative algorithm as a layer in a deep neural network, bridging the gap between traditional iterative algorithms and deep learning approaches. This approach has shown improved performance over baselines in applications such as fusing low-resolution multispectral aerial images with their high-resolution panchromatic counterparts [202], as well as in sparsity-based super-resolution microscopy. The unrolling technique offers several advantages, including the ability to develop efficient, high-performance, and interpretable network architectures from reasonably sized training sets.

The work of Vishal Monga, a prominent researcher in the field, has made significant contributions to the technique of algorithm unfolding. Monga's research has demonstrated how this approach provides a systematic connection between iterative algorithms used in signal processing and deep neural networks, addressing the limitations of deep neural networks, such as their black-box nature and the need for large training sets.

Beyond these specific applications, algorithm unrolling is a promising general technique in deep learning. Compared to traditional black-box deep networks, unrolled networks provide more interpretability by exposing the underlying algorithmic structure, as well as computational efficiency during inference, making them suitable for real-time applications. These advantages are crucial over standard deep learning approaches, which often suffer from a lack of transparency and high computational demands [203].

4 Comparing Deep Learning with Traditional Methods

When comparing DL approaches with traditional methods in the context of SR in satellite imagery, several key differences and advantages emerge.

DL methods, such as CNNs, have gained significant attention and achieved remarkable success in various computer vision tasks, including SR [83, 204]. These methods leverage their ability to learn complex patterns and representations directly from large-scale training data. By training on pairs of LR and HR images, DL models can effectively capture the underlying mapping between the two domains, enabling them to generate high-quality HR outputs.

Unlike traditional methods, which often rely on handcrafted features and explicit modeling of image characteristics, DL methods learn these representations automatically [204]. This eliminates the need for manual feature engineering, allowing the model to adapt and generalize to different satellite image datasets and variations. DL models excel at capturing intricate and nonlinear relationships within the data, enabling them to produce more accurate and visually pleasing HR images.

Furthermore, DL methods have demonstrated superior performance in terms of SR quality. They can recover fine details, enhance texture, and improve overall image sharpness [83, 204]. This is accomplished by leveraging the hierarchical nature of CNN architectures, where lower layers capture low-level features and higher layers capture more complex and abstract information. The ability to capture and leverage multi-level features contributes to the high-quality HR outputs produced by DL models.

Moreover, recent advancements in DL architectures have further improved SR in satellite imagery. For instance, the use of generative adversarial networks (GANs) has shown promising results by incorporating adversarial training to enhance the visual realism of the generated HR [88]. Additionally, attention mechanisms have been employed to selectively focus on important image regions and improve the overall reconstruction quality [205].

However, it is important to note that DL approaches have some limitations. They typically require large amounts of training data to effectively learn the mapping between LR and HR images. Acquiring and labeling such datasets can be time-consuming and expensive. Additionally, training DL models can be computationally intensive, requiring significant computational resources and time.

In contrast, traditional methods often involve explicit mathematical formulations and assumptions about the image characteristics and degradation process. These methods may be more interpretable and computationally efficient, but they may struggle to capture complex relationships and patterns present in the data.

To sum up, DL methods, such as CNNs and GANs, have emerged as powerful approaches for SR in satellite imagery [88, 204, 205]. They can automatically learn representations and mappings from LR to HR images, resulting in improved image quality and detail preservation. While traditional methods have their merits, DL approaches offer significant advantages in terms of accuracy, adaptability, and the ability to learn complex relationships directly from data. There is a comprehensive and qualitative result of the methods mentioned in this literature in Tables 1 and 2.

5 Discussion and Conclusions

We highlight the limitations of LR satellite imagery and the need for enhancing its quality to extract meaningful information for various applications such as urban planning, change monitoring, and population estimation.

The survey categorizes SR techniques into two main categories: SISR and MISR. SISR techniques aim to enhance the resolution of LR images using neural networks and other approaches such as attention mechanisms and generative adversarial networks. However, the one-to-one mapping used in SISR can limit the quality of the reconstruction because the original image contains limited information. SISR can be challenging as it lacks additional reference information and relies solely on a single input image for reconstructing missing image features [185]. Consequently, SISR methods can generate high-resolution (HR) details that are not present in

Table 1 Summary of the qualitative result of the SISR method in this study

SISR	**ISR methods**		**Qualitative results**
	Traditional methods	Interpolation-based	Interpolation-based methods use mathematical interpolation to estimate missing high-frequency details, resulting in smoother transitions, reduced pixelation, and somewhat improved visual consistency compared to the LR input
		Edge-based methods	Edge-based SR techniques focus on preserving and enhancing edge information, leading to significantly sharper delineation of linear features such as roads and buildings, as well as better retention of small-scale details
		Statistical modeling	Statistical modeling methods leverage image statistics and priors to intelligently estimate high-resolution content from low-resolution inputs, producing sharper edges, crisper details, and more natural-looking textures
	Convolutional neural network (CNN)-based methods	Residual networks (ResNets)	ResNets effectively preserve and propagate edge and detail information through their deep architecture, enabling enhanced reconstruction of high-frequency content and reduced artifacts compared to simpler CNN models
		Dense networks (DenseNets)	DenseNets leverage improved feature reuse and multiscale fusion to enhance detail preservation and produce more efficient, robust SR results that can handle noisy or degraded LR inputs
		U-Net architecture	The U-Net's encoder–decoder structure with skip connections allows it to effectively preserve spatial details, object boundaries, and fine-grained textures, resulting in high-quality super-resolution outputs
		MobileNet-based approaches	MobileNet-based models provide efficient, lightweight designs suitable for real-time super-resolution on mobile/embedded devices while still producing visually pleasing, high-quality results

(continued)

Table 1 (continued)

SISR	**ISR methods**		**Qualitative results**
	Generative adversarial network (GAN)-based methods	Super-resolution GAN (SRGAN)	SRGAN leverages the power of generative adversarial networks to produce strikingly realistic and visually appealing super-resolved images, going beyond typical reconstruction-based methods by hallucinating plausible high-frequency details
		ESRGAN (enhanced SRGAN)	Building on SRGAN, ESRGAN further improves image quality through enhanced network architectures, more effective adversarial training, and perceptual loss functions, resulting in sharper edges, more natural textures, and fewer artifacts
		CycleGAN-based approaches	CycleGAN-based super-resolution methods exploit unsupervised cross-domain learning to translate low-resolution images to high-resolution images, without requiring paired training data and can adapt to different imaging modalities or degradation types
	Attention mechanism-based methods for image processing	Self-attention mechanisms	Self-attention mechanisms allow super-resolution networks to model long-range dependencies in the image, capturing global context and enabling more coherent reconstruction of complex structures and textures
		Channel attention mechanisms	Channel attention mechanisms selectively emphasize informative feature channels while suppressing less relevant ones, leading to more discriminative feature representations and improved reconstruction quality, especially for fine details
		Cascading CNN architectures	Cascading CNN architectures leverage a series of progressively more complex super-resolution models, allowing efficient, multiscale processing that can handle diverse degradation levels and produce high-quality results across a wide range of resolutions

Table 2 Summary of the qualitative result of MISR methods in this literature review

MISR	ISR methods	Qualitative results
	Reconstruction-based methods	Sharper, more well-defined edges and boundaries Enhanced preservation of fine details and textures More natural-looking, continuous gradients in complex regions Reduced checkerboard artifacts or other reconstruction-related distortions
	Transformer-based method	Remarkably sharp, well-defined edges Exceptional retention of fine details, small-scale features, and complex textures Highly natural-looking, coherent reconstruction of intricate patterns and gradients Minimal checkerboard artifacts or other reconstruction-related distortions
	Learning-based methods	Enhanced sharpness and clarity of edges, lines, and other high-frequency content, leading to a crisper, more defined appearance Improved retention of fine textures, preserving the richness and complexity of the original high-resolution imagery More natural-looking, less artificial-appearing interpolation of gradients and continuous regions, avoiding the smoothing or distortion common in classical methods Reduced ringing, aliasing, and other reconstruction-related artifacts that can plague traditional super-resolution algorithms

the input image, leading to fake patterns that are unrelated to the original scene, especially when the desired output resolution is high. While some early studies applied SISR algorithms designed for simple images to remote sensing imagery with promising results, the limitations of SISR become evident in scenarios that require accurate guidance. A straightforward remedy to this issue involves augmenting the amount of input information by utilizing multiple LR images simultaneously (MISR methods) [30].

On the other hand, MISR techniques leverage a set of HR images to enhance the resolution of a given LR image. MISR-based image enhancement is advantageous in the field of remote sensing for two main reasons. Firstly, acquiring multiple LR satellite images and a few HR ones is relatively straightforward, enabling accurate evaluation of region-specific reconstructions. Secondly, employing multiple LR images can mitigate problems arising from misalignments and pixel-level discrepancies by incorporating information fusion. In general, MISR is capable of producing more accurate HR reconstructions compared to SISR as it aggregates more information extracted from multiple views of the target region [206]. MISR methods have been widely applied in HR remote sensing and also found extensive application in hyperspectral remote sensing, either independently or in conjunction with sensors possessing multiangular acquisition capabilities [207]. These techniques leverage the inherent data redundancy present in multiple images to enhance aerial imagery. However, there are scenarios where obtaining or acquiring multiple images of the same scene is challenging or not feasible [208].

The review emphasizes the significant role of DL-based methods in super-resolution, particularly in the field of remote sensing. DL algorithms, such as CNNs, can learn the mapping between LR and HR image pairs, enabling the estimation of HR details from LR inputs. These methods have shown promising results in enhancing the quality of remote sensing imagery.

Looking ahead, future research in SR techniques for remote sensing image enhancement is likely to focus on the following areas:

1. Advanced DL Architectures: As DL continues to advance, researchers may explore more sophisticated network architectures and training strategies to further improve the quality of HR reconstructions. This may involve the integration of techniques such as attention mechanisms, generative adversarial networks, and transformer-based models.
2. Incorporation of Spatial and Spectral Information: Remote sensing imagery contains both spatial and spectral information. Ongoing research in the field of remote sensing, such as the development of pan-sharpening models, focuses on effectively leveraging both types of information to enhance the spatial and spectral quality of reconstructed images. Further advancements in this area are expected to contribute to the improvement of super-resolution techniques.
3. Fusion of Multiple Data Sources: Integration of data from multiple sources, such as satellite imagery, aerial photography, and ground-level sensing, can provide richer information for super-resolution. Future techniques may focus on effectively fusing data from multiple sources to improve the quality and accuracy of HR reconstructions.
4. Real-Time Super-resolution: Real-time SR techniques are essential for applications that require immediate access to HR imagery. Future research may explore real-time algorithms that can produce high-quality reconstructions in a computationally efficient manner, enabling their deployment in time-critical scenarios.
5. Evaluation Metrics and Benchmark Datasets: Developing standardized evaluation metrics and benchmark datasets specific to remote sensing SR will enable fair comparisons between different techniques and facilitate progress in the field. Future research may focus on establishing such metrics and datasets to foster advancements in remote sensing image enhancement.
6. Exploration of Advanced DL Architectures: We mentioned that the exploration of advanced DL architectures, such as transformers and generative adversarial networks, can lead to innovative SR solutions tailored for remote sensing applications. These newer DL models may offer enhanced feature extraction, better handling of complex spatial-spectral relationships, and improved image quality.
7. Adaptation to Diverse Applications: As mentioned in the file, DL-based super-resolution can be adapted to different applications, such as object detection, change monitoring, and image analysis, by fine-tuning the models on task-specific datasets. Exploring the versatility of DL-based SR techniques across various remote sensing applications is likely to be an area of future research.

Overall, the document provides valuable insights into the current state of SR techniques for remote sensing image enhancement, and it sets the stage for future research directions in this rapidly evolving field. By addressing the existing limitations and exploring new approaches, researchers can contribute to the development of more accurate and reliable SR methods for remote sensing applications.

References

1. Chen, H., He, X., Qing, L., Wu, Y., Ren, C., Sheriff, R.E., Zhu, C.: Real-world single image super-resolution: a brief review. Inf. Fusion. **79**, 124–145 (2022)
2. Feng, T., Zhao, J.: Review and comparison: building extraction methods using high-resolution images. In: 2009 Second International Symposium on Information Science and Engineering. IEEE (2009)
3. Huang, X., Zhang, L.: A multidirectional and multiscale morphological index for automatic building extraction from multispectral GeoEye-1 imagery. Photogramm. Eng. Remote. Sens. **77**, 721–732 (2011)
4. Huang, G., Liu, Z., Van Der Maaten, L., Weinberger, K.Q.: Densely connected convolutional networks. In: Proceedings of the IEEE Conference on Computer Vision and Pattern Recognition (2017)
5. Rottensteiner, F., Briese, C.: A new method for building extraction in urban areas from high-resolution LIDAR data. In: International Archives of Photogrammetry Remote Sensing and Spatial Information Sciences (2002)
6. Bagan, H., Yamagata, Y.: Landsat analysis of urban growth: how Tokyo became the world's largest megacity during the last 40 years. Remote Sens. Environ. **127**, 210–222 (2012)
7. Dong, L., Shan, J.: A comprehensive review of earthquake-induced building damage detection with remote sensing techniques. ISPRS J. Photogramm. Remote Sens. **84**, 85–99 (2013)
8. Weng, Q.: Remote sensing of impervious surfaces in the urban areas: requirements, methods, and trends. Remote Sens. Environ. **117**, 34–49 (2012)
9. Wang, P., Sertel, E.: Channel–spatial attention-based pan-sharpening of very high-resolution satellite images. Knowl.-Based Syst. **229**, 107324 (2021)
10. Yue, L., Shen, H., Li, J., Yuan, Q., Zhang, H., Zhang, L.: Image super-resolution: the techniques, applications, and future. Signal Process. **128**, 389–408 (2016)
11. Tsagkatakis, G., Aidini, A., Fotiadou, K., Giannopoulos, M., Pentari, A., Tsakalides, P.: Survey of deep-learning approaches for remote sensing observation enhancement. Sensors. **19**, 3929 (2019)
12. Gani, S.R., Spoorthi, H., Rafi, M.: A Review on Super Resolution Algorithms and Applications. (2024)
13. Tsai, R.Y., Huang, T.S.: Multiframe image restoration and registration. In: Multiframe Image Restoration and Registration, pp. 317–339 (1984)
14. Lepcha, D.C., Goyal, B., Dogra, A., Wang, S.H.: An efficient medical image super resolution based on piecewise linear regression strategy using domain transform filtering. Concurr. Comput. Pract. Exp. **34**, e6644 (2022)
15. Wang, P., Bayram, B., Sertel, E.: A comprehensive review on deep learning based remote sensing image super-resolution methods. Earth-Sci. Rev. **232**, 104110 (2022)
16. Anwar, S., Khan, S., Barnes, N.: A deep journey into super-resolution: a survey. ACM Comput. Surv. CSUR. **53**, 1–34 (2020)
17. Jiang, J., Wang, C., Liu, X., Ma, J.: Deep learning-based face super-resolution: a survey. ACM Comput. Surv. CSUR. **55**, 1–36 (2021)
18. Yang, W., Zhang, X., Tian, Y., Wang, W., Xue, J.-H., Liao, Q.: Deep learning for single image super-resolution: a brief review. IEEE Trans. Multimed. **21**, 3106–3121 (2019)

19. Salvetti, F., Mazzia, V., Khaliq, A., Chiaberge, M.: Multi-image super resolution of remotely sensed images using residual attention deep neural networks. Remote Sens. **12**, 2207 (2020)
20. Dong, C., Loy, C.C., He, K., Tang, X.: Learning a deep convolutional network for image super-resolution. In: Computer Vision–ECCV 2014: 13th European Conference (2014)
21. Keys, R.: Cubic convolution interpolation for digital image processing. IEEE Trans. Acoust. Speech Signal Process. **29**, 1153–1160 (1981)
22. Cui, Z., Chang, H., Shan, S., Zhong, B., Chen, X.: Deep network cascade for image super-resolution. In: Computer Vision–ECCV 2014: 13th European Conference (2014)
23. Freedman, G., Fattal, R.: Image and video upscaling from local self-examples. ACM Trans. Graph. TOG. **30**, 1–11 (2011)
24. Huang, J.-B., Singh, A., Ahuja, N.: Single image super-resolution from transformed self-exemplars. In: Proceedings of the IEEE Conference on Computer Vision and Pattern Recognition (2015)
25. Yang, J., Lin, Z., Cohen, S.: Fast image super-resolution based on in-place example regression. In: Proceedings of the IEEE Conference on Computer Vision and Pattern Recognition (2013)
26. Bevilacqua, M., Roumy, A., Guillemot, C., Morel, M.: Low-complexity single-image super-resolution based on nonnegative neighbor embedding. In: Proceedings of the British Machine Vision Conference (2012)
27. Zeyde, R., Elad, M., Protter, M.: On single image scale-up using sparse representations. In: Curves and Surfaces: 7th International Conference, Avignon, France, June 24–30, 2010, Revised Selected Papers. pp. 711–730 (2012). https://doi.org/10.1007/978-3-642-27413-8_47
28. Wu, Q., Zeng, H., Zhang, J., Xia, H.: Multi-image hybrid super-resolution reconstruction via interpolation and multi-scale residual networks. Meas. Sci. Technol. **34**, 075403 (2023)
29. Baker, S., Kanade, T.: Limits on super-resolution and how to break them. IEEE Trans. Pattern Anal. Mach. Intell. **24**, 1167–1183 (2002)
30. Arefin, M.R., Michalski, V., St-Charles, P.-L., Kalaitzis, A., Kim, S., Kahou, S.E., Bengio, Y.: Multi-image super-resolution for remote sensing using deep recurrent networks. In: Proceedings of the IEEE/CVF Conference on Computer Vision and Pattern Recognition Workshops (2020)
31. Lu, T., Wang, J., Zhang, Y., Wang, Z., Jiang, J.: Satellite image super-resolution via multi-scale residual deep neural network. Remote Sens. **11**, 1588 (2019)
32. Bashir, S.M.A., Wang, Y., Khan, M., Niu, Y.: A comprehensive review of deep learning-based single image super-resolution. PeerJ Comput. Sci. **7**, e621 (2021)
33. Aburaed, N., Alkhatib, M.Q., Marshall, S., Zabalza, J., Al Ahmad, H.: 3d expansion of SRCNN for spatial enhancement of hyperspectral remote sensing images. In: 2021 4th International Conference on Signal Processing and Information Security (ICSPIS) (2021)
34. Fadnavis, S.: Image interpolation techniques in digital image processing: an overview. Int. J. Eng. Res. Appl. **4**, 70–73 (2014)
35. Li, X., Orchard, M.T.: New edge-directed interpolation. IEEE Trans. Image Process. **10**, 1521–1527 (2001)
36. Zhang, L., Wu, X.: An edge-guided image interpolation algorithm via directional filtering and data fusion. IEEE Trans. Image Process. **15**, 2226–2238 (2006)
37. Gonzalez, R.C.: Digital image processing. Pearson Education India (2009)
38. Rukundo, O., Cao, H.: Nearest neighbor value interpolation. ArXiv Prepr., ArXiv12111768 (2012)
39. Danahy, E.E., Agaian, S.S., Panetta, K.A.: Algorithms for the resizing of binary and grayscale images using a logical transform. In: Image Processing: Algorithms and Systems V (2007)
40. Zhou, F., Yang, W., Liao, Q.: Interpolation-based image super-resolution using multisurface fitting. IEEE Trans. Image Process. **21**, 3312–3318 (2012)
41. Crow, F.C.: The aliasing problem in computer-generated shaded images. Commun. ACM. **20**, 799–805 (1977)

42. Hou, H., Andrews, H.: Cubic splines for image interpolation and digital filtering. IEEE Trans. Acoust. Speech Signal Process. **26**, 508–517 (1978)
43. Yang, J., Wright, J., Huang, T.S., Ma, Y.: Image super-resolution via sparse representation. IEEE Trans. Image Process. **19**, 2861–2873 (2010)
44. Hurdale, S., Khandekar, P.: Implementation of super-resolution algorithm on FPGA. In: 2015 IEEE International Conference on Computational Intelligence and Computing Research (ICCIC) (2015)
45. Park, S.C., Park, M.K., Kang, M.G.: Super-resolution image reconstruction: a technical overview. IEEE Signal Process. Mag. **20**, 21–36 (2003)
46. Liu, J., Gan, Z., Zhu, X.: Directional bicubic interpolation—a new method of image super-resolution. In: 3rd International Conference on Multimedia Technology (ICMT-13) (2013)
47. Chao, J., Ren, S., Zhang, J., Zhang, Z.: Super-resolution of remote sensing images using support vector regression and bicubic interpolation. In: 2011 4th International Congress on Image and Signal Processing, pp. 1469–1473 (2011)
48. Avidan, S., Shamir, A.: Comparison of interpolation methods for image resizing. ACM Trans. Graph. TOG. **26**, 1–10 (2007)
49. Canny, J.: A computational approach to edge detection. IEEE Trans. Pattern Anal. Mach. Intell., 679–698 (1986)
50. Huang, P.S., Li, Y.: Edge histogram analysis for satellite image classification. IEEE Trans. Geosci. Remote Sens. **39**, 309–315 (2001)
51. Mallat, S., Zhong, S.: Characterization of signals from multiscale edges. IEEE Trans. Pattern Anal. Mach. Intell. **14**, 710–732 (1992)
52. Irani, M., Peleg, S.: Improving resolution by image registration. CVGIP Graph. Models Image Process. **53**, 231–239 (1991)
53. Simoncelli, E.P., Adelson, E.H.: Noise removal via Bayesian wavelet coring. In: Proceedings of 3rd IEEE International Conference on Image Processing (1996)
54. Sun, J., Zheng, N.-N., Shum, H.-Y.: Stereo matching using belief propagation. IEEE Trans. Pattern Anal. Mach. Intell. **25**, 787–800 (2003)
55. Chen, G., Li, Z., Chang, C.: Fusion of hyperspectral and LiDAR remote sensing data for land-cover classification with deep belief networks. IEEE Trans. Geosci. Remote Sens. **52**, 1439–1457 (2014)
56. Zhang, J., Lin, X.: Advances in fusion of optical imagery and LiDAR point cloud applied to photogrammetry and remote sensing. Int. J. Image Data Fusion. **8**, 1–31 (2017)
57. Hunt: Bayesian methods in nonlinear digital image restoration. IEEE Trans. Comput. 100, 219–229 (1977)
58. Qian, W., Titterington, D.M.: Bayesian image restoration: an application to edge-preserving surface recovery. IEEE Trans. Pattern Anal. Mach. Intell. **15**, 748–752 (1993)
59. Portilla, J., Tristan-Vega, A., Selesnick, I.W.: Efficient and robust image restoration using multiple-feature L2-relaxed sparse analysis priors. IEEE Trans. Image Process. **24**, 5046–5059 (2015)
60. Candes, E.J., Plan, Y.: Tight oracle inequalities for low-rank matrix recovery from a minimal number of noisy random measurements. IEEE Trans. Inf. Theory. **57**, 2342–2359 (2011)
61. Celebi, M.E., Smolka, B.: Advances in Low-Level Color Image Processing. Springer (2014)
62. Buades, A., Coll, B., Morel, J.-M.: A review of image denoising algorithms, with a new one. Multiscale Model. Simul. **4**, 490–530 (2005)
63. Tappen, M.F., Freeman, W.T., Adelson, E.H.: Estimating the spatial layout of cluttered rooms. In: Proceedings of the IEEE Computer Society Conference on Computer Vision and Pattern Recognition (CVPR), p. II–555 (2003)
64. Vrigkas, M.I., Nikou, C., Katsaggelos, A.K.: A survey of recent advances in compressive sensing and its applications to signal, speech, and image processing. J. Vis. Commun. Image Represent. **25**, 137–157 (2014)
65. Liebel, L., Körner, M.: Single-image super resolution for multispectral remote sensing data using convolutional neural networks. In: The International Archives of the Photogrammetry, Remote Sensing and Spatial Information Sciences, pp. 883–890 (2016)

66. Tuna, C., Unal, G., Sertel, E.: Single-frame super resolution of remote-sensing images by convolutional neural networks. Int. J. Remote Sens. **39**, 2463–2479 (2018)
67. Liu, D., Xie, S., Li, Y., Zhao, D., El-Alfy, E.-S.M.: Neural Information Processing: 24th International Conference, ICONIP 2017, Guangzhou, China, November 14–18, 2017, Proceedings, Part I. Presented at the (2017)
68. Lei, S., Shi, Z., Zou, Z.: Super-resolution for remote sensing images via local–global combined network. IEEE Geosci. Remote Sens. Lett. **14**, 1243–1247 (2017)
69. Xu, W., Guangluan, X., Wang, Y., Sun, X., Lin, D., Yirong, W.: High quality remote sensing image super-resolution using deep memory connected network. In: IGARSS 2018–2018 IEEE International Geoscience and Remote Sensing Symposium (2018)
70. Jiang, K., Wang, Z., Yi, P., Jiang, J.: A progressively enhanced network for video satellite imagery superresolution. IEEE Signal Process. Lett. **25**, 1630–1634 (2018)
71. Deeba, F., Dharejo, F.A., Zhou, Y., Ghaffar, A., Memon, M.H., Kun, S.: Single image super-resolution with application to remote-sensing image. In: 2020 Global Conference on Wireless and Optical Technologies (GCWOT) (2020)
72. Wang, Z., Jiang, K., Yi, P., Han, Z., He, Z.: Ultra-dense GAN for satellite imagery super-resolution. Neurocomputing. **398**, 328–337 (2020)
73. Ren, C., He, X., Qing, L., Wu, Y., Pu, Y.: Remote sensing image recovery via enhanced residual learning and dual-luminance scheme. Knowl.-Based Syst. **222**, 107013 (2021)
74. Wang, J., Wu, Y., Wang, L., Wang, L., Alfarraj, O., Tolba, A.: Lightweight feedback convolution neural network for remote sensing images super-resolution. IEEE Access. **9**, 15992–16003 (2021)
75. Chang, Y., Luo, B.: Bidirectional convolutional LSTM neural network for remote sensing image super-resolution. Remote Sens. **11**, 2333 (2019)
76. Huan, H., Li, P., Zou, N., Wang, C., Xie, Y., Xie, Y., Xu, D.: End-to-end super-resolution for remote-sensing images using an improved multi-scale residual network. Remote Sens. **13**, 666 (2021)
77. Wang, T., Sun, W., Qi, H., Ren, P.: Aerial image super resolution via wavelet multiscale convolutional neural networks. IEEE Geosci. Remote Sens. Lett. **15**, 769–773 (2018)
78. Ma, W., Pan, Z., Guo, J., Lei, B.: Achieving super-resolution remote sensing images via the wavelet transform combined with the recursive res-net. IEEE Trans. Geosci. Remote Sens. **57**, 3512–3527 (2019)
79. Wagner, L., Liebel, L., Körner, M.: Deep residual learning for single-image super-resolution of multi-spectral satellite imagery. ISPRS Ann. Photogramm. Remote Sens. Spat. Inf. Sci. **4**, 189–196 (2019)
80. Müller, M.U., Ekhtiari, N., Almeida, R.M., Rieke, C.: Super-resolution of multispectral satellite images using convolutional neural networks. ArXiv Prepr., ArXiv200200580 (2020)
81. Zhu, X., Talebi, H., Shi, X., Yang, F., Milanfar, P.: Super-resolving commercial satellite imagery using realistic training data. In: 2020 IEEE International Conference on Image Processing (ICIP) (2020)
82. Haris, M., Shakhnarovich, G., Ukita, N.: Deep back-projection networks for super-resolution. In: Proceedings of the IEEE Conference on Computer Vision and Pattern Recognition (2018)
83. Dong, C., Loy, C.C., He, K., Tang, X.: Image super-resolution using deep convolutional networks. IEEE Trans. Pattern Anal. Mach. Intell. **38**, 295–307 (2015)
84. Dong, C., Loy, C.C., Tang, X.: Accelerating the super-resolution convolutional neural network. In: Computer Vision–ECCV 2016: 14th European Conference (2016)
85. Kim, J., Lee, J.K., Lee, K.M.: Accurate image super-resolution using very deep convolutional networks. In: Proceedings of the IEEE Conference on Computer Vision and Pattern Recognition (2016)
86. Lai, W.-S., Huang, J.-B., Ahuja, N., Yang, M.-H.: Deep Laplacian pyramid networks for fast and accurate super-resolution. In: Proceedings of the IEEE Conference on Computer Vision and Pattern Recognition (2017)

87. Kim, J., Lee, J.K., Lee, K.M.: Deeply-recursive convolutional network for image super-resolution. In: Proceedings of the IEEE Conference on Computer Vision and Pattern Recognition (2016)
88. Ledig, C., Theis, L., Huszár, F., Caballero, J., Cunningham, A., Acosta, A., Aitken, A., Tejani, A., Totz, J., Wang, Z.: Photo-realistic single image super-resolution using a generative adversarial network. In: Proceedings of the IEEE Conference on Computer Vision and Pattern Recognition (2017)
89. Lim, B., Son, S., Kim, H., Nah, S., Mu Lee, K.: Enhanced deep residual networks for single image super-resolution. In: Proceedings of the IEEE Conference on Computer Vision and Pattern Recognition Workshops (2017)
90. Sajjadi, M.S., Scholkopf, B., Hirsch, M.: Enhancenet: single image super-resolution through automated texture synthesis. In: Proceedings of the IEEE international conference on computer vision (2017)
91. Tai, Y., Yang, J., Liu, X.: Image super-resolution via deep recursive residual network. In: Proceedings of the IEEE Conference on Computer Vision and Pattern Recognition (2017)
92. Tai, Y., Yang, J., Liu, X., Xu, C.: Memnet: a persistent memory network for image restoration. In: Proceedings of the IEEE International Conference on Computer Vision (2017)
93. LeCun, Y., Boser, B., Denker, J.S., Henderson, D., Howard, R.E., Hubbard, W., Jackel, L.D.: Backpropagation applied to handwritten zip code recognition. Neural Comput. **1**, 541–551 (1989)
94. Krizhevsky, A., Sutskever, I., Hinton, G.E.: Imagenet classification with deep convolutional neural networks. In: Advances in Neural Information Processing Systems (2012)
95. Zeiler, M.D., Fergus, R.: Visualizing and understanding convolutional networks. In: Computer Vision–ECCV 2014: 13th European Conference, Zurich, Switzerland, September 6–12, 2014, Proceedings, Part I (2014)
96. Sermanet, P., Eigen, D., Zhang, X., Mathieu, M., Fergus, R., LeCun, Y.: Overfeat: integrated recognition, localization and detection using convolutional networks. ArXiv Prepr., ArXiv13126229 (2013)
97. Cheng, G., Han, J., Lu, X.: Remote sensing image scene classification: benchmark and state of the art. Proc. IEEE. **105**, 1865–1883 (2017)
98. Chen, L.-C., Papandreou, G., Kokkinos, I., Murphy, K., Yuille, A.L.: Deeplab: semantic image segmentation with deep convolutional nets, atrous convolution, and fully connected CRFs. IEEE Trans. Pattern Anal. Mach. Intell. **40**, 834–848 (2017)
99. Wang, Q., Wang, S., Chen, M., Zhu, Y.: DARN: distance attention residual network for lightweight remote-sensing image superresolution. IEEE J. Sel. Top. Appl. Earth Obs. Remote Sens. **16**, 714–724 (2022)
100. Kong, D., Gu, L., Li, X., Gao, F.: Multi-scale residual dense network for the super-resolution of remote sensing images. IEEE Trans. Geosci. Remote Sens. **62**, 1 (2024)
101. Patnaik, A., Chaudhary, N., Bhuyan, M., Alfarhood, S., Safran, M.: Remote sensing single-image super-resolution using convolutional block attention residual network with joint adversarial mechanisms. IEEE Access. **12**, 53424 (2024)
102. Wang, X., Girshick, R., Gupta, A., He, K.: Non-local neural networks. In: Proceedings of the IEEE conference on computer vision and pattern recognition (2018)
103. Huang, Z., Cheng, G., Wang, H., Li, H., Shi, L., Pan, C.: Building extraction from multi-source remote sensing images via deep deconvolution neural networks. In: 2016 IEEE International Geoscience and Remote Sensing Symposium (IGARSS) (2016)
104. Wang, K., Zeng, X., Yang, J., Meng, D., Zhang, K., Peng, X., Qiao, Y.: Cascade attention networks for group emotion recognition with face, body and image cues. In: Proceedings of the 20th ACM International Conference on Multimodal Interaction (2018)
105. Hu, J., Li, T., Zhao, M., Wang, F., Ning, J.: A gated content-oriented residual dense network for hyperspectral image super-resolution. Remote Sens. **15**, 3378 (2023)
106. Lin, Z., Li, S., Jiang, Y., Wang, J., Luo, Q.: Feedback multi-scale residual dense network for image super-resolution. Signal Process. Image Commun. **107**, 116760 (2022)

107. Ranneberger, J., Fischer, P.: Brox: U-Net: convolutional networks for biomedical image segmentation. In: International Conference on Medical Image Computing and Computer-Assisted Intervention, pp. 234–241 (2015)
108. Shi, W., Caballero, J., Huszár, F., Totz, J., Aitken, A.P., Bishop, R., Rueckert, D., Wang, Z.: Real-time single image and video super-resolution using an efficient sub-pixel convolutional neural network. In: Proceedings of the IEEE Conference on Computer Vision and Pattern Recognition (2016)
109. Sandler, M., Howard, A., Zhu, M., Zhmoginov, A., Chen, L.-C.: Mobilenetv2: Inverted residuals and linear bottlenecks. In: Proceedings of the IEEE Conference on Computer Vision and Pattern Recognition (2018)
110. Sadeghi-Tehran, P., Angelov, P., Virlet, N., Hawkesford, M.J.: Scalable database indexing and fast image retrieval based on deep learning and hierarchically nested structure applied to remote sensing and plant biology. J. Imaging. **5**, 33 (2019)
111. Yandouzi, M., Grari, M., Idrissi, I., Boukabous, M., Moussaoui, O., Azizi, M., Ghoumid, K., Elmiad, A.K.: Forest fires detection using deep transfer learning. Forest. **13**, 1 (2022)
112. Wang, K., Gou, C., Duan, Y., Lin, Y., Zheng, X., Wang, F.-Y.: Generative adversarial networks: introduction and outlook. IEEECAA J. Autom. Sin. **4**, 588–598 (2017)
113. Wang, X., Sun, L., Chehri, A., Song, Y.: A review of GAN-based super-resolution reconstruction for optical remote sensing images. Remote Sens. **15**, 5062 (2023)
114. Ma, W., Pan, Z., Guo, J., Lei, B.: Super-resolution of remote sensing images based on transferred generative adversarial network. In: IGARSS 2018–2018 IEEE International Geoscience and Remote Sensing Symposium (2018)
115. Jiang, K., Wang, Z., Yi, P., Wang, G., Lu, T., Jiang, J.: Edge-enhanced GAN for remote sensing image superresolution. IEEE Trans. Geosci. Remote Sens. **57**, 5799–5812 (2019)
116. Wang, X., Yu, K., Wu, S., Gu, J., Liu, Y., Dong, C., Qiao, Y., Change Loy, C.: Esrgan: enhanced super-resolution generative adversarial networks. In: Proceedings of the European Conference on Computer Vision (ECCV) Workshops (2018)
117. Lanaras, C., Bioucas-Dias, J., Galliani, S., Baltsavias, E., Schindler, K.: Super-resolution of Sentinel-2 images: learning a globally applicable deep neural network. ISPRS J. Photogramm. Remote Sens. **146**, 305–319 (2018)
118. Pineda, F., Ayma, V., Beltran, C.: A generative adversarial network approach for super-resolution of Sentinel-2 satellite images. In: The International Archives of the Photogrammetry, Remote Sensing and Spatial Information Sciences, pp. 9–14 (2020)
119. Gong, Y., Liao, P., Zhang, X., Zhang, L., Chen, G., Zhu, K., Tan, X., Lv, Z.: Enlighten-GAN for super resolution reconstruction in mid-resolution remote sensing images. Remote Sens. **13**, 1104 (2021)
120. Chen, Y.-Z., Liu, T.-J., Liu, K.-H.: Super-resolution of satellite images based on two-dimensional RRDB and edge-enhanced generative adversarial network. In: 2022 IEEE International Conference on Consumer Electronics (ICCE) (2022)
121. Rakotonirina, N.C., Rasoanaivo, A.: ESRGAN+: further improving enhanced super-resolution generative adversarial network. In: ICASSP 2020-2020 IEEE International Conference on Acoustics, Speech and Signal Processing (ICASSP) (2020)
122. Zhu, J.-Y., Park, T., Isola, P., Efros, A.A.: Unpaired image-to-image translation using cycle-consistent adversarial networks. In: Proceedings of the IEEE International Conference on Computer Vision (2017)
123. Deng, L., Zhang, Y., Wang, X.: High-definition processing of remote sensing images based on CUT-CycleGAN. In: 2021 40th Chinese Control Conference (CCC) (2021)
124. Yuan, Y., Liu, S., Zhang, J., Zhang, Y., Dong, C., Lin, L.: Unsupervised image super-resolution using cycle-in-cycle generative adversarial networks. In: Proceedings of the IEEE conference on computer vision and pattern recognition workshops (2018)
125. Bulat, A., Yang, J., Tzimiropoulos, G.: To learn image super-resolution, use a GAN to learn how to do image degradation first. In: Proceedings of the European Conference on Computer Vision (ECCV) (2018)

126. Zhang, Z., Tian, Y., Li, J., Xu, Y.: Unsupervised remote sensing image super-resolution guided by visible images. Remote Sens. **14**, 1513 (2022)
127. Ghelichkhani, S., Salem, Y.A.M.A.-D., Salahov, H., Adam, F.A.I., Charbatji, A.J., Abdulkareem, M.I.: Generative adversarial networks for artificial satellite image creation and manipulation. In: Adversarial Multimedia Forensics, pp. 125–143. Springer (2023)
128. Chen, L., Liu, H., Yang, M., Qian, Y., Xiao, Z., Zhong, X.: Remote sensing image super-resolution via residual aggregation and split attentional fusion network. IEEE J. Sel. Top. Appl. Earth Obs. Remote Sens. **14**, 9546–9556 (2021)
129. Gao, L., Sun, H.-M., Cui, Z., Du, Y.-B., Sun, H.-B., Jia, R.-S.: Super-resolution reconstruction of single remote sensing images based on residual channel attention. J. Appl. Remote. Sens. **15**, 016513–016513 (2021)
130. Dai, T., Cai, J., Zhang, Y., Xia, S.-T., Zhang, L.: Second-order attention network for single image super-resolution. In: Proceedings of the IEEE/CVF Conference on Computer Vision and Pattern Recognition (2019)
131. Vaswani, A., Shazeer, N., Parmar, N., Uszkoreit, J., Jones, L., Gomez, A.N., Kaiser, Ł., Polosukhin, I.: Attention is all you need. In: Advances in neural information processing systems (2017)
132. Qu, Y., Baghbaderani, R.K., Qi, H., Kwan, C.: Unsupervised pansharpening based on self-attention mechanism. IEEE Trans. Geosci. Remote Sens. **58**, 7835–7847 (2020)
133. Xu, Y., Luo, W., Hu, A., Xie, Z., Xie, X., Tao, L.: TE-SAGAN: an improved generative adversarial network for remote sensing super-resolution images. Remote Sens. **14**, 2425 (2022)
134. Hu, J., Shen, L., Sun, G.: Squeeze-and-excitation networks. In: Proceedings of the IEEE Conference on Computer Vision and Pattern Recognition (2018)
135. Li, L., Zhou, Z., Wang, B., Miao, L., An, Z., Xiao, X.: Domain adaptive ship detection in optical remote sensing images. Remote Sens. **13**, 3168 (2021)
136. Wang, Q., Shen, F., Cheng, L., Jiang, J., He, G., Sheng, W., Jing, N., Mao, Z.: Ship detection based on fused features and rebuilt YOLOv3 networks in optical remote-sensing images. Int. J. Remote Sens. **42**, 520–536 (2021)
137. Hu, J., Zhi, X., Shi, T., Zhang, W., Cui, Y., Zhao, S.: PAG-YOLO: a portable attention-guided YOLO network for small ship detection. Remote Sens. **13**, 3059 (2021)
138. Qin, C., Wang, X., Li, G., He, Y.: An improved attention-guided network for arbitrary-oriented ship detection in optical remote sensing images. IEEE Geosci. Remote Sens. Lett. **19**, 1–5 (2022)
139. Chen, Y., Wang, J., Zhang, Y., Liu, Y.: Arbitrary-oriented ship detection based on Kullback-Leibler divergence regression in remote sensing images. Earth Sci. Inform. **16**, 3243–3255 (2023)
140. Zhang, F., Cai, N., Cen, G., Li, F., Wang, H., Chen, X.: Image super-resolution via a novel cascaded convolutional neural network framework. Signal Process. Image Commun. **63**, 9–18 (2018)
141. Wei, W., Feng, G., Zhang, Q., Cui, D., Zhang, M., Chen, F.: Accurate single image super-resolution using cascading dense connections. Electron. Lett. **55**, 739–742 (2019)
142. Lan, R., Sun, L., Liu, Z., Lu, H., Su, Z., Pang, C., Luo, X.: Cascading and enhanced residual networks for accurate single-image super-resolution. IEEE Trans. Cybern. **51**, 115–125 (2020)
143. Zhang, W., Zhao, W., Li, J., Zhuang, P., Sun, H., Xu, Y., Li, C.: CVANet: cascaded visual attention network for single image super-resolution. Neural Netw. **170**, 622–634 (2024)
144. Zhang, W.: A Novel MRI image super-resolution reconstruction algorithm based on image representation and sparse dictionary learning. In: 2016 International Conference on Inventive Computation Technologies (ICICT) (2016)
145. Lu, T., Li, S., Fang, L., Ma, Y., Benediktsson, J.A.: Spectral–spatial adaptive sparse representation for hyperspectral image denoising. IEEE Trans. Geosci. Remote Sens. **54**, 373–385 (2015)

146. Elad, M., Hel-Or, Y.: A fast super-resolution reconstruction algorithm for pure translational motion and common space-invariant blur. IEEE Trans. Image Process. **10**, 1187–1193 (2001)
147. Mnih, V., Heess, N., Graves, A.: Recurrent models of visual attention. In: Advances in Neural Information Processing Systems (2014)
148. Märtens, M., Izzo, D., Krzic, A., Cox, D.: Super-resolution of PROBA-V images using convolutional neural networks. Astrodynamics. **3**, 387–402 (2019)
149. Deudon, M., Kalaitzis, A., Arefin, M.R., Goytom, I., Lin, Z., Sankaran, K., Michalski, V., Kahou, S.E., Cornebise, J., Bengio, Y.: Highres-net: multi-frame super-resolution by recursive fusion. ArXiv Prepr., ArXiv190412928 (2019)
150. Dorr, F.: Satellite image multi-frame super resolution using 3D wide-activation neural networks. Remote Sens. **12**, 3812 (2020)
151. Molini, A.B., Valsesia, D., Fracastoro, G., Magli, E.: Deepsum: deep neural network for super-resolution of unregistered multitemporal images. IEEE Trans. Geosci. Remote Sens. **58**, 3644–3656 (2019)
152. Molini, A.B., Valsesia, D., Fracastoro, G., Magli, E.: Deepsum++: Non-local deep neural network for super-resolution of unregistered multitemporal images. In: IGARSS 2020–2020 IEEE International Geoscience and Remote Sensing Symposium (2020)
153. Deudon, M., Kalaitzis, A., Goytom, I., Arefin, M.R., Lin, Z., Sankaran, K., Michalski, V., Kahou, S.E., Cornebise, J., Bengio, Y.: Highres-net: recursive fusion for multi-frame super-resolution of satellite imagery. ArXiv Prepr., ArXiv200206460 (2020)
154. Nguyen, N.L., Anger, J., Davy, A., Arias, P., Facciolo, G.: Self-supervised multi-image super-resolution for push-frame satellite images. In: Proceedings of the IEEE/CVF Conference on Computer Vision and Pattern Recognition (2021)
155. Razzak, M.T., Mateo-García, G., Lecuyer, G., Gómez-Chova, L., Gal, Y., Kalaitzis, F.: Multi-spectral multi-image super-resolution of Sentinel-2 with radiometric consistency losses and its effect on building delineation. ISPRS J. Photogramm. Remote Sens. **195**, 1–13 (2023)
156. Cornebise, J., Oršolić, I., Kalaitzis, F.: Open high-resolution satellite imagery: the worldstrat dataset–with application to super-resolution. In: Advances in Neural Information Processing Systems, pp. 25979–25991 (2022)
157. Kowaleczko, P., Tarasiewicz, T., Ziaja, M., Kostrzewa, D., Nalepa, J., Rokita, P., Kawulok, M.: A real-world benchmark for sentinel-2 multi-image super-resolution. Sci. Data. **10**, 644 (2023)
158. An, T., Zhang, X., Huo, C., Xue, B., Wang, L., Pan, C.: TR-MISR: multiimage super-resolution based on feature fusion with transformers. IEEE J. Sel. Top. Appl. Earth Obs. Remote Sens. **15**, 1373–1388 (2022)
159. Caballero, J., Ledig, C., Aitken, A., Acosta, A., Totz, J., Wang, Z., Shi, W.: Real-time video super-resolution with spatio-temporal networks and motion compensation. In: Proceedings of the IEEE Conference on Computer Vision and Pattern Recognition (2017)
160. Jo, Y., Oh, S.W., Kang, J., Kim, S.J.: Deep video super-resolution network using dynamic upsampling filters without explicit motion compensation. In: Proceedings of the IEEE Conference on Computer Vision and Pattern Recognition (2018)
161. He, K., Zhang, X., Ren, S., Sun, J.: Deep residual learning for image recognition. In: Proceedings of the IEEE Conference on Computer Vision and Pattern Recognition (2016)
162. Woo, S., Park, J., Lee, J.-Y., Kweon, I.S.: Cbam: convolutional block attention module. In: Proceedings of the European conference on computer vision (ECCV) (2018)
163. Dosovitskiy, A., Beyer, L., Kolesnikov, A., Weissenborn, D., Zhai, X., Unterthiner, T., Dehghani, M., Minderer, M., Heigold, G., Gelly, S.: An image is worth 16x16 words: transformers for image recognition at scale. ArXiv Prepr., ArXiv201011929 (2020)
164. Luo, Z., Li, Y., Cheng, S., Yu, L., Wu, Q., Wen, Z., Fan, H., Sun, J., Liu, S.: Bsrt: Improving burst super-resolution with swin transformer and flow-guided deformable alignment. In: Proceedings of the IEEE/CVF Conference on Computer Vision and Pattern Recognition (2022)

165. Yang, F., Yang, H., Fu, J., Lu, H., Guo, B.: Learning texture transformer network for image super-resolution. In: Proceedings of the IEEE/CVF Conference on Computer Vision and Pattern Recognition (2020)
166. Chen, X., Wang, X., Zhou, J., Qiao, Y., Dong, C.: Activating more pixels in image super-resolution transformer. In: Proceedings of the IEEE/CVF Conference on Computer Vision and Pattern Recognition (2023)
167. Chen, H., Wang, Y., Guo, T., Xu, C., Deng, Y., Liu, Z., Ma, S., Xu, C., Xu, C., Gao, W.: Pre-trained image processing transformer. In: Proceedings of the IEEE/CVF Conference on Computer Vision and Pattern Recognition (2021)
168. Dudhane, A., Zamir, S.W., Khan, S., Khan, F.S., Yang, M.-H.: Burstormer: burst image restoration and enhancement transformer. In: 2023 IEEE/CVF Conference on Computer Vision and Pattern Recognition (CVPR) (2023)
169. Song, Q., Xiu, M., Nie, Y., Hu, M., Liu, C.: CoT-MISR: marrying convolution and transformer for multi-image super-resolution. Multimed. Tools Appl., 1–13 (2024)
170. Anand, S., Sharma, R.: Pansharpening and spatiotemporal image fusion method for remote sensing. Eng. Res. Express. **6**, 022201 (2024)
171. Huang, W., Xiao, L., Wei, Z., Liu, H., Tang, S.: A new pan-sharpening method with deep neural networks. IEEE Geosci. Remote Sens. Lett. **12**, 1037–1041 (2015)
172. Liu, J., Zhou, C., Fei, R., Zhang, C., Zhang, J.: Pansharpening via neighbor embedding of spatial details. IEEE J. Sel. Top. Appl. Earth Obs. Remote Sens. **14**, 4028–4042 (2021)
173. Zhong, J., Yang, B., Huang, G., Zhong, F., Chen, Z.: Remote sensing image fusion with convolutional neural network. Sens. Imaging. **17**, 1–16 (2016)
174. Roweis, S.T., Saul, L.K.: Nonlinear dimensionality reduction by locally linear embedding. Science. **290**, 2323–2326 (2000)
175. Bevilacqua, M., Roumy, A., Guillemot, C., Morel, M.-L.A.: Super-resolution using neighbor embedding of back-projection residuals. In: 2013 18th International Conference on Digital Signal Processing (DSP) (2013)
176. Yang, C.-Y., Ma, C., Yang, M.-H.: Single-image super-resolution: a benchmark. In: Computer Vision–ECCV 2014: 13th European Conference, Zurich, Switzerland, September 6–12, 2014, Proceedings, Part IV (2014)
177. Jiang, J., Hu, R., Wang, Z., Han, Z., Dong, S.: Manifold regularized sparse support regression for single image super-resolution. In: 2013 IEEE International Conference on Acoustics, Speech and Signal Processing (ICASSP) (2013)
178. Kang, L.W., Hsu, C.C., Zhuang, B., Lin, C.W., Yeh, C.H.: Learning-based joint super resolution and deblocking for a highly compressed image. IEEE Trans. Multimed. **17**, 921–934 (2015). https://doi.org/10.1109/TMM.2015.2434216
179. Doshi, A., Chaudhuri, S.: Chaudhuri: satellite image super-resolution using deep learning: a comprehensive survey. Remote Sens. **14** (2022)
180. Yang, M.C., Wang, Y.C.F.: A self-learning approach to single image super resolution. IEEE Trans. Multimed. **15**, 498–508 (2013). https://doi.org/10.1109/TMM.2012.2232646
181. Xiong, Z., Xu, D., Sun, X., Wu, F.: Example-based super-resolution with soft information and decision. IEEE Trans. Multimed. **15**, 1458–1465 (2013). https://doi.org/10.1109/TMM.2013.2264654
182. He, L., Qi, H., Zaretzki, R.: Beta Process Joint Dictionary Learning for Coupled Feature Spaces with Application to Single Image Super-Resolution. Presented at the (2013). https://doi.org/10.1109/CVPR.2013.51
183. Timofte, R., De Smet, V., Van Gool, L.: A+: Adjusted anchored neighborhood regression for fast super-resolution. In: Computer Vision–ACCV 2014: 12th Asian Conference on Computer Vision, Singapore, Singapore, November 1–5, 2014, Revised Selected Papers, Part IV, p. 12 (2015)

184. Mishra, D., Hadar, O.: Accelerating neural style-transfer using contrastive learning for unsupervised satellite image super-resolution. IEEE Trans. Geosci. Remote Sens. **61**, 1 (2023)
185. Li, Y., Wang, Y., Tian, Y., Xia, Z.: A comprehensive survey of deep learning in single-image super-resolution. IEEE Trans. Neural Netw. Learn. Syst. **32**, 2434–2454 (2021)
186. Zhang, K., Li, X., Van Nguyen, T., Ghanem, B., Ahuja, N.: Deep fusion of multiple motion-compensated frames for video super-resolution. In: Proceedings of the IEEE/CVF Conference on Computer Vision and Pattern Recognition, pp. 10345–10354 (2021)
187. Gao, F., Masek, J., Schwaller, M., Hall, F.: On the blending of the Landsat and MODIS surface reflectance: predicting daily Landsat surface reflectance. IEEE Trans. Geosci. Remote Sens. **44**, 2207–2218 (2006)
188. Hilker, T., Wulder, M.A., Coops, N.C., Linke, J., McDermid, G., Masek, J.G., Gao, F., White, J.C.: A new data fusion model for high spatial-and temporal-resolution mapping of forest disturbance based on Landsat and MODIS. Remote Sens. Environ. **113**, 1613–1627 (2009)
189. Roy, D.P., Ju, J., Lewis, P., Schaaf, C., Gao, F., Hansen, M., Lindquist, E.: Multi-temporal MODIS–Landsat data fusion for relative radiometric normalization, gap filling, and prediction of Landsat data. Remote Sens. Environ. **112**, 3112–3130 (2008)
190. Zhu, X., Chen, J., Gao, F., Chen, X., Masek, J.G.: An enhanced spatial and temporal adaptive reflectance fusion model for complex heterogeneous regions. Remote Sens. Environ. **114**, 2610–2623 (2010)
191. Wang, Q., Atkinson, P.M.: Spatio-temporal fusion for daily Sentinel-2 images. Remote Sens. Environ. **204**, 31–42 (2018)
192. Kwan, C., Budavari, B., Gao, F., Zhu, X.: A hybrid color mapping approach to fusing MODIS and Landsat images for forward prediction. Remote Sens. **10**, 520 (2018)
193. Jia, D., Cheng, C., Song, C., Shen, S., Ning, L., Zhang, T.: A hybrid deep learning-based spatiotemporal fusion method for combining satellite images with different resolutions. Remote Sens. **13**, 645 (2021)
194. Shi, W., Guo, D., Zhang, H.: A reliable and adaptive spatiotemporal data fusion method for blending multi-spatiotemporal-resolution satellite images. Remote Sens. Environ. **268**, 112770 (2022)
195. Ma, Y., Wang, Q., Wei, J.: Spatiotemporal fusion via conditional diffusion model. IEEE Geosci. Remote Sens. Lett. **21**, 1–5 (2024)
196. Laina, I., Rupprecht, C., Belagiannis, V., Tombari, F., Navab, N.: Deeper depth prediction with fully convolutional residual networks. In: 2016 Fourth international conference on 3D vision (3DV) (2016)
197. Guo, Y.: Progressive fusion network for depth-guided image super-resolution. In: Proceedings of the IEEE/CVF Conference on Computer Vision and Pattern Recognition, pp. 5855–5864 (2020)
198. Ma, C.: Depth-aware CNN for RGB-D segmentation. In: Proceedings of the European Conference on Computer Vision, pp. 239–255 (2018)
199. Li, B.: Depth-attentional features for single-image super-resolution. In: Proceedings of the IEEE/CVF Conference on Computer Vision and Pattern Recognition, pp. 305–314 (2019)
200. Godard, C.: Digging into self-supervised monocular depth estimation. In: Proceedings of the IEEE/CVF International Conference on Computer Vision, pp. 3828–3838 (2019)
201. Gao: surface normal guided depth map super-resolution. In: Proceedings of the IEEE/CVF Conference on Computer Vision and Pattern Recognition. pp. 9096–9105 (2020)
202. Lohit, S., Liu, D., Mansour, H., Boufounos, P.T.: Unrolled projected gradient descent for multi-spectral image fusion. In: ICASSP 2019-2019 IEEE International Conference on Acoustics, Speech and Signal Processing (ICASSP) (2019)
203. Yang, G., Zhou, M., Yan, K., Liu, A., Fu, X., Wang, F.: Memory-augmented deep conditional unfolding network for pan-sharpening. In: Proceedings of the IEEE/CVF Conference on Computer Vision and Pattern Recognition (2022)
204. Chen, X., Yi, Z., Pu, H., Sun, Z., Wang, X.: A simple baseline for unsupervised representation learning from pixel-wise correlations. ArXiv Prepr. ArXiv. (2020)

205. Zhang, B., Li, Y., Zhang, S., Wang, X., Huang, H., Sun, J.: Non-local neural networks. Proc. IEEE Conf. Comput. Vis. Pattern Recognit. (2018)
206. Kawulok, M., Benecki, P., Hrynczenko, K., Kostrzewa, D., Piechaczek, S., Nalepa, J., Smolka, B.: Deep learning for fast super-resolution reconstruction from multiple images. In: Real-Time Image Processing and Deep Learning 2019 (2019)
207. Bioucas-Dias, J.M., Plaza, A., Camps-Valls, G., Scheunders, P., Nasrabadi, N., Chanussot, J.: Hyperspectral remote sensing data analysis and future challenges. IEEE Geosci. Remote Sens. Mag. **1**, 6–36 (2013)
208. Simoes, M., Bioucas-Dias, J., Almeida, L.B., Chanussot, J.: A convex formulation for hyperspectral image superresolution via subspace-based regularization. IEEE Trans. Geosci. Remote Sens. **53**, 3373–3388 (2014)

Multi-Image Super-Resolution Using Graph Neural Networks

Tomasz Tarasiewicz and Michal Kawulok

1 Introduction

In the digital age, the clarity and detail of images play a pivotal role across various domains, including remote sensing, medical imaging, surveillance, and more. The quality of an image, largely determined by its resolution, affects not only the visual appeal but also the accuracy and reliability of the information it conveys. Despite technological advancements, achieving *high-resolution* (HR) images directly from imaging sensors often faces practical constraints, such as physical limitations of the sensors, bandwidth and storage issues, and environmental factors that can degrade image quality.

Super-resolution reconstruction (SRR) is a set of techniques designed to enhance the spatial resolution of images beyond the capabilities of the imaging devices themselves. SRR has emerged as a crucial area of research, addressing the gap between the existing hardware limitations and the growing demand for high-quality imagery. This pursuit is not merely an academic endeavor but a necessity for practical applications that rely on detailed visual information for critical decision-making.

The evolution of SRR encompasses various methodologies, with *single-image super-resolution* (SISR) and *multi-image super-resolution* (MISR) standing out as two primary branches. SISR focuses on generating an HR image from a single *low-resolution* (LR) input, tackling the challenge of inferring missing details. In contrast, MISR leverages multiple LR images of the same scene captured at different points in time, exploiting the additional information available to produce a more detailed HR output. Both approaches have their unique advantages and challenges, contributing differently to the field of SRR.

T. Tarasiewicz · M. Kawulok (✉)
Department of Algorithmics and Software, Silesian University of Technology, Gliwice, Poland
e-mail: tomasz.tarasiewicz@polsl.pl; michal.kawulok@polsl.pl

M. Kawulok et al. (eds.), *Super-Resolution for Remote Sensing*, Unsupervised and Semi-Supervised Learning, https://doi.org/10.1007/978-3-031-68106-6_4

Moreover, the integration of deep learning techniques has revolutionized SRR, offering unprecedented capabilities for learning complex patterns and relationships within image data. Among these advancements, *graph neural networks* (GNNs) have shown promise in capturing relational information between elements in an image, offering a novel perspective on addressing SRR challenges.

This chapter explores the applications of GNNs for MISR, comparing them to leading methods. It starts with an overview of SRR, its history, and the key distinctions between SISR and MISR, alongside significant models in the field. Following this, it introduces graph theory fundamentals and influential graph-based neural networks. The discussion then moves to two specific GNN models for MISR, MagNet and MagNet++, detailing their architecture and data preparation processes. It proceeds to describe the datasets utilized in experiments, the training approach, loss functions, and metrics for evaluating model performance. The results section presents both quantitative and qualitative findings, leading to a conclusion on the role of GNNs in MISR. The final section offers a summary and future research directions for GNNs in this domain.

2 Super-Resolution Reconstruction

This section focuses on the methodologies and innovations that define the current landscape of SRR. The distinction between SISR and MISR encapsulates a broad spectrum of techniques and challenges in enhancing image quality. While SISR aims to extract detailed information from a single LR image, leveraging deep learning to infer missing details, MISR utilizes multiple images of the same scene to reconstruct higher-quality outputs, capitalizing on the additional information spread across the image set. The progression from traditional approaches to the incorporation of advanced neural network architectures, including *convolutional neural networks* (CNNs), *generative adversarial networks* (GANs), and more recently, GNNs, marks a significant evolution in the quest for achieving HR imagery. This section aims to provide a comprehensive overview of the significant strides made in SISR and MISR, highlighting key models and methodologies that have shaped the field.

2.1 Single-Image Super-Resolution

SISR has been an area of extensive exploration in the past years, with *convolutional neural networks* (CNNs) being central to the majority of the state-of-the-art solutions [100]. CNNs have been instrumental in advancing the performance of SISR solutions by effectively modeling the relationship between LR and HR images through feature representation and nonlinear mapping.

The first CNN proposed for SR, the *super-resolution convolutional neural network* (SRCNN) [19], was composed of just three convolutional layers. Subsequent advancements in the field gave rise to SRResNet [52], which notably introduced the incorporation of residual blocks to simplify training and enhance super-resolution performance. This design was inspired by the success of ResNet, the pioneering architecture that initially introduced the concept of deep residual learning [34]. Following this, the *multi-scale deep super-resolution* (MDSR) network was developed, offering an innovative approach by integrating multiple scaling factors into a single model, allowing it to handle different upsampling tasks without switching models [54]. Building on these advancements, the *residual channel attention network* (RCAN) [116] was proposed, which further enhanced SISR by using a channel attention mechanism to weigh the importance of different channels and refine the feature representations, proving particularly beneficial for capturing intricate image details. Additionally, more complex models of much larger capacities, like the *enhanced deep SR network* (EDSR) [54], showcased the potential to model the LR-to-HR relationship even more effectively.

In addition to traditional CNNs, the application of GAN models to SISR has witnessed notable advancements. SRGAN [52], which utilized SRResNet as its generator and paired it with a single discriminator network, showcases the power of adversarial training in super-resolution, particularly in producing images with high perceptual quality. Building upon the foundation laid by SRGAN, ESRGAN [97] introduced architectural and loss function modifications, leveraging a robust adversarial loss and integrating residual-in-residual dense blocks. This resulted in images with enhanced details and sharper textures, setting new standards in the field. It is important to note that while GANs are particularly effective in reconstructing images of high perceptual quality, they do not necessarily recover the actual high-frequency information [82]. Despite this, GANs have been utilized in remote sensing [101] or medical imaging [30] applications, and it has been demonstrated that certain constraints on the adversarial loss can increase the reliability of the reconstruction outcome [43]. Another recent development in SISR is the use of vision transformers, which dynamically adjust the size of the feature maps, thus reducing the model complexity [60]. While this approach has shown potential, its broader implications and effectiveness in the context of SRR are still being explored.

2.2 *Multi-Image Super-Resolution*

The concept of MISR has its roots in early video processing techniques. In the late 1980s and early 1990s, the idea of using multiple LR frames to reconstruct a higher-resolution frame was first proposed [91]. This technique was initially known as multi-frame image restoration, and it was primarily used to enhance the quality of video footage.

The process exploited the natural movement in a video sequence, which caused each frame to capture slightly different information about the scene. By aligning and combining these frames, it was possible to extract more details than could be seen in any single frame, resulting in a higher-resolution output. This concept was revolutionary at the time, as it provided a means to enhance the resolution of video footage without requiring any improvements in hardware. As computational power increased and algorithms became more sophisticated, the concept of MISR was extended to still images, resulting in the development of new techniques. MISR has since evolved into a complex field of study that involves various subdisciplines, including image registration [67], image fusion [72], and machine learning.

Compared to SISR, its multi-image counterpart often achieves higher accuracy in reconstructing images [69, 109]. While SISR methods generate missing high-frequency details from a single image, MISR models leverage the information contained in multiple LR images of the same scene. This multi-image approach provides richer data, enabling the fusion of information from different images to enhance resolution.

MISR has witnessed significant progress during recent years, largely driven by the PROBA-V Super-Resolution competition [64], an initiative of the European Space Agency in 2018 that ran for 8 months. The primary objective of this competition was to super-resolve satellite images with a GSD of 300 meters to a finer, 100-m GSD resolution; thus, it aimed to upsample the input images by a factor of three. GSD refers to the distance between two consecutive pixel centers measured on the ground, indicating the spatial resolution of an image [53]. The data for this challenge was sourced from the PROBA-V satellite, with each scene composed of a varying number of observations. It introduced challenges such as obscured pixels from obstructions like clouds or their shadows and uncaptured pixels for which the data was not observed. Even though the challenge concluded in 2019, the servers remain operational, allowing for so-called postmortem evaluations. Current results from these ongoing assessments are recorded and showcased on a dedicated postmortem leaderboard.

Early in the deep learning era of MISR, EvoNet [42] emerged, applying convolutional neural networks for SISR combined with an evolutionary fusion strategy for multi-image integration [41]. It was succeeded by DeepSUM [68], a pioneering end-to-end deep learning model for MISR and the winner of the PROBA-V challenge. Although powerful, DeepSUM's architecture is constrained to a fixed number of LR inputs and necessitates extensive training. HighRes-Net [18] subsequently addressed DeepSUM's limitations by introducing a flexible model capable of handling variable input image counts. Its unique recursive fusion mechanism for latent representations marked a significant departure from earlier models. For a detailed exploration of HighRes-Net's architecture and contributions, refer to Sect. 2.3. Similarly, RAMS [77] model, while operational on a fixed number of LR images like DeepSUM, incorporated attention mechanisms [4], setting a new direction for MISR research. The MISR-GRU [75] model brought a different perspective by utilizing recurrent neural networks (RNNs) [76] to perform a fusion of information along the temporal dimension. It employed gated recurrent units (GRUs) to adaptively capture temporal dependencies, showcasing its potential in addressing MISR's inherent challenges related to handling temporal variations. Another groundbreaking model was PIUNET [92], focusing on permutation invariance and uncertainty in MISR tasks. Further innovations were witnessed with

TR-MISR [2], which built upon HighRes-Net's foundation. Integrating visual transformers [93] into MISR showcased the capability of attention mechanisms in this domain. Detailed insights into TR-MISR's and PIUNET's approaches are elaborated in the subsequent section. It is worth noting that both of these models currently lead in the postmortem PROBA-V Super-Resolution competition, underscoring the efficacy of these state-of-the-art techniques in addressing MISR challenges.

Multitemporal fusion based on CNNs has been applied to both burst-image super-resolution [10] and video super-resolution [39], with a comprehensive overview provided by Liu et al. [56]. 3D CNNs have shown effectiveness in addressing video SRR challenges [45]. Moreover, contemporary methods are merging SISR and MISR techniques to enhance video streams more effectively [36, 80]. Such methodologies have also found applications in satellite imagery enhancement, particularly in processing satellite image bursts. These strategies typically leverage the consistent and known temporal frequency of input frames. They are often designed to handle moving objects by estimating motion fields, distinguishing them from multi-image fusion techniques that work with an unordered collection of images without specific timestamps, rather than a chronological sequence [71].

2.3 *Prominent Models in SRR*

Super-resolution, encompassing both its single- and multi-image variants, has seen the emergence of numerous architectures. Many of these have contributed to advancing the field, introducing novel methodologies, or refining existing ones. This section explores models that have substantially influenced the landscape of SRR and, particularly, the direction of the research.

2.3.1 FSRCNN

The *fast super-resolution convolutional neural network* (FSRCNN) [21], a landmark model in deep learning-based SISR domain, serves as the foundational architecture for the MagNet model, which is the central focus of the discussion in this chapter. By understanding the intricate design and operational principles of FSRCNN, readers will gain insights into the advancements and efficiencies that the MagNet model inherits and builds upon. FSRCNN stands out for its tailored architecture designed for computational efficiency without compromising reconstruction quality. This section describes FSRCNN's innovative approach, which eliminates the explicit bicubic interpolation step found in its predecessor, SRCNN, thereby facilitating faster processing and setting a new benchmark for SISR models.

The model's architecture is structured into several phases. Its design ensures efficient feature representation and has set a benchmark for subsequent SISR models. Conceptually, FSRCNN can be viewed as being composed of three primary blocks:

Feature Extraction At the heart of FSRCNN is its feature extraction phase, which employs a single convolution layer. The aim here is to capture the low-level features of the input LR image. This step ensures that the foundational features, which are essential for subsequent stages, are adequately represented.

Shrinking and Nonlinear Mapping Post feature extraction, FSRCNN introduces a shrinking layer which reduces the number of feature maps, ensuring a compact representation. This is followed by a series of convolutional layers whose main goal is to capture more complex features that could potentially enhance the granularity and richness of the output image. The motivation behind this block is to refine the features before the expansion phase, ensuring that the model has a comprehensive latent set of features to work with during the upsampling process.

Expansion and Deconvolution The final block of FSRCNN involves the expansion of the refined features back to a higher-dimensional space, setting the stage for the deconvolution process. The deconvolution layer [110, 111], or transposed convolutional layer [23, 94], then upsamples the feature maps to produce the HR output. This design choice avoids the explicit bicubic interpolation used in traditional SR methods like SRCNN, resulting in a significant boost in computational efficiency.

The genius of FSRCNN lies in its ability to maintain a balance. While it adopts a lightweight structure, it does not compromise the quality of super-resolved images.

2.3.2 DeepSUM

DeepSUM [68] marked a significant stride in the domain of MISR. Central to its design is a combination of CNN-based SISR techniques with an evolutionary fusion mechanism, enabling the model to assimilate details from multiple images effectively. A standout feature of DeepSUM is the RegNet submodule consisting of a series of shared 2D convolutional layers and a *global dynamic convolutional* (GDC) layer. Instead of attempting to register the LR images directly, DeepSUM first employs its SISR component to upsample these images. RegNet then takes center stage, ensuring alignment of these upsampled versions based predominantly on their spatial shifts. Specifically, for each upsampled image, except for a reference one, RegNet learns a specific filter, which, when convolved with the image by the GDC, aligns it to the reference.

DeepSUM's proficiency was unambiguously showcased when it emerged victorious in the PROBA-V Super-Resolution competition, thus attesting to its efficacy in challenging real-world settings. However, one aspect of DeepSUM that warrants attention is its two-phased training approach. Initially, the RegNet submodule undergoes training, post which the overarching model is trained. While this sequential approach was pivotal in achieving the desired outcomes, it introduced additional complexity, particularly concerning training duration and computational requirements.

2.3.3 HighRes-Net

HighRes-Net [18] introduced an innovative approach to MISR by addressing the limitation of handling a fixed number of input images. One of its standout features is the formulation of shared representations and the embedding mechanism.

The backbone of HighRes-Net's feature extraction process is an architecture of its embedding block, represented as emb_θ, which encompasses a convolutional layer followed by two residual blocks, each activated by a *parametric rectified linear unit* (PReLU) [35] function. This architecture ensures consistent and robust feature extraction and is uniformly applied across all LR images independently. To derive a common representation for the set of LR images, the model computes a reference image by taking the median values across the entire set of N LR frames. The value of a reference pixel at position (x, y) is defined as:

$$\text{ref}(x, y) = \text{median}(\text{LR}_0(x, y), \ldots, \text{LR}_{N-1}(x, y)). \tag{1}$$

This computed reference serves as a composite representation of the scene, synthesized from multiple vantage points. Each LR frame is then cohesively embedded with this reference. The resulting embedded states, originally denoted by s_0^i, capture the distinct features of each frame in comparison to the reference:

$$s_0^i = \text{emb}_\theta([\text{LR}_i, \text{ref}]). \tag{2}$$

After the embedding phase, HighRes-Net unveils its recursive fusion mechanism. This procedure operates on the latent representations, amalgamating the information from multiple frames in a hierarchical manner. In instances where the number of LR frames does not align with a power of two, the model compensates by padding the set with zero-valued views, ensuring the fusion process remains consistent. The culmination of this fusion is a super-resolved image, enriched by the collective information from all the input LRs.

Through its unique approach of shared representation, embedding, and recursive fusion, HighRes-Net has solidified its place as a noteworthy contribution to the evolution of MISR techniques.

2.3.4 RAMS

The residual attention multi-image super-resolution network (RAMS) [77] represents a significant advancement in MISR, integrating attention mechanisms to refine the fusion process. Recognizing that different parts of the input images contribute variably to the super-resolved output, RAMS employs attention maps to amplify the most crucial regions, ensuring better feature utilization.

Central to RAMS is its series of attention blocks dispersed among its super-resolution layers. These blocks generate attention maps that gauge the interdependencies between different regions of the input images. By dynamically modulating the feature maps, RAMS ensures the subsequent network layers prioritize the emphasized regions, resulting in sharper and more accurate outputs.

However, a limitation of RAMS, similar to DeepSUM, is its reliance on 3D convolutional layers in the attention blocks, which fixes the number of input images it can process. Despite this, RAMS' introduction of attention mechanisms into MISR has not only elevated the standard for super-resolved images but also set a precedent for future models blending attention with super-resolution.

2.3.5 PIUNET

PIUNET [92], a significant advancement in the MISR domain, tackled two complex challenges: permutation invariance and uncertainty estimation. Recognizing the potential inconsistencies arising from different orders of input images, PIUNET ensured a consistent super-resolved output regardless of the sequence of its input images. This design is especially crucial in real-world scenarios where the order of image acquisition might differ, making consistent output a priority.

Furthermore, PIUNET brought an added dimension of interpretability to MISR by providing a measure of the uncertainty associated with its predictions. With an uncertainty map accompanying the super-resolved image, it highlights regions where the model is confident and areas where it is less certain. This not only aids in understanding the model's decision-making but also offers a practical metric to assess the reliability of the super-resolved outputs in various parts of the image.

The architecture of PIUNET skillfully integrates these features, producing a model that excels in the quality of its outputs while also offering insightful meta-information about its predictions. Its effectiveness is evident from its ranking in the PROBA-V challenge, where it occupies the second-best position, highlighting PIUNET's capability and its importance in guiding future MISR research.

2.3.6 TR-MISR

TR-MISR [2], building upon the foundation laid by HighRes-Net, introduced the capabilities of transformers to the MISR domain. The central innovation in TR-MISR is its shift from the recursive fusion mechanism of its predecessor, HighRes-Net, to a transformer-based fusion. This change reflects the growing appreciation for the potential of attention mechanisms in deep learning, especially in tasks requiring detailed feature fusion.

Transformers, originally designed for natural language processing tasks, have been adopted across various domains due to their ability to handle long-range dependencies and intricate relationships between data points. In the context of MISR, transformers allow TR-MISR to effectively merge information from multiple LR images, capturing subtle details and relationships between features across these images.

By using self-attention mechanisms, TR-MISR can dynamically weigh the importance of features from different input images, leading to a more refined and contextually aware super-resolved output. This ability to adjust the focus on different parts of the input images based on their relevance allows TR-MISR to produce superior super-resolved images.

Highlighting its effectiveness, TR-MISR currently leads the postmortem leaderboard of the PROBA-V Super-Resolution competition, marking its position as one of the foremost models in the field of MISR. This achievement showcases the value of transformer architectures in pushing the boundaries in areas like multi-image super-resolution and provides a new standard for future research in this domain.

3 Graph Neural Networks

A recent yet less explored area within super-resolution pertains to the use of GNNs. These networks have demonstrated considerable promise across various applications due to their inherent capability to model complex relational information in graph-structured data [48]. Super-resolution tasks fundamentally revolve around understanding and leveraging intricate relational information present in images [69]. While traditional methods, mainly CNNs, have predominantly focused on spatial dependencies, the true richness of relationships in image data goes beyond just spatial interactions. GNNs, with their ability to capture not just spatial, but more intricate and higher-order dependencies, are poised as potentially more suitable for super-resolution tasks. Their broader relational modeling capabilities can provide an edge in tasks where understanding nuanced relationships is paramount.

The advent of GNNs can be traced back to 2014 with the introduction of the *graph convolutional network* (GCN) [15]. GCNs represent a significant evolution in the realm of neural networks, adapting the foundational principles of traditional CNNs to suit graph-structured data. Unlike CNNs, which are optimized for grid-like data such as images, GCNs are designed to handle data that exists in non-grid formats, where entities have complex interrelationships. In essence, GCNs capture and process the relational information inherent in graphs. They achieve this by employing a form of convolution that respects the topology of the graph, allowing for the effective propagation of information across connected nodes. This unique capability has made GCNs particularly valuable in scenarios where understanding the relationships between data points is crucial, e.g., node classification [46], graph classification [114], recommendation systems [108], or relational reasoning [79].

Incorporating the *graph attention networks* (GATs) [95] into the realm of GNNs marked a significant advancement. GATs introduced an attention mechanism that assigns different weights to various edges in a graph, diverging from the conventional uniform information aggregation method in traditional GCNs. This allowed GATs to process varying contributions from different connected nodes based on their relative importance. The unique attention mechanism not only enables the model to highlight salient features but also devalues less important ones, leading to more expressive and discriminative feature representations. This selective aggregation of information, driven by the attention mechanism, enhances GATs' capability to capture informative nodes effectively, finding applications in tasks like person reidentification [5], action recognition [107], or prompt-driven object detection [96].

Recurrent graph networks (RGNs) are another variant of GNNs that process graph-structured data by recursively updating node representations. This recursive nature allows the network to capture deeper and more intricate relationships within the graph. RGNs have been particularly beneficial for video analysis [55], where the temporal sequence of frames can be modeled as a graph, and the recurrent nature helps in capturing the temporal dependencies. Additionally, RGNs have shown promise in modeling social relationships [27], capturing intricate patterns and dynamics within social networks.

A significant milestone in the development of GNNs for computer vision tasks is the introduction of SplineCNN. This model presents a generalization of the convolution operation that respects the local structure and spatial characteristics of arbitrary graphs. The distinctive feature of SplineCNN is its use of continuous B-spline kernels for convolution, effectively applying spline-based filters on graphs and efficiently utilizing spatial information of each node. By parametrizing the filters as splines, the model can better adapt to the characteristics of the graph data. These spline-based filters ensure smooth transformations, which are more expressive and robust against over-smoothing, a prevalent issue in traditional GCNs. SplineCNNs have been effectively used in the task of image graph classification, graph node classification and 3D shape recognition on point clouds [25].

While the use of GNNs in super-resolution is still in its nascent phase, the introduction of the *cross-scale internal graph neural network* (IGNN) [104]—which integrates traditional convolutional layers with the graph-based ones—highlights the immense promise of these networks for resolution enhancement tasks. The design of IGNN is particularly tailored for the SISR problem, proving especially advantageous for images that exhibit repetitive patterns across varying scales, akin to self-exemplar techniques like the one proposed by Huang et al. [37]. It establishes an internal graph structure from the input image and taps into both local and global correlations, employing a cross-scale approach to boost reconstruction quality. However, the application of GNNs for MISR remains an open area of research, presenting a potential avenue for groundbreaking discoveries and advancements in the domain.

Further enriching the landscape of GNNs in SRR is the introduction of the inter-layer feature-representation-based GNN for image super-resolution, LSGNN [84]. This model, similarly to IGNN, is tailored specifically to tackle the SISR problem. While many CNNs for SISR have primarily focused on broader and deeper architecture designs, they often overlook the detailed information inherent in the image itself and the potential relationships between features captured at different stages of a network. The LSGNN model addresses these gaps by emphasizing the importance of understanding the interdependence between the extracted features of different layers. It introduces a layer feature graph representation learning module that captures this interdependence, enabling the extraction of deeper and more fine-grained image detail features.

In the multifaceted domain of machine learning, neural networks have been instrumental in facilitating substantial progress across a wide range of fields, from image and speech recognition to natural language processing. These networks draw their design inspiration from the neural structure of the human brain, providing them with remarkable abilities to identify patterns and make data-driven predictions. However, traditional neural networks often face challenges when dealing with data that is not regularly structured, as is the case with images or time series, and often overlook the relational information inherent in the input data. This is the intersection where GNNs showcase their strength.

3.1 The Concept of Graphs

A fundamental understanding of graphs is essential for grasping the intricacies of GNNs. In the realm of data structures, a graph is defined as a collection of entities, termed as *nodes* or *vertices*, which are interconnected by links known as edges [33, 103]. The power of graphs lies in their ability to represent intricate relationships and systems, from social networks and molecular formations to the vast interconnectedness of web pages [118].

Within the domain of graph theory, various classifications of graphs exist based on their structural properties and the kind of relationships they depict [70]. Notable among them are:

- **Undirected graphs**: These are graphs where edges do not have a direction. An edge between vertex A and vertex B is identical to an edge between vertex B and vertex A.
- **Directed graphs (digraphs)**: In these graphs, edges have a clear direction. An edge from vertex A to vertex B does not reciprocate an edge from vertex B to vertex A.
- **Weighted graphs**: In such graphs, every edge is assigned a specific weight or cost. This weight can represent various metrics, such as distance, cost, or any domain-specific value.
- **Unweighted graphs**: Contrary to weighted graphs, all edges in these graphs are identical and do not carry any specific weight.
- **Cyclic graphs**: These graphs contain cycles, which are closed paths where the starting and ending vertices are the same.
- **Acyclic graphs**: Such graphs are devoid of any cycles. A special kind of acyclic graph, termed a tree, ensures a unique path between any two vertices.

- **Connected graphs**: In these graphs, a path exists between every pair of vertices.
- **Disconnected graphs**: These are graphs where certain vertices might not be reachable from others.
- **Bipartite graphs**: The vertices of these graphs can be divided into two distinct sets where no two vertices within the same set share an edge.
- **Complete graphs**: In these graphs, a unique edge connects every pair of distinct vertices.

Mathematically, a graph $\mathcal{G}$ can be defined as a tuple $\mathcal{G} = (\mathcal{V}, \mathbf{H}, \mathcal{E}, U)$, where:

- $\mathcal{V}$ is a set of vertices (or nodes). Each vertex represents an entity in the graph. Formally, $\mathcal{V} = \{v_1, v_2, \ldots, v_n\}$ where n is the number of vertices.
- $\mathbf{H}$ is a matrix representing the features of each node in $\mathcal{V}$. For a given node v_i, its feature vector is represented as $\boldsymbol{h_i} \in \mathbb{R}^F$, where F is the number of features in each node; thus, $\mathbf{H} \in \mathbb{R}^{n \times F}$.
- $\mathcal{E}$ is a set of edges that connect pairs of vertices. Each edge represents a relationship or connection between two vertices. Formally, $\mathcal{E} \subseteq \mathcal{V} \times \mathcal{V}$, and an edge e_{ij} exists if there is a connection from vertex v_j to vertex v_i. $\mathcal{E}$ can also be represented as a *sparse adjacency matrix* $\mathbf{A} \in \mathbb{B}^{n \times n}$.
- U is a set of attributes associated with the edges. Here, each edge e_{ij} has an attribute from the set U and such attribute is denoted as u_{ij}. Since those attributes can take different forms for different applications, such as single values (attention coefficients [12]), vectors (relative position to another node [24]), or even matrices (multifaceted attributes in knowledge graphs [79]), their dimensionality is not explicitly defined at this point.

If the graph is undirected, then the order of vertices in the edge pair does not matter, that is, $e_{ij} = e_{ji}$ what implies $u_{ij} = u_{ji}$. If the graph is directed, then the order is significant; thus, e_{ij} and e_{ji} are distinct edges.

This understanding of graphs is not only fundamental but also intimately tied to the essence of this research.

3.2 Importance of Graph Neural Networks

GNNs represent a pivotal shift in the realm of neural networks. They are a specialized form of neural networks tailored to efficiently handle and process graph-structured data [13, 31]. This differs from traditional neural networks like CNNs or RNNs, which excel when the data can be structured in a regular form, such as images or time series. These structured data types inherently possess a local order and continuity in their relationships, which these networks leverage to capture patterns and dependencies. However, when the data deviates from this structural format, as is the case with graph data, these traditional networks encounter difficulties [7].

Graph-structured data does not adhere to the spatial or temporal consistency of images or time series data. Instead, graphs are non-Euclidean structures where nodes and their corresponding edges can represent diverse and complex systems [118]. The edges connecting the nodes can symbolize a myriad of relationships, from friendships in a social network [66] and hyperlink connections in the World Wide Web [11] to protein interactions in a biological network [6]. The fundamental challenge is to process this irregularly structured data, extract meaningful features, and identify complex patterns that capture the inherent relationships within the graph [28, 32].

GNNs rise to this challenge by working directly with data in a graph format, allowing them to exploit the unique properties and the richness of relationships that graphs offer [13, 78]. Unlike traditional neural networks that may lose relational information when applied to graph data, GNNs are designed to natively preserve and manipulate the relationships between the nodes [46]. They are capable of learning the topology of the graph and the attributes of the nodes and edges [26], which equips them with the ability to infer the intricate and often nonlinear relationships within the data [7].

This capability of GNNs opens the door to enhanced performance across a broad array of tasks that involve relational or structured data. In essence, they are adept at tasks that require inference about relationships, relational reasoning, and the propagation of relational information. These tasks span the spectrum from node classification, link prediction, graph classification, and even complex system dynamics prediction [12].

The power of GNNs to handle and learn from graph-structured data signifies a paradigm shift in the realm of deep learning and artificial intelligence [105]. By processing data in its native graph format, they are not only better equipped to handle the data's complexity but also to unlock insights from the intricate relationships within the graph. It is this transformative capability of GNNs that introduces a new era in machine learning, paving the way for novel applications and research directions.

3.3 Fundamentals of Graph Neural Networks

The operational philosophy of GNNs centers on principles known as *message passing* or *neighborhood aggregation* [26]. This methodology, derived from the fundamental characteristics of graphs, forms the backbone of GNNs, enabling them to process and extract information from graph-structured data in an iterative and layered manner.

In essence, the message-passing framework postulates that a node's representation can be effectively informed by aggregating information from its immediate neighborhood. This is formalized in a two-step process—a *message function* and an *update function* [78].

Message Function

Initially, each node v_i in the graph is assigned a feature vector $h_i^{(0)} \in \mathbb{R}^F$, derived from its attributes. This feature vector serves as the initial state of the node. As the network progresses through each layer of computation, the nodes perform the message-passing steps, updating their states based on the aggregated messages from their neighbors and their current states. A node v_j is considered a neighbor to v_i if there exists an edge e_{ij} between them, meaning $\mathbf{A}(i, j) = 1$. The capability to perform these operations repeatedly, often referred to as "stacking layers" in GNNs, allows the nodes to gradually incorporate information from an increasingly larger neighborhood [46]. After several iterations, each node in the graph holds a feature representation that captures not only its attributes but also the contextual information from its extended neighborhood. This process is pivotal to learning in GNNs, with the final node representations serving as powerful feature vectors for downstream tasks.

Each node v_i computes a message $\boldsymbol{m}_i$ based on its current state, $\boldsymbol{h}_i$, and the state of its neighboring nodes. This can be expressed mathematically as:

$$\boldsymbol{m}_i = \bigoplus_{j \in \mathcal{N}(i)} \phi(\boldsymbol{h}_i, \boldsymbol{h}_j, u_{ij}), \tag{3}$$

where $\bigoplus$ represents a differentiable and permutation invariant aggregation function (e.g., sum, product, mean), $\mathcal{N}(i)$ is the neighborhood of node v_i, $\boldsymbol{h}_i$ and $\boldsymbol{h}_j$ are the features of nodes v_i and v_j, u_{ij} is the edge attribute from node v_i to v_j, and ϕ is the message function.

Update Function

Each node v_i updates its state based on its current state and the aggregated messages from its neighbors. This is represented as:

$$\boldsymbol{h}'_i = \gamma(\boldsymbol{h}_i, \boldsymbol{m}_i), \tag{4}$$

where $\boldsymbol{h}'_i$ is the updated feature of node v_i, and γ is the update function.

3.4 *Influential GNN Architectures*

GNNs have evolved into a diverse array of architectures and methodologies, each tailored to process and understand graph-structured data in its unique way [7]. Among the vast landscape of GNN architectures, certain types and specific models have been identified as especially influential, not only in the broader research community but also in shaping the direction and methodologies of this research. The differences between these GNN architectures have been primarily defined by how information from a node's neighbors is aggregated and how features describing their type of connection are utilized—the defining aspect of graph neural processing [106].

In this section, emphasis is placed on two fundamental types of GNNs: GCNs and their attention-augmented variant, GATs. Also, SplineCNN is examined as a specific GNN architecture that is particularly impactful for graphs where spatial relationships are meaningful. While the foundation for processing graph data has been provided by GCNs and GATs, a novel approach has been introduced by SplineCNN that has significantly influenced the research perspective. Each of these architectures has offered unique insights and methodologies for handling graph-structured data, and their contributions to the field have been instrumental in guiding the approaches adopted in this research.

3.4.1 Graph Convolutional Networks

GCNs [15] operate under the principle that nodes closer to each other in the graph space should exhibit similar features. This intuition is materialized by implementing a convolution operation in the graph domain. The convolution is achieved by applying a propagation rule at each layer, mathematically expressed as:

$$\boldsymbol{h}_i' = \sum_{j \in \mathcal{N}(i)} \frac{\mathbf{W}\boldsymbol{h}_j}{\sqrt{\tilde{d}_i \tilde{d}_j}}, i \in \mathcal{N}(i). \quad (5)$$

Here, $\tilde{d}_i$ and $\tilde{d}_j$ indicate node degrees, i.e., a total number of edges from nodes v_i and v_j, respectively. By incorporating a normalization factor based on node degrees, GCNs ensure that feature updates remain stable across layers. It is worth noting that GCNs assume self-connections for each node; hence, $i \in \mathcal{N}(i)$. Additionally, $\mathbf{W} \in \mathbb{R}^{F' \times F}$ is the learnable weight matrix which plays a pivotal role in transforming F input features to F' output features. As the network progresses in depth, a node's representation begins to encapsulate features from more distant nodes, effectively capturing information from its multi-hop neighborhood.

3.4.2 Graph Attention Networks

GATs introduced the concept of attention mechanisms to the world of graph networks [95]. Rather than simply averaging the features of neighbor nodes, GATs perform a weighted average where the weights are learned through the data itself. This ability to adaptively assign importance to neighbors results in a more expressive and flexible model. The attention-guided computation of features for node v_i in GATs can be formulated as:

$$\boldsymbol{h}_i' = \gamma \sum_{j \in \mathcal{N}(i)} \alpha_{ij} \mathbf{W}\boldsymbol{h}_j, \quad (6)$$

for which the attention coefficient α_{ij} is a softmax-normalized weight across all choices of v_j in the neighborhood of node v_i. It can be formulated as:

$$\alpha_{ij} = \frac{\exp\left(\text{LeakyReLU}\left(\boldsymbol{a}^T[\mathbf{W}\boldsymbol{h}_i||\mathbf{W}\boldsymbol{h}_j]\right)\right)}{\sum_{k\in\mathcal{N}(i)}\exp\left(\text{LeakyReLU}\left(\boldsymbol{a}^T[\mathbf{W}\boldsymbol{h}_i||\mathbf{W}\boldsymbol{h}_k]\right)\right)} \tag{7}$$

where $\boldsymbol{a} \in \mathbb{R}^{2F'}$ is a learnable weight vector serving as an attention mechanism in GATs, and the weight matrix $\mathbf{W} \in \mathbb{R}^{F' \times F}$, similarly to GCNs, is used as a higher-level feature extractor. The $||$ operator stands for concatenation operation, and $\cdot^T$ is a transposition. To achieve a higher level of nonlinearity and stabilize the training, the calculation of attention coefficients α_{ij} employs the *leaky rectified linear unit* (LeakyReLU) activation function [62].

3.4.3 Spline Convolutional Neural Networks

SplineCNN [25] represents an important advancement in the field of GNNs, leveraging the power of B-spline basis functions to perform convolution operations on graph-structured data. The key innovation of SplineCNN is the use of continuous kernel functions, defined using these B-spline basis functions, which allows for more flexible and efficient handling of irregularly structured data.

The convolution operation in the SplineCNN takes into account the relative positions of each node with respect to its neighbors. This is achieved by using spatial relation vectors, or pseudo-coordinates, to define the kernel functions. These pseudo-coordinates are contained in the set of edge attributes U where the attributes of the edge e_{ij} are denoted by $\boldsymbol{u}_{ij} \in \mathbb{R}^D$ for a D-dimensional scenario. The ability to utilize spatial information is a key advantage of SplineCNN. This feature allows for the processing of graph-structured data in a way that is more similar to the processing of image data in traditional CNNs.

A pivotal aspect of the design and success of SplineCNN is its encoding of spatial relationships through edge attributes. This design philosophy draws parallels with the fundamental operations of traditional 2D convolutions. Just as a sliding window in traditional convolutions identifies neighboring pixels based on their relative positions without knowledge of global position, SplineCNN discerns the relationships among nodes through the encoded spatial relationships in edge attributes. The graph structure inherently encodes not just the presence of a neighbor but also its direction and distance. This spatial encoding becomes especially pivotal for spline-based convolutions. Unlike traditional convolutional filters that are fixed, the relative spatial information embedded in the edge attributes facilitates a more fluid and adaptive convolutional operation in SplineCNN, making it adept at capturing intricate spatial relationships inherent in graph data.

SplineCNN has been successfully applied to a range of tasks, as demonstrated in the original paper [25]. The authors showcased the effectiveness of SplineCNN in image graph classification, graph node classification, and shape correspondence on meshes. In these applications, the architecture, with its spline-based convolution operator, proved particularly proficient.

SplineCNN operates by aggregating information from neighboring nodes, with each node's contribution weighted by a kernel function $g_\theta : [a_1, b_1] \times \ldots \times [a_D, b_D] \rightarrow R^F$. It is crucial to mention that it performs a normalization step before the convolution operation. Specifically, the edge attributes $\boldsymbol{u}_{ij}$ are normalized and then rescaled such that they align with the span of the kernel, i.e., $[a_1, b_1] \times \ldots \times [a_D, b_D]$. This ensures that the convolution is performed in a consistent space, leading to more meaningful feature aggregations. The computation of F' output features for the ith node can be mathematically expressed as:

$$\boldsymbol{h}'_i = \frac{1}{|\mathcal{N}(i)|} \Big\Vert_{s=1}^{F'} \sum_{j \in \mathcal{N}(i)} \boldsymbol{h}_j^T g_{\theta_s}(\boldsymbol{u}_{ij}) \tag{8}$$

where $\Vert$ stands for a concatenation operation. The g_{θ_s} function is fundamentally based on D open B-spline basis functions of degree m, which can be denoted as $((N_{1,i}^m)_{1 \leq i \leq k_D}, \ldots, (N_{D,i}^m)_{1 \leq i \leq k_D})$ and uniformly positioned based on equidistant knot vectors [73] with $\mathbf{k} = \{k_1, \ldots, k_D\}$ defining the number of basis functions for each dimension:

$$g_{\theta_s}(\boldsymbol{u}) = \Big\Vert_{r=1}^{F} \sum_{p \in \mathcal{P}} w_{s,p,r} \cdot B_p(\boldsymbol{u}). \tag{9}$$

Here, p is a D-element tuple of basis functions taken from the Cartesian product of $\mathcal{P} = (N_{1,i}^m)_i \times \cdots \times (N_{D,i}^m)_i$. Each tuple p has assigned a learnable parameter $w_{s,p,r}$ for rth input feature and sth output feature. $B_p(\boldsymbol{u})$ is the product of the basis functions in p at position $\boldsymbol{u}$:

$$B_p(\boldsymbol{u}) = \prod_{d=1}^{D} N_{d,p_d}^m(\boldsymbol{u}(d)) \tag{10}$$

Following the theoretical overview of SplineCNN, it is beneficial to explain its operations visually. A one-dimensional context provides an intuitive foundation before progressing to the more complex two-dimensional scenario. In both cases, the basis functions have a constant degree $m = 2$.

In the one-dimensional context, Fig. 1 depicts the positioning of individual B-spline basis functions along the continuum. When these functions are combined, they produce a consolidated spline-based surface. The interval $[a_1, b_1]$ marks the kernel's definition range, where the sum of all uniform basis functions equals one. This configuration provides a consistent backdrop for understanding the convolution operation in SplineCNN.

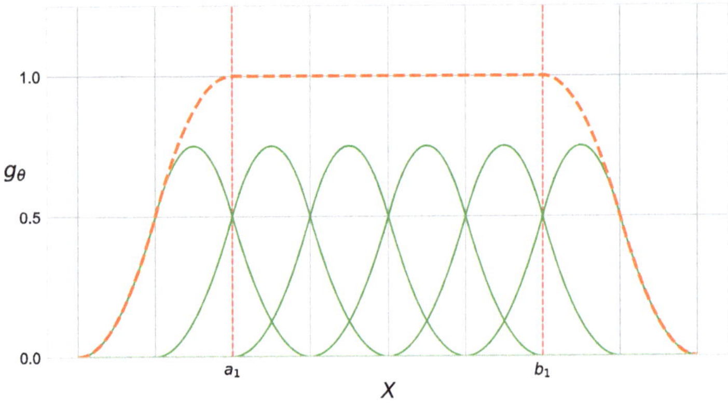

Fig. 1 Positioning of B-spline basis functions in a 1D context. The orange line indicates a surface created by summing the $k_1 = 6$ spline bases of degree $m = 2$, and $[a_1, b_1]$ denote the lower and upper bounds of an interval in which the value of a spline surface equals one

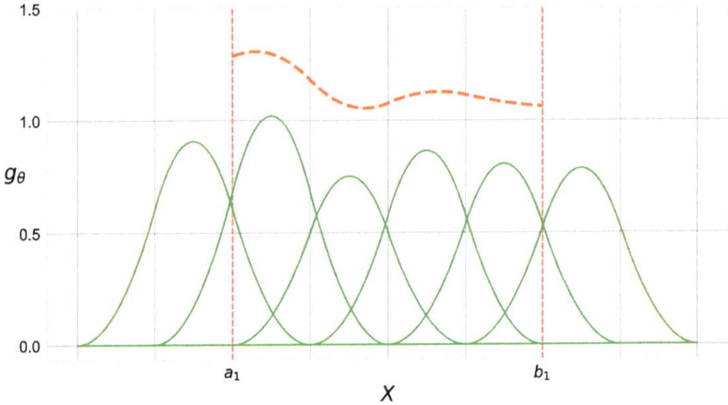

Fig. 2 1D basis functions multiplied by their corresponding weights. The continuous kernel is defined as a sum of the altered bases over the interval $[a_1, b_1]$

Building upon this, Fig. 2 shows the outcome when each basis function is multiplied by its corresponding weight, $w_{s,p,r}$. The result is a fully defined kernel within the $[a_1, b_1]$ span. Notably, the kernel's profile deviates from the unity observed in Fig. 1 due to the influence of the weights. This modification allows the kernel to discern intricate patterns in the data, offering flexibility in the convolution operation.

Transitioning to the two-dimensional scenario ($D = 2$), Fig. 3 illustrates the spatial arrangement of B-spline basis functions in a plane, culminating in the formation of a continuous 2-dimensional kernel. This kernel operates within a 2D span, defined by $[a_1, b_1] \times [a_2, b_2]$. Unlike the 1D setting where each basis function is associated with an individual weight, in the 2D context, each weight corresponds

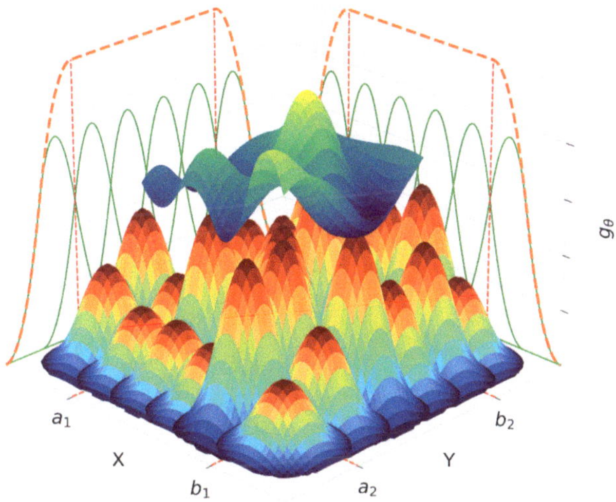

Fig. 3 Creation of a continuous 2D kernel using B-spline basis functions. Here, each product B_p of basis functions in p is multiplied by its corresponding weight w_p and summed within a given interval to create the continuous kernel

to a pair of basis functions, one from each dimension. Thus, with 6 basis functions for each dimension, a total of 36 weights are defined.

Lastly, Fig. 4 demonstrates the application of the 2D kernel to a specific node in a graph. This visualization provides insight into how SplineCNN processes graph-structured data, drawing parallels to traditional CNNs' handling of image data. Through these illustrations, the capabilities and potential of SplineCNN in managing graph-structured datasets with irregularly located nodes become evident.

4 GNNs for MISR

Super-resolution, particularly within the scope of MISR, is an ill-posed problem in optimization theory [71]. Addressing this challenge necessitates maximizing the information encapsulated in the input data to achieve high-quality outputs. A balance between effective data representation and algorithmic design is pivotal for harnessing this information.

This section focuses on the data creation process, explaining the transformation of a stack of input images into a single, unified graph. Also, the proof-of-concept GNN architecture developed for MISR, MagNet [87], along with its successor, MagNet++ [85], designed specifically to address the limitations identified in the original model, are described.

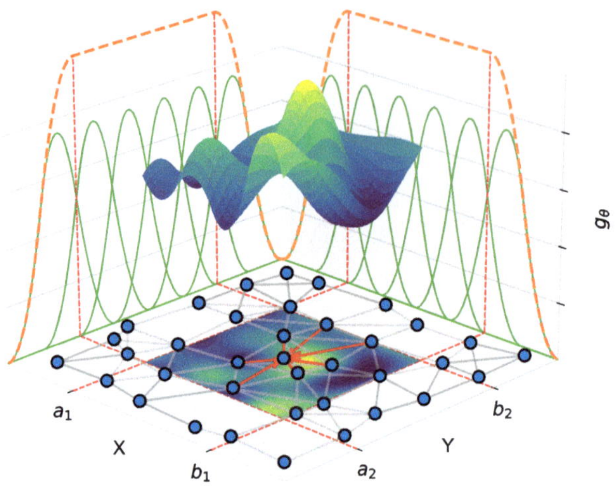

Fig. 4 Application of the 2D spline-based kernel to a specific node in a graph. The red arrows indicate which neighbors within the kernel interval propagate their information to the center node

4.1 Converting a Stack of LR Images into a Single Graph

This section elucidates the transition from conventional image-based data structures to a graph representation, marking not merely a change in format but a fundamental shift in data representation. Initially, each pixel is transformed into a node on a mutual 2D plane, a step referred to as node positioning. Subsequently, these node positions are adjusted to account for shifts between LR images. Finally, nodes are interconnected to fabricate the graph structure. This unique approach amalgamates multiple LR images into a single integrated graph, ensuring a lossless transition while preserving all original information.

4.1.1 Node Positioning

Establishing a meaningful position for each node within the eventual graph is paramount for preserving the inherent spatial relationships of the original images. At this preliminary stage, the challenge lies in mapping pixels to nodes on a unified plane, ensuring that the relative positions of these nodes reflect the spatial coherence of the original image.

To construct a graph, the primary components are initially defined. As introduced in Sect. 3.1, a graph $\mathcal{G}$ can be represented as $\mathcal{G} = (\mathcal{V}, \mathbf{H}, \mathcal{E}, U)$. At this initial stage, two of these components are defined: the set of vertices $\mathcal{V}$ and the matrix of F_{in} features for each node, $\mathbf{H} \in \mathbb{R}^{n \times F_{in}}$. The vertices are derived from the pixels of the

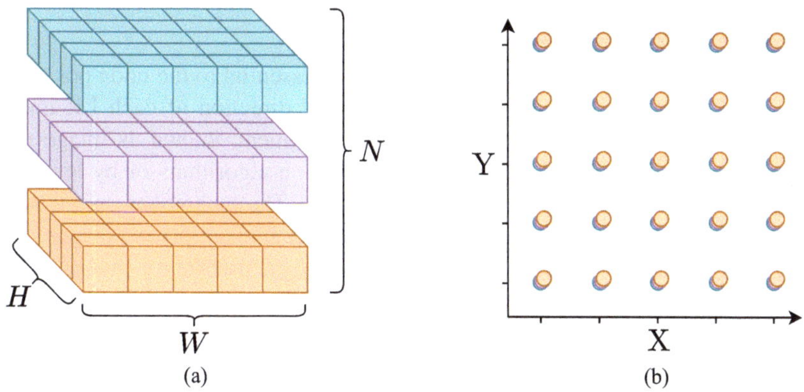

Fig. 5 Illustration of the process of converting a stack of LR images into a single graph. (**a**) A stack of $N = 3$ input images of dimensions 5×5 (**b**) Conversion of pixels to nodes on a mutual 2-dimensional plane

LR images. Meanwhile, the input features, **H**, depend on the number of channels in the LR images. For instance, in the case of RGB images, each pixel has three channels (red, green, and blue); hence, each node in the graph would have three input features. Generally, if the LR images have F_{in} channels, each node would have the same number of input features.

As visualized in Fig. 5, every pixel from the given set of LR images, shown in subfigure (A), is represented as a node on this plane. Its position on the plane, depicted in subfigure (B), directly corresponds to its original position within its respective image. Specifically, if a node corresponding to the ith pixel is denoted as v_i, its position on the plane can be defined by a function $loc : \mathcal{V} \rightarrow \mathbb{N}^2$ and the resulting node position $loc(v_i) = (x_i, y_i)$ corresponds to indices of a pixel from an image in a matrix form; thus, it maintains the spatial integrity between the graph and the corresponding LR image. Hence, x_i and y_i are constrained by the image dimensions: $x_i \in \{0, \ldots, W-1\}$ and $y_i \in \{0, \ldots, H-1\}$, where W and H represent the width and height of the LR images, respectively.

Given N input images, each of size $W \times H$, there are in total $n = N \times W \times H$ nodes on this plane. However, it is crucial to note that at each occupied position on the plane, there are N vertices, each from a different image. Considering the nature of images and their grid-like structure, the closest horizontal neighbors of a pixel at (x, y) would be at $(x \pm 1, y)$, and the vertical neighbors would be at $(x, y \pm 1)$.

By this definition, at this initial phase, the node positions are direct representations of the pixel positions in the LR images without any adjustments. Subsequent steps involve refining these positions and establishing connections between these nodes to form a cohesive graph.

4.1.2 Displacement Calculation

The process of determining displacement vectors, essential to the node positioning strategy, seeks to identify both the magnitude and direction of each LR image's deviation concerning a reference image. This alignment, known as image registration, can be achieved using traditional registration algorithms or by leveraging neural networks specifically designed for this task. The techniques employed for registration can be broadly categorized into full-pixel and sub-pixel methods.

Full-pixel registration offers a direct methodology, translating pixels to align images while preserving their inherent values [14]. This approach is computationally efficient but might overlook the detailed nuances present at sub-pixel levels. On the other hand, sub-pixel registration adjusts pixel values to achieve more refined alignments, though at the potential cost of introducing noise or distortions [89].

Several traditional methods have been employed for image registration. Cross-correlation [49] serves as a foundational technique. Feature detection, such as SIFT [57], is crucial for facilitating feature matching methods [119]. Discrete Fourier transform (DFT)-based techniques also offer refined alignments [29]. In the realm of deep learning, models like RegNet [68] and ShiftNet [18] have emerged, along with convolutional networks designed specifically for geometric transformations in image registration [17]. The choice between these methodologies largely depends on the application's specific needs, available computational resources, and the desired level of precision.

In many MISR models, the inherent challenge is that they lack direct input information about sub-pixel shifts. Instead, these models must deduce such granular details independently in their feed-forward pass, often missing out on the nuanced information. In contrast, in this research, an efficient sub-pixel image translation registration by cross-correlation has been selected [29]. This choice was influenced by its notable advantages of relatively low computation time and high accuracy.

Regardless of the chosen method, the primary objective remains consistent: to determine the sub-pixel displacement for each LR image, ensuring the nodes' accurate positioning in the graph. Formally, for an ensemble of N images, where the image at index $i = 0$ is chosen as the reference, the displacement vectors $\mathbf{u_i}$ for each of the $N - 1$ remaining images are computed. Each vector $\mathbf{u_i}$ is a two-dimensional representation capturing the horizontal and vertical shifts, given by $\mathbf{u_i} = [\bar{x_i}, \bar{y_i}]$, where i ranges from 0 to $N - 1$, and $\bar{x_0} = \bar{y_0} = 0$. Building on this, the adjusted position of the node v_j, accounting for the displacement of its associated image, can be represented as $loc'(v_j) = loc(v_j) + \mathbf{u}_{img(v_j)}$. Here, $img(v_j)$ is a function that maps a node v_j to its associated image index; hence, $img : \mathcal{V} \rightarrow \{0, \ldots, N - 1\}$. Given $loc(v_j) = (x_j, y_j)$, this translates to:

$$\text{loc'}(v_j) = \begin{bmatrix} x_j + \bar{x}_{\text{img}(v_j)} \\ y_j + \bar{y}_{\text{img}(v_j)} \end{bmatrix} = \begin{bmatrix} x'_j \\ y'_j \end{bmatrix}, \quad j \in \{1, \ldots, n\}. \tag{11}$$

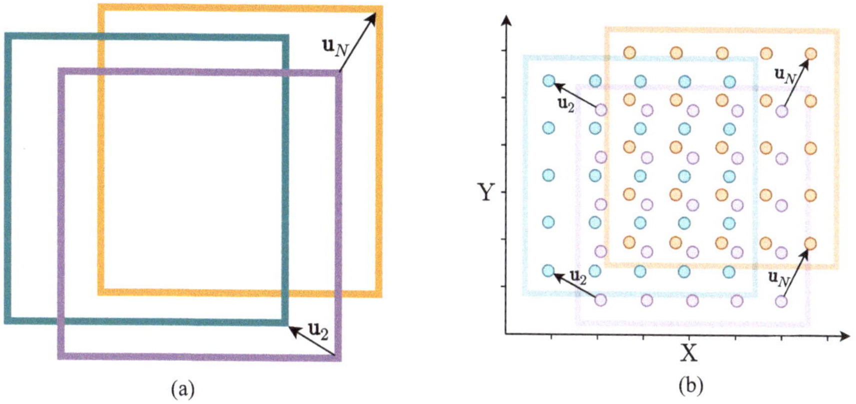

Fig. 6 Illustration of the process of computing shift vectors and adjusting node positions based on these shifts. (**a**) Calculated shift vectors using image registration algorithms. (**b**) Node positions adjusted using the computed shift vectors

This mapping ensures that the spatial relationships between nodes are consistent with the displacements of their respective LR images.

As depicted in Fig. 6, the left image (subfigure A) visualizes the resulting shift vectors derived from the image registration algorithms. Meanwhile, the right image (subfigure B) illustrates how these vectors have been applied to adjust the nodes' positions within the graph.

It is important to note that in more complex scenarios, where images exhibit rotational deviations relative to each other, the procedure can be extended to compute a distinct shift vector for each node using algorithms such as optical flow [113]. This approach accommodates rotational and other complex transformations, determining spatial displacements on a per-node level. However, this method is not adopted in this research as the employed datasets exhibit only translations between the input images. Additionally, optical flow is sensitive to noise, potentially causing significant errors in images with high temporal variations, and is computationally more demanding [117]. Hence, this research focuses on addressing translations alone, utilizing the described displacement calculation mechanism to ascertain and compensate for spatial deviations between different LR images.

4.1.3 Graph Construction

In constructing the graph, the first critical step, post-node positioning, is the creation of edges, effectively binding the individual nodes into an integrated graph structure. As visualized in Fig. 7, the establishment of connections is primarily dictated by proximity, employing the concept of Euclidean distance. Formally, for any two nodes v_i and v_j with positions $loc'(v_i) = (x'_i, y'_i)$ and $loc'(v_j) = (x'_j, y'_j)$, the Euclidean distance $d : V, V \rightarrow \mathbb{R}$ between them is defined as:

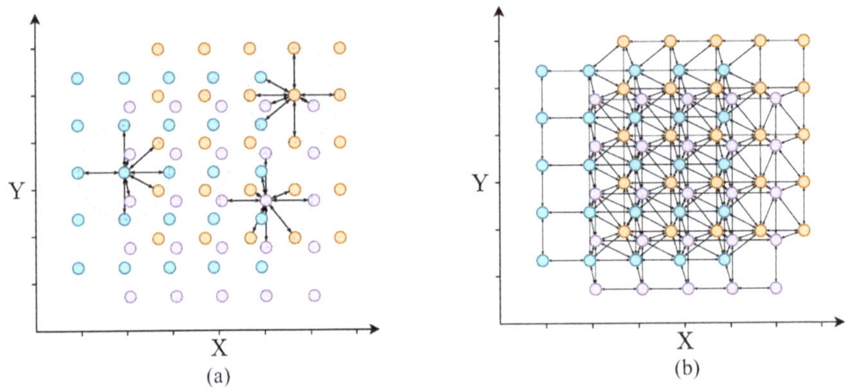

Fig. 7 The illustration provides a depiction of the graph construction process. The left subfigure (**a**) demonstrates the method of connecting nodes that fall within a specified radius, while the right (**b**) displays the completed graph, integrating both nodes and edges. (**a**) Connecting procedure on two arbitrary nodes for radius $r = 1$. (**b**) Fully prepared graph

$$d(v_i, v_j) = \sqrt{(x'_j - x'_i)^2 + (y'_j - y'_i)^2}. \tag{12}$$

Using this distance metric, the graph's adjacency matrix $\mathbf{A}$ and the edge set $\mathcal{E}$ are updated as:

$$\mathbf{A}(i, j) = \begin{cases} 1 & \text{if } d(v_i, v_j) \leq r \\ 0 & \text{otherwise} \end{cases}. \tag{13}$$

Here, an edge e_{ij} is present if $\mathbf{A}(i, j) = 1$ and absent otherwise. The threshold r, set to one in this context, marks the boundary for node connection. Nodes within or on the circumference of a circle with radius r centered around any given node are regarded as neighbors.

By setting $r = 1$, the connections predominantly align with the pixel width, naturally preserving the spatial relationships inherent within individual images. Such a choice is both intuitive and computationally efficient: A broader radius would compound the number of connections, amplifying the computational burden.

With the foundational graph in place, enriched with the set of connections $\mathcal{E}$, edge attributes U are added not only to boost the graph's descriptive potential but also to fulfill the input requirements of the spline-based convolution highlighted in Sect. 3.4.3. These attributes capture the spatial offsets between paired nodes, providing essential spatial relationship data that the spline-based convolution needs to operate efficiently.

For any two nodes v_i and v_j interconnected by edge e_{ij}, the attribute $\boldsymbol{u}_{ij}$ records their spatial displacement:

$$\boldsymbol{u}_{ij} = loc'(v_j) - loc'(v_i) = \begin{bmatrix} x'_j - x'_i \\ y'_j - y'_i \end{bmatrix} = -\boldsymbol{u}_{ji}. \tag{14}$$

With these edge attributes, the graph's depth transcends mere structural connections, shedding light on relative spatial orientations between nodes. These attributes grant crucial context, particularly during graph processing or analysis, as the relative node positions can be inferred directly from edge data.

In conclusion, these steps come together to create a structured graph that accurately represents the original set of LR images. This representation not only maintains the inherent spatial dynamics and displacements but also enriches the spatial relational details embedded within. Unlike the conventional matrix-based methods, where images are simply stacked in a sequence without adding any new information beyond raw pixel values, this graph-based approach is a significant advancement in preserving and enhancing the contextual relationships among image elements. It enriches the input data with more descriptive information, facilitating a more informed and nuanced analysis without altering the existing data.

4.1.4 Benefits of Graph Data Representation for MISR

In theory, the graph-based data representation method offers numerous advantages that make it particularly suitable for MISR tasks. These potential benefits, including *permutation invariance*, the ability to handle *varying quantities of inputs* or *data heterogeneous in size*, and *flexible relationship modeling*, are discussed in this section.

Permutation Invariance

Permutation invariance in MISR signifies that the sequence of input LR images should not affect the super-resolution results. Regrettably, not every MISR approach inherently embodies this attribute. Many conventional methods necessitate a static ordering of input images, inadvertently inducing dependence on the specific sequence of LR inputs.

While contemporary techniques, particularly PIUNET and TR-MISR, do ensure permutation invariance, they do so through distinct architectural strategies. For instance, PIUNET employs self-attention mechanisms in the temporal dimension and shared convolutional layers. Conversely, TR-MISR utilizes a transformer-based fusion mechanism to ensure input order does not affect outcomes.

However, the graph-based representation inherently guarantees permutation invariance by placing each pixel from all LR images on a unified two-dimensional plane, irrespective of their origin. Achieving this at the data-creation level eliminates the requirement for special architectural decisions to tackle this issue.

Varying Number of Input Images

A distinctive feature of the graph-based representation is its adaptability to various numbers of input images. In contrast, some traditional methods, such as RAMS, DeepSUM, and, to a certain extent, HighRes-Net, demonstrate limitations. Specifically, RAMS and DeepSUM can only handle the exact number of LR images during both training and inference as was available during their respective training sessions. Meanwhile, HighRes-Net frequently resorts to padding its input matrix with blank images to round up to the nearest power of two.

In the graph-based approach, the accommodation of varied numbers of LR images is straightforward and does not demand architectural adjustments. Regardless of the input count, they all form a single graph, with the only variable being the graph's density. However, it is worth noting that while this model demonstrates scalability, an exponential rise in complexity accompanies each additional input image due to the drastically increasing number of connections that need to be stored in the memory.

Handling of Heterogeneous Inputs

When it comes to real-world SRR tasks, there is often a mix of input images varying in aspects like resolution, size, or orientation, especially in remote sensing applications [51, 86]. Traditional MISR methods might grapple with this diversity, necessitating extra preprocessing steps to create a uniform dataset. This can be cumbersome and resource-intensive. In stark contrast, the graph-based approach can natively process a mix of different numbers, sizes, and orientations of LR images, assuming a proper node positioning procedure. This inherent flexibility means that whether the inputs are uniformly oriented LR images or a mix of varying resolutions and orientations, the graph-based model can process them without problems, positioning it as a versatile choice for diverse scenarios.

Flexible Modeling of Relationships

Graphs are renowned for their ability to capture relationships between entities [13], an attribute that is highly beneficial in the context of MISR. In the adopted graph-based representation, nodes are interconnected through edges, which can represent various relationships, from spatial proximity to pixel value similarity. Such dynamic connection modeling enables the discernment and utilization of intricate interpixel relations, potentially augmenting the super-resolution results.

In the research presented, these relationships are characterized as relative displacements between pairs of pixels. While this serves as the primary relational feature for the current focus, in more advanced scenarios, other relational characteristics can be encoded. This is reminiscent of the approaches in more complex models, where, for instance, the significance of a node relative to its neighbors is encoded, hinting at the versatility of the graph-based approach.

4.2 *MagNet: A Comprehensive Analysis*

MagNet [87], developed to support the propositions put forth in this thesis, pioneers in utilizing GNN for MISR. In the briskly advancing domain of image super-resolution, MagNet introduces a unique application of GNNs for MISR, carving out a novel path for research in this domain.

The architecture of the MagNet model seamlessly integrates GNNs with the framework of the FSRCNN [22] model. Even though FSRCNN was originally conceptualized as a SISR architecture, its structure lends itself efficiently to MagNet's needs. The graph data representation in MagNet essentially translates the multi-image super-resolution task into a single-graph super-resolution. This means the model operates similarly to a single-image scenario, yet it processes a graph composed of multiple images. By harnessing this graph-centric representation, MagNet adeptly merges insights from various LR images, enhancing its overall performance in the MISR challenge.

As MagNet entered the MISR domain, its functionality and prowess were put to the test through an array of meticulously planned experiments. These tests did more than just highlight its operational efficiency; they emphasized its capability and prospective contributions in managing the intricate facets of multiple-image super-resolution.

It is pivotal to note that MagNet's present incarnation serves primarily as a proof of concept, illustrating the practicality of using GNNs in MISR. The initial rounds of evaluation were predominantly anchored on simulated data, extracted from a diverse pool of standard benchmark datasets. The primary reason for initially using simulated data, as detailed in Sect. 5, describing the employed datasets, was to assess the suitability of the architecture for the MISR problem in a controlled and more manageable testing environment.

However, leaning solely on simulated data is but a preliminary phase in MagNet's developmental journey. The inherent limitations of simulated data make it imperative to test the model in real-world scenarios. This transition to real-world data testing, detailed further in Sect. 7.2, is expected to shed more light on the model's potential and areas needed for refinement.

In essence, while the simulated data-centric proof-of-concept phase was instrumental in unveiling MagNet's potential, the move towards real-world data promises a more comprehensive understanding of its capabilities, suggesting an imminent evolution in the world of MISR.

4.2.1 Dissecting the Architecture of MagNet

As depicted in Fig. 8, the architecture of MagNet is influenced by the SISR network FSRCNN, discussed in Sect. 2.3.1. The method of data representation, which merges multiple LR images into a solitary graph, makes this similarity possible. This unified graph can be perceived as a "single super-image," possessing richer information than a conventional LR image used in SISR. MagNet employs spline-based convolutions, as introduced in SplineCNN (Sect. 3.4). This forms a composite model integrating traditional convolution methods with a message-passing technique inherent to GNNs. This hybrid design empowers the model to harness both the features and connection weights present in graph nodes and the spatial information often disregarded in standard GNNs, due to their relational-centric scenarios.

The very first layer in MagNet, serving as the feature extraction component, employs spline-based convolution kernels with a configuration of $\mathbf{k} = \{3, 3\}$. In this setup, the convolution utilizes two sets of B-spline basis functions, one for each spatial dimension, with each set having three basis functions. This configuration dictates the density of the weights within the kernel's spatial span and not the span itself. The spatial extent of this kernel, specifically from -1 to 1 in both dimensions, adheres to the connection radius described in Sect. 4.1.1. Thus, it ensures that the

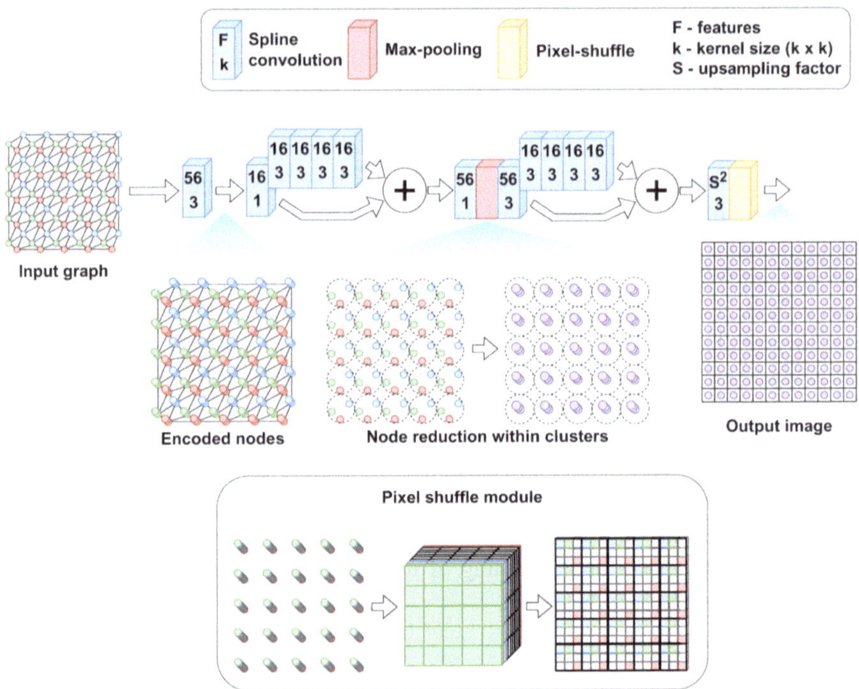

Fig. 8 The architecture of MagNet

convolution is influenced only by the nodes within this defined spatial range around each node. With this convolution, the input channels of each node are transformed to produce 56 features. The choice of outputting 56 features from this layer draws its inspiration directly from the FSRCNN architecture.

Subsequent to this, MagNet integrates a shrinking layer to streamline the architecture. This layer efficiently consolidates the previously mentioned 56 features, reducing them to a more compact 16 features, once again, a design choice influenced by FSRCNN. To achieve this feature reduction, the architecture employs 1×1 ($\mathbf{k} = \{1, 1\}$) spline-based kernels. Notably, these kernels incorporate only a single pair of B-spline basis functions—one for each spatial dimension. Despite being influenced by just a singular weight for each node, this kernel design is ingeniously capable of assimilating information from different neighboring nodes, a characteristic that distinctively sets it apart from conventional 1×1 convolutional layers. After this, the architecture includes a convolutional block made up of four spline-based convolutional layers, interlinked with a skip connection to counteract vanishing gradient issues in deep neural networks [34].

In the next phase, the feature count is increased back to 56 through a single 1×1 spline-based convolutional layer. This is followed by a max-pooling operation applied to predefined node clusters. To understand the cluster formation, one can refer to the function introduced in Sect. 4.1, denoted as $loc(v_i)$. This function yields the discrete location of a node v_i. Considering multiple LR images, for each discrete position (x, y) where $x \in \{0, \ldots, W-1\}$ and $y \in \{0, \ldots, H-1\}$, there are N nodes, each corresponding to one of the N LR images. These nodes are assembled into a cluster based on their mutual spatial location, as defined by $loc(v_i)$.

By employing this clustering strategy, the architecture generates $W \cdot H$ clusters, and each cluster encompasses N nodes. Here is where the transformation becomes significant: The max-pooling operation consolidates each of these clusters, which consists of N nodes, into a single representative node. Consequently, the total number of nodes is reduced from $N \cdot W \cdot H$ to just $W \cdot H$.

This means that post max-pooling, for every discrete position (x_i, y_i) given by the function $loc(v_i) = (x_i, y_i)$, there exists only one corresponding node. Effectively, the architecture has transformed the graph into a regular, grid-like structure. This uniformity mirrors a standard image grid, with each node denoting a pixel-like entity at a particular spatial location. Such a structure not only makes further processing more streamlined but also enhances MagNet's scalability and adaptability, ensuring that the model remains efficient irrespective of the number of input images.

After the max-pooling step, the graph assumes a regular, grid-like structure. Before transitioning into a matrix format, MagNet conducts further spline-based convolutions on this grid-like graph, incorporating a residual block and a single spline-based convolution modifying the feature depth to $\mathcal{S}^2$, with $\mathcal{S}$ being the upsampling factor of the model. Upon the conclusion of these convolutional processes, the graph is converted into a matrix form, preserving the spatial integrity of each pixel according to the $loc(v_i)$ function, and preparing the data for the pixel-shuffle operation [81]. The operation works by reshuffling the channels of the input tensor to the spatial dimensions. For an input tensor with shape $[H, W, F \cdot \mathcal{S}^2]$, where F is the number of features, the pixel-shuffle operation outputs a tensor with shape $[H \cdot \mathcal{S}, W \cdot \mathcal{S}, F]$. In the context of MagNet, given that the matrix representation post convolutions is of shape $[H, W, \mathcal{S}^2]$, the pixel-shuffle operation transforms it into $[H \cdot \mathcal{S}, W \cdot \mathcal{S}, 1]$. This effectively expands the spatial dimensions by a factor of $\mathcal{S}$ and produces a single channel, resulting in a representation resembling a high-resolution single-channel image.

In summary, the MagNet architecture, through its unique incorporation of graph-based techniques and spline-based convolutions, presents a distinct approach to addressing the MISR challenge. It effectively consolidates multiple LR images into a cohesive graph structure, applies convolutions to harness rich node information, and subsequently employs max-pooling and pixel-shuffling to produce a higher-resolution image.

4.2.2 Limitations of MagNet

The MagNet architecture, while being a promising approach to SRR tasks, poses certain challenges that need consideration. Central to these challenges is the application of max-pooling before the upsampling step. The primary objective of max-pooling in this context is to transform the graph into a more structured, grid-like configuration, essential for its conversion into a matrix form suitable for the pixel shuffle operation. However, while this transformation preserves the spatial boundaries, it reduces the graph's node count, potentially diminishing the amount of information embedded in the graph.

Consequently, the following pixel shuffle operation, applied to this now simplified matrix, might not fully utilize the complete spectrum of information present in the graph's state before the max-pooling. In light of these limitations, the next section introduces an evolved architecture, MagNet++, which aspires to retain the foundational strengths of MagNet while addressing its weaknesses, aiming for an optimized super-resolution output.

4.3 Graph-Based Upsampling in MagNet++

Built on the foundations of the MagNet model, a new and refined architecture named MagNet++ was designed and introduced in [85]. While retaining core elements such as the feature extraction layer, shrinking layer, and a convolutional block with skip connections—all leveraging spline-based convolutions—its primary distinction lies in its upscaling technique. This method is optimized for the use of input node information and is crafted to reduce the likelihood of information loss, thereby exploiting the unique aspects of this data representation more effectively.

In MagNet, the upscaling process involved the max-pooling of nodes corresponding to the same pixel position of the LR images. This operation reduced the number of nodes by a factor of N, transforming the nodes into a grid-like configuration. The transformed data was processed by a convolutional block, followed by a pixel shuffle operation to enhance the spatial resolution of the image.

MagNet++, on the other hand, embraces a different approach. Instead of transforming the graph into a matrix before resolution enhancement, a fully graph-based approach is adopted here. This method utilizes a bipartite graph structure. To construct it, initially, the connections within the original graph are entirely removed, rendering it a null graph—a type of graph consisting of nodes but lacking any edges. Next, a new null grid-like graph is crafted having of $W \cdot H \cdot \mathcal{S}^2$ nodes, with a distance between each pair of neighboring nodes (vertically or horizontally) equal to $\mathcal{S}^{-1}$. This new graph is then overlaid onto the original, now edge-less one, so that their centers are aligned. The connection strategy utilizes one-way directed edges, wherein the original nodes, retaining their information, connect exclusively to the new nodes. The chosen radius, $r = \sqrt{2}$, ensures that each new node establishes a connection to the original nodes; thus, it avoids rendering a disconnected graph. A smaller radius, such as previously used $r = 1$, could risk nodes positioned centrally between original pixels (across both dimensions) being too distant to form any connection. This could be particularly problematic in scenarios with a low count of input images or minimal displacements between them. These new edges ensure a one-way flow of information. Consequently, the newly overlaid nodes act solely as targets, receiving information from the original nodes without initiating any connections of their own. Visualization of this process can be seen in Fig. 9, which provides a comprehensive schematic representation of the graph transformation and connection establishment.

In this bipartite graph structure, a single spline-based convolutional layer activated by the PReLU function is applied to gather information from the source nodes. Next, the overlaid grid-like graph is separated from the original one and transformed into a tensor of shape $[\mathcal{S} \cdot W \times \mathcal{S} \cdot H]$. With this method, information loss from

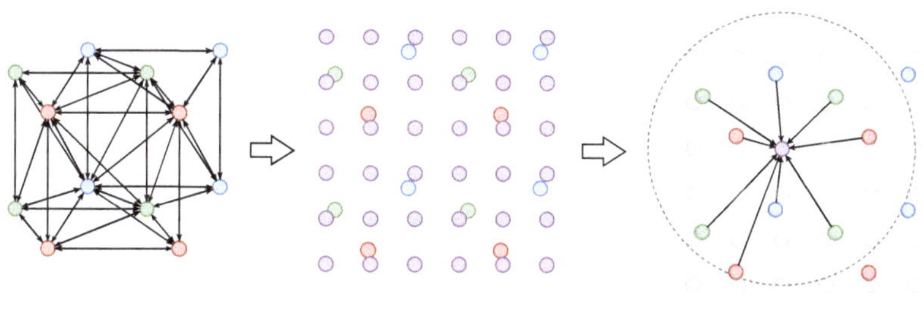

Fig. 9 Schematic representation of the graph transformation and connection establishment for the bipartite upsampling procedure. The original nodes (red, green and blue) establish directed edges to the newly overlaid nodes (purple), ensuring a one-way flow of information in the new bipartite structure

the input graph is potentially reduced, addressing a notable limitation, discussed in Sect. 4.2.2, of the MagNet architecture.

4.3.1 MagNet++ Architecture

With the return of a single rectangular tensor from the bipartite upscaling operation, standard 2D convolutional layers can be applied instead of the spline-based ones to reduce the computational overhead of the model. The data is further processed through a single residually connected convolutional block, followed by a convolutional layer with a single kernel, producing the final super-resolved image. The architecture of MagNet++ is shown in Fig. 10.

A noteworthy advantage of MagNet++ is its flexibility with the shape and completeness of the input LR images. Unlike many models that necessitate rectangular input, MagNet++ is capable of processing non-rectangular images, which provides greater adaptability in various application contexts. Furthermore, it also permits the processing of images with specific nodes removed, such as masked nodes. This functionality is particularly beneficial in cases like remote sensing applications, where certain nodes might depict unwanted or noise-inducing elements, such as clouds or uncaptured pixels. By introducing these new features, MagNet++ demonstrates a significant stride forward in the field of super-resolution imaging, offering a promising prospect for further advancements.

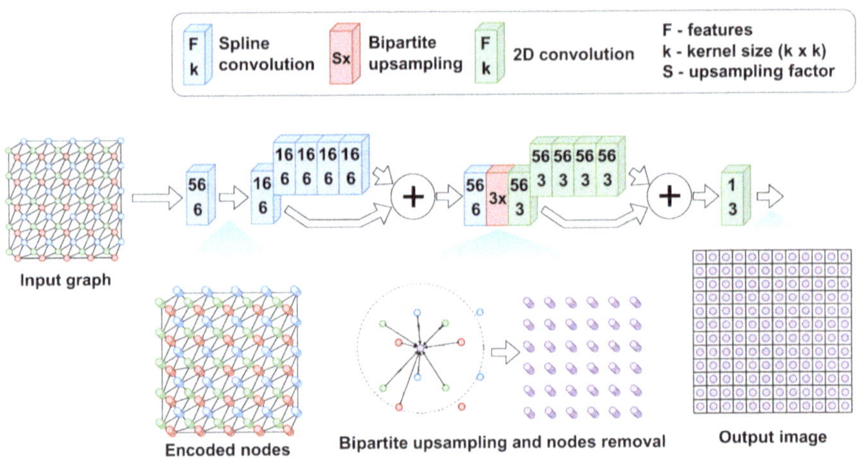

Fig. 10 The architecture of MagNet++

5 Data Description and Simulation

High-quality data is fundamental to the success of any machine learning or deep learning project. When it comes to super-resolution tasks, the choice of data and its preparation become even more critical due to the challenges inherent in this domain. This section focuses on the datasets used in this research, specifically detailing the process of data simulation, which played a key role in training and evaluating the developed models.

5.1 Simulated Dataset

This research strategically incorporates both simulated and real-world datasets. The simulated dataset affords advantages in generating a plethora of training examples from a limited set of HR images. It provides a controlled environment where specific features of the LR images, such as degradation levels, can be manipulated. Furthermore, it permits emulation of certain real-world conditions, particularly minor shifts between the input images that are crucial for MISR.

A simulated dataset was meticulously developed to focus on sub-pixel shifts between LR images. This dataset is instrumental in assessing model capabilities to identify such minor translations, thereby ensuring high fidelity in the super-resolution process. The choice to simulate solely these sub-pixel shifts aligns with the study's goal to isolate and understand the impact of such minute discrepancies on the performance of MISR models, without the interference of broader global variations like brightness, contrast, noise, or temporal variations inherent to the real-world scenarios.

In the context of this research, a pivotal decision concerned the downsampling factor for simulating LR images. Opting for a $3\times$ downsampling factor, the resultant LR images are three times smaller compared to their HR versions. This configuration mirrors the PROBA-V Super-Resolution challenge and its corresponding dataset employed in this study (Sect. 5.2). Notably, numerous state-of-the-art models, which were employed and evaluated in this work, were originally tailored for this challenge and its specific upsampling factor. To preserve the original training conditions of the state-of-the-art models, the upsampling factor, S, was consistently set at three for retraining and evaluation in this study.

The simulated dataset sources images from established databases. Specifically, the DIV2K [1] database was employed for training the models, while others, namely, BSDS100 [3], historical [50], Manga109 [65], Set5 [9], Set14 [112], and Urban100 [50], were utilized for benchmarking purposes.

5.1.1 Training Dataset

The DIV2K dataset was chosen as the primary resource for training and validation purposes. DIV2K is widely recognized in the image processing domain for its diverse and extensive collection of high-quality images. The original dataset comprises 1000 RGB images that display a range of scenes, objects, and textures.

For this study, all images from the DIV2K dataset, as well as those from all other included datasets, were converted to grayscale. This conversion was undertaken to simplify the research process, concentrating on the core problem of image super-resolution without the added complexity of color information.

Additionally, a preprocessing step was introduced to ensure uniformity in image sizes, a vital aspect for facilitating batch creation and collation during the training phase. The smallest image size in the dataset was identified, and all larger images were divided into nonoverlapping sub-images of this determined dimension. In this research, the chosen size was 222×222 pixels. As a result of this cropping strategy, the effective number of images in the dataset was expanded from the original 1000 to 1409.

An 80:20 split was then applied to the preprocessed DIV2K dataset, with 80% of the images reserved for training and the remaining 20% designated for validation. This commonly adopted ratio ensures a substantial majority of data aids in discerning core features and patterns, while a significant portion is preserved to evaluate the model's capabilities and adaptability on previously unseen data during the validation stage, thus mitigating overfitting risks.

5.1.2 Benchmark Datasets

In addition to DIV2K, several other datasets were used for benchmark purposes. These datasets—BSDS100, historical, Manga109, Set5, Set14, and Urban100—are well known in the super-resolution community and widely employed for model evaluation, especially in SISR [20, 44, 90].

Each benchmark dataset offers unique image characteristics, providing a diverse testing ground. Notably, the numbers in the names of datasets like BSDS100, Manga109, Set5, Set14, and Urban100 indicate the number of images they contain. BSDS100 comprises natural images with complex textures and structures. The historical dataset, distinctively, consists of 10 images that showcase older, classical styles and textures. Manga109 is characterized by manga-style images demanding the preservation of sharp edges. Set5 and Set14, although smaller in size, offer a diverse and broadly used collection of images for benchmarking in the super-resolution community and image processing in general. Urban100 features images of urban scenes, predominantly with man-made structures and patterns. Using these varied datasets ensures that the performance of the models is scrutinized across different image types and conditions, thereby measuring their robustness and adaptability—factors vital for effective deployment in real-world SRR tasks.

5.1.3 The Process of Generating Simulated Datasets

The process of generating simulated LR images involved two crucial steps: *image shifting* and *downsampling*. In the initial stage, every HR image was subjected to a series of shifts, resulting in nine distinct shifted versions for each. These shifts aimed to emulate the variability frequently observed in real-world situations where factors such as changes in perspective or environmental variables might modify a scene's representation. Shift vectors applied to each HR image independently were derived from a uniform distribution within the range of $[-1.2, 1.2]$ for both dimensions individually. This range was selected based on calculations and observations from the real-world PROBA-V dataset. Specifically, sub-pixel shifts between LR images rarely surpass the interval $[-0.4, 0.4]$. Given the chosen downsampling rate of three, the range for HR images effectively matches this observed interval for the LR images. Following the shifting operations, the HR images were downsampled $3\times$ to produce the respective LR counterparts.

From this method, nine LR images were produced for each original HR image, intending to introduce foundational intricacies associated with MISR tasks. The entire procedure of generating multiple LR images from a singular HR example is detailed in Algorithm 1, and the resulting images are depicted in Fig. 11.

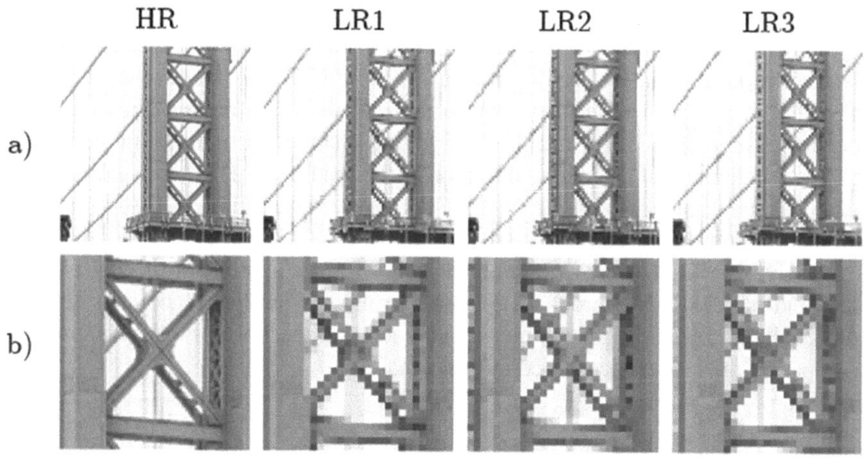

Fig. 11 Illustration of the images synthesized for the simulated dataset. The first row (**a**) presents the full-sized images, while the second row (**b**) offers magnified segments, emphasizing the sub-pixel shift variations between them

Algorithm 1 Algorithm used to simulate multiple LR images

```
Initialize empty set LRs
Set number of shifts n ← 9
Set shift range r ← [−1.2, 1.2]
Set downscale factor d ← 3
for i ← 1 to n do
    x_shift ← random number within r
    y_shift ← random number within r
    shift_vector ← (x_shift, y_shift)
    shifted_image ← shift HR by shift_vector
    LR ← downscale shifted_image by factor d
    Add LR to the set LRs
end for
return LRs
```

5.2 *Real-World Dataset*

Real-world data is pivotal in this research, enabling the testing and validation of super-resolution models under genuine conditions. To achieve this, the PROBA-V multi-image super-resolution dataset [64] was chosen.

Several factors guided the selection of the PROBA-V dataset. Firstly, it offers real-world images captured by the PROBA-V satellite, creating an authentic and challenging environment for developing and testing super-resolution models. This dataset incorporates the inherent complexities typical of satellite imagery, including data gaps and variations in image quality due to atmospheric conditions like cloud coverage. These characteristics make it a true reflection of real-world scenarios.

Moreover, the PROBA-V dataset provides an element that the simulated dataset lacks: temporal variation. As the images in the PROBA-V dataset were captured at different points in time, they expose the models to the natural alterations that occur between subsequent captures, including changes in lighting, movement of objects, and potential growth or alteration of the landscape. This temporal variation is an indispensable factor in real-world SRR tasks, making the PROBA-V dataset a crucial tool for this research.

Lastly, the PROBA-V MISR challenge has been a significant driving force in the field of MISR. Numerous state-of-the-art models have emerged as a result of this challenge, providing innovative techniques and benchmarks for evaluating and comparing the models developed in this research.

5.2.1 PROBA-V MISR Dataset Structure

The PROBA-V MISR dataset, provided for the challenge, consists of satellite data from 74 regions around the globe, which were hand-selected at different points in time within a 30-day window. A total of 1450 scenes are contained in this dataset, of which 1160 scenes are allocated for training and 290 for testing; however, the latter set is devoid of HR target images.

Each scene consists of exactly one HR (100 m GSD) image of shape 384 × 384, which is accompanied by multiple LR (300 m GSD) of shape 128 × 128. Additionally, each image has a corresponding quality map. Their role is to indicate which pixels in the image are obscured (e.g., by clouds, cloud shadows, ice, water) and which ones remain clear.

Inclusion criteria for an image in this dataset dictate that at least 75% of its pixels must be clear for 100-m resolution images and 60% for 300-m resolution images. The number of LR images per scene can vary from 9 to 32, with an average count of 19. The original test set, encompassing 290 scenes, is not accompanied by HR reference images and is specifically structured for evaluations on the competition's servers. For the sake of enabling local evaluations, the original training subset was divided into three distinct subsets: training, validation, and test, adhering to an 80:10:10 ratio, respectively. Through this division, comprehensive evaluations of the models at different developmental stages were facilitated.

5.2.2 Spectral Bands of PROBA-V Dataset

The PROBA-V MISR dataset includes two spectral bands: *near-infrared* (NIR) and *red* (RED). The data consists of radiometrically and geometrically corrected *top-of-atmosphere reflectances* for these spectral bands.

The NIR band, with wavelengths between 780 and 2500 nm, is often used in remote sensing and satellite imaging to study vegetation, water bodies, and atmospheric properties. Images in the NIR band can reveal details about plant health, water content, and other features that are not visible in the RGB spectrum. The RED band, with wavelengths approximately between 620 and 750 nm, is particularly sensitive to chlorophyll and can provide information about vegetation's photosynthetic activity. This makes it useful for monitoring plant growth, assessing crop yields, and tracking changes in forest cover, among other applications. In this study, both NIR and RED subsets, comprising 566 and 594 scenes, respectively, were independently utilized for training, validation, and testing of the models.

5.3 *Handling Real-World Challenges*

The PROBA-V MISR dataset embodies various real-world challenges, underscoring the imperative for advanced super-resolution models. The models developed must not only replicate real-world dynamics but also adapt to myriad inconsistencies to ensure their practical relevance.

A significant challenge in the dataset is the presence of areas obscured by clouds, cloud shadows, ice, and water. To assist in managing this, the dataset provides clearance masks for both LR and HR images. During the training phase, these masks

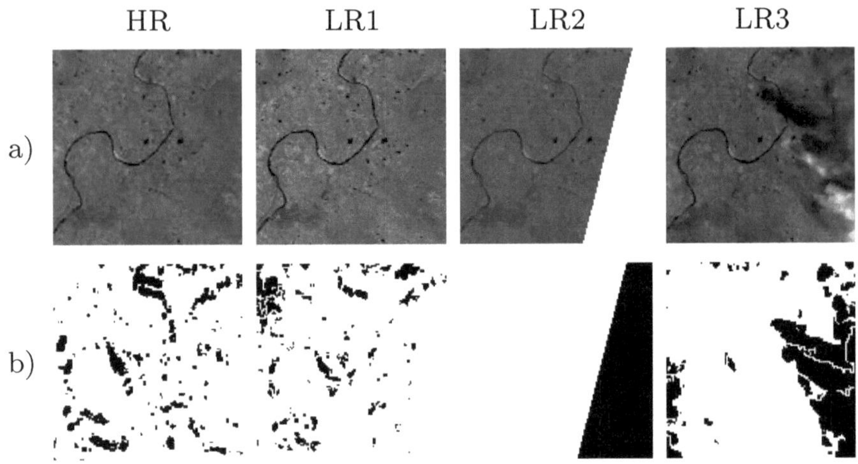

Fig. 12 Illustrative samples from the PROBA-V MISR dataset. The top row (**a**) displays selected images, while the bottom row (**b**) presents the corresponding clearance maps, highlighting the areas of clarity (white) versus obscurity (black)

play an invaluable role, allowing the exclusion of obscured or irrelevant pixels. During the super-resolution reconstruction phase, these masks ensure a better input by excluding the irrelevant pixels.

Another nuance in the dataset is its data storage method. The image files use a 16-bit depth, but the actual image content occupies only 14 bits. Consequently, areas without captured content reach the maximum of the 16-bit range. This specific representation is not complexity in itself but a feature of the dataset. Super-resolution models need to recognize this aspect, ensuring these pixels are properly managed to prevent distortions.

Geometric intricacies, inherent in satellite imagery, appear as shifts and rotations due to factors like satellite position and Earth's rotation. The PROBA-V MISR dataset addresses this by utilizing the Plate Carrée projection [83]. While this projection significantly rectifies geometric distortions using elevation data and the sensor's metadata, minor shifts—primarily at the subpixel level—persist between the LR images. These shifts, however, can be seen as an advantage. By offering varied views of the same scene, they can potentially enrich the super-resolution process. Models should capitalize on these shifts, extracting enhanced information and producing more detailed outputs. To visually comprehend the nuances of the PROBA-V MISR dataset, Fig. 12 showcases representative images from the dataset alongside their respective clearance maps.

5.4 Importance of the PROBA-V MISR Dataset

The PROBA-V MISR dataset's utilization in this research is seen as instrumental for the development and evaluation of super-resolution models under authentic real-world conditions. Characterized by its broad geographical coverage and diverse atmospheric conditions, this dataset can be considered a pivotal resource for super-resolution studies. Further advancements in the field have been significantly catalyzed by the PROBA-V MISR challenge, with an array of innovative techniques and cutting-edge models being introduced. By engaging with this dataset, existing knowledge from these developments was built upon, and further contributions to this dynamic field were facilitated.

Although the PROBA-V dataset is not the only resource available for MISR, other notable datasets have been introduced, such as *MuS2* [47] and the *semi-simulated Sentinel-2 dataset* [88]. MuS2 was designed for super-resolving multiple Sentinel-2 images, using WorldView-2 as the HR reference. However, it faces challenges due to the distinct sensor characteristics of the Sentinel-2 and WorldView satellites, leading to subtle discrepancies between the low- and high-resolution images. Moreover, while the semi-simulated Sentinel-2 dataset approximates real-world conditions better than typical simulated datasets, it does not capture the full authenticity of purely real-world data. A distinguishing factor of these datasets is their orientation towards multispectral MISR, which presents a different set of challenges and considerations. Given these considerations, the PROBA-V dataset was favored for this research. This choice was also influenced by the state-of-the-art MISR models specifically tailored for PROBA-V, which were pivotal for comparison in this study. Additionally, the PROBA-V dataset ensures a uniform sensor profile for both its resolution scales. Coupled with its intrinsic real-world challenges and the unbiased evaluation potential via the challenge's servers, the PROBA-V dataset stands out as the principal dataset for this study.

5.5 Dataset Comparison

Understanding the differences between datasets is crucial for appreciating the challenges and opportunities they present. To shed light on the distinctions between the simulated and PROBA-V datasets, a comparative table has been developed below (Table 1).

Given the outlined characteristics of each dataset, the subsequent section details how this data was employed for training the super-resolution models. The methodology adopted and the metrics chosen for assessment are also discussed to provide a comprehensive view of the experimental approach.

Table 1 Comparison of the datasets used in this research

Parameter	Simulated	PROBA-V (NIR)	PROBA-V (RED)
Training examples	1409	510	535
Test examples	338	56	59
Images/scene	9	9–32	9–32
LR size	74	128	128
HR size	222	384	384
LR shifts	✓	✓	✓
Global LR differences	✗	✓	✓
Local variations	✗	✓	✓

6 Training Methodology and Evaluation Metrics

The methodology adopted for training significantly influences a model's performance. This section describes the training processes utilized for the discussed GNN and various state-of-the-art CNN models. The training was conducted on the simulated and the real-world datasets, each presenting its own unique challenges. Notably, for each architecture, four independent models were trained, one for each simulated dataset and one for each PROBA-V band subset. The procedures followed while handling the training for each dataset, the choice of hyperparameters, and the techniques used to overcome various challenges are discussed in detail in the subsequent sections.

6.1 Loss Function

In the training phase of deep learning models, the loss function plays a pivotal role in optimizing the model to produce outputs closely resembling the target images. To train the models, the *corrected PSNR* (cPSNR) is used as a loss function. cPSNR is a specialized version of the *peak signal-to-noise ratio* (PSNR) [74, 102] and was originally introduced for the PROBA-V competition to address the unique challenges posed by its corresponding dataset in MISR tasks. A key feature of this metric is the exclusion of concealed pixels, which is particularly useful for real-world satellite imagery to avoid reconstructing obscured areas, such as clouds, that can introduce inaccuracies. Additionally, cPSNR incorporates brightness correction and full-pixel registration, ensuring that image quality assessment remains consistent regardless of variations in brightness and positional discrepancies between the compared images.

6.1.1 cPSNR Computation

The foundation of cPSNR lies in its emphasis on image registration. Given the inherent complexities of the multi-image super-resolution (MISR) scenario, where a series of mutually shifted LR images is provided without a precise HR image reference, it would be unjust to penalize the model for minor registration deviations. To account for this, cPSNR permits full-pixel shifts, aligning the super-resolved and HR images and adjusting for possible displacements during the super-resolution phase.

For cPSNR computation, the HR images are divided into patches labeled as $HR_{i,j}$. These patches are created by shifting the HR image within a $[-3, 3]$ range in both spatial dimensions, resulting in 49 distinct patches. Concurrently, the super-resolved image is cropped to obtain a single, central patch. It is important to note that each patch has a diminished effective size due to a 3-pixel exclusion on all sides. This exclusion ensures that border effects, resulting from shifting, do not influence the computation. The focus is primarily on comparing patches of these shifted HR images with the corresponding central patch of the super-resolved image.

Firstly, the brightness correction term b is calculated by subtracting the average intensities of the cropped super-resolved image for each HR patch independently, considering their subsets $\bar{H}R_{i,j}$ including only clear (unobstructed) pixels with indices denoted as x and y:

$$b = \frac{1}{|\bar{H}R_{i,j}|}\left(\sum_{x,y\in\bar{H}R_{i,j}} HR_{i,j}(x,y) - SR(x,y)\right). \tag{15}$$

Next, the corrected mean squared error (cMSE) between the cropped super-resolved image and each HR patch is computed, with a correction for brightness. The cMSE accounts for the average intensity difference between the super-resolved and HR images and, analogously to the brightness correction term, is calculated only over clear pixels:

$$\text{cMSE}(HR_{i,j}, SR) = \frac{1}{|\bar{H}R_{i,j}|}\sum_{x,y\in\bar{H}R_{i,j}} (HR_{i,j}(x,y) - (SR(x,y)+b))^2. \tag{16}$$

Finally, the cPSNR is computed by finding the maximum value calculated for different HR patches:

$$\text{cPSNR}(HR, SR) = \max_{i,j\in 0,\ldots,6}\left\{-10\log_{10}(\text{cMSE}(HR_{i,j}, SR))\right\}. \tag{17}$$

In the following section, a detailed discussion of the training methodologies employed for the models is provided.

6.2 Training Details of Super-Resolution Models

During this study, for each super-resolution architecture, three models were trained—one for each of the datasets: simulated, PROBA-V NIR, and PROBA-V RED. The specific training configurations adopted for these models are detailed in Table 2.

Hyperparameters for the state-of-the-art models—HighRes-Net, RAMS, PIUNET, and TR-MISR—were adopted without alteration based on their corresponding papers and repositories. This adherence to original specifications encompasses preprocessing as well: While images for certain models underwent *rescaling*, adjusting image values to lie between 0 and 1, others utilized *standardization*. The latter was conducted using the overall mean and standard deviation of all training set images. Furthermore, the Adam optimizer was consistently employed for all models.

In terms of loss function optimization, most models were trained to improve the cPSNR value. However, RAMS and PIUNET took a different path, aiming to minimize the cL1 loss function. This cL1 is an adapted version of the L1 loss, adjusted with brightness and shift corrections analogous to the transition from PSNR to cPSNR. Additionally, PIUNET introduces the *uncertainty loss* component during training. Further insights into this specific aspect of PIUNET can be found in [92].

A notable divergence in the adopted approach was observed in terms of training duration. Instead of following epoch counts as outlined in the corresponding literature for each model, a consistent training period of 1000 epochs was established for

Table 2 Training parameters for super-resolution models across different datasets

Dataset	Model	Learning rate	LR patch size	LR images	Batch size	Loss Function	Preprocessing
Simulated	HighRes-Net	0.0007	64 × 64	9	32	cPSNR	Rescaling
	RAMS	0.0005	32 × 32	9	32	cL1	Standardization
	PIUNET	0.0001	32 × 32	9	24	cL1 with uncertainty	Standardization
	TR-MISR	0.0005	64 × 64	9	24	cPSNR	Rescaling
	MagNet	0.001	32 × 32	9	32	cPSNR	Rescaling
	MagNet++	0.001	32 × 32	9	32	cPSNR	Rescaling
PROBA-V	HighRes-Net	0.0007	64 × 64	9-32	32	cPSNR	Rescaling
	RAMS	0.0005	32 × 32	9	32	cL1	Standardization
	PIUNET	0.0001	32 × 32	9	24	cL1 with uncertainty	Standardization
	TR-MISR	0.0005	64 × 64	9-24	4	cPSNR	Rescaling
	MagNet	0.0005	32 × 32	9-16	16	cPSNR	Rescaling
	MagNet++	0.0005	32 × 32	9-16	16	cPSNR	Rescaling

all models. To counter potential biases, model weights that demonstrated optimal performance on a validation subset were chosen, regardless of when this peak performance occurred during the training process. This approach was devised to address potential overfitting challenges that could arise from the extended training duration.

Compared to PROBA-V, the simulated datasets' uniform structure of always having nine LR images per scene allowed for the batch sizes to be increased for certain models. By maximizing the batch size within hardware memory limits, more efficient gradient approximations can be achieved, potentially accelerating convergence and improving generalization.

When it came to validation, a shift was made from the patch-based evaluation to using entire images from the validation subset. This strategic deviation ensured a holistic assessment of the models, gauging their capability to super-resolve entire images, which is more aligned with practical applications. It should be highlighted that the validation metric employed consistently for all models was cPSNR.

6.3 Evaluation

The post-training assessment spanned both simulated and PROBA-V datasets, focusing on the adaptability of models to diverse data conditions.

For the simulated datasets, evaluations were based on image dimensions up to 160×160 for LR inputs and 480×480 for HR images. Images larger than these sizes were either cropped or split into nonoverlapping patches. This method aimed to gauge the models' adaptability beyond training constraints and their scalability for diverse image sizes. In the context of PROBA-V, evaluations engaged with original-sized images—128×128 for LR images and 384×384 for HR references—offering a hands-on evaluation of the models' super-resolution abilities on entire images.

The prudent selection of evaluation metrics, particularly for satellite imagery, is pivotal to ensure that super-resolution techniques align with real-world applications [8]. In this research, the metrics employed include the cPSNR, previously discussed in Sect. 6.1.1, and four additional metrics: *structural similarity index measure* (SSIM) [98], *learned perceptual image patch similarity* (LPIPS) [115], *mean gradient error* (MGE) [58], and *the blur effect* (TBE) [16]. Except for TBE, which is a non-reference metric calculated solely from the super-resolved image, other metrics use their "corrected" version, denoted by the "c" prefix. This correction, mirroring the cPSNR's adjustment process, encompasses brightness rectification and full-pixel registration, ensuring consistent, precise model performance comparisons.

6.3.1 Structural Similarity Index Measure

The SSIM [98] is a perceptual metric used to assess super-resolved image quality against HR references. While traditional metrics like PSNR emphasize signal fidelity, SSIM focuses on preserving structural information, reflecting a more human-centric perception.

SSIM operates in a unique manner, moving a window of size $K \times K$ pixel by pixel over the entire image and evaluating the similarity between local regions in the super-resolved image and the corresponding regions in the HR image. It is a location-dependent analysis providing detailed insights into where the super-resolution model excels and where it falls short.

The SSIM index is calculated for each pair of windows x and y, both of size $K \times K$, taken from the super-resolved image and the HR reference image, respectively:

$$SSIM(x, y) = \frac{(2\mu_x\mu_y + c_1)(2\sigma_{x,y} + c_2)}{(\mu_x^2 + \mu_y^2 + c_1)(\sigma_x^2 + \sigma_y^2 + c_2)}. \tag{18}$$

Here μ_x and μ_y are the averages of windows x and y, respectively; σ_x^2 and σ_y^2 are the variances of x and y; $\sigma_{x,y}$ is the covariance of x and y; $c_1 = (k_1L)^2$ and $c_2 = (k_2L)^2$ are two variables to stabilize the division with a weak denominator; L is the dynamic range of the pixel values; and $k_1 = 0.01$ and $k_2 = 0.03$ by default. SSIM values range from 0 to 1, with 1 denoting perfect similarity.

In the context of image super-resolution, SSIM stands out for its perceptual focus, emphasizing structural and contrast changes over pure signal fidelity. This aligns it more with human vision, which prioritizes structural detail. Its sliding window approach also highlights specific areas of dissimilarity for a nuanced evaluation.

However, SSIM is not without limitations. The sliding window approach introduces a higher computational cost. Additionally, while it offers a perceptual quality score, it does not necessarily align perfectly with human judgment in all cases, given the vast complexity of vision and the many factors that influence human perception of image quality [99].

6.3.2 Learned Perceptual Image Patch Similarity

LPIPS [115] is a state-of-the-art metric designed for perceptual similarity assessment, differentiating it from traditional metrics such as PSNR and SSIM that target low-level image features. It uses deep learning to concentrate on high-level structures and details. It employs a neural network trained on human-rated images to estimate perceptual similarity between a pair of images. Notably, the LPIPS score ranges between 0 and 1, with 0 indicating perfect perceptual similarity and 1 signifying maximal perceptual dissimilarity. Capturing both pixel-level accuracy and nuanced visual characteristics, LPIPS has been recognized in recent research [47] as effective for assessing the quality of super-resolved images, especially when combined with PSNR and SSIM. However, its dependence on the training data and increased computational demand are acknowledged limitations.

Comparing SSIM with LPIPS highlights distinct approaches. SSIM measures structural similarity by considering aspects like contrast, luminance, and structure.

In contrast, LPIPS assesses perceptual differences based on human judgment. While LPIPS offers a deep perceptual analysis, SSIM's deterministic approach delivers results without the potential biases from training data and is less computationally intensive. Both metrics serve their purpose: SSIM for structural similarity and LPIPS for deeper perceptual differences. Using both provides a thorough evaluation of super-resolved images in terms of both structure and visual perception.

6.3.3 Mean Gradient Error

MGE [58] is a metric uniquely tailored to evaluate image edge sharpness—a critical component of perceptual quality [61]. While metrics like PSNR, SSIM, and LPIPS provide a broader perspective on image quality, MGE directly contrasts the gradient magnitudes between super-resolved and HR reference images, offering a specialized assessment of detail sharpness.

Edge sharpness delineates objects and communicates essential boundary and texture details. Given its capability to detect changes in edge sharpness, MGE's importance in super-resolution is underscored. Traditionally utilized as a loss function [58, 59], through the evaluation of gradient magnitudes, MGE measures the preservation of edge sharpness in super-resolved outputs. Hence, a lower MGE score is preferable, indicating reduced sharpness differences between the images.

The MGE score between an HR image and a super-resolved image is defined as follows:

$$MGE(HR, SR) = \frac{1}{HW} \sum_{i=1}^{H} \sum_{j=1}^{W} (G(i, j) - \widehat{G}(i, j))^2. \tag{19}$$

In this equation, $G(i, j)$ and $\widehat{G}(i, j)$ represent the gradient magnitude at pixel location (i, j) in the HR and super-resolved images, respectively, with H and W indicating the total number of pixels in the vertical and horizontal dimensions. The gradient magnitude at a given pixel location is calculated using the formula:

$$G(i, j) = \sqrt{G_x^2(i, j) + G_y^2(i, j)}. \tag{20}$$

Here, G_x and G_y represent the horizontal and vertical gradients, respectively. These gradients are obtained by convolving the image Y with the Sobel operators [38], which are well-established edge detection filters.

6.3.4 The Blur Effect

TBE [16] metric is designed to offer a no-reference evaluation of image sharpness by quantifying perceived blur. Unlike traditional metrics that require a reference image,

TBE operates independently, eliminating the need for correction procedures such as full-pixel registration and brightness adjustments. Its relevance and applicability have been demonstrated in other super-resolution studies, including those by Maier et al. [63] and Karmakar et al. [40].

TBE's foundation is based on the concept that blurring significantly affects perceived image quality, originating from various factors like camera focus errors, motion blur, atmospheric effects, or compression artifacts. Designed to capture perceptual aspects of the blur, TBE evaluates local contrasts, especially in edge regions, which are most perceptually significant. The following steps are involved in the computation of TBE:

1. **Edge detection**: Edges within the image are identified using specific techniques. Such edges, marking boundaries between distinct regions, are vital for perceiving sharpness.
2. **Local contrast assessment**: The local contrast around detected edges is evaluated. Sharp edges show pronounced intensity changes over short distances, whereas blurred edges present gradual transitions.
3. **Spatial distribution analysis**: The distribution of local contrasts across the image is analyzed. This analysis provides a holistic view of the image's sharpness.
4. **Metric aggregation**: Local contrast measurements and their spatial distribution insights are combined into a single metric value, indicating the overall perceived blur in the image.

MGE and TBE have been incorporated to offer a well-rounded sharpness assessment. While MGE centers on edge structures through gradient magnitude comparisons, TBE focuses on the broader aspect of a perceived blur. MGE is sensitive to registration quality, introducing potential variability, whereas TBE's non-reliance on references minimizes such concerns. Given TBE's alignment with human perceptions of blur and validation against human judgments, using both metrics ensures a comprehensive and robust evaluation for super-resolution model assessment.

7 Experimental Results and Discussion

This section presents a comprehensive evaluation of the discussed GNN models, as well as various state-of-the-art models for MISR, namely, HighRes-Net, RAMS, PIUNET, and TR-MISR. The evaluation was conducted on both the simulated and the real-world PROBA-V datasets. Each model was evaluated on the test subset of each dataset. The quality of the super-resolved images was assessed based on several reconstruction quality metrics, as discussed in Sect. 6. Each metric offers a unique perspective on the quality of the super-resolved images, thereby enabling a comprehensive evaluation of the models' performance.

In addition to the state-of-the-art deep learning models, a multi-image version of bicubic interpolation was employed as a baseline for comparison. This version operates by interpolating each LR image independently before averaging the stack of upsampled images to generate the final super-resolved image. This process aligns with the methodology delineated in [87], allowing for a methodical comparison of bicubic interpolation with more advanced MISR models. The following sections focus on the results of these evaluations, starting with the experiment on the simulated dataset.

7.1 Results on the Simulated Dataset

The super-resolution models were evaluated on the simulated data and their performance metrics for each test benchmark are presented in Table 3. To facilitate a direct comparison of model performance, Table 4 provides the overall mean scores, consolidating results obtained on the entire simulated dataset.

In evaluating the performance of super-resolution models on the simulated dataset, both MagNet and MagNet++ have shown notable capabilities, outperforming the bicubic interpolation and the HighRes-Net model. The results underscore the potential application and effectiveness of GNNs in MISR tasks, even though, at their current state, they do not outperform the leading MISR architectures, such as RAMS, TR-MISR, and PIUNET.

Interestingly, the implementation of a more sophisticated upsampling method in the MagNet++ did not yield the anticipated performance boost, with the model slightly underperforming in comparison to its predecessor, MagNet. This unexpected outcome may suggest that the simulated dataset's simplicity, lacking more complex relationships between pixels and temporal distortions inherent for real-world scenarios, might not fully leverage potentially more advanced capabilities of MagNet++.

Visual examinations of super-resolution outputs serve as critical adjuncts to quantitative metrics, offering intuitive insights into model performance. The high-resolution target images utilized in this qualitative assessment are depicted in Fig. 13 with red rectangles denoting their specific regions on which this visual analysis is conducted. These super-resolved regions are shown in Fig. 14, providing a better understanding of each model's strengths and weaknesses.

In the analyzed super-resolution outputs, high-frequency details, initially absent in each LR image, were effectively reintroduced by the models. However, subtle discrepancies in their outputs became apparent, especially in regions featuring complex textures. These variations, often emerging as disruptions in texture or irregular pixel transitions, compromised the visual fidelity of the super-resolved images.

This was particularly evident in high-frequency repeatable patterns such as the parallel lines on the bird's wings in the BSDS100 example or the fur in the image from the Set14 dataset. In these examples, while both MagNet and MagNet++

Table 3 Super-resolution model performance on simulated datasets, showcasing mean quality metric scores for each benchmark in the test subset. The best and second-best scores are highlighted in bold and underlined, respectively

		Model						
Metric	Dataset	Bicubic	HighRes-Net	RAMS	PIUNET	TR-MISR	MagNet	MagNet++
cPSNR	BSDS100	25.46	29.11	31.27	**31.42**	29.67	30.32	30.22
	Manga109	25.56	32.11	**34.81**	34.35	33.35	33.62	33.84
	Set14	25.96	29.31	31.70	**31.72**	30.16	30.67	30.61
	Set5	28.91	33.77	35.10	**36.41**	34.79	34.13	33.80
	Urban100	23.11	27.43	**29.59**	28.74	27.87	27.84	27.66
	Historical	22.10	26.53	28.50	**28.72**	26.99	27.47	27.52
cSSIM	BSDS100	0.737	0.878	0.921	**0.925**	0.885	0.901	0.896
	Manga109	0.852	0.959	0.975	**0.976**	0.966	0.968	0.969
	Set14	0.770	0.884	0.927	**0.930**	0.899	0.890	0.886
	Set5	0.865	0.954	0.967	**0.974**	0.964	0.959	0.956
	Urban100	0.739	0.887	**0.927**	0.915	0.894	0.885	0.878
	Historical	0.704	0.885	0.920	**0.922**	0.889	0.895	0.897
cLPIPS	BSDS100	0.418	0.088	**0.058**	0.066	0.085	0.090	0.095
	Manga109	0.284	0.027	**0.017**	0.023	0.025	0.026	0.027
	Set14	0.358	0.065	**0.043**	0.051	0.060	0.063	0.066
	Set5	0.250	0.029	0.020	**0.016**	0.026	0.032	0.038
	Urban100	0.392	0.071	**0.048**	0.074	0.071	0.090	0.100
	Historical	0.475	0.078	**0.054**	0.061	0.076	0.094	0.096
cMGE	BSDS100	0.080	0.028	**0.017**	**0.017**	0.025	0.024	0.025
	Manga109	0.110	0.022	**0.012**	0.014	0.016	0.016	0.016
	Set14	0.078	0.026	**0.015**	0.018	0.021	0.019	0.021
	Set5	0.053	0.012	**0.006**	**0.006**	0.008	0.009	0.011
	Urban100	0.172	0.053	**0.033**	0.044	0.048	0.058	0.065
	Historical	0.152	0.049	0.037	**0.035**	0.046	0.046	0.044
TBE	BSDS100	0.384	0.302	**0.297**	0.301	0.304	0.304	0.304
	Manga109	0.412	**0.313**	0.318	0.326	0.320	0.316	0.318
	Set14	0.426	**0.339**	**0.339**	0.345	0.342	0.343	0.341
	Set5	0.474	**0.391**	0.395	0.397	0.397	0.395	0.394
	Urban100	0.406	0.318	**0.317**	0.328	0.321	0.322	0.323
	Historical	0.352	0.273	**0.271**	0.278	0.278	0.280	0.280

effectively captured the textures, other models, particularly HighRes-Net and TR-MISR, struggled with the task, often misrepresenting the true direction and nuance of these patterns.

Table 4 The overall mean metric scores across all simulated benchmark datasets combined. The best scores are highlighted in bold, and the second-best scores are underlined

Model	cPSNR	cSSIM	cLPIPS	cMGE	TBE
Bicubic	24.56	0.783	0.355	0.129	0.405
HighRes-Net	29.60	0.913	0.057	0.036	**0.314**
RAMS	**31.96**	**0.945**	**0.038**	**0.021**	**0.314**
PIUNET	31.49	0.941	0.052	0.026	0.323
TR-MISR	30.38	0.920	0.055	0.031	0.318
MagNet	30.60	0.921	0.064	0.034	0.317
MagNet++	30.59	0.917	0.070	0.038	0.318

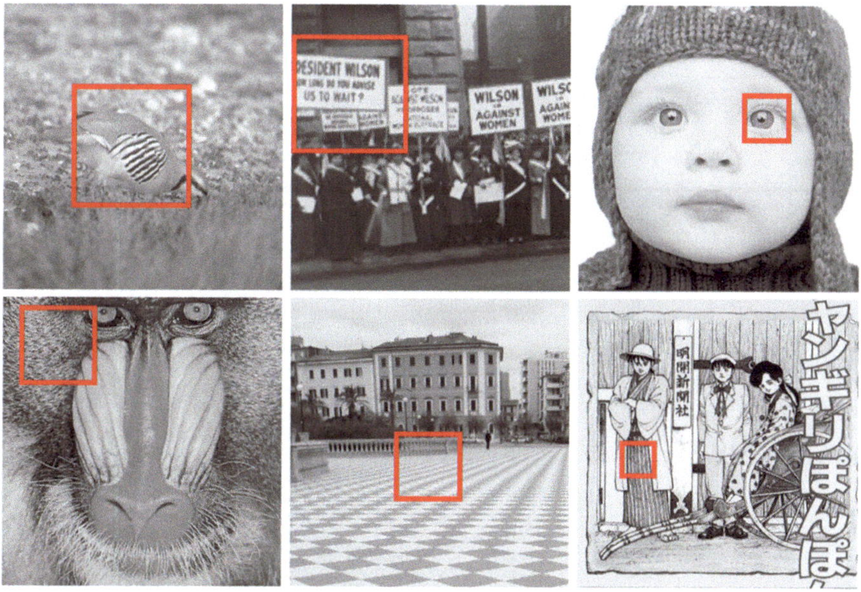

Fig. 13 High-resolution target images utilized in this qualitative assessment. Each image originates from a distinct benchmark dataset, respectively (from top left to bottom right): BSDS100, historical, Set5, Set14, Urban100, and Manga109. The red rectangles highlight the regions of interest examined in the qualitative analysis

7.2 Real-World Evaluation: The PROBA-V Dataset

In this section, a detailed analysis of the performance of the models on both spectral subsets of the PROBA-V datasets is presented, along with a discussion on the visual quality of super-resolved images. The aim is to ascertain whether the conclusions drawn from the simulated dataset are upheld in a real-world, remote-sensing scenario and to potentially uncover new insights into the performance of these models under more complex conditions.

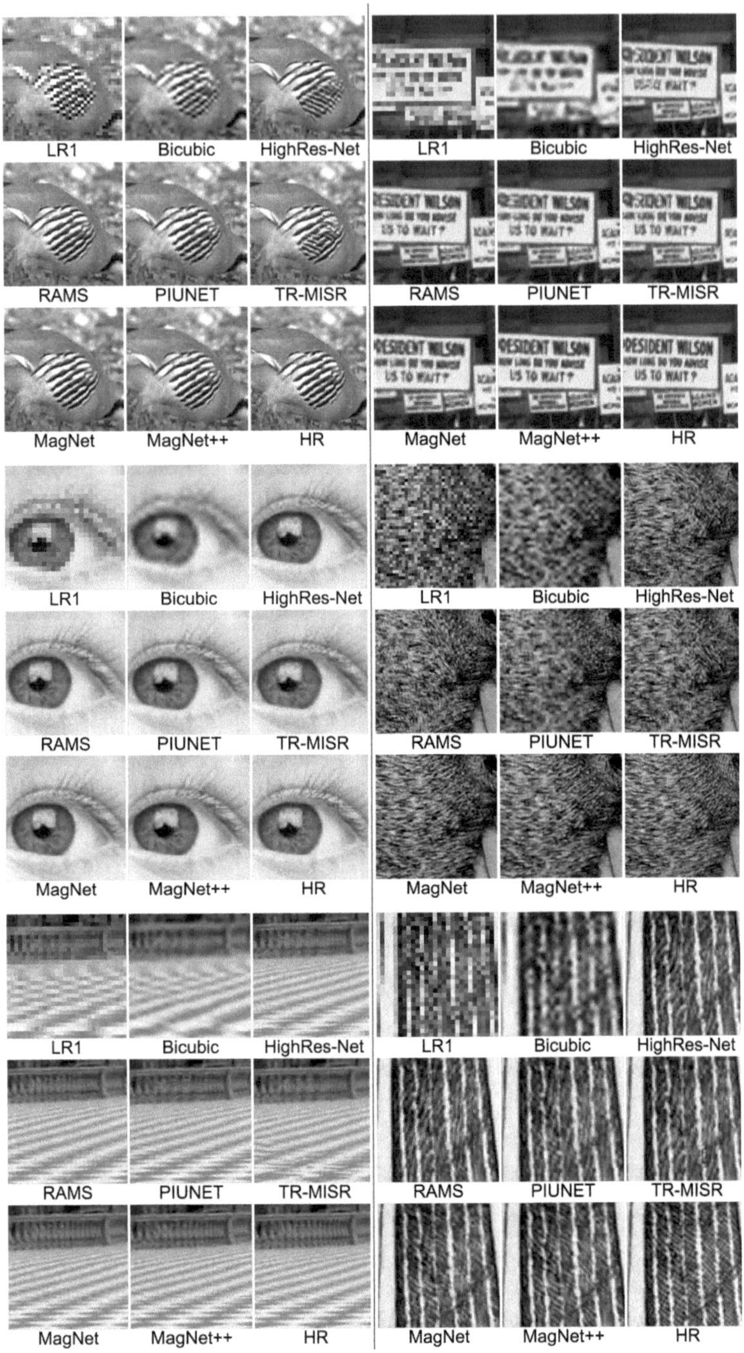

Fig. 14 Examples of super-resolved simulated images

Table 5 Performance metrics obtained by all tested methods on the PROBA-V dataset. The best scores are highlighted in bold, while the second-best scores are underlined for each spectral band independently

Band	Model	cPSNR	cSSIM	cLPIPS	cMGE	TBE
NIR	Bicubic	32.895	0.8602	0.2839	0.0137	0.4679
	HighRes-Net	34.760	0.9036	0.1523	0.0074	0.3472
	RAMS	35.648	**0.9148**	0.1571	**0.0065**	0.3382
	PIUNET	**35.735**	0.9144	0.1669	0.0068	0.3565
	TR-MISR	34.946	0.9058	**0.1367**	0.0071	0.3400
	MagNet	35.166	0.8992	0.2213	0.0072	**0.3307**
	MagNet++	35.500	0.9056	0.2151	0.0067	0.3364
RED	Bicubic	35.420	0.8980	0.3128	0.0082	0.4573
	HighRes-Net	37.389	0.9285	0.1614	0.0043	0.3424
	RAMS	38.492	0.9411	0.1601	**0.0033**	0.3206
	PIUNET	**38.629**	**0.9430**	0.1706	0.0035	0.3462
	TR-MISR	38.218	0.9382	**0.1418**	0.0035	0.3275
	MagNet	37.705	0.9235	0.2270	0.0043	**0.3202**
	MagNet++	37.872	0.9280	0.2248	0.0041	0.3282

7.2.1 Performance Metrics

The PROBA-V satellite mission offers a unique dataset for assessing super-resolution models in real-world conditions, particularly within the remote sensing domain. In this set of experiments, each model was tasked with super-resolving a scene from nine LR input images. The experiments spanned across both NIR and RED spectral bands of the PROBA-V dataset. This setup not only tests the models' capacity to enhance the spatial resolution but also allows for a direct comparison of their effectiveness across different spectral information, which is crucial for applications in earth observation and environmental monitoring. The performance scores for both PROBA-V spectral bands are presented in Table 5.

An analysis of the performance metrics reveals noteworthy observations. Across both NIR and RED spectral bands, different models exhibited varying degrees of success, with the best and second-best scores highlighted for each metric. In the real-world scenario presented by the PROBA-V dataset, MagNet++ demonstrated its strengths, particularly when compared to the simulated data results. While MagNet outperformed MagNet++ in every metric within simulated environments, the more sophisticated bipartite upsampling method incorporated in MagNet++ proved to be better suited for the complex tasks of super-resolution in real-world settings. This finding suggests that MagNet++'s approach to managing the variability and discrepancies among input LR images is more effective, highlighting the importance of model choice and adaptation to the specific challenges posed by real-world data. Despite its strengths, it is critical to acknowledge that MagNet++, while outperforming MagNet in the real-world dataset, still lags behind leading models such as RAMS, PIUNET, and TR-MISR in most metrics. Despite these observations, the simpler architecture of MagNet++ when compared to the state-of-the-art models and the juvenile state of GNNs in MISR tasks underscore a promising potential for graph-based models in this domain, indicating untapped capabilities amidst complex real-world super-resolution challenges.

7.3 Qualitative Analysis on NIR and RED Subsets

Figures 15 and 16 visually compare models' performance on the NIR and RED subsets of the PROBA-V dataset. The same region was used for a direct comparison, allowing for an immediate evaluation of each model's super-resolution capabilities. Bicubic interpolation, as expected, yields images that are less sharp and lack the fine details present in other super-resolved images, showcasing its limitation as a traditional interpolation method.

In the first NIR example, the HR image of a river shoreline shows high-frequency details (roughness and irregularity) that none of the models fully captured. This difficulty in replicating such fine details is also observed in other images, indicating the challenge of MISR in real-world scenarios.

Despite not achieving close reconstructions to the HR images certain trends are noticeable. All models face challenges in reconstructing scenes from specific time points due to temporal variations among the LR images, which is a core issue in real-world MISR. This results in visible artifacts in many reconstructed images, particularly noticeable in RED band examples featuring dynamically changing cultivated areas. MagNet and MagNet++ show satisfactory performance in managing these scenarios by producing images without any artifacts, unlike other models where grid-like distortions are evident in high temporal variation regions. Notably, PIUNET, a sophisticated state-of-the-art model, also avoids such artifacts, however, MagNet and MagNet++ achieve this with simpler, graph-based architectures. This effectiveness suggests promising potential for GNN-centric approaches in MISR.

8 Summary and Conclusions

This work aimed to explore the potential of GNNs in the field of super-resolution reconstruction. Initially, it provided an introduction to super-resolution, distinguishing between SISR and MISR, and discussed the challenges specific to MISR, especially the temporal variations among input images. Various methodologies within both MISR and SISR were reviewed, with an emphasis on those that significantly influenced the field. Fundamental concepts of graphs and GNNs were explained, focusing on architectures like GCN, GAT, and SplineCNN.

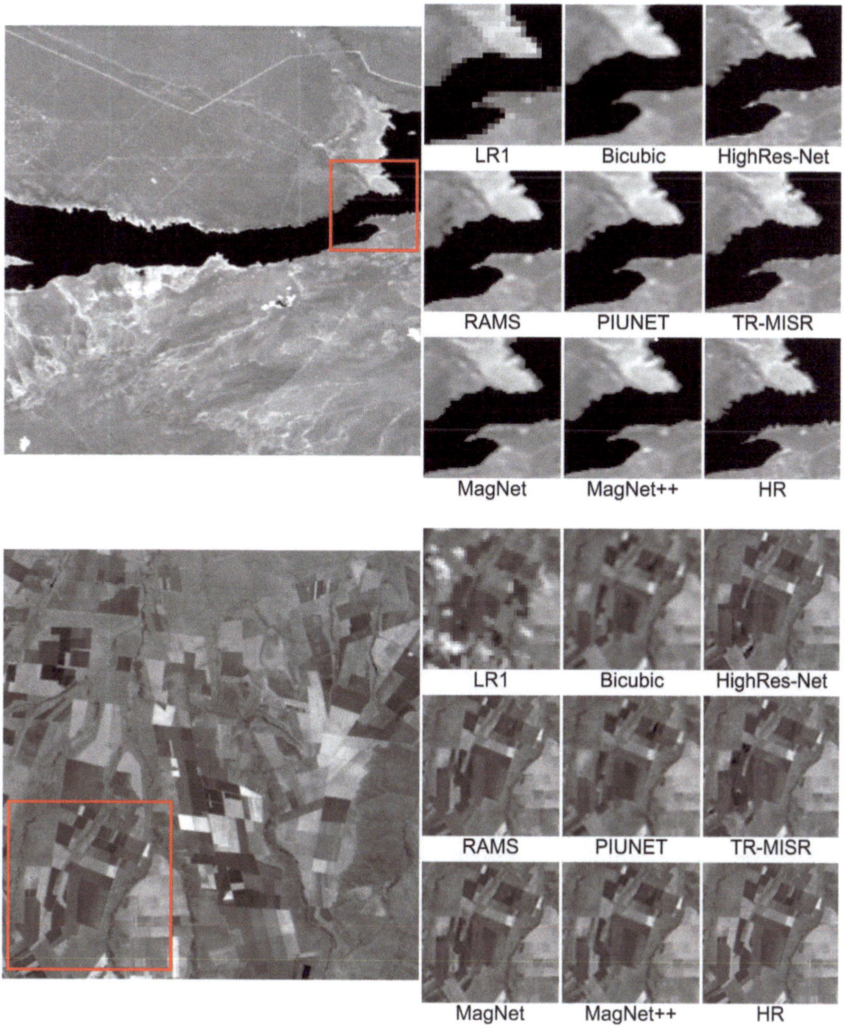

Fig. 15 Visual comparisons of different models on the imgset0625 (top) and imgset0878 (bottom) NIR examples of the PROBA-V dataset. The images are cropped to the region of interest centered at the same location for all models indicated as the red rectangle on the HR image (leftmost)

The methodology section detailed how multiple LR images were represented as a graph, explaining the rationale behind such representation. Also, this section described in detail two GNNs for MISR, namely, MagNet and MagNet++. The motivation for using simulated data for initial testing was explained, followed by the choice of the PROBA-V dataset for real-world MISR evaluation.

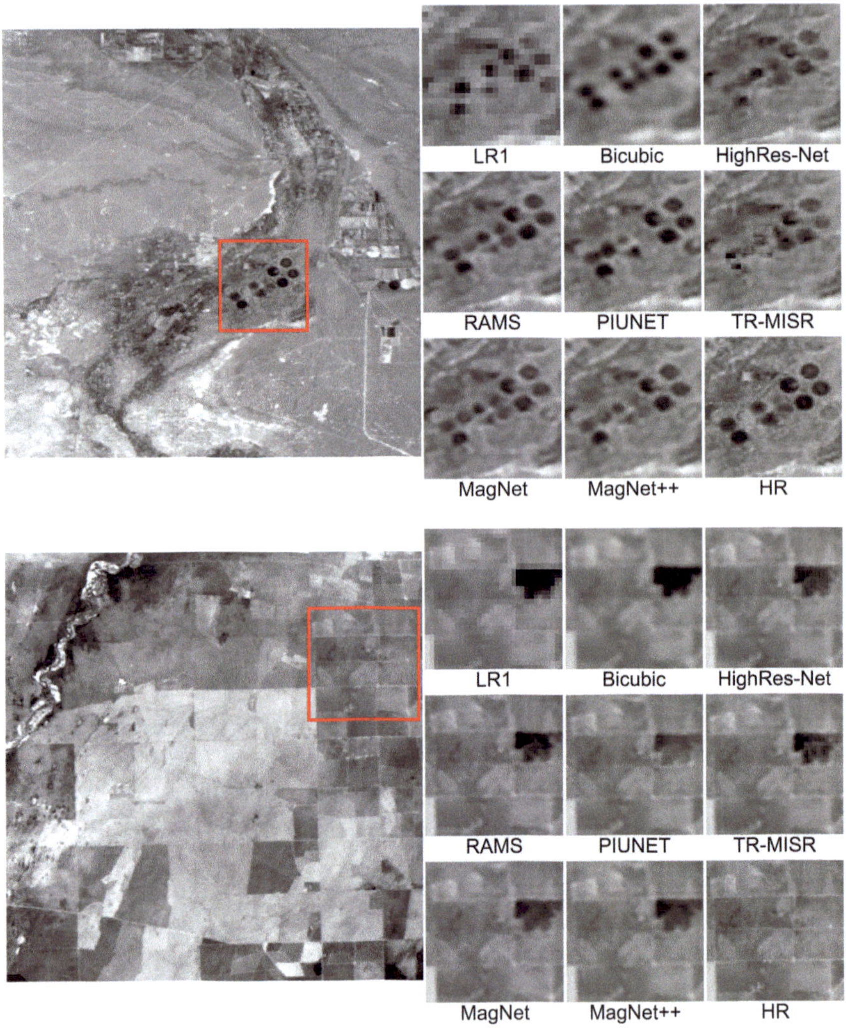

Fig. 16 Visual comparisons of different models on the imgset0132 (top) and imgset0263 (bottom) RED examples of the PROBA-V dataset. The images are cropped to the region of interest centered at the same location for all models indicated as the red rectangle on the HR image (leftmost)

The comparative analysis across simulated and real-world datasets reveals the GNN models' capabilities in the context of MISR, positioning them relative to leading methods. Quantitative assessments underscore the competitive performance of GNNs, particularly in handling the temporal variability that challenges real-world super-resolution tasks. Despite not outperforming all state-of-the-art models, GNNs show promise for future advancements in the MISR domain.

8.1 Future Research Directions in GNNs for MISR

The exploration into GNNs for super-resolution reveals several promising yet unexplored areas within GNN architectures and their application in MISR. Future research directions worth pursuing include:

1. **Dynamic graph structures**: Investigating GNNs that can adapt their graph structure dynamically in response to the specific features of input images. This adaptability could enhance the model's ability to handle diverse temporal and spatial variations more effectively.
2. **Cross-domain feature transfer**: Exploring methods for transferring features across different spectral bands or even modalities (e.g., optical to radar) using GNNs could open new avenues in multispectral and multimodal super-resolution, offering richer information extraction from LR images.
3. **Attention mechanisms in GNNs**: While attention mechanisms have shown promise in other domains, their integration into GNNs for MISR is still an unexplored direction. Researching how these mechanisms can be tailored to prioritize relevant features and relationships in the graph could lead to significant improvements in reconstruction quality.
4. **GNNs for nonrigid transformations**: Extending GNN architectures to accommodate nonrigid transformations in LR images, such as those caused by lens distortions, varying atmospheric conditions, or moving objects, remains a largely unexplored area. Developing models that can account for these transformations could drastically improve MISR's applicability, especially in remote sensing scenarios.
5. **Scalability and efficiency of GNNs**: As GNNs become more complex, their scalability and computational efficiency become crucial. Research into more efficient graph convolution operations or novel GNN architectures that maintain high performance while reducing computational overhead is necessary.
6. **Unsupervised and semi-supervised Learning**: Given the scarcity of labeled high-resolution images for certain applications, advancing unsupervised and semi-supervised learning techniques within GNN frameworks for MISR could leverage unlabeled data more effectively, broadening the applicability of GNNs in super-resolution tasks.

By focusing on these areas, future research can unlock new potentials in GNNs, pushing the boundaries of what is achievable in super-resolution reconstruction and beyond, particularly in challenging MISR scenarios.

Acknowledgments This work was funded by the National Science Centre, Poland, under research grant 2019/35/B/ST6/03006.

References

1. Agustsson, E., Timofte, R.: NTIRE 2017 challenge on single image super-resolution: Dataset and study. In: Proceedings of the IEEE CVPR Workshops, pp. 126–135 (2017)
2. An, T., Zhang, X., Huo, C., Xue, B., Wang, L., Pan, C.: TR-MISR: Multiimage super-resolution based on feature fusion with transformers. IEEE J-STARS **15**, 1373–1388 (2022)
3. Arbelaez, P., Maire, M., Fowlkes, C., Malik, J.: Contour detection and hierarchical image segmentation. IEEE TPAMI **33**(5), 898–916 (2010)
4. Bahdanau, D., Cho, K., Bengio, Y.: Neural machine translation by jointly learning to align and translate (2016)
5. Bao, L., Ma, B., Chang, H., Chen, X.: Masked graph attention network for person re-identification. In: 2019 IEEE/CVF Conference on Computer Vision and Pattern Recognition Workshops (CVPRW), pp. 1496–1505 (2019). https://doi.org/10.1109/CVPRW.2019.00191
6. Barabasi, A.L., Oltvai, Z.: Network biology: understanding the cell's functional organization. Nat. Rev. Genet. **5**, 101–13 (2004). https://doi.org/10.1038/nrg1272
7. Battaglia, P.W., Hamrick, J.B., Bapst, V., Sanchez-Gonzalez, A., Zambaldi, V., Malinowski, M., Tacchetti, A., Raposo, D., Santoro, A., Faulkner, R., Gulcehre, C., Song, F., Ballard, A., Gilmer, J., Dahl, G., Vaswani, A., Allen, K., Nash, C., Langston, V., Dyer, C., Heess, N., Wierstra, D., Kohli, P., Botvinick, M., Vinyals, O., Li, Y., Pascanu, R.: Relational inductive biases, deep learning, and graph networks (2018)
8. Benecki, P., Kawulok, M., Kostrzewa, D., Skonieczny, L.: Evaluating super-resolution reconstruction of satellite images. Acta Astronaut. **153**, 15–25 (2018). https://doi.org/10.1016/j.actaastro.2018.07.035. https://www.sciencedirect.com/science/article/pii/S0094576518300109
9. Bevilacqua, M., Roumy, A., Guillemot, C., Morel, M.L.A.: Low-complexity single-image super-resolution based on nonnegative neighbor embedding. In: Proceedings of the BMVC (2012)
10. Bhat, G., Danelljan, M., Van Gool, L., Timofte, R.: Deep burst super-resolution. In: Proceedings of the IEEE/CVF Conference on Computer Vision and Pattern Recognition (CVPR), pp. 9209–9218 (2021)
11. Brin, S., Page, L.: The anatomy of a large-scale hypertextual web search engine. Comput. Netw. **30**, 107–117 (1998). http://www-db.stanford.edu/~backrub/google.html
12. Brody, S., Alon, U., Yahav, E.: How attentive are graph attention networks? (2022)
13. Bronstein, M.M., Bruna, J., LeCun, Y., Szlam, A., Vandergheynst, P.: Geometric deep learning: going beyond euclidean data. IEEE Signal Process. Mag. **34**(4), 18–42 (2017). https://doi.org/10.1109/msp.2017.2693418. https://doi.org/10.1109%2Fmsp.2017.2693418
14. Brown, L.G.: A survey of image registration techniques. ACM Comput. Surv. **24**(4), 325–376 (1992). https://doi.org/10.1145/146370.146374
15. Bruna, J., Zaremba, W., Szlam, A., LeCun, Y.: Spectral networks and locally connected networks on graphs. In: International Conference on Learning Representations (2014)
16. Crété-Roffet, F., Dolmière, T., Ladret, P., Nicolas, M.: The blur effect: perception and estimation with a new no-reference perceptual blur metric. Human Vision and Electronic Imaging **12**, EI-6492 (2007). https://doi.org/10.1117/12.702790
17. DeTone, D., Malisiewicz, T., Rabinovich, A.: Deep image homography estimation. CoRR **abs/1606.03798** (2016). http://arxiv.org/abs/1606.03798
18. Deudon, M., Kalaitzis, A., et al.: HighRes-net: recursive fusion for multi-frame super-resolution of satellite imagery. arXiv preprint arXiv:2002.06460 (2020)
19. Dong, C., Loy, C.C., He, K., Tang, X.: Learning a deep convolutional network for image super-resolution. In: Proceedings of the ECCV, pp. 184–199. Springer, Berlin (2014)
20. Dong, C., Loy, C.C., He, K., Tang, X.: Image super-resolution using deep convolutional networks. IEEE Trans. Pattern Anal. Mach. Intell. **38**(2), 295–307 (2015)
21. Dong, C., Loy, C.C., Tang, X.: Accelerating the super-resolution convolutional neural network. CoRR **abs/1608.00367** (2016). http://arxiv.org/abs/1608.00367

22. Dong, C., Loy, C.C., Tang, X.: Accelerating the super-resolution convolutional neural network. In: Proceedings of the ECCV, pp. 391–407 (2016)
23. Dumoulin, V., Visin, F.: A guide to convolution arithmetic for deep learning (2018)
24. Fey, M., Lenssen, J.E.: Fast graph representation learning with PyTorch Geometric. In: ICLR Workshop on Representation Learning on Graphs and Manifolds (2019)
25. Fey, M., Lenssen, J.E., Weichert, F., Müller, H.: SplineCNN: fast geometric deep learning with continuous B-spline kernels. In: Proceedings of the IEEE CVPR, pp. 869–877 (2018)
26. Gilmer, J., Schoenholz, S.S., Riley, P.F., Vinyals, O., Dahl, G.E.: Neural message passing for quantum chemistry, In: International Conference on Machine Learning. PMLR, New York (2017)
27. Goel, A., Ma, K.T., Tan, C.: An end-to-end network for generating social relationship graphs. In: Proceedings of the IEEE/CVF Conference on Computer Vision and Pattern Recognition (2019)
28. Grover, A., Leskovec, J.: node2vec: Scalable feature learning for networks. In: Proceedings of the 22nd ACM SIGKDD International Conference on Knowledge Discovery and Data Mining (2016)
29. Guizar-Sicairos, M., Thurman, S.T., Fienup, J.R.: Efficient subpixel image registration algorithms. Opt. Lett. **33**(2), 156–158 (2008)
30. Gupta, R., Sharma, A., Kumar, A.: Super-resolution using gans for medical imaging. Procedia Comput. Sci. **173**, 28–35 (2020). https://doi.org/10.1016/j.procs.2020.06.005. https://www.sciencedirect.com/science/article/pii/S1877050920315076. International Conference on Smart Sustainable Intelligent Computing and Applications under ICITETM2020
31. Hamilton, W.L., Ying, R., Leskovec, J.: Inductive representation learning on large graphs (2018)
32. Hamilton, W.L., Ying, R., Leskovec, J.: Representation learning on graphs: methods and applications (2018)
33. Harary, F.: Graph Theory. Addison-Wesley, Reading, MA (1969)
34. He, K., Zhang, X., Ren, S., Sun, J.: Deep residual learning for image recognition. In: Proceedings of the IEEE Conference on Computer Vision and Pattern Recognition (2015)
35. He, K., Zhang, X., Ren, S., Sun, J.: Delving deep into rectifiers: surpassing human-level performance on imagenet classification. CoRR **abs/1502.01852** (2015). http://arxiv.org/abs/1502.01852
36. He, Z., Li, J., Liu, L., He, D., Xiao, M.: Multiframe video satellite image super-resolution via attention-based residual learning. IEEE Trans. Geosci. Remote Sens. **60**, 1–15 (2022). https://doi.org/10.1109/TGRS.2021.3072381
37. Huang, J.B., Singh, A., Ahuja, N.: Single image super-resolution from transformed self-exemplars. In: Proceedings of the IEEE CVPR, pp. 5197–5206 (2015)
38. Kanopoulos, N., Vasanthavada, N., Baker, R.: Design of an image edge detection filter using the sobel operator. IEEE J. Solid State Circuits **23**(2), 358–367 (1988). https://doi.org/10.1109/4.996
39. Kappeler, A., Yoo, S., Dai, Q., Katsaggelos, A.K.: Video super-resolution with convolutional neural networks. IEEE TCI **2**(2), 109–122 (2016)
40. Karmakar, J., Kumar, A., Nandi, D., Mandal, M.K.: A Novel Super-Resolution Reconstruction from Multiple Frames via Sparse Representation, pp. 33–45. Springer Singapore, Germany (2020). https://doi.org/10.1007/978-981-15-2854-5_4. https://lens.org/048-290-652-634-19X. https://link.springer.com/chapter/10.1007/978-981-15-2854-5_4
41. Kawulok, M., Benecki, P., Kostrzewa, D., Skonieczny, L.: Evolving imaging model for super-resolution reconstruction. In: Proceedings of the GECCO, pp. 284–285. ACM, New York (2018)
42. Kawulok, M., Benecki, P., Piechaczek, S., Hrynczenko, K., Kostrzewa, D., Nalepa, J.: Deep learning for multiple-image super-resolution. IEEE GRSL **17**(6), 1062–1066 (2020)
43. Kim, D.W., Chung, J.R., Kim, J., Lee, D.Y., Jeong, S.Y., Jung, S.W.: Constrained adversarial loss for generative adversarial network-based faithful image restoration. ETRI J. **41**(4), 415–425 (2019)

44. Kim, J., Kwon Lee, J., Mu Lee, K.: Accurate image super-resolution using very deep convolutional networks. In: Proceedings of the IEEE CVPR, pp. 1646–1654 (2016)
45. Kim, S.Y., Lim, J., Na, T., Kim, M.: Video super-resolution based on 3D-CNNs with consideration of scene change. In: Proceedings of the IEEE International Conference on Image Processing (ICIP), pp. 2831–2835. IEEE, New York (2019)
46. Kipf, T.N., Welling, M.: Semi-supervised classification with graph convolutional networks (2017)
47. Kowaleczko, P., Tarasiewicz, T., Ziaja, M., Kostrzewa, D., Nalepa, J., Rokita, P., Kawulok, M.: A real-world benchmark for sentinel-2 multi-image super-resolution. Sci. Data **10**(1), 644 (2023). https://doi.org/10.1038/s41597-023-02538-9
48. Krzywda, M., Lukasik, S., Gandomi, A.H.: Graph neural networks in computer vision—architectures, datasets and common approaches. In: 2022 International Joint Conference on Neural Networks (IJCNN). IEEE, New York (2022). https://doi.org/10.1109/ijcnn55064.2022.9892658. https://doi.org/10.1109%2Fijcnn55064.2022.9892658
49. Kuglin, C.: The phase correlation image alignment method. In: Proceedings of the International Conference on Cybernetics Society pp. 163–165 (1975)
50. Lai, W.S., Huang, J.B., Ahuja, N., Yang, M.H.: Fast and accurate image super-resolution with deep Laplacian pyramid networks. IEEE TPAMI **41**(11), 2599–2613 (2018)
51. Lanaras, C., Bioucas-Dias, J., Galliani, S., Baltsavias, E., Schindler, K.: Super-resolution of Sentinel-2 images: learning a globally applicable deep neural network. ISPRS J. Photogramm. Remote Sens. **146**, 305–319 (2018)
52. Ledig, C., Theis, L., Huszár, F., et al.: Photo-realistic single image super-resolution using a generative adversarial network. In: Proceedings of the CVPR, vol. 2, p. 4 (2017)
53. Lillesand, T.: Remote Sensing and Image Interpretation. Wiley, New York (2004). https://books.google.pl/books?id=8fiIPwAACAAJ
54. Lim, B., Son, S., Kim, H., Nah, S., Mu Lee, K.: Enhanced deep residual networks for single image super-resolution. In: Proceeding of the IEEE/CVF Conference on Computer Vision and Pattern Recognition Workshops, pp. 136–144 (2017)
55. Ling, H., Gao, J., Kar, A., Chen, W., Fidler, S.: Fast interactive object annotation with curve-gcn. In: Proceedings of the IEEE/CVF Conference on Computer Vision and Pattern Recognition (2019)
56. Liu, H., Ruan, Z., Zhao, P., Dong, C., Shang, F., Liu, Y., Yang, L., Timofte, R.: Video super-resolution based on deep learning: a comprehensive survey. Artif. Intell. Rev. **55**(8), 5981–6035 (2022)
57. Lowe, D.G.: Distinctive image features from scale-invariant keypoints. Int. J. Comput. Vision **60**(2), 91–110 (2004). https://doi.org/10.1023/B:VISI.0000029664.99615.94
58. Lu, Z., Chen, Y.: Single image super resolution based on a modified u-net with mixed gradient loss (2019)
59. Lu, Z., Chen, Y.: Dense u-net for super-resolution with shuffle pooling layer (2021)
60. Lu, Z., Li, J., Liu, H., Huang, C., Zhang, L., Zeng, T.: Transformer for single image super-resolution. In: Proceedings of the IEEE/CVF Conference on Computer Vision and Pattern Recognition, pp. 457–466 (2022)
61. Ma, Z., Liao, R., Tao, X., Xu, L., Jia, J., Wu, E.: Handling motion blur in multi-frame super-resolution. In: 2015 IEEE Conference on Computer Vision and Pattern Recognition (CVPR), pp. 5224–5232 (2015). https://doi.org/10.1109/CVPR.2015.7299159
62. Maas, A.L.: Rectifier nonlinearities improve neural network acoustic models (2013)
63. Maier, R., Stückler, J., Cremers, D.: Super-resolution keyframe fusion for 3D modeling with high-quality textures. In: 2015 International Conference on 3D Vision, pp. 536–544 (2015). https://doi.org/10.1109/3DV.2015.66
64. Märtens, M., Izzo, D., Krzic, A., Cox, D.: Super-resolution of PROBA-V images using convolutional neural networks. Astrodynamics **3**(4), 387–402 (2019)
65. Matsui, Y., Ito, K., Aramaki, Y., Fujimoto, A., et al.: Sketch-based manga retrieval using Manga109 dataset. Multimed. Tools Appl. **76**(20), 21811–21838 (2017)

66. Mislove, A., Marcon, M., Gummadi, K.P., Druschel, P., Bhattacharjee, B.: Measurement and analysis of online social networks. In: Proceedings of the 7th ACM SIGCOMM Conference on Internet Measurement, IMC '07, p. 29–42. Association for Computing Machinery, New York (2007). https://doi.org/10.1145/1298306.1298311
67. Mistry, D., Banerjee, A.: Review: Image registration. Int. J. Graphics and Image Processing(ISSN 2249–5452), **II**, 18–22 (2012)
68. Molini, A.B., Valsesia, D., Fracastoro, G., Magli, E.: DeepSUM: Deep neural network for super-resolution of unregistered multitemporal images. IEEE TGRS **58**(5), 3644–3656 (2020)
69. Nasrollahi, K., Moeslund, T.B.: Super-resolution: a comprehensive survey. Mach. Vis. Appl. **25**(6), 1423–1468 (2014)
70. Newman, M.E.J.: Networks: An Introduction. Oxford University, Oxford (2010)
71. Nguyen, N.L., Anger, J., Davy, A., Arias, P., Facciolo, G.: Self-supervised multi-image super-resolution for push-frame satellite images. In: Proceedings of the IEEE/CVF Conference on Computer Vision and Pattern Recognition (CVPR), pp. 1121–1131 (2021)
72. Omar, Z., Stathaki, T.: Image Fusion: An Overview, pp. 306–310 (2014). https://doi.org/10.1109/ISMS.2014.58
73. Piegl, L., Tiller, W.: The NURBS Book, 2nd edn. Springer, New York (1996)
74. Ponomarenko, N., Ieremeiev, O., Lukin, V., Egiazarian, K., Carli, M.: Modified image visual quality metrics for contrast change and mean shift accounting. In: 2011 11th International Conference the Experience of Designing and Application of CAD Systems in Microelectronics (CADSM), pp. 305–311 (2011)
75. Rifat Arefin, M., et al.: Multi-image super-resolution for remote sensing using deep recurrent networks. In: Proceedings of the IEEE CVPR Workshops, pp. 206–207 (2020)
76. Rumelhart, D.E., McClelland, J.L.: Learning Internal Representations by Error Propagation, pp. 318–362 (1987)
77. Salvetti, F., Mazzia, V., Khaliq, A., Chiaberge, M.: Multi-image super resolution of remotely sensed images using residual attention deep neural networks. Remote Sens. **12**(14), 2207 (2020)
78. Scarselli, F., Gori, M., Tsoi, A.C., Hagenbuchner, M., Monfardini, G.: The graph neural network model. IEEE Trans. Neural Netw. **20**(1), 61–80 (2009). https://doi.org/10.1109/TNN.2008.2005605
79. Schlichtkrull, M., Kipf, T.N., Bloem, P., van den Berg, R., Titov, I., Welling, M.: Modeling relational data with graph convolutional networks (2017)
80. Shen, H., Qiu, Z., Yue, L., Zhang, L.: Deep-learning-based super-resolution of video satellite imagery by the coupling of multiframe and single-frame models. IEEE Trans. Geosci. Remote Sens. **60**, 1–14 (2022). https://doi.org/10.1109/TGRS.2021.3121303
81. Shi, W., Caballero, J., Huszár, F., Totz, J., Aitken, A.P., Bishop, R., Rueckert, D., Wang, Z.: Real-time single image and video super-resolution using an efficient sub-pixel convolutional neural network. In: Proceedings of the IEEE Conference on Computer Vision and Pattern Recognition (2016)
82. Singla, K., Pandey, R., Ghanekar, U.: A review on single image super resolution techniques using generative adversarial network. Optik **266**, 169607 (2022). https://doi.org/10.1016/j.ijleo.2022.169607. https://www.sciencedirect.com/science/article/pii/S0030402622009032
83. Snyder, J.: Flattening the Earth: Two Thousand Years of Map Projections. University of Chicago, Chicago (1997). https://books.google.pl/books?id=0UzjTJ4w9yEC
84. Tang, S., Yao, K., Liang, J., Wang, Z., Liang, J.: Graph neural networks with interlayer feature representation for image super-resolution. In: Proceedings of the Sixteenth ACM International Conference on Web Search and Data Mining, WSDM '23, pp. 652–660. Association for Computing Machinery, New York (2023). https://doi.org/10.1145/3539597.3570436
85. Tarasiewicz, T., Kawulok, M.: Graph-based representation for multi-image super-resolution. In: Vento, M., Foggia, P., Conte, D., Carletti, V. (eds.) Graph-Based Representations in Pattern Recognition. GbRPR 2023. Lecture Notes in Computer Science, vol 14121. Springer, Cham (2023). https://doi.org/10.1007/978-3-031-42795-4_14

86. Tarasiewicz, T., Nalepa, J., Farrugia, R.A., Valentino, G., Chen, M., Briffa, J.A., Kawulok, M.: Multitemporal and multispectral data fusion for super-resolution of sentinel-2 images. IEEE Trans. Geosci. Remote Sens. (TGRS). https://doi.org/10.1109/TGRS.2023.3311622
87. Tarasiewicz, T., Nalepa, J., Kawulok, M.: A graph neural network for multiple-image super-resolution. In: Proceedings of the IEEE ICIP, pp. 1824–1828 (2021)
88. Tarasiewicz, T., Nalepa, J., Kawulok, M.: Semi-simulated training data for multi-image super-resolution. In: IGARSS 2022—2022 IEEE International Geoscience and Remote Sensing Symposium, pp. 481–484 (2022). https://doi.org/10.1109/IGARSS46834.2022.9884565
89. Thevenaz, P., Ruttimann, U., Unser, M.: A pyramid approach to subpixel registration based on intensity. IEEE Trans. Image Process. **7**(1), 27–41 (1998). https://doi.org/10.1109/83.650848
90. Timofte, R., De, V., Gool, L.V.: Anchored neighborhood regression for fast example-based super-resolution. In: 2013 IEEE International Conference on Computer Vision, pp. 1920–1927 (2013). https://doi.org/10.1109/ICCV.2013.241
91. Tsai, R.Y., Huang, T.S.: Multiframe image restoration and registration (1984). https://api.semanticscholar.org/CorpusID:59796060
92. Valsesia, D., Magli, E.: Permutation invariance and uncertainty in multitemporal image super-resolution. IEEE TGRS **60**, 1–12 (2022)
93. Vaswani, A., Shazeer, N., Parmar, N., Uszkoreit, J., Jones, L., Gomez, A.N., Kaiser, L., Polosukhin, I.: Attention is all you need. In: Advances in Neural Information Processing Systems (2023)
94. Vedaldi, A., Lenc, K.: Matconvnet: convolutional neural networks for matlab. In: Proceedings of the 23rd ACM International Conference on Multimedia, pp. 689–692 (2016)
95. Veličković, P., Cucurull, G., Casanova, A., Romero, A., Lio, P., Bengio, Y.: Graph attention networks. In: 6th International Conference on Learning Representations (2017)
96. Wang, P., Wu, Q., Cao, J., Shen, C., Gao, L., van den Hengel, A.: Neighbourhood watch: Referring expression comprehension via language-guided graph attention networks. In: Proceedings of the IEEE/CVF Conference on Computer Vision and Pattern Recognition, pp. 1960–1968 (2018)
97. Wang, X., Yu, K., Wu, S., Gu, J., Liu, Y., Dong, C., Loy, C.C., Qiao, Y., Tang, X.: ESRGAN: enhanced super-resolution generative adversarial networks. CoRR **abs/1809.00219** (2018). http://arxiv.org/abs/1809.00219
98. Wang, Z., Bovik, A., Sheikh, H., Simoncelli, E.: Image quality assessment: from error visibility to structural similarity. IEEE Trans. Image Process. **13**(4), 600–612 (2004). https://doi.org/10.1109/TIP.2003.819861
99. Wang, Z., Bovik, A.C., Sheikh, H.R., Simoncelli, E.P.: Image quality assessment: from error visibility to structural similarity. IEEE Trans. Image Process. **13**(4), 600–612 (2004)
100. Wang, Z., Chen, J., Hoi, S.C.H.: Deep learning for image super-resolution: A survey. IEEE Trans. Pattern Anal. Mach. Intell. **43**(10), 3365–3387 (2021). https://doi.org/10.1109/TPAMI.2020.2982166
101. Wang, Z., Jiang, K., Yi, P., Han, Z., He, Z.: Ultra-dense GAN for satellite imagery super-resolution. Neurocomputing **398**, 328–337 (2020)
102. Welstead, S.T.: Fractal and wavelet image compression techniques (1999). https://api.semanticscholar.org/CorpusID:118961255
103. West, D.B.: Introduction to Graph Theory, 2 edn. Prentice Hall, New York (2000)
104. Wu, B., Li, X., Cai, H., Chen, C.W.: Cross-scale internal graph neural network for image super-resolution. In: Proceedings of the IEEE/CVF International Conference on Computer Vision, pp. 15738–15747 (2021)
105. Wu, Z., Pan, S., Chen, F., Long, G., Zhang, C., Yu, P.S.: A comprehensive survey on graph neural networks. IEEE Trans. Neural Networks Learn. Syst. **32**(1), 4–24 (2021). https://doi.org/10.1109/tnnls.2020.2978386. https://doi.org/10.1109%2Ftnnls.2020.2978386
106. Xu, K., Hu, W., Leskovec, J., Jegelka, S.: How powerful are graph neural networks? (2019)
107. Yan, S., Xiong, Y., Lin, D.: Spatial temporal graph convolutional networks for skeleton-based action recognition. In: Proceedings of the AAAI Conference on Artificial Intelligence, vol. 32(1) (2018)

108. Ying, R., He, R., Chen, K., Eksombatchai, P., Hamilton, W.L., Leskovec, J.: Graph convolutional neural networks for web-scale recommender systems. In: Proceedings of the 24th ACM SIGKDD International Conference on Knowledge Discovery & Data Mining. ACM, New York (2018). https://doi.org/10.1145/3219819.3219890. https://doi.org/10.1145%2F3219819.3219890
109. Yue, L., Shen, H., Li, J., Yuan, Q., Zhang, H., Zhang, L.: Image super-resolution: the techniques, applications, and future. Signal Process. **128**, 389–408 (2016)
110. Zeiler, M.D., Krishnan, D., Taylor, G.W., Fergus, R.: Deconvolutional networks. In: 2010 IEEE Computer Society Conference on Computer Vision and Pattern Recognition, pp. 2528–2535 (2010). https://doi.org/10.1109/CVPR.2010.5539957
111. Zeiler, M.D., Taylor, G.W., Fergus, R.: Adaptive deconvolutional networks for mid and high level feature learning. In: 2011 International Conference on Computer Vision, pp. 2018–2025 (2011). https://doi.org/10.1109/ICCV.2011.6126474
112. Zeyde, R., Elad, M., Protter, M.: On single image scale-up using sparse-representations. In: Proceedings of the International Conference on Curves and Surfaces, pp. 711–730 (2010)
113. Zhai, M., Xiang, X., Lv, N., Kong, X.: Optical flow and scene flow estimation: a survey. Pattern Recogn. **114**, 107861 (2021). https://doi.org/10.1016/j.patcog.2021.107861. https://www.sciencedirect.com/science/article/pii/S0031320321000480
114. Zhang, M., Cui, Z., Neumann, M., Chen, Y.: An end-to-end deep learning architecture for graph classification. In: McIlraith, S.A., Weinberger, K.Q. (eds.) Proceedings of the Thirty-Second AAAI Conference on Artificial Intelligence, (AAAI-18), the 30th innovative Applications of Artificial Intelligence (IAAI-18), and the 8th AAAI Symposium on Educational Advances in Artificial Intelligence (EAAI-18), New Orleans, Louisiana, USA, February 2–7, 2018, pp. 4438–4445. AAAI Press, New York (2018). https://www.aaai.org/ocs/index.php/AAAI/AAAI18/paper/view/17146
115. Zhang, R., Isola, P., Efros, A.A., Shechtman, E., Wang, O.: The unreasonable effectiveness of deep features as a perceptual metric. In: Proceedings of the IEEE/CVF CVPR (2018)
116. Zhang, Y., Li, K., Li, K., Wang, L., Zhong, B., Fu, Y.: Image super-resolution using very deep residual channel attention networks. In: V. Ferrari, M. Hebert, C. Sminchisescu, Y. Weiss (eds.) Proceedings of the ECCV, pp. 294–310. Springer, Cham (2018)
117. Zhao, W., Sawhney, H.S.: Is super-resolution with optical flow feasible? In: A. Heyden, G. Sparr, M. Nielsen, P. Johansen (eds.) Computer Vision—ECCV 2002, pp. 599–613. Springer, Berlin (2002)
118. Zhou, J., Cui, G., Hu, S., Zhang, Z., Yang, C., Liu, Z., Wang, L., Li, C., Sun, M.: Graph neural networks: A review of methods and applications (2021)
119. Zitova, B., Flusser, J.: Image registration methods: a survey. Image Vis. Comput. **21**(11), 977–1000 (2003)

Effectiveness Analysis of Example-Based Machine Learning and Deep Learning Methods for Super-resolution Hyperspectral Images

Ediz Polat and Murat Simsek

1 Introduction

Hyperspectral imagery collects image data simultaneously in hundreds of narrow and adjacent spectral bands providing a more detailed spectral profile for each pixel. Because of having a large quantity of spectral information, hyperspectral images are used in many applications including remote sensing, mineral identification, environmental studies, food processing, surveillance, and object classification. The additional spectral information allows for better discrimination between different materials or conditions, enabling more precise analysis and interpretation of the data. Although they have high spectral resolution, hyperspectral images have low spatial resolution due to hardware limitation. Because of this drawback, many techniques have been developed in the literature for improving hyperspectral spatial resolution such as super-resolution [1–3].

Super-resolution refers to the process of enhancing the resolution or level of detail in an image. This technique is particularly useful when dealing with low-resolution images, and it aims to generate a higher-resolution version of the image by improving the clarity and sharpness of details. In the hyperspectral image context, the super-resolution methods fuse low spatial resolution of hyperspectral images and high spatial resolution of panchromatic or multispectral images. The basic idea behind the super-resolution techniques is to use combination of low resolution images of the same scene to generate higher resolution images. It is

E. Polat (✉)
School of Electrical Engineering and Computer Science, University of North Dakota, Grand Forks, ND, USA
e-mail: ediz.polat@und.edu

M. Simsek
Artificial Intelligence Engineering Department, Ostim Technical University, Ankara, Turkiye
e-mail: murat.simsek@ostimteknik.edu.tr

M. Kawulok et al. (eds.), *Super-Resolution for Remote Sensing*, Unsupervised and Semi-Supervised Learning, https://doi.org/10.1007/978-3-031-68106-6_5

a reconstruction process of the original image with high resolution, given many observed images at lower resolution.

In general, super-resolution methods can be broadly categorized into two categories, namely, single-image super-resolution and multi-image super-resolution. The approach for the first category involves enhancing the resolution of a single low-resolution image [4, 5]. The approach can be computationally more efficient than approaches that involve processing multiple images. The approach can be further categorized into several subcategories. Multi-image super-resolution approach utilizes multiple low-resolution images of the same scene to generate a higher-resolution output by capturing different views or aspects of the same scene [6, 7]. This allows for the extraction of complementary information that may not be present in a single image, leading to improved overall resolution and quality. The use of multiple images can help decrease the effects of noise in individual images. By averaging or combining information, these methods can produce a more robust and higher accuracy in the reconstruction of high-resolution output especially in satellite and remote imaging [8], where multiple captures of the same location are available.

In this research, we investigate the superiority of the methods in these two categories, single-image and multi-image super-resolution, for hyperspectral images. We further divide the methods into three groups. The first group includes multi-image and interpolation-based methods also known as pan-sharpening methods (intensity hue saturation (IHS) and principle component analysis (PCA)). The second group includes example-based machine learning methods, especially, dictionary learning-based sparse representation methods (K-singular value decomposition (K-SVD), online dictionary learning (ODL), and Bayesian). And the third group includes deep learning-based method (convolutional neural network (CNN)). The framework in [9] is also extended in this research to include most popular techniques in the current literature. Thus, more algorithms with extensive experiments are implemented to provide broader perspective on the super-resolution of hyperspectral images.

2 Super-resolution Image Reconstruction Methods

The main objective of super-resolution image methods is to enhance the resolution of a lower-resolution image such that the resulting image becomes more detailed in terms of high-frequency components. There are a few common approaches to achieve super-resolution images. Each method has its strengths and weaknesses, and the choice often depends on the specific requirements of the application. Neural network-based approaches, especially deep learning models, have gained popularity due to their ability to learn complex mappings and generate high-quality super-resolved images.

2.1 Interpolation-Based Methods

Interpolation is a mathematical and computational technique used to estimate values that fall between known values. In the context of super-resolution images, interpolation is often employed to estimate pixel values at inserted pixel coordinates based on the values of neighboring pixels at lower-resolution image pixel coordinates. One of the most common interpolation methods is the bicubic interpolation that involves fitting a cubic polynomial to a set of neighboring pixels and using this polynomial to estimate pixel values at higher resolutions. Bicubic interpolation is relatively computationally more expensive than simpler methods like bilinear interpolation; however it generally produces smoother and higher-quality results [10–12]. Another simple interpolation method is bilinear method that computes the new super-resolution image pixel value as a weighted average of the surrounding pixels in a 2×2 or more neighborhood. While computationally less intensive than bicubic interpolation, bilinear interpolation may result in smoother but less detailed enhanced images [12, 13]. Nearest-neighbor interpolation is another simple method where each pixel in the super-resolution image is assigned with the value of the nearest pixel in the low-resolution image. Nearest-neighbor interpolation is computationally efficient but tends to produce blocky and less visually appealing results [14, 15].

2.2 Frequency Domain Methods

Frequency domain methods involve transforming the image into the frequency domain, manipulating the high-frequency components using different methods, and then transforming it back to the spatial domain. One of these methods includes interpolation in Fourier domain in which several interpolation techniques are applied to the image directly in the frequency domain to estimate the pixel values at new frequencies [16]. Another method is sub-pixel registration where multiple low-resolution images taken at slightly different positions are aligned in the frequency domain which results in enhanced high-frequency details [17]. Phase information of the frequency domain image could also be combined with amplitude information to reconstruct a high-resolution image. The use of prior information for the expected frequency content of super-resolution image can also be incorporated in a Bayesian framework (i.e., using prior distributions of the image regions) to guide the super-resolution process. Multi-scale analysis in the frequency domain methods is also used for super-resolution images [18, 19]. The process involves decomposing the low-resolution image into different frequency bands, enhancing the resolution at each scale, and then combining the results. This approach is often used in conjunction with wavelet transforms [20]. Deep learning techniques have also been applied to super-resolution in the frequency domain [21, 22]. Convolutional neural networks (CNNs) can learn complex mappings from low-frequency information to high-frequency details. While they can be effective in certain scenarios, they are often used in combination with spatial domain methods or deep learning-based approaches to achieve better overall results.

2.3 Example-Based Methods

Example-based learning methods often yield high predictive accuracy, especially when trained on large and diverse datasets. They learn patterns from examples and can make accurate predictions for new, unseen data. Most of these methods are versatile and applicable to various types of data, from images and text to numerical and categorical data [23–27]. As long as labeled examples are available, one can train a model to learn the underlying patterns. In many cases, example-based models are interpretable. It is easy to understand how the model makes predictions by examining the importance of features or the coefficients associated with them. This interpretability is crucial in applications where understanding the decision-making process is essential. These methods can easily incorporate new examples to adapt to changing data patterns. This makes them suitable for scenarios where the underlying patterns evolve over time. Example-based methods excel in classification tasks, where the goal is to categorize input data into predefined classes or labels. This makes them widely used in image processing applications.

2.4 Neural Networks and Deep Learning-Based Methods

Neural networks and deep learning-based image super-resolution is a technique to enhance the resolution and quality of images. Several comprehensive surveys can be found in [28–31]. The goal is to generate a higher-resolution version of a given low-resolution image while preserving important details. A large dataset of high-resolution images paired with their low-resolution counterparts is used for training the model. The low-resolution images are typically created by downscaling the high-resolution images. Convolutional neural networks (CNNs) are commonly used for image super-resolution [32–34]. The network is trained to learn the mapping between low-resolution and high-resolution image pairs. The architecture may include layers like convolutional layers, activation functions, and upsampling layers. During training, a loss function is defined to measure the difference between the generated high-resolution image and the actual high-resolution image. Common loss functions include mean squared error (MSE) or perceptual loss, which takes into account perceptual differences. The neural network is trained on the prepared dataset using backpropagation and optimization techniques. The goal is for the network to learn the underlying patterns in the data that allow it to effectively upscale low-resolution images.

2.5 *Image Feature-Based Methods*

Image feature-based super-resolution is another approach to enhancing the resolution of images; however it differs from deep learning-based methods. Instead of relying on neural networks, feature-based super-resolution focuses on extracting and manipulating specific features within an image to achieve higher resolution [35, 36]. Feature-based super-resolution involves identifying and extracting relevant features from the low-resolution image. Features can include edges, textures, and other distinctive patterns [37, 38]. Once features are extracted, interpolation techniques are often used to estimate the high-resolution values of these features. Additionally, registration methods may be employed to align corresponding features between the low-resolution and high-resolution images. Mathematical mappings or transformations are applied to the extracted features to upscale and enhance their resolution. These transformations aim to preserve the structural details present in the original image. The high-resolution features obtained from the previous steps are combined to form the final high-resolution image. This combination process may involve blending or merging the enhanced features in a way that maintains coherence and realism. Some feature-based super-resolution methods incorporate post-processing techniques to further refine the output and reduce artifacts. One common technique used in feature-based super-resolution is sparse representation, where the low-resolution image is represented as a linear combination of high-resolution basis functions. This representation is then used to estimate the high-resolution image. While deep learning-based super-resolution methods have gained popularity due to their ability to learn complex mappings, feature-based approaches are often used in scenarios where computational efficiency and interpretability are crucial.

3 Multi-image, Interpolation-Based and Pan-sharpening for Hyperspectral Super-resolution

Hyperspectral images contain high spectral resolution that captures information across a wide range of spectral bands often tens to hundreds. Each band represents a specific range of wavelengths across the electromagnetic spectrum. This allows for detailed analysis of the spectral signatures of materials. They can be used in tasks that require detailed spectral information, such as environmental monitoring, agriculture, mineralogy, and target detection. Spatial resolution can vary, but it might not be as high as panchromatic images. The emphasis is on capturing a broad range of spectral information. They tend to be data-heavy due to the large number of spectral bands, requiring more storage and processing resources. On the other hand, panchromatic images contain a single spectral band, typically capturing grayscale information. They lack the detailed spectral information found in hyperspectral images. They consist of a single band, capturing intensity values without distinguishing between different wavelengths. Panchromatic images are commonly used for general visual interpretation and mapping and often combined with multispectral data to enhance spatial resolution. Typically they have higher spatial resolution, capturing fine details in terms of shape and structure.

Pan-sharpening is a process of enhancing the spatial resolution of a multispectral (MS) image by fusing it with a higher-resolution panchromatic image. A comprehensive review for pan-sharpening methods can be found in [39, 40]. The goal is to create a high-resolution color image that retains both the spectral characteristics of the multispectral data and the fine spatial details from the panchromatic data. One of the most common methods in this context is intensity-hue-saturation (IHS) method where high-resolution panchromatic image and low-resolution multispectral images are combined using IHS color space to reconstruct super-resolution image. Another method is principal component analysis (PCA) where PCA is used to find the principal components of panchromatic and multispectral images. The first few principal components, which contain most of the variance, are sharpened using the panchromatic image, and the result is transformed back to get high-resolution image. In our framework, we employ two pan-sharpening methods including IHS and PCA that are commonly used for hyperspectral image super-resolution. The details for these two methods are given in the following sections.

3.1 Hyperspectral Intensity-Hue-Saturation (IHS)

The intensity-hue-saturation (IHS) method is one of the computationally efficient and popular techniques used in remote sensing, particularly for fusing high-resolution panchromatic images with lower-resolution multispectral images. Due to its simplicity, it can process a large volume of data relatively in a short time [41]. A similar approach can be considered for hyperspectral super-resolution by decomposing the hyperspectral data into different components. The IHS effectively transforms multispectral image from Red-Green-Blue (RGB) color domain into spatial (I) and spectral (H, S) information into three components:

- Intensity (I): Represents the grayscale or brightness information of the image
- Hue (H): Represents the color information
- Saturation (S): Represents the richness of the color

The main steps of the method can be given as follows:

- Intensity: Extracting high-resolution intensity information from a hyperspectral image. This could involve using a higher-resolution band or applying a super-resolution technique specific to intensity.
- Hue and saturation: Retaining the spectral information by keeping the original hyperspectral bands for hue and saturation. In most cases, it is needed to resample or enhance the spatial resolution of these bands to match the intensity component.

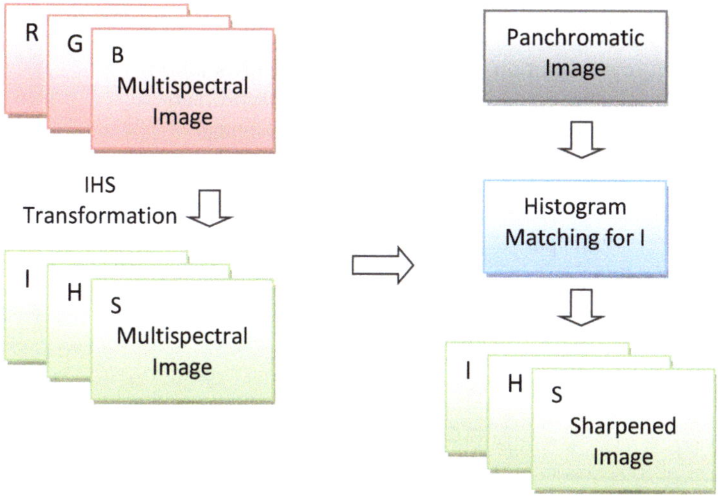

Fig. 1 Overview of IHS method for super-resolution of hyperspectral images

- Fusion: Combining the high-resolution intensity component with the spatially enhanced hue and saturation components. A histogram matching algorithm is performed for high-resolution image. The histogram matching is a process of transforming pixel values of the source image to obtain a specified histogram. After the process, the resulting image histogram matches the specified image histogram. Then, intensity (I) component of hyperspectral image is replaced with high-resolution image's intensity (I) component. Figure 1 demonstrates the overview of the method.
- Inverse transformation: Transforming the fused image back to the original hyperspectral space.

3.2 *Hyperspectral Principal Component Analysis (PCA)*

Principal component analysis (PCA) is a statistical technique used for dimensionality reduction and data compression. The main idea behind PCA is to transform a dataset into a new coordinate system, where the axes are the principal components of the data. These principal components are orthogonal (uncorrelated) and capture the maximum variance in the data. The eigenvectors represent the directions of maximum variance, and the eigenvalues indicate the magnitude of variance along those directions. The output of PCA is a transformed dataset where the new variables (principal components) are uncorrelated and capture the most significant information in the original data. PCA is widely used in various fields, including image processing, signal processing, and machine learning, to reduce the complexity of data while retaining essential features [42].

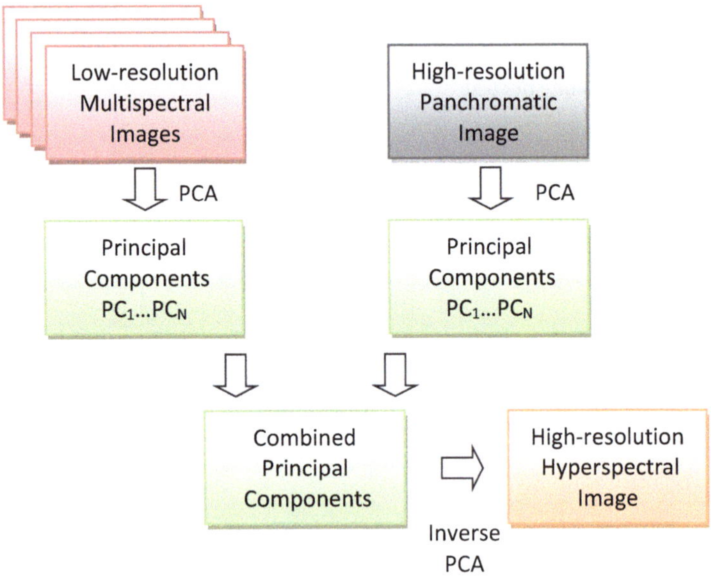

Fig. 2 Overview of PCA method for hyperspectral images

In the context of super-resolution, PCA can be applied to reduce the dimensionality of the image data while retaining its essential features. This reduction in dimensionality can help in efficient processing and reconstruction of high-resolution images [43, 44]. The idea is to represent the image using its principal components, which are the key patterns or features in the data. By keeping the most significant components and discarding the less important ones, one can achieve compression and denoising. Then, through the super-resolution process, one can reconstruct a higher-resolution image using the retained information. In hyperspectral-PCA, the consecutive bands of hyperspectral images are highly correlated, and usually they carry almost the same information. The transformation is used to remove the correlation among the neighboring bands. It is based on the mathematical principle known as eigenvalue decomposition of the covariance matrix of hyperspectral image bands. Initially, hyperspectral bands are transformed into their principal components (PCs). The most important one is the first principal component (PC_1). PC_1 contains most of the spatial data variance between all the bands. Figure 2 shows the overview of the method. In this method, initially the principle components of hyperspectral image and high-resolution image are obtained. Then, hyperspectral image's most informative component PC_1 is replaced by high-resolution image's most informative component PC_1. Finally, backward transformation is performed to obtain new high-resolution hyperspectral image.

4 Example-Based Hyperspectral Super-resolution

Example-based methods can produce high-quality and visually pleasing results by learning patterns from a dataset of high-resolution and corresponding low-resolution image pairs. The model understands how to generate fine details from low-resolution inputs. These methods often give good results at preserving the structural information in the image. The learned models understand the contextual relationships between different image elements and enhance details while maintaining overall structure. Example-based super-resolution methods are adaptable to different types of images, including natural scenes, medical images, satellite imagery, and more. As long as a diverse dataset is available for training, the model can learn to enhance various visual features. The models can capture and generate realistic textures and details, making the enhanced images visually appealing. Once trained, example-based super-resolution models can generate high-resolution images efficiently. This is valuable for real-time applications, such as video upscaling or enhancing images for real-time systems. Example-based methods, when well-designed, can minimize artifacts in the enhanced images. Artifacts, such as blurring or distortion, are undesirable in super-resolution tasks, and these methods attempt to reduce or eliminate them. Trained models often generalize well to unseen data, allowing for the enhancement of new images not seen during training. This transferability is crucial for the practical application of super-resolution methods. Example-based learning methods for image super-resolution offer high-quality enhancement, preservation of structural information, adaptability, realism, computational efficiency, and the ability to learn both global and local features, making them valuable in numerous visual processing applications. In our framework, we employed dictionary learning methods for sparse representation in hyperspectral super-resolution image context. The details are given in the following sections.

4.1 Dictionary Learning for Sparse Representation

Dictionary learning methods, particularly in the context of sparse coding and signal representation, offer several advantages. The main idea is to represent data using a sparse linear combination of dictionary elements. This leads to compact and efficient representations, especially in situations where the data has sparse features or can be well-represented using a small subset of dictionary elements. Dictionary learning methods adapt to the characteristics of the input data. The dictionary is learned directly from the data, allowing it to capture specific patterns and structures present in the dataset. Dictionaries in these methods can be overcomplete, meaning they have more elements than the dimensionality of the input data. This overcompleteness can capture a rich set of features and variations in the data, enabling more expressive representations. Sparse coding using learned dictionaries can be robust to noise. By representing data sparsely, the impact of noise is minimized, and the underlying structure of the signal can be preserved. Dictionary learning can be performed in an unsupervised manner, where the dictionary is learned directly from the input data without the need for labeled examples. This makes it suitable for scenarios where labeled data is scarce or unavailable. By learning an appropriate dictionary, these methods effectively perform dimensionality reduction. The data can be represented in a lower-dimensional space using a sparse set of coefficients, making it computationally more efficient and easier to analyze. The learned dictionary serves as a basis for feature extraction. The atoms in the dictionary represent meaningful features, and the sparse coefficients indicate the presence and strength of these features in the input data. The learned dictionary can provide insights into the underlying structure of the data. Each atom in the dictionary corresponds to a distinctive feature, and analyzing these atoms can reveal important characteristics of the data. The learned and adaptive dictionaries play important role on the representation performance. Current methods for obtaining adaptive dictionaries usually involve optimization in terms of task to be performed such as representation or classification. Such adaptive methods work by iteratively alternating between sparse coding the input data given current dictionary and updating atoms in the dictionary to better fit the data. In our framework, K-singular value decomposition (K-SVD), online dictionary learning (ODL), and Bayesian dictionary learning algorithms are employed to build adaptive dictionaries in this study. G-SOMP+ [45] algorithm is also used for sparse approximation. It is a generalization of simultaneous orthogonal matching pursuit algorithm. It imposes non-negativity constraint and takes advantage of spatial structure of the scene. The details are given in the following sections.

4.2 K-SVD Dictionary Learning-Based Sparse Representation

K-SVD is designed to find a dictionary that allows sparse representation of input data. Sparse coding involves representing data as a linear combination of a few dictionary elements with sparse coefficients. The K-SVD algorithm iteratively updates the dictionary and the sparse codes to minimize the reconstruction error. It alternates between fixing the sparse codes and updating the dictionary in a way that improves the overall representation. K-SVD adapts the dictionary to the specific characteristics of the input data. It learns a set of atoms (dictionary elements) that are most effective in representing the training data in a sparse manner. K-SVD allows the creation of overcomplete dictionaries, meaning the number of atoms in the dictionary can be greater than the dimensionality of the input data. This overcompleteness contributes to capturing a rich set of features and variations. The iterative nature of K-SVD helps in making the learned dictionary and sparse codes robust to noise. It can effectively handle noisy data and produce sparse representations that emphasize the underlying structure. K-SVD is often employed in an unsupervised learning setting, where the dictionary is learned directly from the input data without the need for labeled examples. This makes it versatile and applicable to various domains. The atoms in the learned dictionary can be interpreted as meaningful features present in the data. Analyzing these atoms provides insights into the characteristics and structure of the input data.

Using K-SVD-based dictionary learning for super-resolution image tasks involves training a dictionary on pairs of low-resolution and high-resolution image patches. In the context of super-resolution, the low-resolution and high-resolution image patches are used as training pairs. The goal is to learn a dictionary that captures the relationship between these patches, allowing for the enhancement of low-resolution images. The learned dictionary is used to sparsely represent the low-resolution image patches. This means that each patch is represented as a linear combination of a few dictionary elements with sparse coefficients. The K-SVD algorithm iteratively updates the dictionary and the sparse codes to minimize the reconstruction error. During each iteration, the dictionary atoms are updated to better represent the characteristics of the training patches. To minimize the reconstruction error, orthogonal matching pursuit (OMP) or basic pursuit algorithms could be used. In our framework, G-SOMP+ derived from OMP is used. K-SVD adapts the dictionary to the specific structures and features present in the image patches. This adaptability is crucial for capturing the relevant details and textures that contribute to high-resolution features. The dictionary can be overcomplete, allowing it to capture a rich set of features and variations in the image patches. This overcompleteness contributes to the ability to represent diverse high-resolution details. K-SVD-based dictionary learning is applied to enhance the resolution of low-resolution images. By using the learned sparse representation, the algorithm can generate high-resolution patches from their low-resolution counterparts. The iterative nature of K-SVD and the sparse representation can help reduce artifacts in the super-resolved images. It aims to generate visually pleasing and realistic high-resolution results. K-SVD can handle noisy input images by learning a dictionary that emphasizes the underlying structures while minimizing the impact of noise. This is particularly beneficial in real-world scenarios where images may have noise. Once trained on a dataset of low- and high-resolution patches, the learned dictionary can be applied to new, unseen low-resolution images for super-resolution. This transferability is essential for practical applications (Fig. 3).

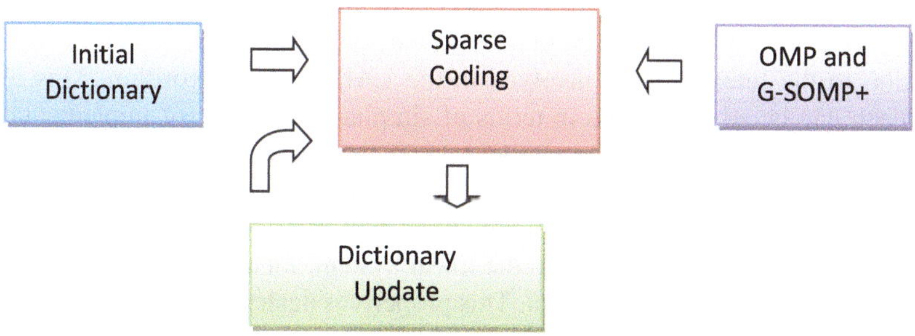

Fig. 3 Overview of K-SVD

4.3 ODL-Based Sparse Representation

Online dictionary learning (ODL) is an optimization algorithm based on stochastic approximations [46]. It can be used with large datasets of training samples. It has been shown that ODL can achieve faster performance and better dictionaries. Online learning allows the algorithm to adapt its dictionary in real time as new data becomes available. This is especially beneficial when dealing with dynamic or changing datasets. The adaptability of online learning can lead to dictionaries that generalize well to various types of data. This flexibility is crucial in scenarios where the input data distribution may change over time. Online learning, by updating the dictionary incrementally, can help prevent overfitting to specific patterns in the data. It allows the algorithm to maintain a more generalized dictionary that captures essential features without getting too specialized. The online nature of the learning process can make it more scalable to large datasets or continuous streams of data. Traditional batch learning approaches may struggle with the sheer volume of information, whereas online learning can handle it more efficiently.

The algorithm consists of two main components including sparse coding and dictionary update similar to K-SVD and others. Sparse coding aims to represent data using a small set of non-zero coefficients, leading to a more efficient representation. This can be advantageous in terms of storage, processing, and memory requirements. The sparse coding problem with fixed dictionary is a $\ell 1$-regularized linear least-squares problem. To solve this problem, for all possible values of regularization parameter, a Cholesky-based implementation of the LARS-Lasso algorithm [47] is used. It provides faster solution with higher accuracy. On the other hand, the dictionary update uses block-coordinate descent with warm restarts. It benefits from being parameter-free and does not require any learning and rate-tuning process. Cholesky decomposition allows the LARS-Lasso algorithm to factorize the data matrix efficiently. This can result in faster computation compared to methods that involve iterative optimization techniques. Cholesky decomposition is numerically stable when solving linear systems. This stability is beneficial in the context of the LARS-Lasso algorithm, as it helps avoid numerical issues that can arise in ill-conditioned matrices. Cholesky-based implementations often have lower memory requirements compared to some iterative optimization methods. This can be advantageous, especially when dealing with large datasets where memory efficiency is crucial. Cholesky decomposition provides a direct solution to the linear system involved in the LARS-Lasso algorithm. This direct approach can be advantageous in terms of simplicity and ease of implementation. Cholesky decomposition can be parallelized efficiently, taking advantage of parallel processing architectures. This can lead to further speedup when implementing the LARS-Lasso algorithm on parallel computing platforms. Cholesky decomposition provides a deterministic solution to the linear system, meaning that the same input will always produce the same output. This property is desirable in many applications for reproducibility.

4.4 Bayesian Sparse Representation

Bayesian statistics provides a framework for reasoning about uncertainty. Instead of providing a single point estimate, Bayesian methods model the entire probability distribution of possible parameter values. This is particularly useful when dealing with limited data or when there is uncertainty in the model parameters. In the context of sparse representation, Bayesian methods can be applied to model the uncertainty associated with the sparse coefficients. Instead of obtaining a deterministic solution, Bayesian sparse representation provides a probabilistic representation of the sparse coefficients, taking into account uncertainty and variability in the data. The Bayesian sparse representation problem can be formulated as a probabilistic model. This model incorporates a prior distribution on the sparse coefficients, representing prior beliefs about their distribution, and combines it with a likelihood function that captures the relationship between the observed data and the sparse coefficients. Bayesian inference is used to update the prior distribution based on the observed data, resulting in a posterior distribution over the sparse coefficients. This posterior distribution reflects the updated beliefs about the coefficients given the observed data. Bayesian sparse representation naturally incorporates regularization through the choice of prior distributions. The hyperparameters of these priors can be tuned to control the sparsity and encourage certain characteristics in the sparse coefficients. Bayesian sparse representation allows for model selection by considering different sparse representations and evaluating their probabilities given the observed data. This can help in choosing the most appropriate model complexity for a given task.

Bayesian sparse representation for hyperspectral image super-resolution fuses a low-resolution hyperspectral image with a high-resolution image in a multi-stage process [48]. A formulation is performed using a Bayesian model that describes the relationship between the low-resolution and high-resolution images. This involves defining a likelihood function that captures the observed low-resolution image given the high-resolution image and introducing prior distributions that encode sparsity assumptions. A utilization of sparse coding techniques is also performed to represent the high-resolution image in a sparse manner. This involves finding a set of coefficients that best capture the essential features of the image. Bayesian sparse coding introduces a probabilistic framework for estimating these coefficients [49]. Definition of prior distributions is needed on the sparse coefficients. These priors express the beliefs or assumptions about the characteristics of the high-resolution image, encouraging sparsity and guiding the reconstruction process. Applying Bayesian inference is also required to update the prior distributions based on the observed low-resolution image. The goal is to obtain the posterior distribution over the sparse coefficients, incorporating information from both the observed data and the prior beliefs. Sampling from the posterior distribution is then preformed using methods like Markov Chain Monte Carlo (MCMC), or optimization of the posterior distribution is performed to find the most likely sparse coefficients given the observed data. Finally, reconstruction of the high-resolution image is performed using the estimated sparse coefficients and the dictionary of basis functions. This involves multiplying the sparse coefficients by the basis functions and summing them up to generate the high-resolution image.

5 Deep Learning for Hyperspectral Super-resolution

Deep learning-based methods have been extensively used in creating super-resolution images from low-resolution images. The most common methods are super-resolution convolutional neural network (SRCNN), enhanced deep super-resolution (EDSR), and super-resolution generative adversarial network (SRGAN). SRCNN utilizes convolutional neural networks (CNNs) to directly learn the mapping between low-resolution and high-resolution image patches. Its advantage is being simple and efficient. EDSR takes the concept further with an even deeper architecture, optimizing the reconstruction process for enhanced performance. Higher depth allows it to capture more complex features and details, making it more powerful in terms of super-resolution quality. SRGAN introduces the magic of generative adversarial networks (GANs) to the super-resolution game. It uses a generator and a discriminator in a competitive manner. In addition to more details, it also creates more realistic textures. SRGANs bring a touch of artistry to the super-resolution by introducing adversarial training.

In our analysis framework, we used the most simple and efficient method of CNN in super-resolution context, also known as super-resolution convolutional neural network (SRCNN) [50]. It was one of the early deep learning approaches to single-image super-resolution and demonstrated significant improvements over traditional multi-image methods. One of the key advantages of SRCNN is its ability to learn end-to-end mappings from low-resolution to high-resolution images, without the need for manual/handcrafted features or intermediate processing steps. This makes it highly effective and efficient for super-resolution tasks. Even though it is a deep neural network, SRCNN is computationally efficient compared to some other super-resolution methods as it has been demonstrated in this paper. It achieves high-quality results with relatively low computational overhead, making it practical for real-time or large-scale image processing applications.

The low-resolution image is divided into smaller patches. A dataset of corresponding high-resolution and low-resolution patch pairs is created. The network needs several pairs to learn. The training process consists of training the network using the high- and low-resolution patch pairs. The network learns the mapping between low- and high-resolution features during this phase. A CNN architecture is then designed for the network. A series of convolutional layers for feature extraction is required. These layers capture and highlight important patterns in the image. During training, the network is optimized to minimize the difference between the predicted high-resolution image and the ground-truth high-resolution image. This is typically done using a loss function such as mean squared error (MSE) or mean absolute error (MAE). The final layer produces the high-resolution image. The overview of the method is given in Fig. 4 where given a low-resolution image,

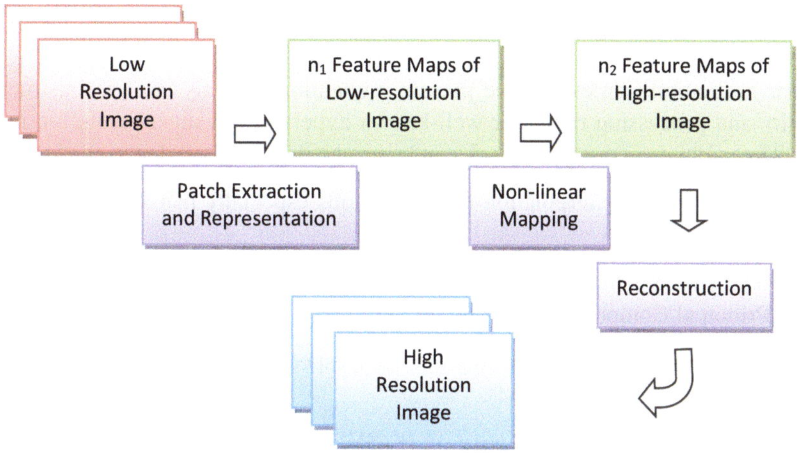

Fig. 4 Overview of CNN deep learning-based super-resolution method

the first convolutional layer extracts a set of feature maps. The second layer maps these feature maps nonlinearly to high-resolution patch representations. The last layer incorporates the estimations within a spatial region to reconstruct the high-resolution image.

In more improved networks, residual learning is a technique used in deep neural networks to address the problem of vanishing gradients and improve the training of very deep networks [51]. In traditional deep neural networks, as the network depth increases, it becomes increasingly difficult to train the network effectively. One of the main challenges is the vanishing gradient problem, where gradients become extremely small as they propagate backward through many layers during training. This can lead to slower convergence and difficulties in learning meaningful representations, especially in very deep networks. Residual learning is introduced with residual blocks. Each residual block contains skip connections, allowing the network to learn the residual (difference) between the low- and high-resolution images. This helps in mitigating the vanishing/exploding gradient problem. Instead of directly learning complex mappings from inputs to outputs, residual networks focus on learning the residuals and the difference between the desired output and the input. This can simplify the learning task, as the network only needs to learn to adjust or refine the input features rather than reconstruct them from scratch. With residual connections, it becomes feasible to design and train extremely deep neural networks. This has led to the development of state-of-the-art architectures with hundreds or even thousands of layers, such as ResNet, which have achieved remarkable performance on various tasks including image classification, object detection, and segmentation.

6 Experimental Results

In our evaluation framework, we present a quantitative analysis of reconstruction quality and processing times for well-known hyperspectral super-resolution methods. The methods in our framework can be grouped into three categories as follows:

1. Multi-image and interpolation-based methods especially pan-sharpening methods:
   - Intensity Hue Saturation (IHS)
   - Principal Component Analysis (PCA)
2. Example-based machine learning methods, especially, dictionary learning-based sparse representation methods:
   - K-Singular Value Decomposition (K-SVD)
   - Online Dictionary Learning (ODL)
   - Bayesian Dictionary Learning
3. Deep learning-based methods:
   - Convolutional Neural Network (CNN)

The main reason for the selection of these groups and methods is the popularity, simplicity, and output quality.

In our analysis framework, two hyperspectral image databases are used. The first one is collected by Airborne Visible/Infrared Imaging Spectrometer (AVIRIS) over San Diego, CA. The spatial resolution is 3.5 m per pixel. It has 224 spectral bands in the wavelengths ranging from 370 to 2510 nm. Figure 5 shows two example images from the AVIRIS dataset. The second database is the CAVE database [52]

Fig. 5 Example images from AVIRIS dataset collected over San Diego airport. Left: RGB image. Right: Hyperspectral image

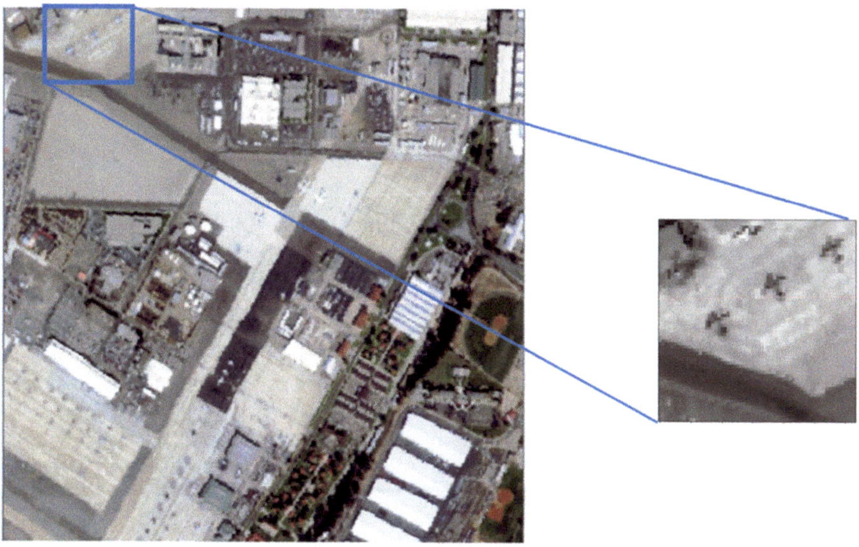

Fig. 6 An example of an image region used in the CNN deep learning method

containing 32 different image sets. Each image set is in the spatial dimension of 512×512 pixels and acquired at a wavelength interval of 10 nm in the range of 400–700 nm. The SPAMS framework [46] is used in the dictionary learning stage of sparse representation for the ODL method. The pan-sharpening framework [53] is also modified by us for hyperspectral images and used in the evaluation framework.

The CNN model takes both high-resolution and low-resolution hyperspectral images as input. An initial step involves using a transpose convolution layer to upsample the low-resolution input, effectively increasing its size by a factor of 8. Subsequently, the outputs of the input layers are concatenated. This mixture includes the image obtained by upsampling the low-resolution image and the high-resolution image. The concatenated image undergoes processing through three consecutive convolution layers. Each convolution layer is utilized for extracting feature maps and is coupled with leaky rectified linear unit (ReLU) activation functions. These steps represent a feature extraction process aimed at obtaining learnable features. The output layer of the model is responsible for predicting the high-resolution hyperspectral image. Transpose convolution layers are employed to transform the low-resolution input into a higher-resolution output. This output is trained by comparing it to the actual high-resolution hyperspectral image. The learning parameters of the model are specified, and the training process is initiated using training data. The model is trained for a defined 10 epochs and 34 batches, and at the end of the training, it is evaluated on test data. Figure 6 shows an image region used in CNN model. Figure 7 shows downsampled image, low-resolution, high-resolution, and reconstructed images.

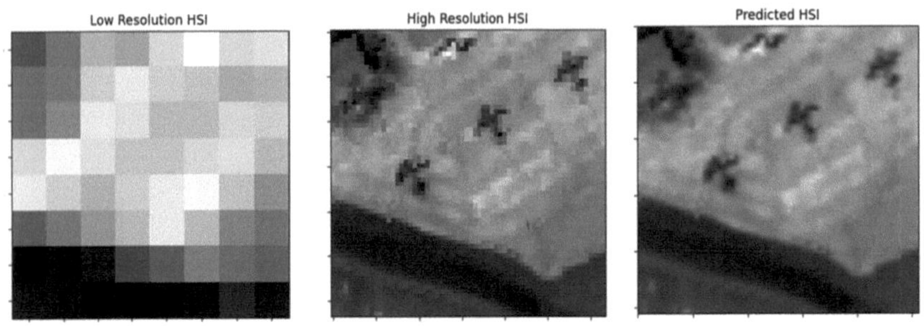

Fig. 7 Resulting hyperspectral images using CNN deep learning method. Left: Downsampled version of the image (low resolution image). Middle: High-resolution image. Right: Predicted super-resolution image

Table 1 RMSE values and processing times of dictionary-based sparse representation methods and CNN deep learning method using San Diego Airport AVIRIS hyperspectral image database

Method	RMSE	Processing time (s)
K-SVD	2.230	52
ODL	1.77	41
Bayesian	2.7024	55
CNN	0.0519	0.2311

Table 1 shows that the CNN deep learning-based method has the lowest RMSE reconstruction value and processing time (excluding training time) among all the dictionary learning-based sparse representation methods. The RMSE value is measured between the original high-resolution hyperspectral images and the reconstructed super-resolution image. RMSE value is based on the average difference between the bands of reference images and the resulting image. The ideal value of RMSE is 0 that means the super-resolution image is reconstructed without any error. The formulation for RMSE value is given as follows:

$$\mathrm{RMSE} = \sqrt{\frac{\sum \left\| I - \hat{I} \right\|^2}{L \times M \times N}}$$

where I is the original ground truth image and $\hat{I}$ is the estimated super-resolution image as matrices in $R^{L \times M \times N}$. The images are in the spatial dimension of $M \times N$ and spectral dimension of L.

All of the experiments are conducted using a laptop computer with the following specifications: Intel® Core™ i7-10750H CPU, 2.60 GHz, and 12 GB RAM. The CNN deep learning algorithm runs on NVIDIA GTX1650 GPU, and the training time is not included in the result table.

CNNs offer several advantages in the context of image super-resolution. CNNs are proficient at learning hierarchical features from images. In the case of super-resolution, they can automatically learn and extract low-level to high-level features, capturing intricate details that contribute to enhancing image resolution. CNNs use shared weights and biases across different parts of the image, which helps in preserving spatial hierarchies and patterns. This parameter sharing is particularly beneficial in image super-resolution tasks as it allows the network to generalize well to different image regions. CNNs utilize local receptive fields to capture local dependencies in an image. This is crucial for super-resolution, where understanding the context and dependencies of neighboring pixels is essential for generating realistic high-resolution images. The nonlinear activation functions used in CNNs, such as rectified linear unit (ReLU), introduce nonlinearity into the model, enabling it to learn complex mappings from low-resolution to high-resolution features. This helps in capturing nonlinear relationships in the image data. CNNs facilitate end-to-end learning, allowing the model to directly learn the mapping between low-resolution and high-resolution images. This end-to-end approach simplifies the training process and enables the network to optimize its parameters for the specific super-resolution task. Pre-trained CNN models on large datasets can be fine-tuned for specific super-resolution tasks. Transfer learning leverages the knowledge gained from one task to improve performance on another related task, leading to more efficient and effective training. CNNs can benefit from data augmentation techniques, which involve applying transformations (e.g., rotations, flips) to the training data. This helps the model generalize better and handle variations in the input data, contributing to improved super-resolution performance. CNN operations can be parallelized efficiently, making them suitable for hardware acceleration (e.g., GPUs). This parallel processing capability accelerates training and inference times, making CNNs practical for real-time or near-real-time super-resolution applications.

In our evaluation framework, IHS and PCA are also used for further analysis of hyperspectral super-resolution images. IHS method is able to process large volume of image data relatively in a short time. IHS separates the intensity (luminance) component from the color information (hue and saturation). This separation can be beneficial for super-resolution, as it allows for independent processing of luminance and color components. By focusing on the luminance information, fine features and structures can be brought out more clearly, contributing to an improved perception of sharpness. The reconstructed image is improved in terms of spatial resolution; however, the method causes spectral distortion. Since IHS separates intensity from color, the color information (hue and saturation) remains largely unaffected during the super-resolution process applied to the intensity component. This helps in preserving the original color appearance of the image. Processing only the intensity component helps maintain color consistency in the final output. The color information remains largely unchanged, avoiding potential artifacts or distortions that could arise if color components were processed independently. By processing the intensity component separately from the color components, computational efficiency can be improved. Different algorithms and techniques can be applied to the intensity component without the need to perform complex operations on the entire color space.

The PCA method is specifically used to remove highly correlated data in the consecutive bands of hyperspectral images that carry almost the same information. As a result, PCA helps to reduce the dimensionality of the data by transforming the original image into a set of principal components. This reduction in dimensionality can make subsequent processing more computationally efficient. PCA can be effective in denoising images by capturing the most important features and suppressing noise. This is particularly beneficial in super-resolution, where the goal is often to enhance fine details that may be obscured by noise in lower-resolution images. PCA extracts the most significant features of the data. In the context of super-resolution, these features may correspond to important structures and details in the image. By focusing on these principal components, PCA can contribute to the enhancement of relevant image information. PCA decorrelates the data by transforming it into uncorrelated components. This can be advantageous in super-resolution, where correlated pixel values in the original image may hinder the extraction of high-frequency details. PCA can help in decorrelating the data and highlighting important information. The eigenvalues associated with the principal components provide a measure of the importance of each component. This ranking can guide the selection of components to include in the reconstruction process, allowing for a more tailored and efficient super-resolution approach.

In addition to IHS and PCA, the example-based machine learning methods with dictionary learning and sparse-based representations (K-SVD, ODL, and Bayesian) are also used in our framework for hyperspectral super-resolution. It is obvious that these methods require more computation because of their learning-based structure. There is a trade-off between computation and output quality of reconstruction. In these methods, a sparse combination of basic elements (atoms) is obtained from a dictionary. Since the dictionary is learned from training data, these methods in this group provide more accurate reconstruction than IHS- and PCA- based methods. An extensive visual comparison using CAVE database is given in Fig. 8.

Figure 8 shows the resulting super-resolution hyperspectral images for "Balloons" dataset in CAVE database. It can be easily seen from the images in Fig. 8 that the reconstruction quality is relatively poor for IHS and PCA methods. The main reason for relatively poor result in IHS transformation is that the panchromatic image is a histogram-matched image. It is transformed by a constant value so that the resulting image exhibits similar mean and variance with intensity component I (Intensity) in IHS. On the other hand, because the histogram-matched panchromatic image and the intensity component I in IHS do not usually have the same local mean, a large spectral distortion can be seen as color changes. The reason for this negative effect is that the spectral response of the I component in IHS (obtained using multispectral bands via transformation) can be quite different from that of panchromatic image. This difference causes local spatial distortions and space-varying radiance offsets [54].

K-SVD is known for its ability to provide sparse representations of images. In the context of super-resolution, this sparsity can be beneficial for efficiently representing high-frequency details and fine structures in the image. K-SVD learns an overcomplete dictionary from the input data. In the context of super-resolution,

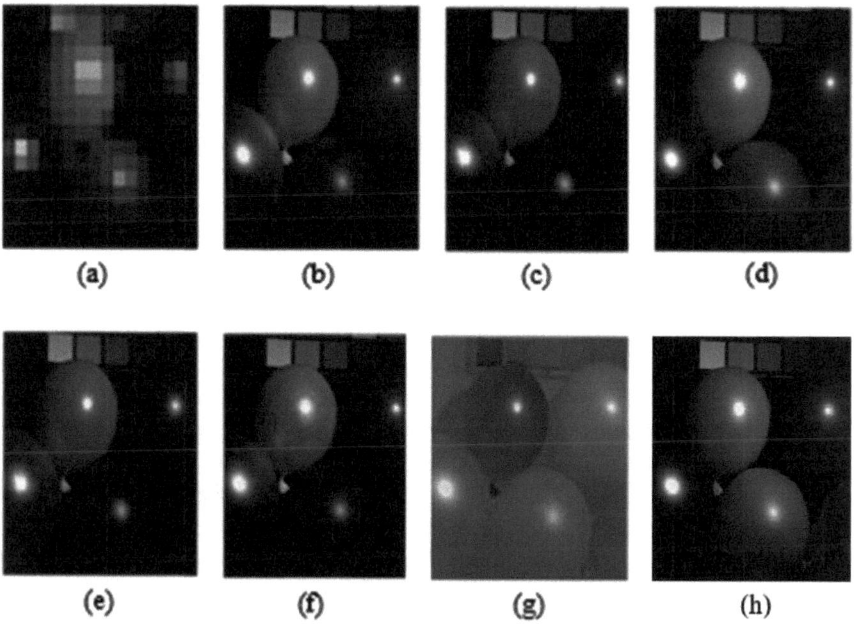

Fig. 8 Resulting super-resolution hyperspectral images for Balloons dataset in CAVE database. (**a**) Low-resolution hyperspectral image, (**b**) high-resolution multispectral image, (**c**) resulting image using K-SVD, (**d**) using ODL, (**e**) using Bayesian, (**f**) using IHS, (**g**) using PCA, (**h**) using CNN

this dictionary can capture essential features and patterns present in high-resolution images, enabling more accurate reconstruction of details during the super-resolution process. K-SVD adapts its dictionary to the specific characteristics of the input images. This adaptability is advantageous for super-resolution tasks, where different images may have varying levels of detail, textures, and structures. The learned dictionary can be more effective in representing the specific features of a given image. The overcomplete dictionary learned by K-SVD can efficiently handle redundant information in the image. This is important in super-resolution, where redundant or correlated information in lower-resolution images may need to be reconstructed with increased fidelity in the high-resolution output. K-SVD is capable of capturing and preserving the underlying structure of the input images. In super-resolution, this can contribute to maintaining the integrity of object boundaries, edges, and other important structural details. K-SVD can perform well even with a limited amount of training data. In scenarios where obtaining a large dataset for training is challenging, K-SVD's ability to learn a dictionary from a smaller set of examples can be advantageous. K-SVD can be extended to handle nonlinear transformations in image super-resolution. This flexibility makes it suitable for addressing complex relationships between low-resolution and high-resolution image patches. K-SVD is computationally efficient, and its iterative optimization process allows for scalable implementation. This makes it suitable for real-time or near-real-time applications, where efficiency is crucial.

ODL allows the dictionary to adapt to changing data distributions over time. In super-resolution, where the characteristics of images may vary, adapting the dictionary online can capture new patterns and structures that emerge in the data. Super-resolution tasks often involve large datasets, and ODL allows for incremental updates, making it computationally more efficient than batch dictionary learning. It enables continuous learning and updating without the need to process the entire dataset at once. The incremental nature of ODL makes it suitable for real-time or near-real-time applications. As new data arrives, the dictionary can be updated, allowing the super-resolution system to adapt dynamically to changing input. ODL can be more resource-efficient compared to batch learning, as it does not require storing and processing the entire dataset simultaneously. This can be particularly advantageous in scenarios where memory or computational resources are limited. Super-resolution images may exhibit variations in content, lighting conditions, or other factors. Online dictionary learning can adapt to these variations by updating the dictionary based on the most recent data, ensuring that the learned representations remain relevant and effective. ODL supports continual learning scenarios where the super-resolution system needs to adapt to new information without retraining on the entire dataset. This is valuable in applications where the system is deployed over an extended period, encountering diverse images over time. ODL can be less sensitive to the initial dictionary compared to batch learning methods. This is advantageous when the super-resolution system is applied to different types of images or when the initialization may not be optimal for all scenarios.

Bayesian dictionary learning approaches allow the incorporation of prior information about the data distribution. In super-resolution, this can be advantageous for modeling the statistical properties of high-resolution images, leading to more informed and context-aware super-resolution. Bayesian methods naturally handle uncertainty through probability distributions. This is beneficial in super-resolution, where uncertainties in the data, noise levels, or model parameters can be effectively modeled, leading to more robust and reliable results. Bayesian dictionary learning provides a natural way to introduce regularization through prior distributions. This regularization can help prevent overfitting and improve generalization, which is crucial in super-resolution tasks to avoid amplifying noise or artifacts. Bayesian methods allow for the estimation of hyperparameters from the data itself. This adaptability is advantageous in super-resolution, where the optimal hyperparameters may vary across different images or datasets. Bayesian model selection techniques can be employed to automatically select the appropriate complexity of the model. This is important in super-resolution, where selecting an appropriate dictionary size or complexity is critical for capturing relevant image structures without introducing unnecessary details. Bayesian methods provide a natural framework for modeling noise and outliers. In super-resolution, where low-resolution images may be affected by noise, outliers, or artifacts, Bayesian approaches can improve the robustness of the model by explicitly modeling these uncertainties. Bayesian dictionary learning

Table 2 RMSE values of IHS, PCA, dictionary-based sparse representations (K-SVD, ODL, and Bayesian), and CNN deep learning methods for CAVE hyperspectral image database

	RMSE value					
Image dataset	K-SVD	ODL	Bayesian	IHS	PCA	CNN
Balloons	1.53	1.44	2.1	9.1	11.1	0.16
Beads	3.66	3.26	5.4	5.8	9.8	0.18
CD	4.92	4.25	5.3	5.8	7.5	0.15
Cloth	2.44	2.11	4.0	5.6	6.2	0.15
Clay	1.68	1.54	2.7	2.9	9.3	0.13
Faces	1.23	1.07	1.9	4.1	12.3	0.16

enables the computation of probabilistic estimates for the high-resolution image. Instead of providing a single deterministic solution, Bayesian methods can produce probability distributions over possible high-resolution images, providing a more nuanced representation of uncertainty. Bayesian approaches facilitate joint inference of model parameters, including dictionary elements and regularization parameters. This joint inference can lead to more coherent and consistent results in super-resolution. Bayesian sparse coding can be incorporated into the dictionary learning process, allowing for more accurate and sparse representations of the data. This is particularly relevant in super-resolution, where the goal is often to capture and enhance fine details efficiently.

The quantitative analysis of all the methods described in earlier sections is given in Table 2 for CAVE hyperspectral image database that includes Balloons, Beads, CD, Cloth, Clay, and Faces image datasets.

The quality of the reconstructed super-resolution image depends on the accurate learning of dictionary in sparse representation. The parameter number of atoms in the dictionary is directly proportional to RMSE value between original and reconstructed super-resolution images. A generalization of simultaneous orthogonal matching pursuit (G-SOMP+) algorithm [45] is used for sparse approximation. In G-SOMP+ algorithm, it is reported that RMSE values are the lowest when the number of dictionary atoms selected in each iteration is chosen between 15 and 25. It is chosen as 20 in our experiments to process 8x8 image patches.

Table 2 shows that CNN deep learning method provides the best RMSE value resulting high-quality super-resolution images. CNN are commonly used and have shown success in hyperspectral image super-resolution, the choice of architecture, training strategy, and dataset characteristics also play significant roles in achieving optimal results. This method provides the advantages over other methods in terms of local and global feature learning using multiple bands in the hyperspectral context, parameter sharing through convolutional layers, translation-invariant features, spatial hierarchy learning, and nonlinearity. Table 2 also shows that the dictionary learning and sparse representation-based methods (K-SVD, ODL, and Bayesian) perform better in terms of RMSE value comparing to IHS and PCA. The lowest RMSE value is obtained using ODL among these two groups. As demonstrated with RMSE values, the ODL algorithm outperforms K-SVD and Bayesian dictionary learning algorithms in terms of creating a better dictionary and processing time. As a result, better dictionary provides better reconstruction of super-resolution hyperspectral images with smaller RMSE value.

The ODL algorithm performs better between the two groups as expected for the following reasons:

- It enables efficient learning of dictionaries for sparse coding.
- It uses LARS for the sparse coding step that provides faster solution with higher accuracy.
- It uses a block-coordinate approach for the dictionary update with better restart. It benefits from being parameter-free, it does not require any learning and rate tuning process.
- It solves a large-scale matrix factorization problem using efficient algorithms.
- It can handle potentially infinite datasets.
- It can adapt to dynamic training sets.
- It can be dramatically faster than batch algorithms.

The graphical demonstration of the results is also given in Fig. 9. For each method, the image datasets and their RMSE values are shown in different colors.

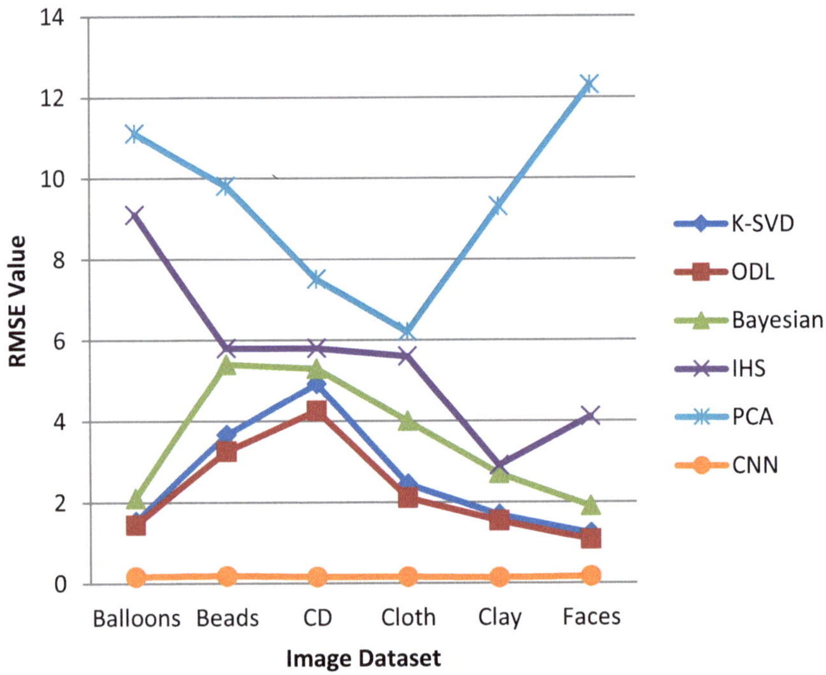

Fig. 9 RMSE values of the methods for the super-resolution reconstruction of hyperspectral image datasets in CAVE database

Table 3 The processing times of the methods for the super-resolution reconstruction of hyper-spectral image datasets in CAVE database

	Processing time (s)					
Image dataset	K-SVD	ODL	Bayesian	IHS	PCA	CNN
Balloons	26	21	31	7	7	0.15
Beads	27	22	31	6	7	0.07
CD	27	24	33	6	6	0.07
Cloth	28	22	32	5	7	0.08
Clay	23	23	33	6	6	0.08
Faces	24	22	35	8	9	0.17

Table 3 shows the performances of the methods in terms of processing times. Due to less complexity and simplicity of the algorithms in IHS and PCA, the processing times are shorter comparing to dictionary learning for sparse representations algorithms. The processing time for dictionary learning in sparse representation depends on several factors, including the size of the dataset, the dimensionality of the data, the sparsity level desired, and the specific algorithm used for dictionary learning. Dictionary learning is a computationally intensive task, and its processing time can vary based on these factors. The availability of hardware acceleration, such as graphics processing unit (GPU) support, can greatly speed up the processing time. Many deep learning libraries and frameworks provide GPU support, which can be leveraged for faster dictionary learning.

The processing time for CNN-based super-resolution depends on several factors including the architecture of the CNN; deeper networks with more parameters generally require more computation. GPUs are often used to accelerate the computations involved in deep learning tasks. Utilizing a high-performance GPU can significantly reduce processing times compared to using a CPU. The batch size, or the number of images processed simultaneously during each iteration, can impact processing time. Larger batch sizes may improve parallelism and GPU utilization, but they also require more memory. Various optimization techniques, such as model pruning, quantization, and model compression, can be applied to reduce the number of parameters and improve processing speed without sacrificing the performance. Some deep learning frameworks and libraries support parallel processing, allowing you to distribute computations across multiple devices or processors. This can significantly speed up the training and inference process. Given all these factors, CNN deep learning-based algorithm provides the best processing time as shown in the table.

The graphical demonstration of the processing times is also given in Fig. 10.

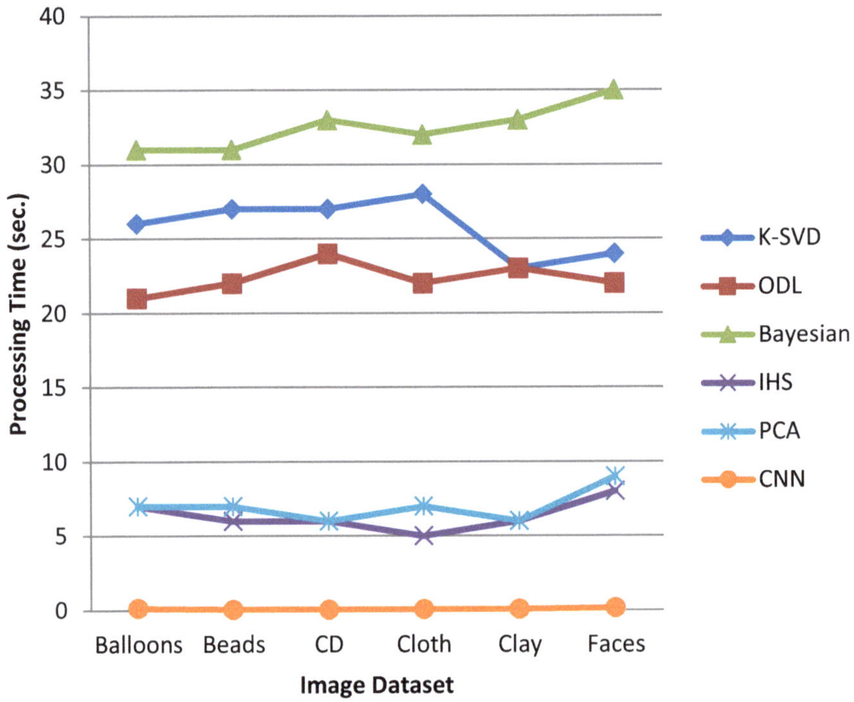

Fig. 10 Processing times of the methods for the super-resolution reconstruction of hyperspectral image datasets in CAVE database

7 Conclusion

We present an effectiveness analysis evaluation framework for super-resolution of hyperspectral images. Three different groups of methods are used for their simplicity, efficiency, and reconstruction quality. The first group includes multi-image and interpolation-based methods also known as pan-sharpening methods (IHS and PCA). Second group consists of example-based machine learning methods, especially, dictionary learning-based sparse representation methods (K-SVD, ODL, and Bayesian). Finally, the third group includes deep learning-based method (CNN). The quantitative analysis of the methods is based on two measures, namely, RMSE values between original and reconstructed super-resolution images, and processing times. The IHS transformation separates the intensity (high-resolution) component from the hue and saturation components (lower-resolution). This simplifies the process of enhancing only the intensity component for higher resolution. The IHS method suffers from blurring of details, as it primarily focuses on enhancing the intensity information while leaving the hue and saturation components at their original resolutions. The improvement in spatial resolution is limited by the original resolution of the hue and saturation components. The reconstruction results show that the quality is limited for this method. PCA takes advantage of the correlation between different bands, allowing for the integration of multispectral information to enhance spatial resolution. It also effectively captures the principal components that contribute most to the variance in the data. While PCA focuses on maximizing variance, it also does not preserve the original spectral characteristics as other methods. Dictionary learning methods provide interpretable and sparse representations, making it easier to understand the underlying features captured in the dictionary. They are effective for datasets with limited training samples or in scenarios where training data might be scarce. The process of learning a dictionary and performing sparse coding are computationally expensive as it can be easily seen from the processing time results. CNN-based deep networks captures complex, nonlinear relationships in the data, allowing for potentially more accurate super-resolution. The reconstructions results show that CNN-based method provides the best quality with the lowest RMSE error. CNNs, especially with modern architectures and hardware acceleration, are more computationally efficient during both training and inference compared to dictionary learning methods. The processing time result shows that the use of hardware acceleration such as GPU provides better processing times.

References

1. Lepcha, D.C., Goyal, B., Dogra, A., Goyal, V.: Image super-resolution: a comprehensive review, recent trends, challenges and applications. Inf. Fusion. **91**, 230–260 (2023)
2. Yue, L., Shen, H., Li, J., Yuan, Q., Zhang, H., Zhang, L.: Image super-resolution: the techniques, applications, and future. Signal Process. **128**, 389–408 (2016)
3. Yang, W., Zhang, X., Tian, Y., Wang, W., Xue, J.-H., Liao, Q.: Deep learning for single image super-resolution: a brief review. IEEE Trans. Multimed. **21**(12), 3106–3121 (2019). https://doi.org/10.1109/TMM.2019.2919431
4. Yang, C.Y., Ma, C., Yang, M.H.: Single-image super-resolution: a benchmark. In: Fleet, D., Pajdla, T., Schiele, B., Tuytelaars, T. (eds.) Computer Vision – ECCV 2014. ECCV 2014. Lecture Notes in Computer Science, vol 8692. Springer, Cham (2014)
5. Lim, B., Son, S., Kim, H., Nah, S., Mu Lee, K.: Enhanced deep residual networks for single image super-resolution. In: Proceedings of the IEEE Conference on Computer Vision and Pattern Recognition (CVPR) Workshops, July, 2017 (2017)
6. Kawulok, M., Benecki, P., Piechaczek, S., Hrynczenko, K., Kostrzewa, D., Nalepa, J.: Deep learning for multiple-image super-resolution. IEEE Geosci. Remote Sens. Lett. **17**(6), 1062–1066 (2020). https://doi.org/10.1109/LGRS.2019.2940483
7. Salvetti, F., Mazzia, V., Khaliq, A., Chiaberge, M.: Multi-image super resolution of remotely sensed images using residual attention deep neural networks. Remote Sens. **12**(14), 2207 (2020). https://doi.org/10.3390/rs12142207
8. Arefin, M.R., Michalski, V., St-Charles, P. L., Kalaitzis, A., Kim, S., Kahou, S.E., Bengio, Y.: Multi-image super-resolution for remote sensing using deep recurrent networks. In: Proceedings of the IEEE/CVF Conference on Computer Vision and Pattern Recognition (CVPR) Workshops, June, 2020 (2020)
9. Simsek, M., Polat, E.: Performance evaluation of pan-sharpening and dictionary learning methods for sparse representation of hyperspectral super-resolution. SIViP. **15**, 1099–1106 (2021). https://doi.org/10.1007/s11760-020-01836-8

10. Liu, J., Gan, Z., Zhu, X.: Directional bicubic interpolation—a new method of image super-resolution. In: Proceedings of 3rd International Conference on Multimedia Technology (ICMT-13), pp. 463–470 (2013)
11. Tang, Z., Deng, M., Xiao, C., Yu, J.: Projection onto convex sets super-resolution image reconstruction based on wavelet bi-cubic interpolation. In: Proceedings of 2011 International Conference on Electronic & Mechanical Engineering and Information Technology, Harbin, China, pp. 351–354 (2011). https://doi.org/10.1109/EMEIT.2011.6022950
12. Khaledyan, D., Amirany, A., Jafari, K., Moaiyeri, M.H., Khuzani, A.Z., Mashhadi, N.: Low-cost implementation of bilinear and bicubic image interpolation for real-time image super-resolution. In: 2020 IEEE Global Humanitarian Technology Conference (GHTC), Seattle, WA, USA, pp. 1–5 (2020). https://doi.org/10.1109/GHTC46280.2020.9342625
13. Suresha, D., Prakash, H.N.: Single picture super resolution of natural images using N-Neighbor Adaptive Bilinear Interpolation and absolute asymmetry based wavelet hard thresholding. In: 2016 2nd International Conference on Applied and Theoretical Computing and Communication Technology (iCATccT), Bangalore, India, pp. 387–393 (2016). https://doi.org/10.1109/ICATCCT.2016.7912029
14. Timofte, R., De Smet, V., Van Gool, L.: Proceedings of the IEEE International Conference on Computer Vision (ICCV), pp. 1920–1927 (2013)
15. Mishra, D., Majhi, B., Kumar Sa, P., Dash, R.: Development of robust neighbor embedding based super-resolution scheme. Neurocomputing. **202**, 49–66 (2016)
16. Anbarjafari, G., Demirel, H.: Image super resolution based on interpolation of wavelet domain high frequency subbands and the spatial domain input image. ETRI J. **32**, 390–394 (2010)
17. Chen, K., Li, L., Liu, H., Li, Y., Tang, C., Chen, J.: SwinFSR: stereo image super-resolution using SwinIR and frequency domain knowledge. In: Proceedings of the IEEE/CVF Conference on Computer Vision and Pattern Recognition (CVPR) Workshops, June 2023, pp. 1764–1774 (2023)
18. Fritsche, M., Gu, S., Timofte, R.: Frequency separation for real-world super-resolution. In: 2019 IEEE/CVF International Conference on Computer Vision Workshop (ICCVW), Seoul, Korea (South), pp. 3599–3608 (2019). https://doi.org/10.1109/ICCVW.2019.00445
19. Zhou, Y., Deng, W., Tong, T., Gao, Q.: Guided frequency separation network for real-world super-resolution. In: Proceedings of the IEEE/CVF Conference on Computer Vision and Pattern Recognition (CVPR) Workshops, June 2020 (2020)
20. Moca, V.V., Bârzan, H., Nagy-Dăbâcan, A., et al.: Time-frequency super-resolution with superlets. Nat. Commun. **12**, 337 (2021). https://doi.org/10.1038/s41467-020-20539-9
21. Shi, W., Caballero, J., Huszar, F., Totz, J., Aitken, A.P., Bishop, R., Rueckert, D., Wang, Z.: Real-time single image and video super-resolution using an efficient sub-pixel convolutional neural network. In: Proceedings of the IEEE Conference on Computer Vision and Pattern Recognition (CVPR), June 2016 (2016)
22. Chen, Y., Jiang, G., Jiang, Z., Yu, M., Ho, Y.-S.: Deep light field super-resolution using frequency domain analysis and semantic prior. IEEE Trans. Multimed. **24**, 3722–3737 (2022). https://doi.org/10.1109/TMM.2021.3106775
23. Freeman, W.T., Jones, T.R., Pasztor, E.C.: Example-based super-resolution. IEEE Comput. Graph. Appl. **22**(2), 56–65 (2002). https://doi.org/10.1109/38.988747
24. Salvador, J.: Example-Based Super-Resolution. Academic/Elsevier (2017)
25. Timofte, R., De Smet, V., Van Gool, L.: Anchored neighborhood regression for fast example-based super-resolution. In: Proceedings of the IEEE International Conference on Computer Vision (ICCV), December 2013 (2013)
26. Timofte, R., Rothe, R., Van Gool, L.: Seven ways to improve example-based single image super resolution. In: Proceedings of the IEEE Conference on Computer Vision and Pattern Recognition (CVPR), June, 2016 (2016)
27. Trinh, D.-H., Luong, M., Dibos, F., Rocchisani, J.-M., Pham, C.-D., Nguyen, T.Q.: Novel example-based method for super-resolution and denoising of medical images. IEEE Trans. Image Process. **23**(4), 1882–1895 (2014). https://doi.org/10.1109/TIP.2014.2308422

28. Liu, H., Ruan, Z., Zhao, P., et al.: Video super-resolution based on deep learning: a comprehensive survey. Artif. Intell. Rev. **55**, 5981–6035 (2022)
29. Jiang, J., Wang, C., Liu, X., Ma, J.: Deep learning-based face super-resolution: a survey. Assoc. Comput. Machinery. **55**(1) (2023)
30. Wang, P., Bayram, B., Sertel, E.: A comprehensive review on deep learning based remote sensing image super-resolution methods. Earth Sci. Rev. **232** (2022)
31. Wang, X., Yi, J., Guo, J., Song, Y., Lyu, J., Xu, J., Yan, W., Zhao, J., Cai, Q., Min, H.: A review of image super-resolution approaches based on deep learning and applications in remote sensing. Remote Sens. **14**(21), 5423 (2022)
32. Arun, P.V., Buddhiraju, K.M., Porwal, A., Chanussot, J.: CNN-based super-resolution of hyperspectral images. IEEE Trans. Geosci. Remote Sens. **58**(9), 6106–6121 (2020). https://doi.org/10.1109/TGRS.2020.2973370
33. Tian, C., Xu, Y., Zuo, W., Zhang, B., Fei, L., Lin, C.-W.: Coarse-to-fine CNN for image super-resolution. IEEE Trans. Multimed. **23**, 1489–1502 (2021). https://doi.org/10.1109/TMM.2020.2999182
34. Su, H., Wang, A., Zhang, T., Qin, T., Du, X., Yan, X.-H.: Super-resolution of subsurface temperature field from remote sensing observations based on machine learning. Int. J. Appl. Earth Obs. Geoinf. **102** (2021)
35. Huang, Z.-X., Jing, C.-W.: Super-resolution reconstruction method of remote sensing image based on multi-feature fusion. IEEE Access. **8**, 18764–18771 (2020). https://doi.org/10.1109/ACCESS.2020.2967804
36. Xin, J., Li, J., Jiang, X., Wang, N., Huang, H., Gao, X.: Wavelet-based dual recursive network for image super-resolution. IEEE Trans. Neural Netw. Learn. Syst. **33**(2), 707–720 (2022). https://doi.org/10.1109/TNNLS.2020.3028688
37. Liu, J., Zhang, W., Tang, Y., Tang, J., Wu, G.: Residual feature aggregation network for image super-resolution. In: Proceedings of the IEEE/CVF Conference on Computer Vision and Pattern Recognition (CVPR), June 2020 (2020)
38. Chen, Y., Liu, L., Phonevilay, V., et al.: Image super-resolution reconstruction based on feature map attention mechanism. Appl. Intell. **51**, 4367–4380 (2021)
39. Javan, F.D., Samadzadegan, F., Mehravar, S., Toosi, A., Khatami, R., Stein, A.: A review of image fusion techniques for pan-sharpening of high-resolution satellite imagery. ISPRS J. Photogramm. Remote Sens. **171**, 101–117 (2021). https://doi.org/10.1016/j.isprsjprs.2020.11.001
40. Loncan, L., et al.: Hyperspectral pansharpening: a review. IEEE Geosci. Remote Sens. Mag. **3**(3), 27–46 (2015). https://doi.org/10.1109/MGRS.2015.2440094
41. Tu, T.-M., Huang, P.S., Hung, C.-L., Chang, C.-P.: A fast intensity-hue-saturation fusion technique with spectral adjustment for IKONOS imagery. IEEE Geosci. Remote Sens. Lett. **1**(4), 309–312 (2004). https://doi.org/10.1109/LGRS.2004.834804
42. Rodarmel, C., Shan, J.: Principal component analysis for hyperspectral image classification. Surv. Land Inf. Syst. **62**(2) (2002)
43. Zhang, H., Zhang, L., Shen, H.: A super-resolution reconstruction algorithm for hyperspectral images. Signal Process. **92**(9), 2082–2096 (2012). https://doi.org/10.1016/j.sigpro.2012.01.020
44. Zhang, L., Su, H., Shen, J.: Hyperspectral dimensionality reduction based on multiscale superpixelwise kernel principal component analysis. Remote Sens. **11**(10), 1219 (2019). https://doi.org/10.3390/rs11101219
45. Akhtar, N., Shafait, F., Mian, A.: Sparse spatiospectral representation for hyperspectral image superresolution. In: Computer Vision – ECCV 2014, pp. 63–78 (2014)
46. Mairal, J., Bach, F., Ponce, J., Sapiro, G.: Online dictionary learning for sparse coding. In: International Conference on Machine Learning (ICML) (2009)
47. Efron, B., Hastie, T., Johnstone, I., Tibshirani, R.: Least angle regression. Ann. Stat. **32**(2), 407–499 (2004)
48. Akhtar, N., Shafait, F., Mian, A.: Bayesian sparse representation for hyperspectral image super resolution. In: 2015 IEEE Conference on Computer Vision and Pattern Recognition (CVPR), pp. 3631–3640 (2015)

49. Paisley, J., Carin, L.: Nonparametric factor analysis with beta process priors. In: ICML '09: Proceedings of the 26th Annual International Conference on Machine Learning, pp. 777–784 (2009)
50. Dong, C., Change Loy, C., He, K., Tang, X.: Image super-resolution using deep convolutional networks. IEEE Trans. Pattern Anal. Mach. Intell. **38**(2), 295–307 (2016)
51. He, K., Zhang, X., Ren, S., Sun, J.: Proceedings of the IEEE Conference on Computer Vision and Pattern Recognition (CVPR), pp. 770–778 (2016)
52. Yasuma F., Mitsunaga T., Iso D., Nayar S.: Generalized Assorted Pixel Camera: Post-Capture Control of Resolution, Dynamic Range and Spectrum. Technical report, Department of Computer Science, Columbia University CUCS-061-08 (2008)
53. Huang, W., Xiao, L., Lui, H., Wei, Z.: Hyperspectral imagery super-resolution by compressive sensing inspired dictionary learning and spatial-spectral regularization. Sensors. **15**, 2041–2058 (2015)
54. Alparone, L., Wald, L., Chanussot, J., Thomas, C., Gamba, P., Bruce, L.M.: Comparison of pansharpening algorithms: outcome of the 2006 GRS-S data-fusion contest. IEEE Trans. Geosci. Remote Sens. **45**(10), 3012–3021 (2007)

Synergy of Images: Multi-Image Fusion Empowering Super-Resolution in Remote Sensing

Hailiang Lu, Mercedes E. Paoletti, Lirong Han, Weipeng Jing, Guangsheng Chen, and Juan M. Haut

1 Introduction

As remote sensing (RS) technologies advance rapidly, an increasing number of imaging sensors mounted on satellites, spacecraft, unmanned aerial vehicles (UAVs), and mobile terrestrial platforms are being launched and deployed in an effort to better comprehend terrestrial environments, processes, and phenomena [49, 50, 106, 127]. Thus, technological advances in RS, together with open data policies, have made it possible to collect huge amounts of RS images of the terrestrial surface through different Earth observation (EO) campaigns and missions [52, 63, 99, 111, 157, 179], providing a rich and varying information about Earth through various observation modes (active/passive sensors, optical/radar/thermal technology, etc.). This information plays a crucial role in research and decision-making procedures, resulting in significant advances in a wide variety of economic and societal activities, such as environmental monitoring [10], resource management [114], urban planning [174], land surface classification [79], and risk prevention [39].

Among other features (e.g., the data acquisition frequency, object range, etc.), there are two fundamental parameters in RS images, i.e., spatial resolution and spectral resolution. Spatial resolution indicates the mapping of a sensor's pixel or a narrow aperture onto the Earth's surface. The level of spatial resolution during data acquisition has dual implications: On the one hand, it determines the capability

H. Lu · W. Jing · G. Chen
College of Computer and Control Engineering, Northeast Forestry University, Harbin, China
e-mail: luhailiang@nefu.com; jwp@nefu.edu.cn; kjc_chen@163.com

M. E. Paoletti · L. Han · J. M. Haut (✉)
Department of Technology of Computers and Communications, University of Extremadura, Cáceres, Spain
e-mail: mpaoletti@unex.es; lironghan_upc@163.com; juanmariohaut@unex.es

M. Kawulok et al. (eds.), *Super-Resolution for Remote Sensing*, Unsupervised and Semi-Supervised Learning, https://doi.org/10.1007/978-3-031-68106-6_6

to distinguish different features and, on the other hand, the capability to measure their sizes accurately. The former aspect influences classification precision, while the latter impacts the precision of dimensional measurements. Spectral resolution refers to the ability to capture the spectral reflectance or spectral emission of an object or area across multiple wavelengths in a range of the electromagnetic spectrum. By analyzing responses across different wavelength ranges, various features and intricate details within an image can be distinguished. For instance, broader features such as vegetation can be differentiated using wider wavelength bands like visible and near-infrared (VNIR), while more specific distinctions, such as different types of rocks, necessitate a comparison across narrower wavelength bands. Therefore, spectral resolution characterizes the sensor's capacity to define precise spectral intervals, effectively sampling the image in segmented wavelength ranges. This capability enables the determination of the spectral composition of the image. RS images equipped with high spatial and spectral resolutions offer an extensive array of observational data, enabling a more comprehensive understanding of the observed area or objects. This detailed imagery facilitates highly precise and nuanced analyses of ground objects, enhancing accuracy in identification, classification, and interpretation of features or changes within the observed area.

A well-known problem in RS is the trade-off between spatial and spectral resolutions, due to limitations inherent in sensor technology and other factors. This compromise arises as an outcome of the challenges involved in achieving both high spatial and high spectral resolutions simultaneously, necessitating a balance between the two aspects in acquiring imagery [166]. To deal with this, modern satellites often utilize dual sensors: one for capturing panchromatic (PAN) images, attaining exceptional spatial resolution despite being comprised of only a single band, and another for acquiring multispectral (MS) or hyperspectral (HS) images, which are composed of several or hundreds of spectral bands and have a wealth of spectral information but lower spatial resolution. Table 1 illustrates the spatial and spectral resolutions of those images obtained by various optical sensors. The spatial resolution ratio among these sensors is relatively high. For instance, in GeoEye-1, the spatial resolution ratio between PAN and MS is 4, while for PRISMA, it stands at 6. In this context, multiple techniques have been introduced in literature aiming to improve the spatial resolution of MS data. These methods, collectively known as *pansharpening*, constitute an interesting type of super-resolution technique.

As a super-resolution technique, pansharpening attempts to fuse a PAN image with a MS image to produce an enhanced image, exhibiting the high spatial resolution from the former and the superior spectral resolution from the latter. Therefore, it has the potential to substantially improve the spatial resolution of MS images without compromising the spectral information [53]. Figure 1 provides a comparison between conventional single-image super-resolution (SR) techniques (particularly, the EXP interpolator) and a pansharpening technique (the pansharpening neural network [PNN]) [94]. In contrast to the former, the pansharpening technique enhances the reconstruction of both spatial and spectral information more effectively. Furthermore, as pansharpening technology advances, some methods have been extended to solve slightly different problems, such as

Table 1 The spatial and spectral resolutions on various optical sensors

Equipment		Spectral resolution (band)	Spatial resolution (m)	Launch time (year)
Landsat-7, 8, 9	MS	6	30	1999, 2013, 2021
	PAN	1	15	
IKONOS	MS	4	4	1999
	PAN	1	1	
QuickBird	MS	4	2.44	2001
	PAN	1	0.61	
GeoEye-1	MS	4	4	2008
	PAN	1	1	
Pleiades-1A, 1B	MS	4	2	2011, 2012
	PAN	1	0.5	
WorldView-2, 3, 4	MS	8, 8, 4	2, 1.24, 1.24	2009, 2014, 2016
	PAN	1	0.5, 0.31, 0.31	
GF-1, 2	MS	4	8, 4	2013, 2014
	PAN	1	2, 1	
PRISMA	HS	250	30	2019
	PAN	1	5	
EO-1	HS	220	30	2000
	PAN	1	10	
HyMap	HS	125	20	2007
	PAN	1	4	
AVIRIS	HS	224	100	–
	PAN	1	20	

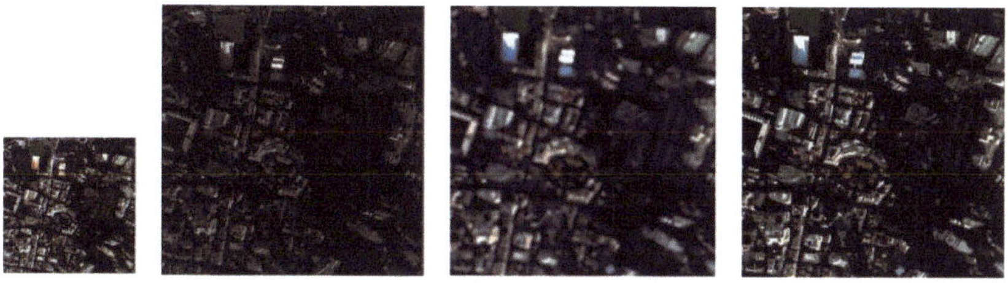

Original *MS* *PAN* Single-image SR (EXP) Pansharpening (PNN)

Fig. 1 A comparison between single-image super-resolution and pansharpening methods (multi-image fusion)

the hyperpansharpening [42, 57, 59] and the synthesis of MS and HS images [150, 161, 182]. In this context, pansharpening is a powerful and versatile tool in RS. It offers significant benefits for improving both spectral and spatial quality of RS images, extracting rich and valuable information while reducing the costs associated with advanced sensor instruments. Moreover, pansharpening techniques have become essential for various RS applications, including land use and cover mapping, deserving further investigation.

Chapter Organization

This chapter provides a comprehensive review of pansharpening methods to explore their feasibility and performance as suitable super-resolution strategies in the context of multi-image fusion strategies. In this regard, the chapter is organized as follows: Firstly, the recently progress made in pansharpening is elaborated in Sect. 2, delving into the literature to provide a detailed discussion of the different types of pansharpening methods, including traditional and machine learning approaches. Also, some publicly available resources are collected in Sect. 3. Furthermore, assessments are conducted in Sect. 4 to investigate the effectiveness of various pansharpening methods using diverse quality evaluation procedures. Finally, some RS applications based on pansharpening are summarized in Sect. 5.

2 Pansharpening as Super-Resolution Technology: An Overview

Pansharpening roots can be traced back to 1980 [120]. With the deployment of the Système Pour l'Observation de la Terre-1 (SPOT-1), which provided low-resolution MS images and high-resolution PAN images, the development of pansharpening methods progressed rapidly [98]. In recent years, the number of contributions to pansharpening has increased considerably. Thus, nowadays an extensive literature is available on the subject of pansharpening. Indeed, to advance the progress of this domain, scholars have conducted in-depth reviews. For instance, Zhang et al. [175] presented an exhaustive examination of various pansharpening methods which are based on PAN and MS images; Vivone [133] provided a summary of the research advancements in MS and HS images fusion, covering fundamental theories and available resources. Current challenges and potential directions for future research are also discussed, for instance, Deng et al. [27] analyzed the evolution of machine learning (ML)-based techniques for pansharpening. In addition, the 12th IEEE workshop on Hyperspectral Image and Signal Processing: Evolution in Remote Sensing (IEEE WHISPERS 2022) organized the 2022 Hyperspectral Pansharpening Challenge [137]. This project aims to merge a PAN image with a HS image to produce a high-resolution image that retains the spectral information of HS data while matching the spatial resolution of PAN images.

2.1 Literature Statistics

To analyze the available literature regarding pansharpening methods for super-resolution, the Web of Science[1] (WOS) and the software tool for constructing and

[1] Source: www.webofscience.com [Accessed January, 30th, 2024].

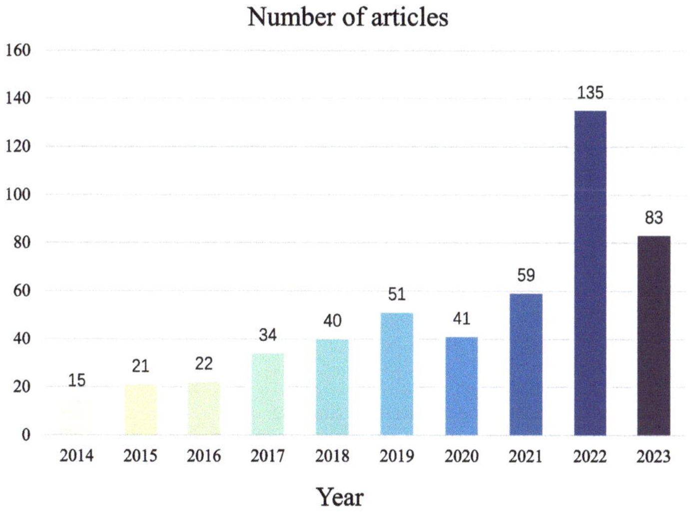

Fig. 2 Number of papers published about pansharpening over the years according to WOS

visualizing bibliometric networks VOSviewer[2] have been selected as the primary analysis tools. In this sense, "pansharpening" and "remote sensing" have been used as keywords to search the relevant literature from the WOS Core Collection. Furthermore, framing the scope of this study to the past decade, i.e., from 2014 to 2023, a total of 501 papers published during this period have been included for subsequent analysis.

The trend of relevant papers published from 2014 to 2023 is illustrated in Fig. 2, where more than 500 papers have been collected and evaluated from WOS. As it is observed, the bar chart illustrates a growing interest in the pansharpening field, showing a consistent year-over-year increase in the number of publications. In particular, the number of published papers reached a peak in 2022, with a total of 135 papers, and although the number of papers decreased in 2023, it still remains at a substantial level of publications.

Furthermore, additional analysis about the impact of the journals in which papers are published is presented. These journals are arranged based on the number of citations, where the top 10 journals have been chosen for statistics. As shown in Table 2, IEEE Transactions on Geoscience and Remote Sensing (IEEE T GEOSCI REMOTE) remains the highest in both the number of papers published and total citations. IEEE Geoscience and Remote Sensing Letters (IEEE GEOSCI REMOTE S) holds the second-highest position, despite having a relatively small number of papers. Of particular note is the IEEE Geoscience and Remote Sensing Magazine (IEEE GEOSC REM SEN M), which has only 5 articles, yet the number of

[2] Source: www.vosviewer.com [Accessed January, 30th, 2024].

Table 2 Journals impact based on the number of published papers

Journal	Published papers	Total citations	Average citations
IEEE T GEOSCI REMOTE[a]	131	1788	13.65
IEEE GEOSCI REMOTE S[b]	55	949	17.25
REMOTE SENS-BASEL[c]	74	527	7.12
IEEE J-STARS[d]	51	402	7.88
IEEE GEOSC REM SEN M[e]	5	270	54
IEEE T IMAGE PROCESS[f]	7	239	34.14
INFORM FUSION[g]	9	149	16.56
ISPRS J PHOTOGRAMM[h]	11	125	11.36
IEEE T NEUR NET LEAR[i]	13	69	23
REMOTE SENS LETT[j]	11	41	3.73

[a] IEEE Transactions on Geoscience and Remote Sensing
[b] IEEE Geoscience and Remote Sensing Letters
[c] Remote Sensing
[d] IEEE Journal of Selected Topics in Applied Earth Observations and Remote Sensing
[e] IEEE Geoscience and Remote Sensing Magazine
[f] IEEE Transactions on Image Processing
[g] Information Fusion
[h] ISPRS Journal of Photogrammetry and Remote Sensing
[i] IEEE Transactions on Neural Networks and Learning Systems
[j] Remote Sensing Letters

citations is remarkably high, demonstrating both the high impact and quality of the publications. In general, these journals have demonstrated significant influence in the field of pansharpening.

Keywords serve as a fundamental summary of a paper, and analyzing them provides significant insight into the research field. By examining keyword co-occurrence, a deep understanding of the relationships among topics within the represented discipline can be gained. In this sense, the co-occurrence map of keywords was generated using VOSviewer and is presented in Fig. 3. The most frequently appearing keywords, such as "pansharpening," are positioned at the center of the map. Simultaneously, pansharpening methods are continuously evolving and being updated over time. Before 2020, traditional methods were mainly used, as indicated by keywords like "component substitution (cs)," "wavelet," and "multiresolution analysis" in the left of the figure. However, in recent years, many approaches based on machine learning and deep learning have emerged, as indicated by keywords like "deep learning," "convolutional neural networks," and "transformer" in the right region of the figure (highlighted in yellow). This suggests that machine learning technologies are playing an increasingly paramount role in the field of pansharpening.

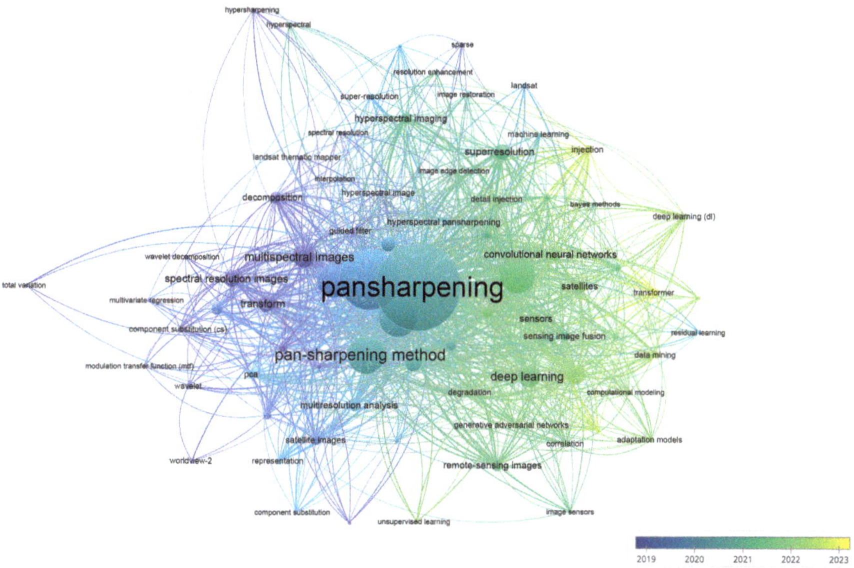

Fig. 3 The map of keyword co-occurrence obtained by VOSviewer

2.2 *Pansharpening Categorization*

Pansharpening techniques can be categorized according to the level at which image fusion is performed. This includes pixel-level, feature-level, and decision-level techniques [47]. Pixel-level fusion involves manipulating the pixel values of images using fusion rules, providing an interesting tool for the management of substantial data volumes and resulting in a high-quality fused image with enhanced accuracy. Nevertheless, the pixel-level fusion tends to exhibit high computational complexity. On the contrary, feature-level fusion concentrates on extracting corner points, textures, shapes, and so on, which are subsequently combined to generate the fused image. Within the decision fusion method and employing classification as an example, diverse classifiers are employed across different sets of data features. The classification results are determined by implementing decision fusion rules, which may include methods like majority voting and joint measurement, among others [110]. Regarding the two pansharpening techniques, the scientific community has usually devoted more effort to the first one, since, despite pixel-level fusion being computationally more demanding, its high accuracy makes it more attractive [65].

Meanwhile, a more detailed categorization has been delineated in various ways. For instance, based on the extraction procedure of spatial details from PAN images, Baronti et al. [14] categorized the prevailing pansharpening methods into two primary groups: multiresolution analysis (MRA)-based methods and component substitution (CS)-based methods. Also, Kaur et al. [61] arranged the existing pansharpening methods into deep learning (DL)-based methods, without learning-

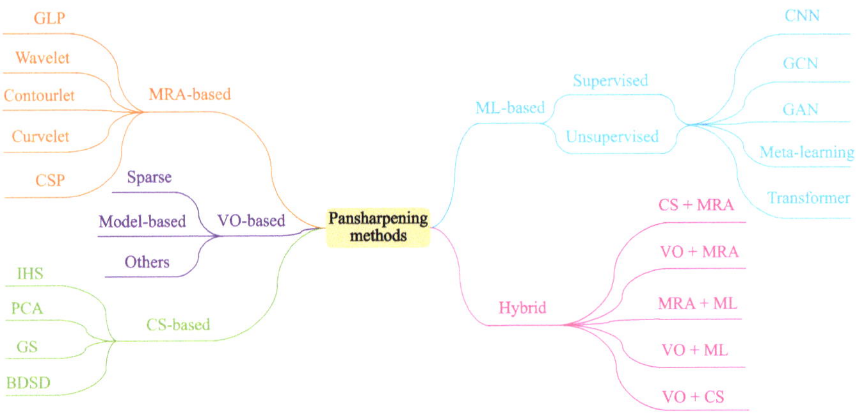

Fig. 4 The categorization of pansharpening methods

based methods, and variational optimization (VO) methods. Finally, Yilmaz et al. [167] classified conventional and state-of-the-art pansharpening methods into six categories: MRA-based, CS-based, color-based (CB), deep learning (DL)-based, VO-based, and hybrid methods. Building upon previous studies, this chapter will concentrate on analyzing the principles of several typical category approaches (Fig. 4).

2.2.1 Component Substitution (CS)-Based Methods

CS-based methods stand out as prevalent pansharpening strategies, the origins of which date back to 1982 [55], often referred to as intensity-hue-saturation (IHS). Currently, a lot of CS-based methods are seamlessly integrated into various RS platforms, such as ENVI [77], Google Earth Engine (GEE) [180], and PCI Geomatica, underscoring their widespread adoption across the RS community. CS-based methods involve the process of substituting high-frequency information from the panchromatic image into the corresponding multispectral bands [160]. Based on the employed transformation strategy, the techniques within this category can be classified into four subclasses, i.e., IHS (intensity-hue-saturation) [18, 46, 55], PCA (principal component analysis) [36, 68], GS (Gram-Schmidt) [5, 145], and BDSD (band-dependent spatial detail) [132]. These subclasses represent distinct approaches that leverage different transformations for pansharpening purposes.

The new understanding of CS-based methods is established by Tu et al. [129] and has been extended by Dou et al. [33]. They provide a unified framework, liberating CS-based methods from intricate processing involving rigid projection transformations. Based on the summary provided by Vivone et al. [136] and Meng et al. [96], the general framework of the new understanding, as illustrated in Fig. 5, can be mathematically described by Eq. (1):

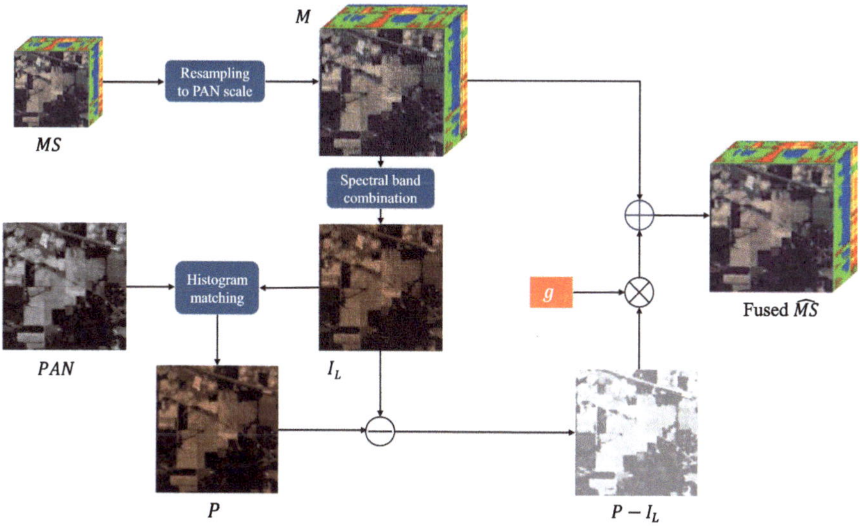

Fig. 5 Diagram of CS-based methods

$$\widehat{MS} = M + g(P - I_L) \tag{1}$$

where $\widehat{MS}$ is the pansharpened image. M presents the MS image which resample to the PAN image scale. g denotes the injection gains which contains the high spatial information. I_L is the component to be substituted, which is obtained by combining spectral bands of the original MS image through a specific transformation. P is the normalized image obtained through histogram matching the original PAN image with I_L.

In this context, the effectiveness of CS-based methods in data fusion is significantly influenced by two key factors, i.e., the selected transformations to obtain I_L and the definition of g. The careful consideration and optimization of these factors play a pivotal role in determining the overall fusion performance. The general expression for I_L can be formulated as:

$$I_L = \sum_{k=1}^{B} w_k M_k \tag{2}$$

where B is the number of bands of M. The weights, denoted as w_k, directly influence the value of I_L and are configured differently in various methods. For instances, Tu et al. [128] assigned weights of $\frac{1}{3}$ to both the red and near-infrared bands and set to $\frac{a}{3}$ and $\frac{b}{3}$ for the green and blue bands separately, where a and b satisfy the condition $a + b = 1$. Garzelli et al. [45] attained the weights for each band through the least square criterion. The same goes for g, which is set to 1 in [128] and set according to the edge map of P and M in [72].

In CS-based methods, inappropriate decomposition solutions consistently led to spectral distortions [19]. However, these methods are typically simple to implement and impose a low computational burden, making them suitable for scenarios where a substantial amount of data needs to be fused [134]. While these methods may not be anticipated to achieve competitive performance compared to other types of methods, such as VO- and ML-based approaches, their simplicity and effectiveness still make them appealing for researchers and worthy of further investigation. In this regard, approaches within this category have undergone continuous updates in recent years. For instance, Wang et al. [146] introduced an enhanced haze-corrected version of the multiplicative-based CS method to address potential distortion issues. Özay et al. [105] developed a new pansharpening method to tackle the spectral distortion. Also, to address the chromatic distortions induced by histogram transformation, Civicioglu et al. [26] proposed the pansharpening using dominant pixels (PDP), which employs a new histogram transformation technique to fully preserve chromatic information.

2.2.2 Multiresolution Analysis (MRA)-Based Methods

As another popular and widely adopted pansharpening technique, MRA-based methods possess a significant capability to model the high frequencies present in source images. Schowengerdt [120] initially implemented pansharpening through the application of a single-level decomposition technology. Afterwards, several decomposition-based methods have been developed to extract spatial structure content from the original PAN image and subsequently incorporate it into the original MS image, such as Laplacian pyramids (LPs) [16, 138], wavelet [91, 104], contourlet [30, 37], and curvelet [31, 124]. Besides, there has been a growing interest in modulation transfer function (MTF)-based MRA methods due to their capability to deliver promising performances [7]. Examples include MTF with generalized Laplacian pyramids (MTF-GLP) [3], MTF-GLP with context-based decision (MTF-GLP-CBD) [8], and full-scale regression-based MTF-GLP (MTF-GLP-FS) [140].

As illustrated in Fig. 6, the present comprehension of MRA-based methods is expanded within a unified fusion framework [4, 151], which can be formulated as:

$$\widehat{MS} = M + g(PAN - PAN_L) \tag{3}$$

where PAN_L refers to the low-pass filtered version of the original PAN image. It is noteworthy that these techniques can be mathematically represented by an equation very similar to CS-based methods. The main difference is the PAN_L, which is used to replace the intensity image (named I_L in Eq. (1)). Hence, the MRA category is defined by the methodology employed to generate the PAN_L image and the injection gains (g). Over the past decades, the efficient approximation of spatial information in original PAN image at various scales and the calculation of injection gains have been extensively explored. For instance, Vivone et al. [140] applied the generalized Laplacian pyramid (GLP) to extract high-frequency information from

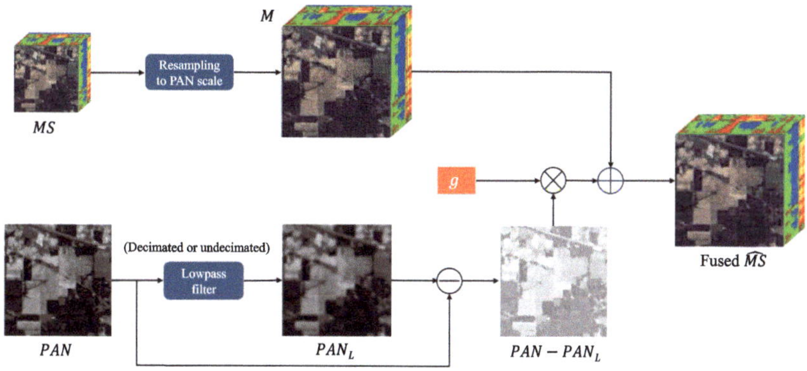

Fig. 6 Diagram of MRA-based methods

the original image. Meanwhile, the injection gains at the full scale were adjusted through regression approaches. Devulapalli et al. [28] employ the curvelet transform for extracting the orientation of spatial details, while the adaptive neuro-fuzzy inference was used to integrate the injection gains from the original MS and PAN images.

Recently, MRA-based pansharpening methods have been further developed. Civicioglu et al. [25] presented the contrast stretching-based pansharpening (CSP) method. This approach employs contrast stretching to produce two pseudo-multispectral images and one pseudo-panchromatic image for the fusion process. Carpentiero et al. [17] introduced an estimation procedure to determine the value of the scale parameter in regression-based pansharpening. Addesso et al. [1] proposed a new MRA-based pansharpening method, which integrates a detail-extraction filter with an enhanced injection scheme. While MRA-based methods generally exhibit strong performance in the spectral aspect, the incorporation of multiscale transformation tends to elevate computational complexity and may lead to spatial distortion [96].

2.2.3 Variational Optimization (VO)-Based Methods

To ensure a balance among spatial and spectral distortions, various VO-based approaches have been introduced. VO-based techniques primarily revolve around solving a variational optimization problem [38], which can be summarized in two steps: One is the construction of an energy function, and the other is the optimization of the constructed function. The earliest VO-based method adopted a Bayesian framework [93]. This approach constructed an estimator and covariance matrix as prior information based on the orthogonality principle and then used a new statistical synthesis formulation to obtain the final fused image. Another classic VO-based method, the so-called P+XS as an abbreviation of PAN and MS images, uses the Euler-Lagrange equations as the energy function combined with gradient

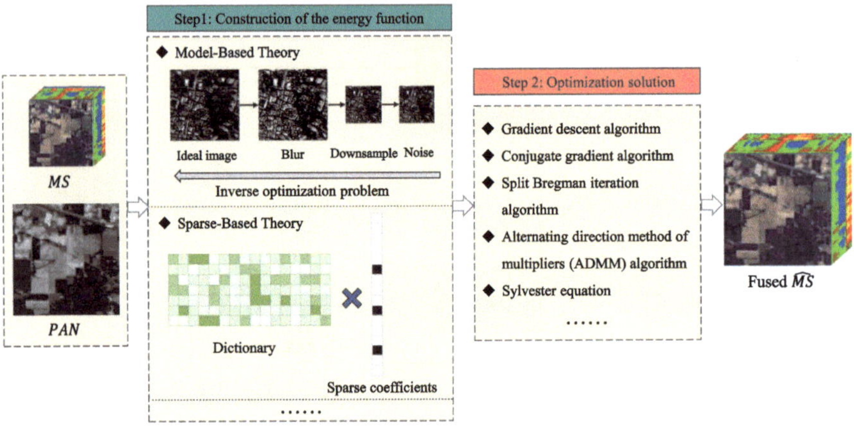

Fig. 7 Diagram of VO-based methods [61, 96, 98]

descent to achieve fast convergence of the optimization [12]. In view of its excellent fusion results, more two-step VO-based methods have been proposed. The primary principles adopted in two-step methods are summarized as shown in Fig. 7, where model-based and sparse representation theory are prominent algorithms used in the first step. For the second step, popular approaches involve an iterative optimization algorithm to solve the energy function. However, alternative methods, such as the Sylvester equation [152], offer interesting ways of circumventing the need for iteration.

Model-based approaches construct the energy function based on the inverse optimization problem and can be formulated as [98]:

$$E(\widehat{MS}) = f_{spectral}(\widehat{MS}, MS) + f_{spatial}(\widehat{MS}, PAN) + f_{prior}(\widehat{MS}) \tag{4}$$

where $\widehat{MS}$ is the fused image. $E(\widehat{MS})$ denotes the constructed energy function. It consists of three components: The spectral fidelity component is built upon the correlation between the $\widehat{MS}$ and the original MS image, aiming to preserve spectral information; the spatial component is formulated by leveraging the relationship between the $\widehat{MS}$ and the original PAN image to enhance spatial information; and finally, the prior component is used to constrain the $\widehat{MS}$. Some interesting methods have been presented by Zhang et al. [176], who introduced an adjustable model-based approach that incorporated blurring, downsampling, and noise operators in the spectral fidelity component. The spatial component is achieved through a linear combination, and a Huber-Markov prior is applied to the prior component. Ultimately, the gradient descent algorithm is employed for iterative solutions to the energy function model. Li et al. [78] developed a new model-based methods based on image restoration strategy, where the point spread function (PSF) and additive zero-mean random noises are used for spectral fidelity. Also, a linear combination in the spatial component is employed. Liu et al. [83] presented a non-convex sparse prior, attributed to differences in spatial and spectral gradients. Furthermore, the proposed optimization method is addressed using the alternating direction method of multipliers (ADMM) algorithm. More model-based methods can be found in [81, 82, 86, 107, 158].

Sparse representation-based methods form the foundation of compressed sensing, integrating the dictionary with estimated sparse coefficients [76]. Within a sparse-based model, an image patch is represented through a linear combination of new atoms derived from a chosen or learned dictionary. For example, the sparse representation details injection (SR-D) model [131] divides the high-resolution details into overlapped patches. A patch represents a set of columns within a dictionary, which can be formulated as:

$$x = D^h \alpha \tag{5}$$

where x is the overlapped patch. D^h identifies the dictionary containing the high spatial resolution details. α denotes the sparse coefficients. In this context, the purpose is to minimize the reconstruction error and achieve an optimal representation of x using the elements of the dictionary, all while following a sparsity constraint on the coefficients α. Upon resolving x, the additive model can be incorporated into the original MS image, enabling the construction of a fused $\widehat{MS}$ image with high spatial resolution. Based on this, improved versions were developed by Fei et al. [40, 41]. Through introducing manifold regularization, the local geometric structure of the original MS image is incorporated to enhance the overall fusion performance.

Other methods based on sparse representation have also been proposed. For instance, the sparse fusion of images (SparseFI) [183] explores sparse representations from original PAN images and its low-resolution versions. Lang et al. [70] developed a sparse representation-based model to address the detail discrepancies between spectral channels by analyzing the structural similarity and detail differences between the PAN image and the original MS image in each band. Tian et al. [126] presented a VO-based method based on gradient sparse representation, while Han et al. [54] introduced a novel pansharpening method based on a spectral dictionary. This method employs subpixel-shift decomposition coupled with joint optimization to address spectral distortions in a patch-based spatial dictionary. Although these methods excel at prioritizing the enhancement of details, achieving remarkable spatial quality, they frequently overlook the comprehensive consideration of spectral constraints and can result in spectral distortion in the fused results [86].

2.2.4 Machine Learning (ML)-Based Methods

As a renowned data-driven technology, machine learning (ML) has the capability to implicitly uncover nonlinear relationships among different feature domains. Indeed, working as universal approximators, ML methods have the ability to establish a direct mapping from low-resolution MS images to high-resolution MS images by referring the PAN image [56]. Leveraging their strong performance in learning from data, ML-based methods have demonstrated significant advantages over traditional pansharpening approaches [27]. Nevertheless, it must be noted that ML algorithms often perform better when learning from labeled and known data, i.e., the so-called supervised learning where the ML method is tuned on representative training samples composed of input features matched with corresponding accurate and meaningful output labels. Therefore, the unavailability of labeled data presents a challenge. Regarding pansharpening, ML-based methods encounter a substantial challenge as there is no high-resolution MS image available to serve as the ground truth [24]. Recently, two common approaches have been developed to tackle this challenge. One strategy is downsampling the PAN and MS images, where these downscaled images serve as input samples for training a ML model, with the original MS image acting as the ground truth. This allows for the full supervised training of the ML models, albeit in a lower-resolution domain. Another approach, known as the full-resolution approach, involves training ML models in the high-resolution domain. Indeed, ML models are trained using the original PAN and MS images as input at their native resolution, without downsampling. Simultaneously, the loss function is designed to achieve unsupervised learning by assessing the spatial and spectral variances between the fused MS image and two input images [87].

With the aspect of reduced resolution, the pansharpening neural network (PNN) [94] represents the initial endeavor to investigate the application of convolutional neural networks (CNNs) in pansharpening. The core idea revolves around a resolution shift, where the original MS image is used as the reference instead of the input. This is achieved by proportionally reducing the resolution of both the MS and PAN images on all training images by employing the MTF filter with a corresponding resolution ratio. In this manner, a sufficient number of samples will be available to train the CNN model during the training process. Figure 8 illustrates the conceptual

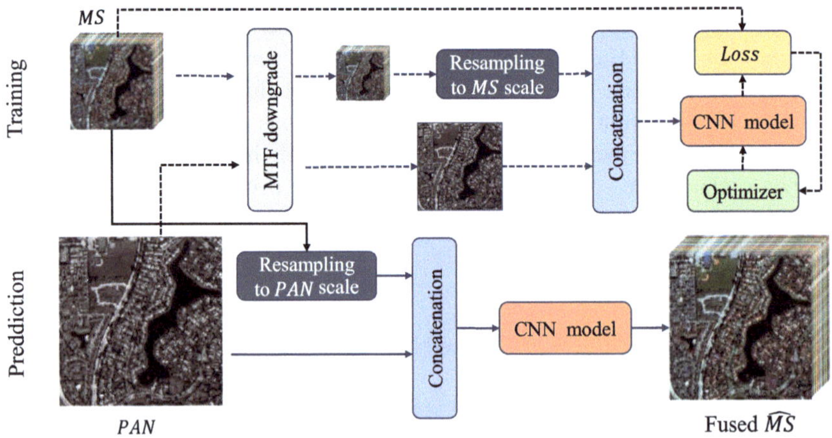

Fig. 8 The overall framework of the PNN

framework introduced by PNN. Upon achieving convergence during model training, the prediction stage allows obtaining the target high spatial resolution MS image by utilizing the original-resolution MS and PAN images as input. Building upon the foundation of PNN, which utilized a simple network with three convolution layers, different types of deep neural networks have been developed. For instance, Scarpa et al. [119] adopted adaptive tuning schemes named A-PNN. Shao et al. [121] introduced a pansharpening method that incorporates a two-branch network with a deeper structure. Additionally, it employs residual learning to comprehensively explore the relationship between original- and high-resolution MS images. He et al. [58] developed a generative adversarial network (GAN)-based fusion network for pansharpening. The self-attention mechanism is also introduced into the fusion network [71, 88]. Other popular used networks for training on the low-resolution strategy include graph convolution network (GCN) [162], transformer [97, 154], and global multiscale context network (GMCNet) [80].

Nevertheless, PNN and its offshoot models often adjust their parameters using simulated datasets, which may constrain their efficacy when deployed on real-world data [20, 90, 144]. Hence, numerous works are currently shifting towards unsupervised strategies at the full-resolution scale. This involves defining suitable loss functions and adopting specific training schemes. Serving as a representative framework operating on ML models at full resolution, Z-PNN [24] is highly versatile and applicable to any deep neural network. A dual-component loss strategy was devised to only employ the original MS and PAN images within the framework, effectively preventing any loss of information. Soon after, a refined version of the Z-PNN was introduced, which named λ-PNN [23]. In this version, a novel ML-based model is introduced to maximize the capabilities of the adopted framework, delivering state-of-the-art performance. Furthermore, the enhancement of spatial and spectral information in the fused image is achieved through further refinement of the two-component loss function. The overall framework of the λ-PNN can be found in Fig. 9, where L_S and L_λ are the spatial and spectral loss terms, respectively. By integrating comprehensive losses with fast target-adaptive strategy during the

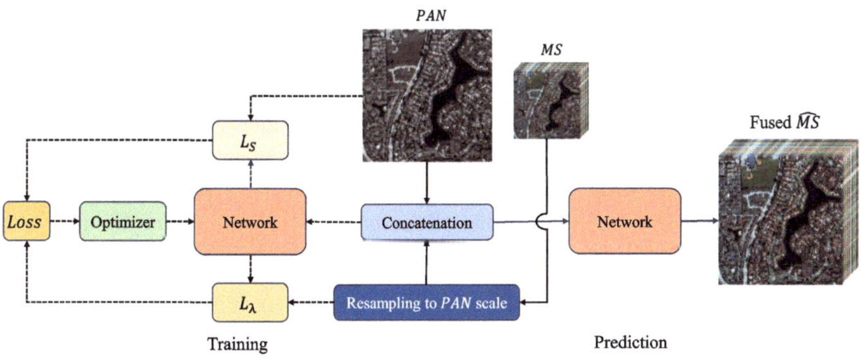

Fig. 9 The overall framework of the λ-PNN

training stage, λ-PNN demonstrates significant potential in practical applications. In addition, numerous outstanding works have been developed at full-resolution domain. For instance, Ni et al. [102] designed a two-stream CNN network with a hybrid spectral loss in an unsupervised manner. The CSFNet [73] introduces two attention blocks designed to address both cross-spectral and spatial scales. High-resolution MS image can be achieved by mapping computed nonlocal similarity from the low-resolution to the high-resolution domain. Xiong et al. [159] introduced an attention-based network composed of two parts: a deep residual network and a spatial texture attention block. By incorporating a loss function calculated at both spatial and frequency scales, the network demonstrated superior performance compared to other ML-based methods. Various techniques are integrated in the recent advancements, including GAN-based [29, 84], meta-learning [144], and transformer [62].

The utilization of ML-based models offers a novel approach to pansharpening, providing impressive fusion results. Nevertheless, whether at full-resolution or a reduced-resolution scale, the expansion of the network architecture results in a significant investment of time and computing resources for training. This poses challenges to the practical implementation of ML-based fusion methods. Hence, there is merit in conducting further research to balance fusion accuracy, managing training time and computational resource consumption, and applying ML-based methods to practical scenarios.

2.2.5 Hybrid

Hybrid pansharpening methods have been developed with the aim of exploiting the advantages of the various methods mentioned above. In the early period, hybrid methods primarily utilized CS and MRA. For instance, to minimize color distortion, the combination of the strengths of both the IHS and wavelet techniques is implemented [103, 178]. Recently, more hybrid methods tend to be combined with the VO-based methods. For instance, Shen et al. [122] introduced a pansharpening method that integrates a deep neural network with a model-based variational optimization. To address the limitation of spatial detail preservation in the traditional SR-D method, Maneshi et al. [92] suggested employing an extended MRA framework with an iterative full-scale injection gain. Wang et al. [147] introduced a two-stage method that relies on both CS and VO strategies. Table 3 enumerates various hybrid methods, often crafted by amalgamating two or more distinct pansharpening techniques. The objective of hybrid techniques is to leverage one method to address the limitations of another, thereby enhancing the quality of the fused image. Therefore, enhancements in any constituent sub-method hold the promise of elevating the overall performance of the hybrid approach.

Table 3 Hybrid methods

Year	Hybrid way (no priority)	Methods	Article
1999	CS and MRA	Wavelet + IHS	Nunez et al. [103]
2005	CS and MRA	Wavelet + IHS	Zhang et al. [178]
2008	CS and MRA	PCA + ATROUS wavelet	Luo et al. [89]
2012	CS and MRA	IHS + curvelet	Valizadeh et al. [130]
2015	VO and MRA	Wavelet + sparse representation	Cheng et al. [21]
2017	MRA and ML	DNN + MRA framework	Azarang et al. [11]
2019	VO and ML	CNN + model-based fusion	Shen et al. [122]
2020	VO and ML	DCNN + model-based fusion	Wu et al. [156]
2020	VO and MRA	SR-D + extended MRA framework	Maneshi et al. [92]
2020	VO, CS and MRA	Variational model + injection model	Yang et al. [165]
2021	VO and ML	Variational model + CNN	Wu et al. [155]
2022	VO and CS	Variational model + IHS	Wang et al. [147]
2023	VO and MRA	Sparse representation + MTF-GLP-CBD	Yilmaz [169]

3 Open Resource

Significant advancements have been made in establishing standardization for pansharpening algorithms and their implementations to date. Nonetheless, achieving complete reproducibility in research remains challenging, primarily due to the absence of benchmark datasets of RS images accessible to this domain. To address this challenge, scholars in recent years have sequentially developed publicly available standardized datasets. At the same time, many classic pansharpening methods were verified based on these datasets. Typically, Gemine Vivone et al. [135] shared a novel benchmark dataset and 9 pansharpening methods of different types were verified, which is named as PAirMax. PAirMax contains a publicly available collection of 9 MS-PAN pairs from various types of sensors and covering different landscapes. PAirMax is primarily employed for conducting pansharpening tests, but supervised deep learning methods are not suitable due to insufficient data for pre-training. Another popular dataset is the large-scale benchmark dataset developed by Meng et al. [98]. This dataset comprises over 2000 pairs of images, captured across diverse scenes by various sensors. Based on this dataset, both traditional and cutting-edge methods are verified. The access addresses for above two datasets are provided in Table 4, along with some publicly available benchmark datasets related to hyperpansharpening (MS + PAN) and HS + MS. Simultaneously, for experimental convenience, Table 5 offers publicly available code resources.

Table 4 Open datasets

Year	Type	Name	Link
2020	Pansharpening	[98]	http://www.escience.cn/people/fshao/database.html
2021	Pansharpening	PAirMax [135]	https://resources.maxar.com/product-samples/pansharpening-benchmark-dataset
2022	Hyperpansharpening	[137]	https://openremotesensing.net/knowledgebase/panchromatic-and-hyperspectral-image-fusion-outcome-of-the-2022-whispers-hyperspectral-pansharpening-challenge/
2023	Hyperpansharpening	[74]	https://aistudio.baidu.com/aistudio/datasetdetail/95831
2013	HS + MS	Hyperion [99]	https://data.nasa.gov/dataset/EO-1-Hyperion/ethf-arwz/data
2015	HS + MS	Indian Pines [15]	https://www.ehu.eus/ccwintco/index.php?title=Hyperspectral_Remote_Sensing_Scenes
2016	HS + MS	Chikusei [170]	https://naotoyokoya.com/Download.html
2023	HS + MS	[74]	https://aistudio.baidu.com/aistudio/datasetdetail/120965

4 Evaluation Methodologies

The precision of the pansharpening process plays a pivotal role in ensuring that the derived images provide trustworthy and high-quality information for various applications in RS. Conducting both quantitative and qualitative evaluations of results from a pansharpening method is an essential step to ascertain the effectiveness of this method. Being a typical inverse problem, the fused image obtained from pansharpening cannot be directly compared with the ground truth in a real RS image. The key challenge is the absence of the reference image, which makes the evaluation process more difficult [51]. To provide a clearer description of the evaluation procedure, two aspects in the following will be discussed. Firstly, a detailed overview of the evaluation process will be presented to enhance understanding. Subsequently, taking pansharpening as an example, classic methods are systematically tested and validated using publicly available datasets to assess their effectiveness and performance.

4.1 Quality Assessment

In addressing the challenges arising from the absence of reference images during the evaluation process, Wald et al. [125, 143] originally devised a protocol

Table 5 Available codes

Year	Type	Name	Link
2020	Pansharpening	Toolbox [136]	https://openremotesensing.net/knowledgebase/a-new-benchmark-based-on-recent-advances-in-multispectral-pansharpening-revisiting-pansharpening-with-classical-and-emerging-pansharpening-methods/
2021	Pansharpening	MQNR [95]	https://github.com/yyxhpkq/MQNR
2022	Pansharpening	MD^3Net [163]	https://github.com/YinsongYan/M3DNet
2022	Pansharpening	Z-PNN [24]	https://github.com/matciotola/Z-PNN
2023	Pansharpening	SFNLR [158]	https://github.com/Jin-liangXiao/SFNLR
2023	Pansharpening	λ-PNN [23]	https://github.com/matciotola/Lambda-PNN
2022	Hyperpansharpening	Pgnet [75]	https://github.com/rs-lsl/Pgnet
2023	Hyperpansharpening and HS + MS	CSSNet [74]	https://github.com/rs-lsl/CSSNet
2011	HS + MS	CNMF [171]	http://naotoyokoya.com/Download.html
2016	HS + MS	Toolbox [170]	http://naotoyokoya.com/Download.html

grounded in the principles of consistency and synthesis. Consistency asserts that the pansharpened image, downgraded to the original MS resolution, ought to closely resemble the original MS image. This essential condition can be validated through a comparison between the original MS image and a degraded version of the pansharpened image. Synthesis holds that the pansharpened image should closely resemble the one obtained from an ideal MS sensor with the spatial resolution matching that of the PAN sensor. This can be verified by conducting fusion on spatially degraded MS and PAN images, aiming to generate a pansharpened image that closely approximates the characteristics of the original MS image. Based on Wald's protocol, two different types of assessment approaches were developed, as shown in Fig. 10.

The reduced-resolution (RR) evaluation firstly performs spatial degradation of the original PAN and MS images according to Wald's protocol. The degraded PAN and MS images are then used as the input of the pansharpening methods. Therefore, the original MS image can be used as the reference to compare with the fusion results obtained from the degraded images. Accordingly, many quality assessment metrics with reference can be adopted in RR evaluation. This approach has been embraced by a diverse range of pansharpening studies, which have systematically assessed the quality of the degraded pansharpened images [9, 32, 168, 172, 177]. Among them, the most commonly used metrics during the evaluation process are:

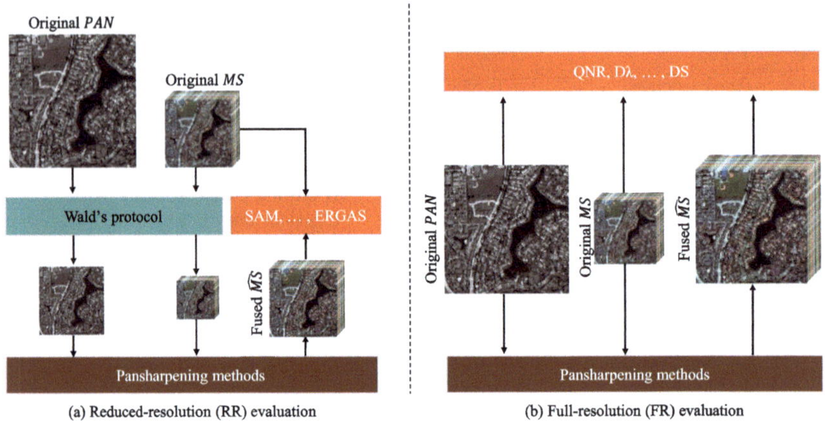

Fig. 10 The two typical evaluation approaches

1. RMSE (root mean square error): This metric provides a comprehensive evaluation of the disparities between two images, which can be used to assess the reconstruction errors.
2. CC (spatial correlation coefficient): CC measures spatial similarity by calculating correlation coefficients between two multispectral images. Despite computational demands, its precision in capturing spatial details makes it a favored method in pansharpening.
3. SAM (spectral angular mapper) [173]: SAM is utilized to evaluate the spectral similarity by calculating the angle of the spectral vectors between the fused image and the reference. It is usually expressed in degrees and is equal to 0 if and only if the fused image is spectrally identical to the reference image. Finally, the global spectral dissimilarity or distortion index is obtained by averaging over the entire scene.
4. SSIM (structural similarity index) [149]: This metric provides a quantitative measure that evaluates the perceived quality or similarity of structural patterns within images, making it a valuable tool in various image processing applications.
5. ERGAS (Erreur Relative Globale Adimensionnelle de Synthése) [115]: A normalized dissimilarity index, which assesses the overall fidelity and accuracy of the pansharpening process concerning a reference multiband image, providing a comprehensive measure of distortion in a test image.
6. $Q2^n$ [44]: $Q2^n$ is an extension derived from the universal image quality index (UIQI) [148]. This metric currently allows the assessment of spectral images with 2^n bands, providing a comprehensive approach for evaluating global image quality.

The fundamental assumption of the RR evaluation is the invariance of image quality across scales, though this alone is insufficient to assert the superiority of a pansharpening methods [27]. Moreover, the choice of downgrading method applied to the original MS and PAN images will also significantly impact the quality of the evaluation. Consequently, as the demand for accuracy persists, evaluation methods at the full scale are continually evolving.

The full-resolution (FR) evaluation stands out for its ability to provide evaluations at the original PAN scale, offering a valuable advantage in maintaining the fidelity of spatial information. This makes the FR approach particularly beneficial in scenarios where preserving original spatial characteristics is crucial for accurate RS applications. As depicted in Fig. 10b, the original MS and PAN images serve not only as inputs for the pansharpening methods but also actively contribute to the evaluation of the fused images, as initially adopted in [181]. And many recent high-quality articles persist in examining quality on the full scale [13, 95, 118].

The three most typical evaluation indicators in FR are:

1. D_S (spatial distortion index) [6]: Two values are calculated using the UIQI formula initially. The first one is derived from the comparison between the original MS image and the PAN image degraded to MS resolution. The second is determined by assessing the fused MS image against the original PAN image. Then the spatial distortion is produced by calculating the average absolute difference across all frequency bands between these values.
2. D_λ (spectral distortion index) [6]: This index is independently computed for both the original MS image and the fused MS image using the UIQI index. The variation in UIQI at two scales can indicate spectral distortion.
3. QNR (quality with no reference) [6]: QNR is commonly employed as a standard practice to validate at FR, which obtains measurements for both spectral and spatial distortion, offering a comprehensive validation. In addition various adaptations of QNR have been created successively, such as HQNR [2] and GQNR [67].

Conducting evaluation both on RR and FR scale can better reflect the performance of the pansharpening method. And it has been widely adopted in recent years. The equations of main metrics at RR and FR evaluation are illustrated in Table 6.

4.2 Experiments

To illustrate the evaluation procedure in more detail, a set of experiments is carried out using publicly available resources. The results of the experiments for various pansharpening methods are validated both at reduced-resolution and full-resolution scales. A complete analysis is provided, incorporating both qualitative and quantitative perspectives.

4.2.1 Experiment Settings

The experiment utilized the publicly available PAirMax [135] datasets, which consist of 9 scenes representing typical instances commonly encountered in pansharpening applications. Captured by various satellite sensors such as GeoEye-1,

Table 6 The equations of commonly used metrics

Equation	Variable description
Reduced-resolution	
$Q2^n = \frac{\lvert\sigma_{z\hat{z}}\rvert}{\sigma_z\sigma_{\hat{z}}} \cdot \frac{2\sigma_z\sigma_{\hat{z}}}{\sigma_z^2+\sigma_{\hat{z}}^2} \cdot \frac{2\lvert\hat{z}\rvert\lvert\overline{\hat{z}}\rvert}{\lvert\hat{z}\rvert^2\lvert\hat{z}\rvert^{2'}}$	M, N: the lines and columns of an image. f, g: the fused and reference images. (x, y): the location of a pixel in an image.
$SAM = \arccos \frac{\sum_{k=1}^{L} p_k q_k}{\sqrt{\sum_{k=1}^{L} p_k^2}\sqrt{\sum_{k=1}^{L} q_k^2}}$	$z, \hat{z}$ are the hypercomplex representation. l: the spatial resolutions of reference.
$RMSE = \sqrt{\frac{\sum_{x=1}^{M}\sum_{y=1}^{N}[f(x,y)-g(x,y)]^2}{MN}}$	h: the spatial resolutions of fused image. p: the spectrum vectors of fused image.
$SSIM = \frac{(2\mu_f\mu_g+c1)(2\sigma_{fg}+c2)}{(\mu_f^2+\mu_g^2+c1)(\sigma_f^2+\Sigma_g^2+c2)}$	q: the spectrum vectors of the reference. L: the number of band in an image.
$ERGAS = 100\frac{h}{l}\sqrt{\frac{1}{L}\sum_{k=1}^{L}(\frac{RMSE(f_k,g_k)}{\mu_g(k)})^2}$	μ: the mean value of an image. σ: image variance.
$CC = \frac{\frac{1}{MN}\sum_{x=1}^{M}\sum_{y=1}^{N}\{[f(x,y)-\mu_f][g(x,y)-\mu_g]\}}{\sqrt{\frac{1}{MN}\sum_{x=1}^{M}\sum_{y=1}^{N}[f(x,y)-\mu_f]^2}\sqrt{\frac{1}{MN}\sum_{x=1}^{M}\sum_{y=1}^{N}[g(x,y)-\mu_g]^2}}$	
Full-resolution	
$D_S = \frac{1}{L}\sum_{k=1}^{L}\lvert UIQI(g_k, P_d) - UIQI(f_k, P)\rvert$	P: the original PAN image. P_d: the degraded PAN image. α, β: the parameters, generally set to 1.
$D_\lambda = \frac{1}{L(L-1)}\sum_{k=1}^{L}\sum_{c=1,c\neq k}^{L}\lvert UIQI(f_k, f_c) - UIQI(g_k, g_c)\rvert$	
$QNR = (1 - D_\lambda)^\alpha(1 - D_S)^\beta$	

WorldView-2, WorldView-3, and WorldView-4, the dataset encompasses scenes depicting urban, natural, and mixed environments. All PAN-MS image pairs in datasets have a 4:1 spatial ratio. MS bands are in visible near-infrared domains, with 4 bands for GeoEye-1 and WorldView-4 and 8 bands for WorldView-2 and WorldView-3.

The experiment employed fusion methods from various pansharpening categories, incorporating a total of 27 methods. These are CS-based (BT-H [85], BDSD [45], C-BDSD [43], BDSD-PC [132], GS [69], GSA [5], C-GSA [116], PRACS [22]), MRA-based (AWLP [104], MTF-GLP [3], MTF-GLP-FS [140], MTF-GLP-HPM [141], MTF-GLP-HPM-H [85], MTF-GLP-HPM-R [139], MTF-GLP-CBD

[8], C-MTF-GLP-CBD [116], MF [117]), VO-based (FE-HPM [142], SR-D [131], PWMBF [108], TV [107], RR [109]), and ML-based (PNN [94], PNN-IDX [94], A-PNN [119], A-PNN-FT [119], and a interpolator [EXP]). The above methods were implemented using the MATLAB toolbox contributed by Vivone et al. [136]. For ML-based methods, the authors commonly provide the weights obtained from their training datasets along with the source code. It is worth noting that the toolbox utilized in the experiment only has pre-trained models suitable for WorldView-2, WorldView-3, and GeoEye-1 sensors and lacks the training model for WorldView-4. In this scenario, similar to the approach outlined in [135], we apply the pre-trained model obtained on GeoEye-1 to the WorldView-4 data. Furthermore, in the A-PNN-FT method, which implements target-adaptive operations, the fine-tuning iterations are configured to 50 to adjust to the new scenario.

The experiments were conducted using the Windows 10 operating system, which is configured with 112 GB of RAM, an Intel Core i9 3.3-GHz CPU, and a Titan RTX GPU. During the experiments, ML-based methods leverage GPU for processing, whereas traditional methods rely on CPU. For the quantitative assessment, classical metrics $Q2^n$, SAM, and SSIM were employed for the RR evaluation, while QNR, D_λ, and D_S index were utilized for the FR evaluation.

4.2.2 Reduced-Resolution Evaluation

For the evaluation of RR, downscaled datasets were generated from FR scenes using the procedure outlined in [135]. The original MS image acts as the reference data. The results from the experiments for all 9 low-resolution scenes, corresponding to the three evaluation metrics $Q2^n$, SAM, and SSIM, are presented in Tables 7, 8, and 9, and the best results are marked in bold. In the two urban areas captured by GeoEye-1 as listed in Table 7, the MRA-based method MTF-GLP-HPM exhibited superiority in terms of the Q_4 and SSIM metrics. The only exception was the SAM in the GE_Tren_Urb dataset, where it did not outperform the ML-based method A-PNN. When considering the comprehensive observation of the four datasets captured by WorldView-4 and WorldView-2 in Tables 7 and 8, various methods perform differently, but A-PNN-FT consistently outperforms others in SAM. This indicates that this method excels in preserving spectral information. The MRA-based method consistently outperforms other methods in the three scenes captured by WorldView-3 (Table 9).

Furthermore, the VO-based method (RR) exhibits the poorest performance on the GeoEye-1 observation scene. Specifically, the Q_4 metric achieved only 0.3863 on the GE_Lond_Urb dataset, a result even inferior to the basic interpolator (EXP). Another similar method is PNN-IDX, which achieved the worst results in the three scenarios observed by WorldView-3 (Table 9). For example, it only attains a score of 0.2814 for Q_8 on W3_Muni_Mix, a substantial decrease compared to the 0.6382 achieved by EXP. To distinctly illustrate the contrasting fusion outcomes of the aforementioned methods, two scenes were chosen for visualization. Figures 11 and 12 display the outcomes for GE_Lond_Urb and W3_Muni_Mix, respectively.

Table 7 Quantitative results of RR evaluation on GE_Lond_Urb, GE_Tren_Urb, and W2_Miam_Mix datasets

	GE_Lond_Urb			GE_Tren_Urb			W2_Miam_Mix		
	Q_4	SAM[°]	SSIM	Q_4	SAM [°]	SSIM	Q_8	SAM [°]	SSIM
GT	1.0000	0.0000	1.0000	1.0000	0.0000	1.0000	1.0000	0.0000	1.0000
EXP	0.5924	4.0029	0.3931	0.5828	7.3827	0.2729	0.5372	10.1668	0.3206
CS-based									
BT-H	0.9093	3.0495	0.8007	0.8922	5.7520	0.7832	0.8564	8.5013	0.7493
BDSD	0.9189	2.8142	0.8208	0.9057	6.7929	0.7985	0.8637	9.6253	0.7442
C-BDSD	0.9176	2.8749	0.8217	0.9072	6.5557	0.8047	0.8645	9.2938	0.7452
BDSD-PC	0.9191	2.8032	0.8209	0.9054	6.7506	0.7982	0.8625	9.1420	0.7460
GS	0.8014	3.9220	0.7460	0.8419	7.0974	0.7377	0.7368	10.8601	0.6377
GSA	0.9142	2.7727	0.8161	0.8988	6.8567	0.7861	0.8425	8.9703	0.7302
C-GSA	0.9091	3.7284	0.7997	0.8991	6.5729	0.7919	0.8407	10.4503	0.6897
PRACS	0.8484	3.1245	0.7135	0.8204	6.7860	0.6614	0.7650	9.3567	0.6355
MRA-based									
AWLP	0.9020	3.1674	0.7860	0.8955	6.9040	0.7774	0.8538	8.5670	0.7408
MTF-GLP	0.9203	2.7531	0.8206	0.9065	6.4088	0.8011	0.8578	8.8611	0.7478
MTF-GLP-FS	0.9200	2.7647	0.8207	0.9033	6.8017	0.7905	0.8417	8.9726	0.7294
MTF-GLP-HPM	**0.9226**	**2.6008**	**0.8258**	**0.9119**	6.5271	**0.8089**	**0.8670**	8.2071	**0.7620**
MTF-GLP-HPM-H	0.9189	3.0366	0.8087	0.8966	5.6987	0.7861	0.8584	8.4659	0.7514
MTF-GLP-HPM-R	0.9220	2.6257	0.8257	0.9084	7.1857	0.7929	0.8493	8.7273	0.7397
MTF-GLP-CBD	0.9198	2.7664	0.8205	0.9031	6.8588	0.7893	0.8412	8.9450	0.7292
C-MTF-GLP-CBD	0.9144	3.7263	0.8023	0.9026	6.5620	0.7941	0.8390	10.3918	0.6897
MF	0.8824	2.7424	0.7686	0.8790	6.6739	0.7594	0.8234	8.7697	0.7080

VO-based									
FE-HPM	0.8951	2.7351	0.7819	0.8902	6.6864	0.7727	0.8323	8.8211	0.7162
SR-D	0.8987	2.9751	0.7541	0.8915	6.0835	0.7642	0.8229	9.1678	0.6890
PWMBF	0.8593	3.3549	0.7452	0.8388	7.8025	0.7158	0.7981	10.8850	0.6774
TV	0.7954	3.9390	0.6221	0.7587	7.2201	0.5397	0.8586	9.4501	0.7315
RR	0.3863	16.8070	0.2111	0.4591	24.4670	0.4090	0.8500	9.5911	0.7247
ML-based									
PNN	0.8568	3.7348	0.7204	0.8808	4.9399	0.7492	0.8606	7.5809	0.7530
PNN-IDX	0.8641	3.3044	0.7190	0.8808	4.9024	0.7505	0.8596	7.7353	0.7504
A-PNN	0.8511	3.4854	0.6934	0.8886	**4.8091**	0.7664	0.8385	7.6758	0.7169
A-PNN-FT	0.8857	2.7906	0.7659	0.8860	4.8552	0.7607	0.8513	**7.5618**	0.7312

Table 8 Quantitative results of RR evaluation on W4_Mexi_Nat, W4_Mexi_Urb, and W2_Miam_Urb datasets

	W4_Mexi_Nat			W4_Mexi_Urb			W2_Miam_Urb		
	Q_4	SAM [°]	SSIM	Q_4	SAM [°]	SSIM	Q_8	SAM [°]	SSIM
GT	1.0000	0.0000	1.0000	1.0000	0.0000	1.0000	1.0000	0.0000	1.0000
EXP	0.7274	2.1751	0.4787	0.5501	6.8951	0.2568	0.5457	9.5731	0.2737
CS-based									
BT-H	0.8901	2.0021	0.8310	0.9132	5.2181	0.8216	0.8966	8.5033	0.7904
BDSD	0.8967	2.0584	0.8400	0.9236	5.6916	0.8407	0.9039	9.3480	0.7908
C-BDSD	0.8944	2.0818	0.8369	0.9226	5.4580	**0.8426**	**0.9090**	8.9095	**0.8054**
BDSD-PC	0.8964	2.0565	0.8399	0.9237	5.6125	0.8410	0.9041	8.9339	0.7922
GS	0.8346	2.3298	0.8298	0.8547	6.4392	0.7724	0.8089	9.3340	0.6939
GSA	0.8977	2.0034	0.8391	0.9192	5.9993	0.8311	0.8862	9.4191	0.7598
C-GSA	0.8975	2.0324	0.8396	0.9154	5.5940	0.8307	0.8843	9.5246	0.7530
PRACS	0.8741	2.1653	0.7822	0.8758	6.5851	0.7625	0.8614	9.6780	0.7227
MRA-based									
AWLP	0.8969	2.1321	0.8247	0.9090	5.8557	0.8099	0.8826	9.0933	0.7637
MTF-GLP	**0.9047**	1.9759	0.8435	0.9237	5.7848	0.8380	0.8934	9.0532	0.7789
MTF-GLP-FS	0.9039	1.9916	0.8406	0.9232	5.9122	0.8358	0.8856	9.3134	0.7618
MTF-GLP-HPM	**0.9047**	1.9877	**0.8451**	**0.9257**	5.9275	0.8404	0.8957	9.2620	0.7806
MTF-GLP-HPM-H	0.8928	1.9729	0.8216	0.9091	5.2514	0.8117	0.8946	8.5635	0.7860
MTF-GLP-HPM-R	0.9038	2.0139	0.8418	0.9251	6.1487	0.8366	0.8872	9.5961	0.7595
MTF-GLP-CBD	0.9039	1.9922	0.8406	0.9231	5.9623	0.8350	0.8851	9.3558	0.7599
C-MTF-GLP-CBD	0.9037	2.0030	0.8371	0.9104	7.5406	0.7822	0.8814	9.4683	0.7506
MF	0.8943	1.9981	0.8306	0.8922	6.1169	0.7858	0.8568	9.2519	0.7238

VO-based									
FE-HPM	0.9024	1.9944	0.8405	0.8997	6.1612	0.7966	0.8646	9.2167	0.7337
SR-D	0.8895	1.9127	0.7875	0.9047	5.4267	0.7998	0.8676	8.8079	0.7337
PWMBF	0.8773	2.3239	0.8034	0.8697	6.6721	0.7708	0.8388	10.0910	0.7150
TV	0.8327	2.1867	0.6706	0.7868	7.1632	0.5798	0.9023	8.8111	0.7930
RR	0.8881	2.0602	0.8104	0.8942	6.0304	0.7913	0.8974	9.0845	0.7835
ML-based									
PNN	0.8838	2.0059	0.7871	0.8884	5.3507	0.7729	0.8924	7.4933	0.7844
PNN-IDX	0.8757	2.1832	0.7816	0.8889	5.4554	0.7701	0.8899	7.6078	0.7825
A-PNN	0.8772	1.9305	0.7726	0.8620	5.5830	0.7221	0.8806	7.4818	0.7620
A-PNN-FT	0.8791	**1.8433**	0.7843	0.8837	**5.0346**	0.7678	0.8851	**7.3300**	0.7737

Table 9 Quantitative results of RR evaluation on W3_Muni_Mix, W3_Muni_Nat, and W3_Muni_Urb datasets

	W3_Muni_Mix			W3_Muni_Nat			W3_Muni_Urb		
	Q_8	SAM [°]	SSIM	Q_8	SAM [°]	SSIM	Q_8	SAM [°]	SSIM
GT	1.0000	0.0000	1.0000	1.0000	0.0000	1.0000	1.0000	0.0000	1.0000
EXP	0.6382	6.0261	0.4981	0.5981	4.2291	0.6305	0.6164	7.7873	0.4510
CS-based									
BT-H	0.9316	4.1272	0.8944	0.8058	2.8129	0.8513	0.9253	5.5560	0.8626
BDSD	0.9321	4.6929	0.8869	0.8068	3.0261	0.8506	0.9235	5.9464	0.8646
C-BDSD	0.9245	4.6817	0.8878	0.7951	3.0831	0.8443	0.9071	6.4571	0.8482
BDSD-PC	0.9322	4.6638	0.8872	0.8073	3.0237	0.8501	0.9241	5.7073	0.8666
GS	0.7889	6.1109	0.7886	0.7124	4.0648	0.8312	0.8126	7.9825	0.7657
GSA	0.9283	4.5476	0.8789	0.8044	3.2656	0.8389	0.9208	5.4690	0.8613
C-GSA	0.9202	5.2856	0.8516	0.8041	3.1755	0.8343	0.9156	6.1941	0.8503
PRACS	0.8613	5.2956	0.7881	0.7203	4.3084	0.7411	0.8550	6.7415	0.7772
MRA-based									
AWLP	0.9225	5.0162	0.8524	0.8093	3.9888	0.8369	0.9109	6.1660	0.8193
MTF-GLP	0.9331	4.4865	0.8866	0.8106	3.1244	0.8443	0.9231	5.4463	0.8636
MTF-GLP-FS	0.9279	4.5296	0.8782	0.8059	3.2340	0.8399	0.9200	5.4407	0.8599
MTF-GLP-HPM	**0.9349**	4.2114	0.8947	**0.8141**	**2.7468**	**0.8540**	**0.9264**	**5.1258**	**0.8705**
MTF-GLP-HPM-H	0.9333	**4.0962**	**0.8961**	0.8074	2.7901	0.8519	0.9251	5.6494	0.8568
MTF-GLP-HPM-R	0.9297	4.4566	0.8830	0.8085	2.9387	0.8495	0.9235	5.1452	0.8660
MTF-GLP-CBD	0.9276	4.5325	0.8778	0.8059	3.2434	0.8395	0.9197	5.4439	0.8595
C-MTF-GLP-CBD	0.9192	5.2943	0.8493	0.8010	3.1756	0.8345	0.9115	6.1972	0.8436
MF	0.9006	4.7023	0.8329	0.8043	2.9378	0.8467	0.8910	5.8893	0.7914

VO-based									
FE-HPM	0.9081	4.6714	0.8432	0.8123	2.7745	0.8524	0.9008	5.8145	0.8104
SR-D	0.9016	4.7611	0.8228	0.7846	3.0034	0.8194	0.8913	6.6353	0.7512
PWMBF	0.9007	5.3236	0.8503	0.7777	3.4502	0.8294	0.8896	6.4463	0.8254
TV	0.8956	5.9394	0.8343	0.6979	4.4365	0.7596	0.8861	7.9770	0.7962
RR	0.8586	5.7944	0.8206	0.7129	3.9698	0.7771	0.8621	8.0900	0.7884
ML-based									
PNN	0.6534	9.3920	0.7399	0.2767	8.6151	0.6939	0.7827	11.6841	0.7021
PNN-IDX	0.2814	17.7024	0.2904	0.1200	14.1356	0.4608	0.4596	21.1385	0.3363
A-PNN	0.7598	7.1208	0.6326	0.5432	4.7176	0.6539	0.7939	9.7334	0.5914
A-PNN-FT	0.8838	4.7886	0.8156	0.7003	3.9368	0.7853	0.8931	6.2204	0.8006

Fig. 11 Fusion results at the RR evaluation on the GE_Lond_Urb dataset

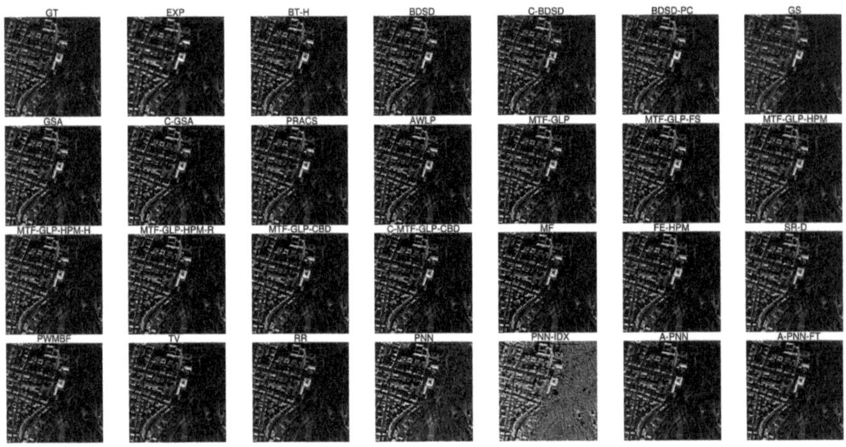

Fig. 12 Fusion results at the RR evaluation on the W3_Muni_Mix dataset

As can be seen, evident spectral distortions are observable in both methods. The VO-based method (RR) exhibits low performance specifically with GeoEye-1 observations. The PNN-IDX method demonstrates inadequate performance when applied to data obtained from WorldView-3. The observation is that the effectiveness of the two approaches is significantly influenced by the type of sensor used. In addition, the PNN method also exhibits significant spectral distortion in the case of W3_Muni_Mix (Fig. 12).

4.2.3 Full-Resolution Evaluation

For the evaluation of FR, all fusion methods utilize the original MS and PAN images as input directly. The evaluation of performance involves measuring the D_λ, D_S, and QNR metrics. In the evaluation experiments for reduced resolution, methods based on MRA and ML demonstrated superior performance. However, FR verification exhibits variations. In general, the VO-based method demonstrates superior performance. Additionally, the MRA-based method exhibits advantages across multiple indicators.

Tables 10 and 11 present the quantitative results for observation scenes captured by GeoEye-1, WorldView-4, and WorldView-2. Among them, VO-based and MRA-based methods stand out for their clear advantages. The VO-based methods (SR-D) demonstrate superior performance in the D_λ metric across various landscapes and sensors, while the MRA-based method (C-MTF-GLP-CBD, MTF-GLP-HPM-H) excels in the QNR metric. In addition, CS-based techniques have also achieved certain results. For instance, BDSD and C-BDSD have achieved the best results in the D_S metric for GE_Tren_Urb and W4_Mexi_Urb datasets, respectively. Simultaneously, the ML-based approach (A-PNN-FT) only attained optimal outcomes for two metrics (D_S and QNR) within a single scene (W4_Mexi_Nat). Regarding the three scenes captured by WorldView-3 (as indicated in Table 12), the SR-D method (VO-based) not only demonstrated superior performance in the D_λ metric but also excelled in the QNR metric of the W3_Muni_Nat dataset. Overall, in the FR quantitative evaluation, the VO-based method (SR-D) demonstrated superior performance in terms of the D_λ metric across all testing datasets, indicating its efficacy in addressing spectral distortion problem. Simultaneously, the MRA-based methods demonstrate strong performance on the QNR metric, suggesting a favorable trade-off between spectral and spatial distortion. Furthermore, it is important to highlight that while the ML-based method's performance may not be exceptional, the A-PNN-FT method exhibits significant enhancement across various metrics when compared to A-PNN. This underscores the crucial role of fine-tuning techniques utilizing testing images in enhancing fusion performance.

For qualitative evaluation, the same observed outcome of the experimental results as in RR evaluation is that the worst results are observed when applying the VO-based method (RR) to scenes acquired by GeoEye-1 and the PNN-IDX method to scenes captured by WorldView-3. Figure 13 illustrates the visualization results of all methods on the W3_Muni_Nat dataset. Due to the absence of high-resolution MS image for FR evaluation, the ground truth (GT) was substituted with the original PAN image when visualizing the fusion results. Similar in RR evaluation, PNN and PNN-IDX once again shows obvious spectral distortion. Except for this, all other methods demonstrate satisfactory performance in qualitative evaluation. Figure 14 illustrates the visual representation of the fusion results obtained by all methods on the W2_Miam_Mix dataset.

Table 10 Quantitative results of FR evaluation on GE_Lond_Urb, GE_Tren_Urb, and W2_Miam_Mix datasets

	GE_Lond_Urb			GE_Tren_Urb			W2_Miam_Mix		
	D_λ	D_S	QNR	D_λ	D_S	QNR	D_λ	D_S	QNR
EXP	0.1036	0.1153	0.7931	0.0820	0.0732	0.8508	0.0645	0.0661	0.8737
CS-based									
BT-H	0.2265	0.0545	0.7314	0.1519	0.0437	0.8110	0.1276	0.0973	0.7875
BDSD	0.2492	0.0425	0.7189	0.2113	**0.0146**	0.7771	0.2415	0.0921	0.6887
C-BDSD	0.2553	0.0265	0.7249	0.2083	0.0195	0.7762	0.2048	0.1135	0.7049
BDSD-PC	0.2492	0.0425	0.7189	0.2083	0.0214	0.7748	0.1808	0.0315	0.7934
GS	0.1899	0.0697	0.7536	0.1759	0.0459	0.7863	0.1584	0.1016	0.7561
GSA	0.2254	0.0792	0.7133	0.1475	0.1010	0.7664	0.1368	0.0748	0.7986
C-GSA	0.2431	0.0883	0.6901	0.1428	0.0990	0.7724	0.1357	0.0745	0.7999
PRACS	0.1566	0.0420	0.8080	0.1010	0.0543	0.8502	0.0807	0.0386	0.8838
MRA-based									
AWLP	0.0686	0.0389	0.8951	0.0478	0.0424	0.9118	0.0444	0.0641	0.8944
MTF-GLP	0.0634	0.0511	0.8887	0.0439	0.0455	0.9126	0.0479	0.0833	0.8727
MTF-GLP-FS	0.0636	0.0530	0.8867	0.0449	0.0495	0.9078	0.0479	0.0750	0.8807
MTF-GLP-HPM	0.0631	0.0471	0.8927	0.0431	0.0309	0.9274	0.0916	0.0568	0.8568
MTF-GLP-HPM-H	0.0625	0.0514	0.8893	0.0442	0.0184	**0.9382**	0.0523	0.1101	0.8434
MTF-GLP-HPM-R	0.0644	0.0510	0.8879	0.0443	0.0359	0.9214	0.0485	0.0633	0.8913
MTF-GLP-CBD	0.0647	0.0535	0.8853	0.0455	0.0515	0.9053	0.0480	0.0725	0.8830
C-MTF-GLP-CBD	0.0806	**0.0188**	**0.9022**	0.0547	0.0496	0.8985	0.0494	0.0479	0.9051
MF	0.0705	0.0387	0.8936	0.0524	0.0334	0.9160	0.0543	0.0747	0.8751
VO-based									
FE-HPM	0.0685	0.0509	0.8840	0.0469	0.0392	0.9157	0.0511	0.0763	0.8765
SR-D	**0.0345**	0.0976	0.8712	**0.0212**	0.1259	0.8556	**0.0299**	0.0508	0.9208
PWMBF	0.1249	0.0816	0.8037	0.0912	0.0654	0.8494	0.1112	0.1347	0.7691
TV	0.1147	0.0893	0.8062	0.0908	0.0399	0.8729	0.0428	0.0245	**0.9338**
RR	0.1364	0.3503	0.5610	0.2312	0.2487	0.5776	0.1288	**0.0171**	0.8563
ML-based									
PNN	0.0897	0.0652	0.8510	0.0494	0.0314	0.9208	0.1074	0.0257	0.8696
PNN-IDX	0.0859	0.1088	0.8146	0.0538	0.0536	0.8955	0.1181	0.0611	0.8280
A-PNN	0.0818	0.1185	0.8094	0.0464	0.0586	0.8977	0.0686	0.0620	0.8736
A-PNN-FT	0.0742	0.0502	0.8793	0.0569	0.0354	0.9098	0.0601	0.0292	0.9124

Another essential consideration relates to the time cost of the fusion algorithm in FR. The efficiency of practical applications is more relative to the time cost; such methods that involve excessively long processing times may become impractical for real-world implementation. Conclusively, we present the computational times associated with all the pansharpening techniques as shown in Table 13. Traditional techniques like CS-based and MRA-based require similar time costs and it is basically not affected by the number of bands. Nevertheless, the VO-based methods typically exhibit longer execution times, particularly with an expanding number of

Table 11 Quantitative results of FR evaluation on W4_Mexi_Nat, W4_Mexi_Urb, and W2_Miam_Urb datasets

	W4_Mexi_Nat			W4_Mexi_Urb			W2_Miam_Urb		
	D_λ	D_S	QNR	D_λ	D_S	QNR	D_λ	D_S	QNR
EXP	0.1114	0.1599	0.7465	0.0828	0.1202	0.8070	0.0496	0.0787	0.8756
CS-based									
BT-H	0.1889	0.1124	0.7200	0.1468	0.0265	0.8306	0.0937	0.0619	0.8502
BDSD	0.2255	0.1472	0.6605	0.1935	0.0150	0.7944	0.1652	0.0693	0.7769
C-BDSD	0.1754	0.0519	0.7818	0.1852	**0.0115**	0.8054	0.1810	0.1137	0.7258
BDSD-PC	0.2258	0.1476	0.6599	0.1935	0.0150	0.7944	0.1430	0.0425	0.8205
GS	0.2826	0.1250	0.6277	0.1728	0.0456	0.7895	0.1257	0.0807	0.8038
GSA	0.2132	0.1592	0.6615	0.1441	0.0776	0.7895	0.0913	0.0605	0.8537
C-GSA	0.2063	0.1545	0.6711	0.1420	0.0754	0.7933	0.0888	0.0556	0.8605
PRACS	0.1656	0.1133	0.7399	0.1100	0.0435	0.8513	0.0680	0.0248	0.9089
MRA-based									
AWLP	0.0877	0.1099	0.8120	0.0465	0.0281	0.9267	0.0273	0.0371	0.9366
MTF-GLP	0.0894	0.1421	0.7812	0.0437	0.0415	0.9167	0.0286	0.0533	0.9196
MTF-GLP-FS	0.0897	0.1428	0.7803	0.0441	0.0429	0.9149	0.0292	0.0469	0.9252
MTF-GLP-HPM	0.0892	0.1435	0.7801	0.0435	0.0378	0.9203	0.0328	0.0470	0.9217
MTF-GLP-HPM-H	0.0895	0.0961	0.8230	0.0443	0.0285	**0.9284**	0.0297	0.0637	0.9085
MTF-GLP-HPM-R	0.0895	0.1427	0.7805	0.0446	0.0381	0.9190	0.0296	0.0418	0.9298
MTF-GLP-CBD	0.0898	0.1424	0.7806	0.0449	0.0432	0.9139	0.0294	0.0456	0.9264
C-MTF-GLP-CBD	0.0915	0.1140	0.8049	0.0551	0.0227	0.9234	0.0325	0.0185	**0.9495**
MF	0.0959	0.1352	0.7819	0.0508	0.0325	0.9184	0.0354	0.0502	0.9162
VO-based									
FE-HPM	0.0945	0.1428	0.7762	0.0499	0.0417	0.9105	0.0319	0.0530	0.9168
SR-D	**0.0533**	0.0774	0.8734	**0.0203**	0.1277	0.8547	**0.0184**	0.0482	0.9343
PWMBF	0.1630	0.1556	0.7067	0.0945	0.0624	0.8490	0.0790	0.0979	0.8309
TV	0.1034	0.1367	0.7741	0.0708	0.0619	0.8717	0.0323	0.0275	0.9411
RR	0.1302	0.2407	0.6604	0.1789	0.2065	0.6515	0.0696	**0.0162**	0.9154
ML-based									
PNN	0.0932	0.0362	0.8740	0.0503	0.0464	0.9056	0.0583	0.0324	0.9112
PNN-IDX	0.0978	0.0551	0.8526	0.0613	0.0782	0.8653	0.0635	0.0419	0.8973
A-PNN	0.0650	0.1015	0.8401	0.0450	0.1106	0.8494	0.0470	0.0565	0.8991
A-PNN-FT	0.0642	**0.0271**	**0.9104**	0.0564	0.0589	0.8880	0.0446	0.0366	0.9205

bands; the time consumption is influenced by the specific fusion approach utilized. Notably, the most impactful approach is the VO-based method (RR). When applied to an image comprising 8 bands, its processing time escalates by almost tenfold, making it the most time-intensive method among all the algorithms. Meanwhile, ML-based methods offer faster processing times due to their utilization of GPU acceleration and pre-trained models.

Table 12 Quantitative results of FR evaluation on W3_Muni_Mix, W3_Muni_Nat, and W3_Muni_Urb datasets

	W3_Muni_Mix			W3_Muni_Nat			W3_Muni_Urb		
	D_λ	D_S	QNR	D_λ	D_S	QNR	D_λ	D_S	QNR
EXP	0.0542	0.1297	0.8231	0.1104	0.1881	0.7223	0.0622	0.1152	0.8297
CS-based									
BT-H	0.1406	0.0623	0.8058	0.1857	0.1769	0.6702	0.1458	0.0902	0.7771
BDSD	0.1820	0.0374	0.7874	0.2584	0.0891	0.6755	0.1757	0.0174	0.8100
C-BDSD	0.1814	0.0646	0.7657	0.2304	0.0892	0.7010	0.2153	0.0678	0.7315
BDSD-PC	0.1642	**0.0238**	0.8159	0.2298	0.1070	0.6878	0.1785	0.0158	0.8086
GS	0.1816	0.0607	0.7687	0.3189	0.1110	0.6055	0.1415	0.0550	0.8113
GSA	0.1479	0.0420	0.8163	0.1839	0.1578	0.6873	0.1664	0.0408	0.7996
C-GSA	0.1557	0.0474	0.8043	0.1826	0.1689	0.6794	0.1724	0.0517	0.7847
PRACS	0.0786	0.0254	0.8980	0.1248	0.0856	0.8003	0.0963	0.0273	0.8790
MRA-based									
AWLP	0.0355	0.0285	**0.9371**	0.0863	0.1469	0.7794	0.0408	0.0358	0.9249
MTF-GLP	0.0355	0.0390	0.9269	0.0907	0.1734	0.7517	0.0394	0.0461	0.9164
MTF-GLP-FS	0.0355	0.0347	0.9310	0.0901	0.1548	0.7690	0.0396	0.0387	0.9233
MTF-GLP-HPM	0.0360	0.0388	0.9266	0.1431	0.1618	0.7182	0.0407	0.0491	0.9122
MTF-GLP-HPM-H	0.0368	0.0521	0.9131	0.0894	0.1802	0.7465	0.0553	0.0896	0.8601
MTF-GLP-HPM-R	0.0359	0.0324	0.9328	0.0891	0.1542	0.7705	0.0408	0.0395	0.9213
MTF-GLP-CBD	0.0356	0.0326	0.9330	0.0901	0.1554	0.7685	0.0397	0.0371	0.9246
C-MTF-GLP-CBD	0.0399	0.0244	0.9366	0.0931	0.1685	0.7540	0.0456	**0.0136**	**0.9414**
MF	0.0424	0.0301	0.9288	0.1057	0.1622	0.7492	0.0470	0.0345	0.9202

VO-based									
FE-HPM	0.0356	0.0291	0.9363	0.0920	0.1676	0.7558	0.0426	0.0409	0.9183
SR-D	**0.0238**	0.0607	0.9170	**0.0617**	0.0971	**0.8472**	**0.0308**	0.0706	0.9008
PWMBF	0.0865	0.0686	0.8509	0.1858	0.1381	0.7018	0.0945	0.0660	0.8458
TV	0.0443	0.0403	0.9172	0.1330	0.1639	0.7249	0.0435	0.0381	0.9200
RR	0.1218	0.0366	0.8460	0.3006	0.2059	0.5554	0.1167	0.0597	0.8306
ML-based									
PNN	0.4641	0.1229	0.4700	0.7955	0.1007	0.1839	0.3814	0.1109	0.5500
PNN-IDX	0.8153	0.5317	0.0865	0.9270	0.3696	0.0460	0.7385	0.4274	0.1497
A-PNN	0.1771	0.2193	0.6425	0.4987	0.2145	0.3938	0.1401	0.1628	0.7199
A-PNN-FT	0.0697	0.0475	0.8861	0.2723	**0.0552**	0.6876	0.0633	0.0209	0.9171

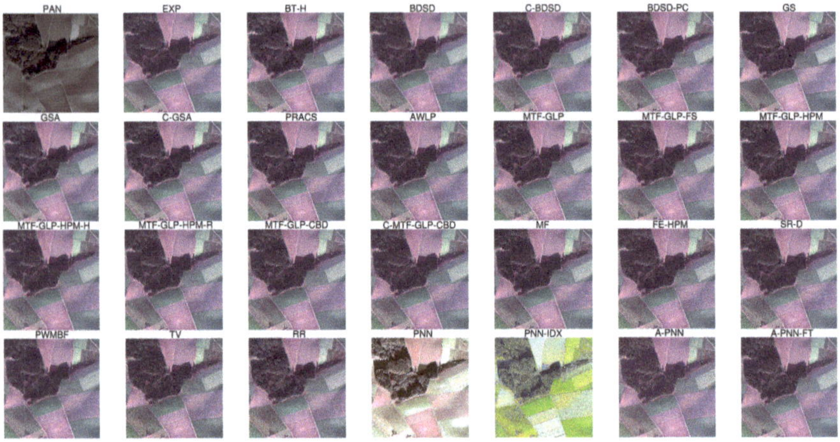

Fig. 13 Fusion results at the FR evaluation on the W3_Muni_Nat dataset

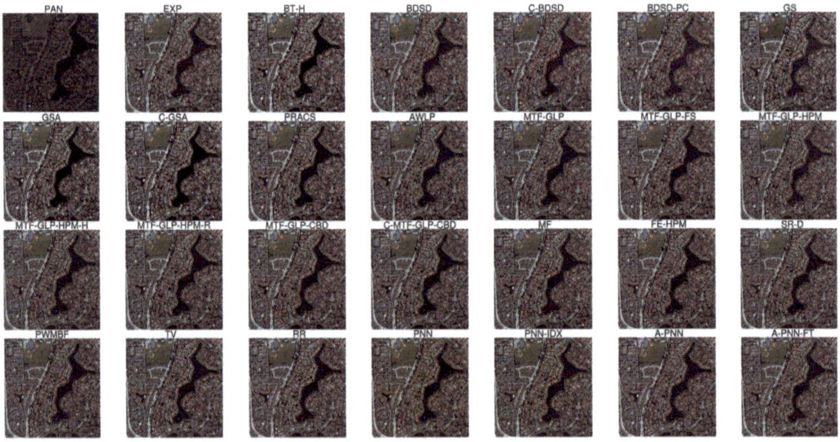

Fig. 14 Fusion results at the FR evaluation on the W2_Miam_Mix dataset

5 Applications

The pansharpening method enables the acquisition of high spatial and spectral resolution RS images in a cost-effective way. Meanwhile, applications relying on RS images are gradually evolving towards finer-resolution scales. Hence, over the past decade, there has been a growing number of applications in RS observations utilizing pansharpening approaches, such as change detection [35], land surface temperature characterization [101], and so on. We have made efforts to gather several representative application scenarios based on the pansharpening method in recent years, as shown in Table 14.

Table 13 The processing durations (measured in seconds) of various methods

		4 bands		8 bands	
		GE_Lond_Urb	GE_Tren_Urb	W2_Miam_Mix	W2_Miam_Urb
CS-based	BT-H	0.88	0.78	0.86	0.87
	BDSD	0.70	0.63	1.11	1.12
	C-BDSD	22.08	22.48	35.39	35.53
	BDSD-PC	0.87	0.60	1.42	1.43
	GS	0.83	0.80	1.55	1.57
	GSA	2.16	2.09	2.98	2.94
	C-GSA	16.51	26.46	21.11	14.81
	PRACS	3.31	3.32	10.53	10.50
MRA-based	AWLP	5.51	5.59	10.97	10.84
	MTF-GLP	1.73	1.72	3.62	3.52
	MTF-GLP-FS	1.36	1.33	2.71	2.65
	MTF-GLP-HPM	1.77	1.74	3.63	3.50
	MTF-GLP-HPM-H	1.84	1.75	3.06	3.03
	MTF-GLP-HPM-R	1.22	1.15	2.45	2.43
	MTF-GLP-CBD	1.45	1.41	3.05	2.93
	C-MTF-GLP-CBD	15.32	27.33	22.49	14.05
	MF	1.52	1.41	2.83	2.77

(continued)

Table 13 (continued)

		4 bands		8 bands	
		GE_Lond_Urb	GE_Tren_Urb	W2_Miam_Mix	W2_Miam_Urb
VO-based	FE-HPM	7.44	7.27	11.13	10.86
	SR-D	267.41	285.97	593.96	636.04
	PWMBF	12.01	11.72	12.40	12.23
	TV	147.02	143.83	311.73	292.31
	RR	358.69	386.29	3286.59	3479.66
ML-based	PNN	6.35	2.61	2.86	2.79
	PNN-IDX	1.41	1.38	2.01	1.96
	A-PNN	1.11	1.04	1.21	1.20
	A-PNN-FT	7.83	5.98	7.19	6.94

Table 14 Some pansharpening-based applications scenarios in the past decade

Article	Year	Data resource	Application areas
[164]	2012	QuickBird, IKONOS, KOMPSAT II	Monitoring coastal barrier island development
[60]	2013	QuickBird	Mapping diseased pine and oak trees
[35]	2013	QuickBird, ALOS	Change detection
[100]	2014	IRS	Mapping and monitoring of the dune
[66]	2015	Landsat-5 TM, World View-2	Vegetation and urban area classification
[153]	2016	QuickBird-2, Worldview-2	Antarctic wildlife monitoring
[112]	2017	CBERS-4, Sentinel-2	Mapping wetland soil cover
[48]	2017	Landsat-8	Crop type differentiation
[113]	2017	Synthetic images	Anomaly detection
[64]	2021	PRISMA	Marine plastic litter detection
[101]	2022	Landsat-8	Land surface temperature characterization
[123]	2023	WorldView-2	Tree aboveground carbon mapping
[34]	2023	Landsat-5, Sentinel-2	Monitoring urban forest cover and LST change

6 Summary

The increasing quantity of multimodal RS images presents both a challenge and an opportunity for Earth observation tasks. The inadequate spatial resolution of multispectral images poses a significant limitation in numerous observational tasks, leading to the distortion of information. To enhance the spatial information of multispectral images, multi-image fusion based on pansharpening becomes a significant topic, which is a crucial branch within super-resolution technology. This chapter provides a thorough introduction to the rapidly expanding field, encompassing a literature analysis, a systematic summary of several widely used pansharpening methods, a compilation of available resources, the evaluation methodologies, and various application scenarios. Furthermore, a detailed analysis of evaluation procedure was conducted, specifically focusing on both reduced resolution and full resolution and delves into the methods, datasets, and metrics used, achieved through experiments performed on publicly available resources.

References

1. Addesso, P., Restaino, R., Vivone, G.: An improved version of the generalized laplacian pyramid algorithm for pansharpening. Remote Sens. **13**(17), 3386 (2021)
2. Aiazzi, B., Alparone, L., Baronti, S., Carlà, R., Garzelli, A., Santurri, L.: Full-scale assessment of pansharpening methods and data products. In: Image and Signal Processing for Remote Sensing XX, vol. 9244, p. 924402. SPIE, New York (2014)
3. Aiazzi, B., Alparone, L., Baronti, S., Garzelli, A., Selva, M.: Mtf-tailored multiscale fusion of high-resolution ms and pan imagery. Photogramm. Eng. Remote Sens. **72**(5), 591–596 (2006)

4. Aiazzi, B., Baronti, S., Lotti, F., Selva, M.: A comparison between global and context-adaptive pansharpening of multispectral images. IEEE Geosci. Remote Sens. Lett. **6**(2), 302–306 (2009)
5. Aiazzi, B., Baronti, S., Selva, M.: Improving component substitution pansharpening through multivariate regression of ms + pan data. IEEE Trans. Geosci. Remote Sens. **45**(10), 3230–3239 (2007)
6. Alparone, L., Aiazzi, B., Baronti, S., Garzelli, A., Nencini, F., Selva, M.: Multispectral and panchromatic data fusion assessment without reference. Photogramm. Eng. Remote Sens. **74**(2), 193–200 (2008)
7. Alparone, L., Baronti, S., Aiazzi, B., Garzelli, A.: Spatial methods for multispectral pansharpening: Multiresolution analysis demystified. IEEE Trans. Geosci. Remote Sens. **54**(5), 2563–2576 (2016)
8. Alparone, L., Wald, L., Chanussot, J., Thomas, C., Gamba, P., Bruce, L.M.: Comparison of pansharpening algorithms: Outcome of the 2006 GRS-S data-fusion contest. IEEE Trans. Geosci. Remote Sens. **45**(10), 3012–3021 (2007)
9. Arienzo, A., Aiazzi, B., Alparone, L., Garzelli, A.: Reproducibility of pansharpening methods and quality indexes versus data formats. Remote Sens. **13**(21), 4399 (2021)
10. Asadzadeh, S., de Oliveira, W.J., de Souza Filho, C.R.: Uav-based remote sensing for the petroleum industry and environmental monitoring: state-of-the-art and perspectives. J. Pet. Sci. Eng. **208**, 109633 (2022)
11. Azarang, A., Ghassemian, H.: A new pansharpening method using multi resolution analysis framework and deep neural networks. In: 2017 3rd International Conference on Pattern Recognition and Image Analysis (IPRIA), pp. 1–6. IEEE, New York (2017)
12. Ballester, C., Caselles, V., Igual, L., Verdera, J., Rougé, B.: A variational model for p+ xs image fusion. Int. J. Comput. Vis. **69**, 43–58 (2006)
13. Bao, K., Meng, X., Chai, X., Shao, F.: A blind full resolution assessment method for pansharpened images based on multistream collaborative learning. IEEE Trans. Geosci. Remote Sens. **60**, 1–11 (2022)
14. Baronti, S., Aiazzi, B., Selva, M., Garzelli, A., Alparone, L.: A theoretical analysis of the effects of aliasing and misregistration on pansharpened imagery. IEEE J. Sel. Top. Sign. Proces. **5**(3), 446–453 (2011)
15. Baumgardner, M.F., Biehl, L.L., Landgrebe, D.A.: 220 band aviris hyperspectral image data set: June 12, 1992 indian pine test site 3. Purdue University Research Repository **10**(7), 991 (2015)
16. Burt, P.J., Adelson, E.H.: The laplacian pyramid as a compact image code. In: Readings in Computer Vision, pp. 671–679. Elsevier, Amsterdam (1987)
17. Carpentiero, M., Vivone, G., Restaino, R., Addesso, P., Chanussot, J.: An optimization procedure for robust regression-based pansharpening. IEEE Trans. Geosci. Remote Sens. **60**, 1–16 (2022)
18. Carper, W., Lillesand, T., Kiefer, R., et al.: The use of intensity-hue-saturation transformations for merging spot panchromatic and multispectral image data. Photogramm. Eng. Remote. Sens. **56**(4), 459–467 (1990)
19. Chen, L., Lai, Z., Vivone, G., Jeon, G., Chanussot, J., Yang, X.: Arbrpn: A bidirectional recurrent pansharpening network for multispectral images with arbitrary numbers of bands. IEEE Trans. Geosci. Remote Sens. **60**, 1–18 (2021)
20. Chen, S., Qi, H., Nan, K.: Pansharpening via super-resolution iterative residual network with a cross-scale learning strategy. IEEE Trans. Geosci. Remote Sens. **60**, 1–16 (2021)
21. Cheng, J., Liu, H., Liu, T., Wang, F., Li, H.: Remote sensing image fusion via wavelet transform and sparse representation. ISPRS J. Photogramm. Remote Sens. **104**, 158–173 (2015)
22. Choi, J., Yu, K., Kim, Y.: A new adaptive component-substitution-based satellite image fusion by using partial replacement. IEEE Trans. Geosci. Remote Sens. **49**(1), 295–309 (2010)
23. Ciotola, M., Poggi, G., Scarpa, G.: Unsupervised deep learning-based pansharpening with jointly-enhanced spectral and spatial fidelity. In: IEEE Transactions on Geoscience and Remote Sensing (2023)

24. Ciotola, M., Vitale, S., Mazza, A., Poggi, G., Scarpa, G.: Pansharpening by convolutional neural networks in the full resolution framework. IEEE Trans. Geosci. Remote Sens. **60**, 1–17 (2022)
25. Civicioglu, P., Besdok, E.: Contrast stretching based pansharpening by using weighted differential evolution algorithm. Expert Syst. Appl. **208**, 118144 (2022)
26. Civicioglu, P., Besdok, E.: Pansharpening of remote sensing images using dominant pixels. Expert Syst. Appl. **242**, 122783 (2024)
27. Deng, L.J., Vivone, G., Paoletti, M.E., Scarpa, G., He, J., Zhang, Y., Chanussot, J., Plaza, A.: Machine learning in pansharpening: A benchmark, from shallow to deep networks. IEEE Geosci. Remote Sens. Mag. **10**(3), 279–315 (2022)
28. Devulapalli, S., Krishnan, R.: Synthesized pansharpening using curvelet transform and adaptive neuro-fuzzy inference system. J. Appl. Remote. Sens. **13**(3), 034519–034519 (2019)
29. Diao, W., Zhang, F., Sun, J., Xing, Y., Zhang, K., Bruzzone, L.: Zergan: zero-reference gan for fusion of multispectral and panchromatic images. IEEE Trans. Neural Networks Learn. Syst. **34**(11), 8195–8209 (2022)
30. Do, M.N., Vetterli, M.: The contourlet transform: an efficient directional multiresolution image representation. IEEE Trans. Image Process. **14**(12), 2091–2106 (2005)
31. Dong, L., Yang, Q., Wu, H., Xiao, H., Xu, M.: High quality multi-spectral and panchromatic image fusion technologies based on curvelet transform. Neurocomputing **159**, 268–274 (2015)
32. Dou, W.: Image degradation for quality assessment of pan-sharpening methods. Remote Sens. **10**(1), 154 (2018)
33. Dou, W., Chen, Y., Li, X., Sui, D.Z.: A general framework for component substitution image fusion: an implementation using the fast image fusion method. Comput. Geosci. **33**(2), 219–228 (2007)
34. Doyog, N.D.: Urban forest cover and LST change monitoring through optical and thermal remote sensing approach. J. Indian Soc. Remote Sens. **51**(12), 2467–2480 (2023)
35. Du, P., Liu, S., Xia, J., Zhao, Y.: Information fusion techniques for change detection from multi-temporal remote sensing images. Inf. Fusion **14**(1), 19–27 (2013)
36. Duran, J., Buades, A.: Restoration of pansharpened images by conditional filtering in the pca domain. IEEE Geosci. Remote Sens. Lett. **16**(3), 442–446 (2018)
37. El-Mezouar, M.C., Kpalma, K., Taleb, N., Ronsin, J.: A pan-sharpening based on the non-subsampled contourlet transform: application to worldview-2 imagery. IEEE J. Sel. Top. Appl. Earth Obs. Remote Sens. **7**(5), 1806–1815 (2014)
38. Fang, F., Li, F., Shen, C., Zhang, G.: A variational approach for pan-sharpening. IEEE Trans. Image Process. **22**(7), 2822–2834 (2013)
39. Farhadi, H., Esmaeily, A., Najafzadeh, M.: Flood monitoring by integration of remote sensing technique and multi-criteria decision making method. Comput. Geosci. **160**, 105045 (2022)
40. Fei, R., Zhang, J., Liu, J., Du, F., Hu, J., Chang, P., Zhou, C., Sun, K.: Weighted manifold regularized sparse representation of featured injected details for pansharpening. Int. J. Remote Sens. **42**(11), 4199–4223 (2021)
41. Fei, R., Zhang, J.S., Liu, J., Du, F., Chang, P., Hu, J.: Manifold regularized sparse representation of injected details for pansharpening. Int. J. Remote Sens. **40**(22), 8395–8417 (2019)
42. Feng, Y., Yan, B., Jeon, S., Yang, X.: A hyperspectral pansharpening method using retrain transformer network for remote sensing images in UAV communications system. Wirel. Netw., 1–14 (2024)
43. Garzelli, A.: Pansharpening of multispectral images based on nonlocal parameter optimization. IEEE Trans. Geosci. Remote Sens. **53**(4), 2096–2107 (2014)
44. Garzelli, A., Nencini, F.: Hypercomplex quality assessment of multi/hyperspectral images. IEEE Geosci. Remote Sens. Lett. **6**(4), 662–665 (2009)
45. Garzelli, A., Nencini, F., Capobianco, L.: Optimal mmse pan sharpening of very high resolution multispectral images. IEEE Trans. Geosci. Remote Sens. **46**(1), 228–236 (2007)
46. Ghahremani, M., Ghassemian, H.: Nonlinear ihs: A promising method for pan-sharpening. IEEE Geosci. Remote Sens. Lett. **13**(11), 1606–1610 (2016)

47. Ghassemian, H.: A review of remote sensing image fusion methods. Inf. Fusion **32**, 75–89 (2016)
48. Gilbertson, J.K., Kemp, J., Van Niekerk, A.: Effect of pan-sharpening multi-temporal landsat 8 imagery for crop type differentiation using different classification techniques. Comput. Electron. Agric. **134**, 151–159 (2017)
49. Gomes, V.C., Queiroz, G.R., Ferreira, K.R.: An overview of platforms for big earth observation data management and analysis. Remote Sens. **12**(8), 1253 (2020)
50. Gruen, A.: Scientific-technological developments in photogrammetry and remote sensing between 2004 and 2008. In: Advances in Photogrammetry, Remote Sensing and Spatial Information Sciences: 2008 ISPRS Congress Book, pp. 39–44. CRC Press, New York (2008)
51. Guan, X., Li, F., Zhang, X., Ma, M., Mei, S.: Assessing full-resolution pansharpening quality: a comparative study of methods and measurements. IEEE J. Sel. Top. Appl. Earth Obs. Remote Sens. **16**, 6860–6875 (2023)
52. Guanter, L., Kaufmann, H., Segl, K., Foerster, S., Rogass, C., Chabrillat, S., Kuester, T., Hollstein, A., Rossner, G., Chlebek, C., et al.: The enmap spaceborne imaging spectroscopy mission for earth observation. Remote Sens. **7**(7), 8830–8857 (2015)
53. Günen, M.A.: Weighted differential evolution algorithm based pansharpening. Int. J. Remote Sens. **42**(22), 8468–8491 (2021)
54. Han, X., Leng, W., Xu, Q., Li, W., Tao, R., Sun, W.: A joint optimization based pansharpening via subpixel-shift decomposition. IEEE Trans. Geosci. Remote Sens. **61**, 1–15 (2023)
55. Haydn, R.: Application of the ihs color transform to the processing of multisensor data and image enhancement. In: Proceedings of the International Symposium on Remote Sensing of Arid and Semi-Arid Lands, Cairo, Egypt, 1982 (1982)
56. He, J., Yuan, Q., Li, J., Xiao, Y., Zhang, L.: A self-supervised remote sensing image fusion framework with dual-stage self-learning and spectral super-resolution injection. ISPRS J. Photogramm. Remote Sens. **204**, 131–144 (2023)
57. He, L., Xi, D., Li, J., Lai, H., Plaza, A., Chanussot, J.: Dynamic hyperspectral pansharpening CNNS. IEEE Trans. Geosci. Remote Sens. **61**, 1–19 (2023)
58. He, L., Zhang, W., Shi, J., Li, F.: Cross-domain association mining based generative adversarial network for pansharpening. IEEE J. Sel. Top. Appl. Earth Obs. Remote Sens. **15**, 7770–7783 (2022)
59. He, L., Zhu, J., Li, J., Plaza, A., Chanussot, J., Li, B.: Hyperpnn: Hyperspectral pansharpening via spectrally predictive convolutional neural networks. IEEE J. Sel. Top. Appl. Earth Obs. Remote Sens. **12**(8), 3092–3100 (2019)
60. Johnson, B.A., Tateishi, R., Hoan, N.T.: A hybrid pansharpening approach and multiscale object-based image analysis for mapping diseased pine and oak trees. Int. J. Remote Sens. **34**(20), 6969–6982 (2013)
61. Kaur, G., Saini, K.S., Singh, D., Kaur, M.: A comprehensive study on computational pansharpening techniques for remote sensing images. In: Archives of Computational Methods in Engineering, pp. 1–18 (2021)
62. Ke, C., Zhang, W., Wang, Z., Ma, J., Tian, X.: Coarse-to-fine cross-domain learning fusion network for pansharpening. IEEE Trans. Geosci. Remote Sens. **61**, 1–14 (2023)
63. Kramer, H.J., et al.: Observation of the Earth and Its Environment: Survey of Missions and Sensors, vol. 10. Springer, Berlin (2002)
64. Kremezi, M., Kristollari, V., Karathanassi, V., Topouzelis, K., Kolokoussis, P., Taggio, N., Aiello, A., Ceriola, G., Barbone, E., Corradi, P.: Pansharpening prisma data for marine plastic litter detection using plastic indexes. IEEE Access **9**, 61955–61971 (2021)
65. Kulkarni, S.C., Rege, P.P.: Pixel level fusion techniques for sar and optical images: a review. Inf. Fusion **59**, 13–29 (2020)
66. Kumar, U., Milesi, C., Nemani, R.R., Basu, S.: Multi-sensor multi-resolution image fusion for improved vegetation and urban area classification. Int. Arch. Photogramm. Remote. Sens. Spat. Inf. Sci. **40**, 51–58 (2015)
67. Kwan, C., Budavari, B., Bovik, A.C., Marchisio, G.: Blind quality assessment of fused worldview-3 images by using the combinations of pansharpening and hypersharpening paradigms. IEEE Geosci. Remote Sens. Lett. **14**(10), 1835–1839 (2017)

68. Kwarteng, P., Chavez, A.: Extracting spectral contrast in landsat thematic mapper image data using selective principal component analysis. Photogramm. Eng. Remote Sens. **55**(1), 339–348 (1989)
69. Laben, C.A., Brower, B.V.: Process for enhancing the spatial resolution of multispectral imagery using pan-sharpening (2000). US Patent 6,011,875
70. Lang, W., Zhao, Z., Fang, S., Cao, Y., Wang, Y.: Sparse representation-based detail-injection method for pan-sharpening. J. Appl. Remote. Sens. **14**(2), 026523–026523 (2020)
71. Lei, D., Huang, J., Zhang, L., Li, W.: Mhanet: A multiscale hierarchical pansharpening method with adaptive optimization. IEEE Trans. Geosci. Remote Sens. **60**, 1–15 (2022)
72. Leung, Y., Liu, J., Zhang, J.: An improved adaptive intensity–hue–saturation method for the fusion of remote sensing images. IEEE Geosci. Remote Sens. Lett. **11**(5), 985–989 (2013)
73. Li, S., Tian, Y., Wang, C., Wu, H., Zheng, S.: Cross spectral and spatial scale non-local attention based unsupervised pansharpening network. IEEE J. Sel. Top. Appl. Earth Obs. Remote Sens. **16**, 4858–4870 (2023)
74. Li, S., Tian, Y., Wang, C., Wu, H., Zheng, S.: Hyperspectral image super-resolution network based on cross-scale non-local attention. IEEE Trans. Geosci. Remote Sens. **61**, 1–15 (2023)
75. Li, S., Tian, Y., Xia, H., Liu, Q.: Unmixing-based pan-guided fusion network for hyperspectral imagery. IEEE Trans. Geosci. Remote Sens. **60**, 1–17 (2022)
76. Li, S., Yang, B.: A new pan-sharpening method using a compressed sensing technique. IEEE Trans. Geosci. Remote Sens. **49**(2), 738–746 (2010)
77. Li, X., Liu, L., Huang, L.: Comparison of several remote sensing image classification methods based on envi. Int. Arch. Photogramm. Remote. Sens. Spat. Inf. Sci. **42**, 605–611 (2020)
78. Li, Z., Leung, H.: Fusion of multispectral and panchromatic images using a restoration-based method. IEEE Trans. Geosci. Remote Sens. **47**(5), 1482–1491 (2008)
79. Liang, L., Zhang, Y., Zhang, S., Li, J., Plaza, A., Kang, X.: Fast hyperspectral image classification combining transformers and simam-based CNNs. IEEE Trans. Geosci. Remote Sens. **61**, 1–19 (2023)
80. Lin, L., Yin, C., Gao, Z.: Pan-sharpening with global multi-scale context network. In: International Conference on Artificial Neural Networks, pp. 375–385. Springer, Berelin (2023)
81. Liu, P.: Pansharpening with spatial hessian non-convex sparse and spectral gradient low rank priors. IEEE Trans. Image Process. **32**, 2120–2131 (2023)
82. Liu, P., Li, Y.: Spectral gradient fidelity and spatial hessian hyper-laplacian sparsity constraints for variational pansharpening. IEEE J. Sel. Top. Appl. Earth Obs. Remote Sens. **15**, 6086–6098 (2022)
83. Liu, P., Xiao, L.: A nonconvex pansharpening model with spatial and spectral gradient difference-induced nonconvex sparsity priors. IEEE Trans. Geosci. Remote Sens. **60**, 1–15 (2021)
84. Liu, X., Liu, X., Dai, H., Kan, X., Plaza, A., Zu, W.: Mun-GAN: A multi-scale unsupervised network for remote sensing image pansharpening. IEEE Trans. Geosci. Remote Sens. **61**, 1–18 (2023)
85. Lolli, S., Alparone, L., Garzelli, A., Vivone, G.: Haze correction for contrast-based multispectral pansharpening. IEEE Geosci. Remote Sens. Lett. **14**(12), 2255–2259 (2017)
86. Lu, H., Yang, Y., Huang, S., Tu, W., Wan, W.: A unified pansharpening model based on band-adaptive gradient and detail correction. IEEE Trans. Image Process. **31**, 918–933 (2021)
87. Luo, S., Zhou, S., Feng, Y., Xie, J.: Pansharpening via unsupervised convolutional neural networks. IEEE J. Sel. Top. Appl. Earth Obs. Remote Sens. **13**, 4295–4310 (2020)
88. Luo, S., Zhou, S., Qi, Y.: Csafnet: Channel similarity attention fusion network for multispectral pansharpening. IEEE Geosci. Remote Sens. Lett. **19**, 1–5 (2020)
89. Luo, Y., Liu, R., Zhu, Y.F.: Fusion of remote sensing image base on the pca+ atrous wavelet transform. Int. Arch. Photogramm. Remote. Sens. Spat. Inf. Sci. **37**(Part B7), 1155–1158 (2008)
90. Ma, J., Yu, W., Chen, C., Liang, P., Guo, X., Jiang, J.: Pan-gan: An unsupervised pansharpening method for remote sensing image fusion. Inf. Fusion **62**, 110–120 (2020)

91. Mallat, S.G.: A theory for multiresolution signal decomposition: the wavelet representation. IEEE Trans. Pattern Anal. Mach. Intell. **11**(7), 674–693 (1989)
92. Maneshi, M., Ghassemian, H., Imani, M.: Sparse representation of injected details for mra-based pansharpening. In: 2020 IEEE India Geoscience and Remote Sensing Symposium (InGARSS), pp. 86–89. IEEE, Berlin (2020)
93. Mascarenhas, N., Banon, G., Candeias, A.: Multispectral image data fusion under a Bayesian approach. Int. J. Remote Sens. **17**(8), 1457–1471 (1996)
94. Masi, G., Cozzolino, D., Verdoliva, L., Scarpa, G.: Pansharpening by convolutional neural networks. Remote Sens. **8**(7), 594 (2016)
95. Meng, X., Bao, K., Shu, J., Zhou, B., Shao, F., Sun, W., Li, S.: A blind full-resolution quality evaluation method for pansharpening. IEEE Trans. Geosci. Remote Sens. **60**, 1–16 (2021)
96. Meng, X., Shen, H., Li, H., Zhang, L., Fu, R.: Review of the pansharpening methods for remote sensing images based on the idea of meta-analysis: practical discussion and challenges. Inf. Fusion **46**, 102–113 (2019)
97. Meng, X., Wang, N., Shao, F., Li, S.: Vision transformer for pansharpening. IEEE Trans. Geosci. Remote Sens. **60**, 1–11 (2022)
98. Meng, X., Xiong, Y., Shao, F., Shen, H., Sun, W., Yang, G., Yuan, Q., Fu, R., Zhang, H.: A large-scale benchmark data set for evaluating pansharpening performance: overview and implementation. IEEE Geosci. Remote Sens. Mag. **9**(1), 18–52 (2020)
99. Middleton, E.M., Ungar, S.G., Mandl, D.J., Ong, L., Frye, S.W., Campbell, P.E., Landis, D.R., Young, J.P., Pollack, N.H.: The earth observing one (EO-1) satellite mission: over a decade in space. IEEE J. Sel. Top. Appl. Earth Obs. Remote Sens. **6**(2), 243–256 (2013)
100. Moosavi, V., Shamsi, S.R.F., Moradi, H., Shirmohammadi, B.: Application of taguchi method to satellite image fusion for object-oriented mapping of barchan dunes. Geosci. J. **18**, 45–59 (2014)
101. Mushore, T.D., Mutanga, O., Odindi, J., Sadza, V., Dube, T.: Pansharpened landsat 8 thermal-infrared data for improved land surface temperature characterization in a heterogeneous urban landscape. Remote Sens. Appl.: Soc. Environ. **26**, 100728 (2022)
102. Ni, J., Shao, Z., Zhang, Z., Hou, M., Zhou, J., Fang, L., Zhang, Y.: LDP-Net: An unsupervised pansharpening network based on learnable degradation processes. IEEE J. Sel. Top. Appl. Earth Obs. Remote Sens. **15**, 5468–5479 (2022)
103. Nunez, J., Otazu, X., Fors, O., Prades, A., Pala, V., Arbiol, R.: Multiresolution-based image fusion with additive wavelet decomposition. IEEE Trans. Geosci. Remote Sens. **37**(3), 1204–1211 (1999)
104. Otazu, X., González-Audícana, M., Fors, O., Núñez, J.: Introduction of sensor spectral response into image fusion methods: application to wavelet-based methods. IEEE Trans. Geosci. Remote Sens. **43**(10), 2376–2385 (2005)
105. Özay, E.K., Tunga, B.: A novel method for multispectral image pansharpening based on high dimensional model representation. Expert Syst. Appl. **170**, 114512 (2021)
106. Pajares, G.: Overview and current status of remote sensing applications based on unmanned aerial vehicles (UAVs). Photogramm. Eng. Remote Sens. **81**(4), 281–330 (2015)
107. Palsson, F., Sveinsson, J.R., Ulfarsson, M.O.: A new pansharpening algorithm based on total variation. IEEE Geosci. Remote Sens. Lett. **11**(1), 318–322 (2013)
108. Palsson, F., Sveinsson, J.R., Ulfarsson, M.O., Benediktsson, J.A.: Model-based fusion of multi-and hyperspectral images using pca and wavelets. IEEE Trans. Geosci. Remote Sens. **53**(5), 2652–2663 (2014)
109. Palsson, F., Ulfarsson, M.O., Sveinsson, J.R.: Model-based reduced-rank pansharpening. IEEE Geosci. Remote Sens. Lett. **17**(4), 656–660 (2019)
110. Pandit, V.R., Bhiwani, R.: Image fusion in remote sensing applications: a review. Int. J. Comput. Appl. **120**(10), 22–32 (2015)
111. Pepe, M., Fregonese, L., Scaioni, M.: Planning airborne photogrammetry and remote-sensing missions with modern platforms and sensors. Eur. J. Remote Sens. **51**(1), 412–436 (2018)
112. Pereira, O.J.R., Melfi, A.J., Montes, C.R.: Image fusion of Sentinel-2 and CBERS-4 satellites for mapping soil cover in the Wetlands of Pantanal. Int. J. Image Data Fusion **8**(2), 148–172 (2017)

113. Qu, Y., Qi, H., Ayhan, B., Kwan, C., Kidd, R.: Does multispectral/hyperspectral pansharpening improve the performance of anomaly detection? In: 2017 IEEE International Geoscience and Remote Sensing Symposium (IGARSS), pp. 6130–6133. IEEE, New York (2017)
114. Rai, P.K., Chandel, R.S., Mishra, V.N., Singh, P.: Hydrological inferences through morphometric analysis of lower kosi river basin of India for water resource management based on remote sensing data. Appl. Water Sci. **8**, 1–16 (2018)
115. Renza, D., Martinez, E., Arquero, A.: A new approach to change detection in multispectral images by means of ergas index. IEEE Geosci. Remote Sens. Lett. **10**(1), 76–80 (2012)
116. Restaino, R., Dalla Mura, M., Vivone, G., Chanussot, J.: Context-adaptive pansharpening based on image segmentation. IEEE Trans. Geosci. Remote Sens. **55**(2), 753–766 (2016)
117. Restaino, R., Vivone, G., Dalla Mura, M., Chanussot, J.: Fusion of multispectral and panchromatic images based on morphological operators. IEEE Trans. Image Process. **25**(6), 2882–2895 (2016)
118. Scarpa, G., Ciotola, M.: Full-resolution quality assessment for pansharpening. Remote Sens. **14**(8), 1808 (2022)
119. Scarpa, G., Vitale, S., Cozzolino, D.: Target-adaptive CNN-based pansharpening. IEEE Trans. Geosci. Remote Sens. **56**(9), 5443–5457 (2018)
120. Schowengerdt, R.A.: Reconstruction of multispatial, multispectral image data using spatial frequency content. Photogramm. Eng. Remote. Sens. **46**(10), 1325–1334 (1980)
121. Shao, Z., Cai, J.: Remote sensing image fusion with deep convolutional neural network. IEEE J. Sel. Top. Appl. Earth Obs. Remote Sens. **11**(5), 1656–1669 (2018)
122. Shen, H., Jiang, M., Li, J., Yuan, Q., Wei, Y., Zhang, L.: Spatial–spectral fusion by combining deep learning and variational model. IEEE Trans. Geosci. Remote Sens. **57**(8), 6169–6181 (2019)
123. Singh, N., Nandy, S., van Leeuwen, L.M: Tree Aboveground Carbon Mapping in an Indian Tropical Moist Deciduous Forest Using Object-Based Image Analysis and Very High Resolution Satellite Imagery. J Indian Soc Remote Sens **52**, 723–734 (2024)
124. Starck, J.L., Candès, E.J., Donoho, D.L.: The curvelet transform for image denoising. IEEE Trans. Image Process. **11**(6), 670–684 (2002)
125. Thomas, C., Ranchin, T., Wald, L., Chanussot, J.: Synthesis of multispectral images to high spatial resolution: a critical review of fusion methods based on remote sensing physics. IEEE Trans. Geosci. Remote Sens. **46**(5), 1301–1312 (2008)
126. Tian, X., Chen, Y., Yang, C., Gao, X., Ma, J.: A variational pansharpening method based on gradient sparse representation. IEEE Signal Process Lett. **27**, 1180–1184 (2020)
127. Toth, C., Jóźków, G.: Remote sensing platforms and sensors: A survey. ISPRS J. Photogramm. Remote Sens. **115**, 22–36 (2016)
128. Tu, T.M., Huang, P.S., Hung, C.L., Chang, C.P.: A fast intensity-hue-saturation fusion technique with spectral adjustment for IKONOS imagery. IEEE Geosci. Remote Sens. Lett. **1**(4), 309–312 (2004)
129. Tu, T.M., Su, S.C., Shyu, H.C., Huang, P.S.: A new look at ihs-like image fusion methods. Inf. Fusion **2**(3), 177–186 (2001)
130. Valizadeh, S.A., Ghassemian, H.: Remote sensing image fusion using combining IHS and Curvelet transform. In: 6th International Symposium on Telecommunications (IST), pp. 1184–1189. IEEE, New York (2012)
131. Vicinanza, M.R., Restaino, R., Vivone, G., Dalla Mura, M., Chanussot, J.: A pansharpening method based on the sparse representation of injected details. IEEE Geosci. Remote Sens. Lett. **12**(1), 180–184 (2014)
132. Vivone, G.: Robust band-dependent spatial-detail approaches for panchromatic sharpening. IEEE Trans. Geosci. Remote Sens. **57**(9), 6421–6433 (2019)
133. Vivone, G.: Multispectral and hyperspectral image fusion in remote sensing: a survey. Inf. Fusion **89**, 405–417 (2023)
134. Vivone, G., Alparone, L., Chanussot, J., Dalla Mura, M., Garzelli, A., Licciardi, G.A., Restaino, R., Wald, L.: A critical comparison among pansharpening algorithms. IEEE Trans. Geosci. Remote Sens. **53**(5), 2565–2586 (2014)

135. Vivone, G., Dalla Mura, M., Garzelli, A., Pacifici, F.: A benchmarking protocol for pansharpening: Dataset, preprocessing, and quality assessment. IEEE J. Sel. Top. Appl. Earth Obs. Remote Sens. **14**, 6102–6118 (2021)
136. Vivone, G., Dalla Mura, M., Garzelli, A., Restaino, R., Scarpa, G., Ulfarsson, M.O., Alparone, L., Chanussot, J.: A new benchmark based on recent advances in multispectral pansharpening: revisiting pansharpening with classical and emerging pansharpening methods. IEEE Geosci. Remote Sens. Mag. **9**(1), 53–81 (2020)
137. Vivone, G., Garzelli, A., Xu, Y., Liao, W., Chanussot, J.: Panchromatic and hyperspectral image fusion: outcome of the 2022 whispers hyperspectral pansharpening challenge. IEEE J. Sel. Top. Appl. Earth Obs. Remote Sens. **16**, 166–179 (2022)
138. Vivone, G., Marano, S., Chanussot, J.: Pansharpening: Context-based generalized Laplacian pyramids by robust regression. IEEE Trans. Geosci. Remote Sens. **58**(9), 6152–6167 (2020)
139. Vivone, G., Restaino, R., Chanussot, J.: A regression-based high-pass modulation pansharpening approach. IEEE Trans. Geosci. Remote Sens. **56**(2), 984–996 (2017)
140. Vivone, G., Restaino, R., Chanussot, J.: Full scale regression-based injection coefficients for panchromatic sharpening. IEEE Trans. Image Process. **27**(7), 3418–3431 (2018)
141. Vivone, G., Restaino, R., Dalla Mura, M., Licciardi, G., Chanussot, J.: Contrast and error-based fusion schemes for multispectral image pansharpening. IEEE Geosci. Remote Sens. Lett. **11**(5), 930–934 (2013)
142. Vivone, G., Simões, M., Dalla Mura, M., Restaino, R., Bioucas-Dias, J.M., Licciardi, G.A., Chanussot, J.: Pansharpening based on semiblind deconvolution. IEEE Trans. Geosci. Remote Sens. **53**(4), 1997–2010 (2014)
143. Wald, L., Ranchin, T., Mangolini, M.: Fusion of satellite images of different spatial resolutions: assessing the quality of resulting images. Photogramm. Eng. Remote. Sens. **63**(6), 691–699 (1997)
144. Wang, D., Zhang, P., Bai, Y., Li, Y.: Metapan: unsupervised adaptation with meta-learning for multispectral pansharpening. IEEE Geosci. Remote Sens. Lett. **19**, 1–5 (2022)
145. Wang, W., Jiao, L., Yang, S.: Novel adaptive component-substitution-based pan-sharpening using particle swarm optimization. IEEE Geosci. Remote Sens. Lett. **12**(4), 781–785 (2014)
146. Wang, W., Liu, H.: An efficient detail extraction algorithm for improving haze-corrected CS pansharpening. IEEE Geosci. Remote Sens. Lett. **19**, 1–5 (2021)
147. Wang, Y., Liu, G., Zhang, R., Liu, J.: A two-stage pansharpening method for the fusion of remote-sensing images. Remote Sens. **14**(5), 1121 (2022)
148. Wang, Z., Bovik, A.C.: A universal image quality index. IEEE Signal Process Lett. **9**(3), 81–84 (2002)
149. Wang, Z., Bovik, A.C., Sheikh, H.R., Simoncelli, E.P.: Image quality assessment: from error visibility to structural similarity. IEEE Trans. Image Process. **13**(4), 600–612 (2004)
150. Wang, Z., Ng, M.K., Michalski, J., Zhuang, L.: A self-supervised deep denoiser for hyperspectral and multispectral image fusion. In: IEEE Transactions on Geoscience and Remote Sensing (2023)
151. Wang, Z., Ziou, D., Armenakis, C., Li, D., Li, Q.: A comparative analysis of image fusion methods. IEEE Trans. Geosci. Remote Sens. **43**(6), 1391–1402 (2005)
152. Wei, Q., Dobigeon, N., Tourneret, J.Y.: Fast fusion of multi-band images based on solving a sylvester equation. IEEE Trans. Image Process. **24**(11), 4109–4121 (2015)
153. Witharana, C., LaRue, M.A., Lynch, H.J.: Benchmarking of data fusion algorithms in support of earth observation based antarctic wildlife monitoring. ISPRS J. Photogramm. Remote Sens. **113**, 124–143 (2016)
154. Wu, K., Yang, X., Nie, Z., Li, H., Jeon, G.: A dual attention transformer network for pansharpening. IEEE Sensors J. **24**(5), 5500–5511 (2023)
155. Wu, Z.C., Huang, T.Z., Deng, L.J., Hu, J.F., Vivone, G.: Vo+ net: An adaptive approach using variational optimization and deep learning for panchromatic sharpening. IEEE Trans. Geosci. Remote Sens. **60**, 1–16 (2021)

156. Wu, Z.C., Huang, T.Z., Deng, L.J., Vivone, G., Miao, J.Q., Hu, J.F., Zhao, X.L.: A new variational approach based on proximal deep injection and gradient intensity similarity for spatio-spectral image fusion. IEEE J. Sel. Top. Appl. Earth Obs. Remote Sens. **13**, 6277–6290 (2020)
157. Wulder, M.A., White, J.C., Loveland, T.R., Woodcock, C.E., Belward, A.S., Cohen, W.B., Fosnight, E.A., Shaw, J., Masek, J.G., Roy, D.P.: The global landsat archive: Status, consolidation, and direction. Remote Sens. Environ. **185**, 271–283 (2016)
158. Xiao, J.L., Huang, T.Z., Deng, L.J., Wu, Z.C., Wu, X., Vivone, G.: Variational pansharpening based on coefficient estimation with nonlocal regression. IEEE Transactions on Geoscience and Remote Sensing (2023)
159. Xiong, Z., Liu, N., Wang, N., Sun, Z., Li, W.: Unsupervised pansharpening method using residual network with spatial texture attention. IEEE Transactions on Geoscience and Remote Sensing (2023)
160. Xu, Q., Li, B., Zhang, Y., Ding, L.: High-fidelity component substitution pansharpening by the fitting of substitution data. IEEE Trans. Geosci. Remote Sens. **52**(11), 7380–7392 (2014)
161. Xu, S., Amira, O., Liu, J., Zhang, C.X., Zhang, J., Li, G.: HAM-MFN: hyperspectral and multispectral image multiscale fusion network with RAP loss. IEEE Trans. Geosci. Remote Sens. **58**(7), 4618–4628 (2020)
162. Yan, K., Zhou, M., Liu, L., Xie, C., Hong, D.: When pansharpening meets graph convolution network and knowledge distillation. IEEE Trans. Geosci. Remote Sens. **60**, 1–15 (2022)
163. Yan, Y., Liu, J., Xu, S., Wang, Y., Cao, X.: Md^3 net: integrating model-driven and data-driven approaches for pansharpening. IEEE Trans. Geosci. Remote Sens. **60**, 1–16 (2022)
164. Yang, B., Kim, M., Madden, M.: Assessing optimal image fusion methods for very high spatial resolution satellite images to support coastal monitoring. GISci. Remote Sens. **49**(5), 687–710 (2012)
165. Yang, Y., Lu, H., Huang, S., Tu, W.: Pansharpening based on joint-guided detail extraction. IEEE J. Sel. Top. Appl. Earth Obs. Remote Sens. **14**, 389–401 (2020)
166. Ye, F., Guo, Y., Zhuang, P.: Pan-sharpening via a gradient-based deep network prior. Signal Process. Image Commun. **74**, 322–331 (2019)
167. Yilmaz, C.S., Yilmaz, V., Gungor, O.: A theoretical and practical survey of image fusion methods for multispectral pansharpening. Inf. Fusion **79**, 1–43 (2022)
168. Yilmaz, V.: A non-dominated sorting genetic algorithm-ii-based approach to optimize the spectral and spatial quality of component substitution-based pansharpened images. Concurrency Comput. Pract. Exper. **33**(5), e6030 (2021)
169. Yilmaz, V.: Adaptive hybrid pansharpening: a novel approach for combining two methods to achieve superior pansharpening performance. Int. J. Remote Sens. **44**(14), 4301–4325 (2023)
170. Yokoya, N., Iwasaki, A.: Airborne hyperspectral data over Chikusei. Space Appl. Lab., Univ. Tokyo, Tokyo, Japan, Tech. Rep. SAL-2016-05-27 **5**, 5 (2016)
171. Yokoya, N., Yairi, T., Iwasaki, A.: Coupled non-negative matrix factorization (CNMF) for hyperspectral and multispectral data fusion: application to pasture classification. In: 2011 IEEE International Geoscience and Remote Sensing Symposium, pp. 1779–1782. IEEE, New York (2011)
172. Yu, L., Liu, D., Mansour, H., Boufounos, P.T.: Fast and high-quality blind multi-spectral image pansharpening. IEEE Trans. Geosci. Remote Sens. **60**, 1–17 (2021)
173. Yuhas, R.H., Goetz, A.F., Boardman, J.W.: Discrimination among semi-arid landscape endmembers using the spectral angle mapper (SAM) algorithm. In: JPL, Summaries of the Third Annual JPL Airborne Geoscience Workshop. AVIRIS Workshop, vol. 1 (1992)
174. Zao, Y., Zou, Z., Shi, Z.: Topology-guided road graph extraction from remote sensing images. In: IEEE Transactions on Geoscience and Remote Sensing (2023)
175. Zhang, K., Zhang, F., Wan, W., Yu, H., Sun, J., Del Ser, J., Elyan, E., Hussain, A.: Panchromatic and multispectral image fusion for remote sensing and earth observation: concepts, taxonomy, literature review, evaluation methodologies and challenges ahead. Inf. Fusion **93**, 227–242 (2023)

176. Zhang, L., Shen, H., Gong, W., Zhang, H.: Adjustable model-based fusion method for multispectral and panchromatic images. IEEE Trans. Syst. Man Cybern. B Cybern. **42**(6), 1693–1704 (2012)
177. Zhang, L., Zhang, J., Ma, J., Jia, X.: SC-PNN: saliency cascade convolutional neural network for pansharpening. IEEE Trans. Geosci. Remote Sens. **59**(11), 9697–9715 (2021)
178. Zhang, Y., Hong, G.: An ihs and wavelet integrated approach to improve pan-sharpening visual quality of natural colour ikonos and quickbird images. Inf. Fusion **6**(3), 225–234 (2005)
179. Zhao, Q., Yu, L., Du, Z., Peng, D., Hao, P., Zhang, Y., Gong, P.: An overview of the applications of earth observation satellite data: impacts and future trends. Remote Sens. **14**(8), 1863 (2022)
180. Zhao, Q., Yu, L., Li, X., Peng, D., Zhang, Y., Gong, P.: Progress and trends in the application of google earth and google earth engine. Remote Sens. **13**(18), 3778 (2021)
181. Zhou, J., Civco, D.L., Silander, J.A.: A wavelet transform method to merge landsat tm and spot panchromatic data. Int. J. Remote Sens. **19**(4), 743–757 (1998)
182. Zhu, C., Deng, S., Zhou, Y., Deng, L.J., Wu, Q.: Qis-gan: A lightweight adversarial network with quadtree implicit sampling for multispectral and hyperspectral image fusion. IEEE Trans. Geosci. Remote Sens. **61**, 1–15 (2023)
183. Zhu, X.X., Bamler, R.: A sparse image fusion algorithm with application to pan-sharpening. IEEE Trans. Geosci. Remote Sens. **51**(5), 2827–2836 (2012)

Unsupervised Pansharpening Using ConvNets

Matteo Ciotola and Giuseppe Scarpa

Acronyms

CNN	Convolutional Neural Network
CC	Correlation Coefficient
CS	Component Substitution
DL	Deep Learning
ERGAS	*Erreur Relative Globale Adimensionnelle de Synthése*
FR	Full Resolution
GAN	Generative Adversarial Metworks
GS	Gram-Schmidt
GT	Ground Truth
IHS	Generalized Intensity-Hue-Saturation
LPF	Low-Pass Filtering
ML	Machine Learning
MRA	Multi-Resolution Analysis
MS	Multispectral image
MTF	Modulation Transfer Function
PAN	Panchromatic Image
PCA	Principal Component Analysis
PSNR	Peak Signal-to-Noise Ratio
RR	Reduced Resolution

M. Ciotola (✉)
Department of Electrical Engineering and Information Technology, University of Naples Federico II, Naples, Italy
e-mail: matteo.ciotola@unina.it

G. Scarpa
Department of Engineering, University of Naples Parthenope, Naples, Italy
e-mail: giuseppe.scarpa@uniparthenope.it

M. Kawulok et al. (eds.), *Super-Resolution for Remote Sensing*, Unsupervised and Semi-Supervised Learning, https://doi.org/10.1007/978-3-031-68106-6_7

SAM	Spectral Angle Mapper
SSIM	Structural Similarity Index
TA	Target Adaptation procedure
TV	Total Variation
VO	Variational Optimization-Based
UIQI	Universal Image Quality Index

1 Introduction

Optical satellites capture valuable information about the Earth's surface by acquiring images in multiple spectral bands. These bands exhibit sensitivity to specific ranges of light wavelengths, facilitating precise characterization of the ground elements in terms of spectral signatures. Such characteristic enables the extraction of varied information, useful for urban planning, natural resource management, agriculture monitoring, disaster response, security, and so forth.

On the downside, these types of systems also come with certain limitations. For instance, acquiring mainly visible and infrared backscattered sunlight, their products are highly conditioned by the weather. Because of clouds, fog, and pollution, the quality of the provided data can be more or less acceptable. Moreover, passive imaging satellites can work only in daylight conditions, which further limits their application scope. Also, the angle at which the Sun illuminates the Earth's surface affects the geometrical and photometric features of the resulting optical images, such as shadows, occlusions, and reflections, with unavoidable impacts on the information extraction processes. Finally, due to technological, physical, and economic constraints, the remote sensing products are bounded in terms of spatial (or geometric), spectral, radiometric, and temporal resolutions. Depending on the goals of any given Earth observation satellite mission, these imaging quality parameters can be balanced in different ways, giving priority to one aspect (e.g. spatial resolution) rather than to another (e.g. spectral resolution), within the frame of a fixed overall bound that calls for trade-off decisions.

One frequently adopted option is to combine the use of multiple sensors with different features, carried by the same satellite platform. It is the case of WorldView-2/-3 [51, 61], GeoEye-1 [25], Pleiades [39], and others, which provide a very high (spatial)-resolution single-band image, the Panchromatic Image (PAN), gathering a wide spectral bandwidth response (roughly the visible near-infrared range), and a lower resolution Multispectral image (MS) whose bands are tuned on disjoint narrower spectral bandwidths. Some PAN-MS pair samples are shown in Fig. 1. The multispectral sensors can provide from a few (GeoEye-1, IKONOS, QuickBird, etc.) up to tens (WorldView-2, WorldView-3, Pleiades Neo, etc.) of bands. In contrast, other satellites, like ESA Sentinel-2, offer only multispectral subsets at varying resolutions, with non-overlapping spectral ranges. Other satellites, e.g. PRISMA [24] and ALI/Hyperion [64], are equipped with hyperspectral sensors instead of multispectral ones, enabling the collection of hundreds of spectral bands.

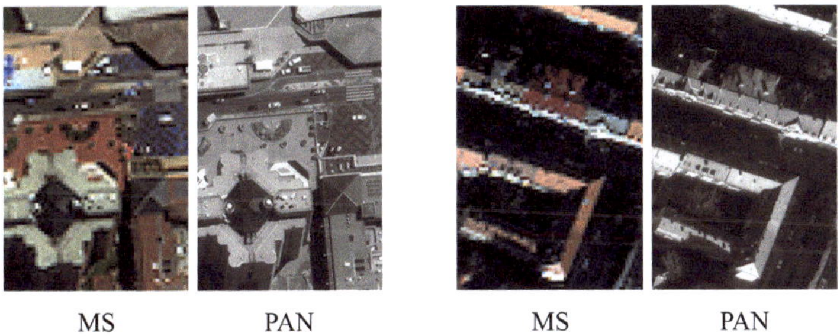

Fig. 1 Examples of PAN-MS pairs from WorldView-3 (for the eight-band MS only three bands are displayed: red, blue, and green)

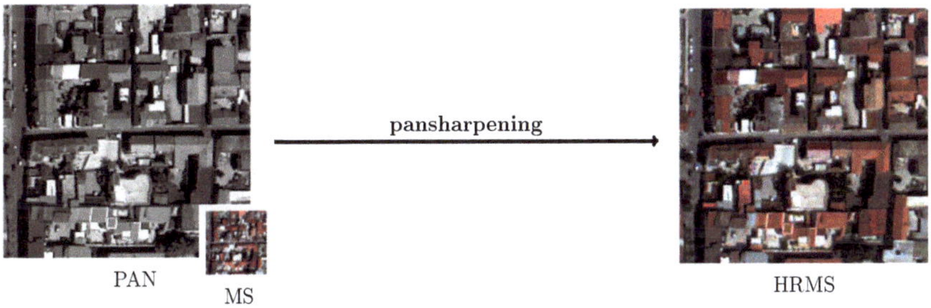

Fig. 2 Panshaprening Problem: PAN and MS image are fused to obtain a unique product at both high spatial and spectral resolution

Within the above-described scenario, data fusion techniques come into play, aiming to raise the lower-resolution bands to their utmost possible resolution. This fusion can be achieved by employing data acquired from the same satellite [33] or from alternative sources possessing superior resolution, albeit characterized by distinct imaging attributes [32]. Among the different fusion opportunities, one of the most frequent is the so-called pansharpening [26, 55, 84], which is the combination of a higher-resolution panchromatic band with a lower-resolution multispectral image acquired by the same satellite system, to produce the high-resolution version of the multispectral input component. Figure 2 clarifies this process. Said in different words, pansharpening can be considered as a super-resolution of the MS input component supported by the spatial guide given by the PAN. Sometimes, the same fusion task can occur for PAN-MS pairs provided by different satellites, with obvious additional problems (non-integer resolution ratios, severe misregistration, temporal changes, etc.). However to a minor extent, misregistration or temporal changes due to fast-moving objects can also occur for PAN-MS pairs from a single satellite, because of the slight temporal delays among the sensors' shooting processes.

Figure 3 shows the spectral response that characterizes each band, including the PAN, for some sample satellites. As can be seen, there are frequencies covered by

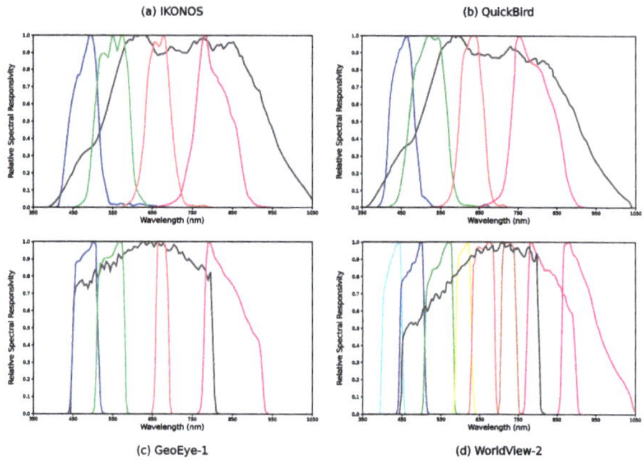

Fig. 3 Relative spectral responsivity of some optical satellites [85]

the MS but not by the PAN and vice versa: the PAN can acquire wavelengths that are not or only partially sampled by the MS bands. In practical scenarios, some objects within a scene might be observable exclusively by either the PAN or the MS, rather than both simultaneously.

The primary challenge associated with pansharpening is the absence of the MS image at full resolution to serve as ground truth. This particular aspect carries noteworthy implications. Generally, an effective pansharpening algorithm is intrinsically linked to the specific satellite in use, necessitating customization for each distinct sensor. A satellite captures images at uniform spatial resolutions, a feature dictated by sensor physics and the distance from the Earth's surface. The lack of ground truths compels the development of protocols to synthesize sample images complemented with corresponding ground truths. Such protocols are useful for pansharpening quality assessment as well as for the development of the algorithms. Unfortunately, the actual image synthesis processes in use for this purpose come at the price of a resolution shift (towards lower resolutions) [91]. Therefore, because of the statistical mismatch among different resolution scales, there is no guarantee that a model, designed, trained (think of deep learning models) and evaluated within a reduced resolution framework, will eventually generalize properly on full-resolution, real-world data. This issue not only accentuates the critical nature of designing a pansharpening algorithm, particularly if it incorporates machine learning but also underscores the primary importance of its evaluation.

2 Pansharpening Methods

The significance of the pansharpening technique in the realm of remotely sensed image fusion is of crucial importance. A vast amount of literature exists on this subject, including a comprehensive book [6] along with various review articles, such as [37], which reinforce its pivotal role. Moreover, dedicated surveys focusing precisely on this topic have populated the literature in recent years [4, 38, 47, 84], validating the growing interest surrounding the object.

Several taxonomies have been applied to pansharpening algorithms. By analysing each case, it is possible to identify the connections and interactions between the various classes involved. The most used classification collects the pansharpening methods in four main classes, characterized by well-distinguished properties: Component Substitution (CS), Multi-Resolution Analysis (MRA), Variational Optimization-Based (VO), and Deep Learning (DL) [86].

2.1 Component Substitution

Component substitution methods are based on a straightforward idea: the replacement of one of the MS components with the PAN. These methods rely upon the projection of the MS image into a new space, where the spatial structure is distinctly separated from the spectral information [78]. This transformation, often carried out using methods like Principal Component Analysis (PCA) [17, 18], Generalized Intensity-Hue-Saturation (IHS) [44, 76], or Gram-Schmidt (GS) [5, 45], permits the enhancement of the MS image by substituting the component containing spatial details with the PAN image. Subsequently, the inverse transformation is applied to bring the data back to their original space [86].

Under the hypothesis of linear transformation and the substitution of a sole component, the CS fusion process can be strongly simplified without the explicit calculation of forward and backward transformations [78, 79]. This leads to a faster implementation that can be generally described, for each $b \in [1, B]$, where b is the b-th band and B is the total number of bands, by

$$\widehat{\mathrm{MS}}_b = \widetilde{\mathrm{MS}}_b + \mathrm{G}_b \cdot (\mathrm{PAN} - \mathrm{I}_L) \tag{1}$$

in which $\widehat{\mathrm{MS}}_b$ is the b-th pansharpened band, $\widetilde{\mathrm{MS}}_b$ is the b-th MS band interpolated to the same scale of the PAN, and $\mathrm{G}_1, \ldots, \mathrm{G}_b, \ldots, \mathrm{G}_B$ are the injection gain matrices. Finally, I_L is the intensity component, defined as

$$\mathrm{I}_L = \sum_{b=1}^{B} \omega_b \widetilde{\mathrm{MS}}_b \tag{2}$$

in which the weight vector $\omega = [\omega_1, \ldots, \omega_b, \ldots, \omega_B]$ is the first row of the forward transformation matrix [78, 79].

Figure 4 illustrates a high-level flowchart depicting the core operations of methods falling under the CS category. The diagram illustrates the process, including interpolation, calculation of the intensity component (as defined in Eq. (2)), spectral alignment between PAN and intensity I_L, and the details injected in agreement with Eq. (1).

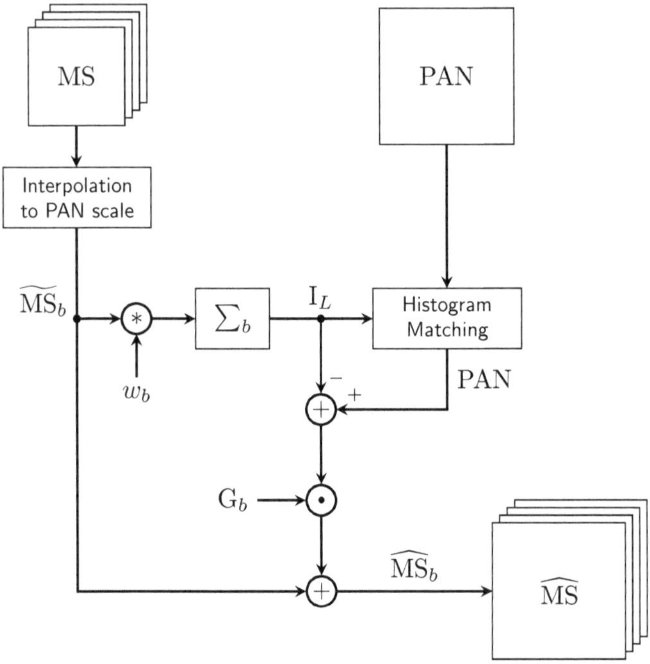

Fig. 4 A flowchart of pansharpening methods belonging to the CS class

Generally, the substitution step usually introduces spectral distortion because of changes in the low spatial frequencies of the MS image. To alleviate this issue, spectral matching procedures are frequently applied before substituting the spatial structure of the MS image with the PAN [84]. CS-based methods get high fidelity in rendering details in the fused products. Furthermore, these methods are usually easy to implement and have a low computational burden [5].

2.2 Multi-Resolution Analysis

The category of Multi-Resolution Analysis (MRA) methods comprises techniques that employ a multiscale decomposition. In most cases, MRA is applied directly to the signals within the original spatial domain. These techniques find their roots in fusion methods that rely on general-purpose decompositions, achieved through simple low-pass filters, or more intricate iterative procedures, such as Laplacian pyramids [16], wavelets [54, 60, 75], curvelets [77], and contourlets [27]. However, as MRA approaches demonstrate significant potential in delivering fused images with remarkable spectral quality, substantial efforts have been dedicated to crafting more specialized pansharpening techniques in the last few years. A pivotal

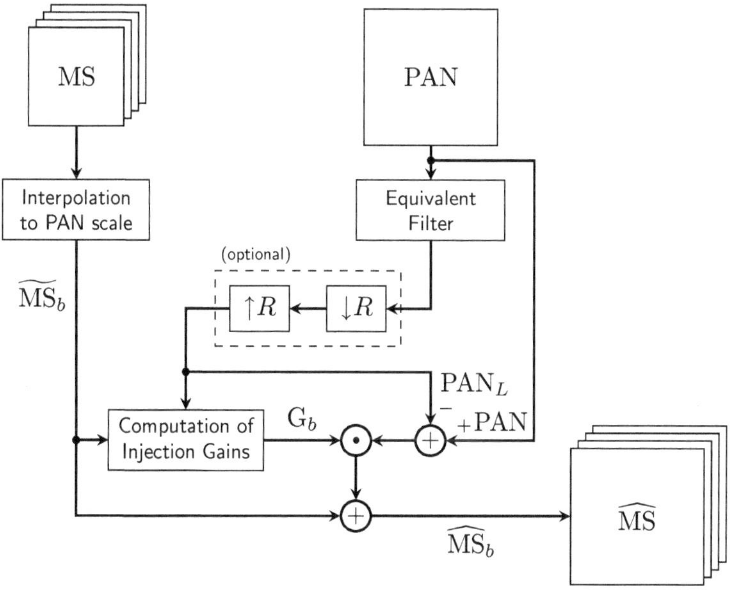

Fig. 5 A flowchart of pansharpening methods belonging to the MRA class

advancement in defining the second-generation era of these algorithms involves information about the acquisition sensor into the decomposition scheme [2, 8, 86].

From a technical standpoint, the MRA methods still rely on the spatial detail injection on the upsampled version of the MS, but their extraction differs from the CS methods. Indeed, MRA algorithms exploit the difference between the PAN and its low-pass spatial version PAN_L. Namely, the fused image is obtained for each $b = 1, \ldots, B$ as follows:

$$\widehat{\mathrm{MS}}_b = \widetilde{\mathrm{MS}}_b + \mathrm{G}_b \cdot (\mathrm{PAN} - \mathrm{PAN}_L) \tag{3}$$

The different approaches belonging to this class are characterized by the way to calculate PAN_L and by employed injection gains $\{\mathrm{G}_b\}_{b=1,\ldots,B}$. In a very general setting, PAN_L is achieved through an iterative decomposition scheme (called *MRA*) that aims at constructing a sequence of 2D signals with progressively reduced information through the repeated application of some analysis operators.

The general fusion process for MRA methods is depicted in Fig. 5. It comprises elements dedicated to interpolating the multispectral (MS) image to match the PAN scale, calculating the low-pass version PAN_L of the PAN image based on the resolution ratio R, determining the band-specific injection gains, $\mathrm{G}b_{b=1,\ldots,B}$, and executing the fusion by Eq. (3). It is noteworthy that a distinction arises when considering whether PAN_L is subjected to decimation-interpolation (as indicated by the dashed box) or not. The former option is typically preferred because it enables compensation for aliasing in the MS image during the fusion process [3]. In conclusion, MRA-based fused products are known for reducing spectral distortions but may occasionally compromise spatial rendering.

2.3 Variational Optimization-Based

Variational optimization-based techniques in the context of pansharpening rely on solving optimization problems. These methods have gained prominence recently due to advancements in convex optimization and inverse problems in imaging, including super-resolution, blind deconvolution, and image restoration.

VO techniques involve developing a model that describes the acquisition process by a sensor or the representation of an image. This model leads to the formulation of an objective function, which is tackled through variational optimization methods to find a suitable solution. In several VO methods, the relationship between the observed PAN image, the MS image, and the high-resolution MS (HRMS) image is defined based on a sensor model. The objective is to estimate the HRMS images from the PAN and LRMS images.

The optimization problem typically involves a cost function with two components: a fidelity term that characterizes the relationship between HRMS images and observed images and a regularization term that incorporates prior beliefs about HRMS images [86].

The seminal VO pansharpening method is the P + XS method [14], which relies on assumptions about LRMS images being subsampled versions of HRMS images, the PAN image being a linear combination of HRMS channels, and the geometric information of an image being in its level sets. To enforce the geometric aspect, a regularization term based on the Total Variation (TV) penalty is introduced. Other VO methods have variations in their assumptions and regularization techniques [62], such as incorporating PAN image information into the TV regularizer [41, 59, 96], assuming low-rank structures in the spectral domain, or relaxing certain assumptions [12]. Bayesian methods also fall into the VO category, assuming Gaussian models [31] and wavelet transformation into the formulation [97] or using sparse representations [48]. Most VO methods involve regularization parameters that require user selection, which remains a challenge in the pansharpening literature, although some articles provide guidelines for parameter settings. In essence, VO techniques in pansharpening leverage optimization and mathematical models to enhance the spatial resolution of multispectral images using panchromatic data, with various approaches and assumptions to achieve this goal [63, 103].

2.4 Deep Learning-Based Pansharpening

In recent years, a paradigm shift from model-based to data-driven approaches has revolutionized all fields of image processing, from computer vision [28, 40, 43, 46,

99] to remote sensing [15, 57, 71, 94]. In pansharpening the first method based on Convolutional Neural Network (CNN) was proposed by Masi et al. in 2016 [55], after which many more followed in a few years [13, 29, 30, 49, 56, 67, 74, 81, 92–95, 98]. It seems safe to say that deep learning is currently the most popular approach for pansharpening. Nonetheless, it suffers from a major problem: the lack of ground truth data for supervised training. Multi-resolution sensors can only provide the original MS-PAN data, downgraded in space or spectrum, never their high-resolution versions, which remain to be estimated.

A widespread solution to this problem is to perform a resolution shift. The PAN-MS data undergo a downsampling process, after which they are used as input samples to train a network where the original MS data play the role of ground truth. By doing so, the network is trained in a fully supervised manner, although in a lower-resolution domain. Eventually, it will be used for pansharpening the original data. Therefore, this solution relies on a sort of scale-invariance assumption: a network trained at low resolution is expected to work equally well at high resolution. That this hypothesis holds up, however, is by no means obvious.

In the literature, this problem is well-known [91], and great attention is devoted to mimicking the sensor Modulation Transfer Function (MTF) to ensure a correct downgrading of data. Even with an ideal scaling process, however, an inherent information gap exists between scales. For example, objects whose typical size amounts to a few pixels at the original resolution will necessarily lose their shape when brought at low resolution. There is no hope that a network trained at reduced resolution will "experience" such tiny geometries. Not surprisingly, networks trained with this approach work very well on reduced-resolution data but show significant quality losses on full-resolution target data [55, 72, 92, 94]. Interestingly, these problems have often been overlooked precisely because, in the absence of ground truth, it is not possible to objectively measure performance at target resolution.

A further limit of deep learning-based pansharpening is the endemic scarcity of remote sensing training data. Due to the high cost of multi-resolution data, networks are usually trained on just a few images which, however large, cannot ensure adequate diversity in terms of geographical position, territorial conformation, atmospheric conditions, acquisition geometry, direction and intensity of light, etc. As a consequence, such networks will hardly generalize to images acquired by sensors not seen in training or even just to different-looking images.

Motivated by these considerations, we propose a new framework for training pansharpening models in the high-resolution domain. Networks are trained using the original PAN-MS pairs as input, at their native resolution, with no downgrading and hence no loss of information. To obviate the lack of a ground truth, a new ad hoc loss is defined, which weights suitably defined indicators of spatial and spectral consistency. These indicators are computed by comparing the pansharpened output with the original PAN and MS components in their respective domains. In addition, to ensure correct operations on images with the most diverse characteristics, notwithstanding the limited datasets available for training, we use the target-adaptive modality proposed originally in [72] which fine-tunes the network on the fly to the target image. Finally, it is worth underlining that the proposed learning framework is fully general and can be used for any deep learning-based pansharpening model. Experiments with three state-of-the-art CNN-based pansharpening models on images acquired by different multi-resolution sensors demonstrate the broad and seamless applicability of this framework, as well as the significant quality improvements ensured by high-resolution training. Moreover, to ensure research reproducibility, we publish online a user-friendly software package (https://github.com/matciotola/Z-PNN) for high-resolution training and testing of pansharpening networks, together with several trained models. In summary, the main innovative contribution of this chapter (of which a significant portion has been adapted and used in [23]) is the proposal of a new fully unsupervised framework which allows training deep learning-based pansharpening models at high resolution. To validate the proposal, we re-train several state-of-the-art methods in the new framework and carry out a wide range of experiments on images acquired by several sensors.

Following this Introduction, in Sect. 3 we account for related work, in Sect. 4 describe the proposed full-resolution training framework, in Sect. 5 present experimental result, and eventually in Sect. 6 draw conclusions.

3 Unsupervised Deep Learning-Based Pansharpening

In recent years there has been a growing awareness that the resolution-shift approach to training pansharpening networks has inherent weaknesses and may cause a performance cap. Starting in 2020, several papers have begun to address this issue and to propose new solutions that carry out training, at least partially, in the high-resolution domain.

The first of these papers [83], in which the training is carried out in a fully supervised modality with a loss that includes both reduced resolution and full-resolution terms. At low resolution, the resolution-shift approach is used, with the original MS acting as ground truth. At high resolution, instead, the output of the MTF-GLP-HPM model-based algorithm [1] takes the role of ground truth. Indeed, this algorithm is known to ensure very good preservation of high-resolution details, which justifies using it as a proxy of the unknown ground truth for the only purpose of improving spatial quality. Needless to say, spatial accuracy cannot be better than that of the auxiliary algorithm, certainly non-optimal. An enhanced version of the method was later proposed in [82], with spatial loss terms relying also on the preservation of spatial gradients. Eventually, both versions provide only minor improvements concerning methods relying on reduced-resolution learning schemes. A further development [21] concerns the fusion of high- and low-resolution spectral bands in Sentinel-2 images, a closely related task.

In [52] a rather complex residual network is proposed, trained at high resolution. Features extracted from the PAN are used in a sequence of fusion units to refine the high-pass details extracted from the upsampled MS. The loss includes spatial and spectral terms, compounding both Euclidean norm and structural similarity (Structural Similarity Index (SSIM)), together with a term depending on a no-reference quality index. Despite the stated goal of overcoming the resolution-shift approach, these loss terms depend heavily on cross-scale consistency indexes, thereby reintroducing a sort of scale invariance assumption. In addition, an MS-to-PAN operator is used (called it $\mathcal{G}$, in Sect. 4) which combines linearly the MS bands through coefficients estimated, again, at low resolution. Experimental results seem promising, but training and test data come from the same scene and do not allow test generalization ability.

A deep CNN, called UPSNet, comprising 28 residual blocks plus two adaptation blocks is proposed in [73].

Loss terms are computed exclusively on high-resolution data, with spatial accuracy pursued by working on the PAN gradients. However, they depend again on some ill-defined pieces of information, such as "greyed" or upsampled versions of the MS. To make up for errors originated by such greyed MS, a further loss is introduced which, however, involves also non-differentiable functions. Despite these shortcomings, good-quality pansharpened images are obtained, although a bit over-smoothed.

A group of recent papers on this topic rely on Generative Adversarial Metworks (GAN). Indeed, GANs seem to fit very well the pansharpening task. The generator may be charged with the task of producing the high-resolution output starting from the available PAN and MS, while two dedicated discriminators validate the quality of results by comparing the panchromatic and low-resolution projections of the output with the original counterparts. None of these processes require a resolution shift. PanGAN [53], PercepPAN [100], and PGMAN [101] all follows this approach, with minor variations. However, despite the elegant formulation, results turn out to be much below expectations, with visible spectral aberrations (PanGAN, PercepPAN) or spatial blurring (PGMAN). Arguably, such poor results may be due to seemingly minor inaccuracies that disrupt the delicate training process of GANs. Such inaccuracies include the use of arbitrary MS-to-PAN linear projections with coefficients estimated on unrepresentative data and imperfect MS interpolation.

Despite their obvious value, these contributions present some common limits and flaws:

- they concern individual pansharpening methods trained at high resolution, not a general training framework;
- rely heavily on potentially detrimental cross-scale processing steps, such as arbitrary forms of interpolation or decimation, or MS-to-PAN conversions;
- generalize poorly to images with characteristics not seen in training, especially if acquired by sensors not represented in the dataset;
- methods and results are hardly reproducible due to the lack of software code online.

On the contrary, we propose a high-resolution training *framework*, applicable to any deep learning-based network, even if designed originally for reduced-resolution training. We minimize cross-scale processing, limited to a single downsizing step for loss computation. Correct operations on the most diverse images are ensured by the target-adaptive modality.

4 Full-Resolution Training Framework

4.1 Problem Statement

In the following, we will use M and P, respectively, to denote multispectral and panchromatic images. A subscript will indicate their spatial scale, with 0 associated with the highest resolution, and a fixed resolution ratio R between scales n and $n+1$, for each n. The relationship between these images is depicted in Fig. 6 where it is also assumed that low-resolution images can be obtained from their higher-resolution versions through a deterministic operator, $\mathcal{D}$, and panchromatic images from the corresponding multispectral ones through another operator, $\mathcal{G}$. This assumption holds with a good approximation for the downscaling operator, $\mathcal{D}$, while MS-to-PAN operators, though often used in applications, are necessarily far from ideal because of the sensors' physical characteristics. Of course, such operators imply a loss of information and hence are not invertible.

In multi-resolution remote sensing, M_1 and P_0 are the only available pieces of information (the MS-PAN pair), and in fact, the goal of pansharpening is to estimate the unknown high-resolution multispectral image M_0 from these spatially and spectrally degraded images

$$\widehat{M}_0 = \phi_0(M_1, P_0) \tag{4}$$

In *deep learning-based* pansharpening, the estimator ϕ_0 is learned from a suitable collection of training data. This would be a standard task if complete training data

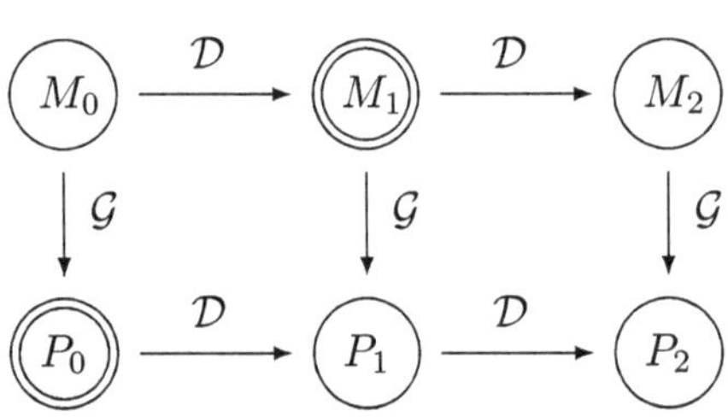

Fig. 6 Images and scales involved in the pansharpening process. The only available pieces of information are the full-resolution panchromatic image, P_0, and the low-resolution multispectral image, M_1, from which the target high-resolution multispectral image, M_0, is estimated. Deterministic (only partially known) operators, $\mathcal{D}$ and $\mathcal{G}$, relate images with their spatially or spectrally downgraded versions

were available, that is, for each training input pair (M_1^i, P_0^i), the corresponding desired output M_0^i was also provided. However, this is not the case; no full-resolution multispectral images are available to be used as ground truth.

Most deep learning-based pansharpening methods proposed thus far [55, 72, 93, 94] have circumvented this problem by means of a domain shift approach, known as Wald's protocol [91], which allows us to assess their synthesis ability. All available images in the dataset are downscaled to the next lower-resolution level

$$M_2^i = \mathcal{D}\left(M_1^i\right), \qquad P_1^i = \mathcal{D}\left(P_0^i\right) \tag{5}$$

For these pairs, the original multispectral images, M_1^i, represent the perfectly known ground truth. Therefore, a conventional training procedure can be used to estimate the network weights, that is, the pansharpening function ϕ_1. This function is eventually used to perform pansharpening at the original scale. A block diagram of this training procedure is shown in the left part of Fig. 7.

Of course, underlying this approach is the assumption that the same network can operate equally well at low resolution and high resolution, that is, $\phi_1 \simeq \phi_0$. This is a convenient approximation, but experimental evidence accumulated over the years proves it to be largely inaccurate. Networks trained under Wald's resolution downgrading protocol work very well on the low-resolution images they have been trained for, but only fairly well [86] on the full-resolution images. In practice, there is a significant domain mismatch between low-resolution and high-resolution pansharpening.

Before proposing our alternative training framework, let us justify intuitively the unsatisfactory behaviour of the resolution shift solution. The fundamental observation is that the network, during the entire training process, never sees the full-resolution data. In particular, the panchromatic images, the only data available at full resolution, are immediately resized, causing an irrecoverable loss of information. To fully realize the importance of this loss, one should also keep in mind that these

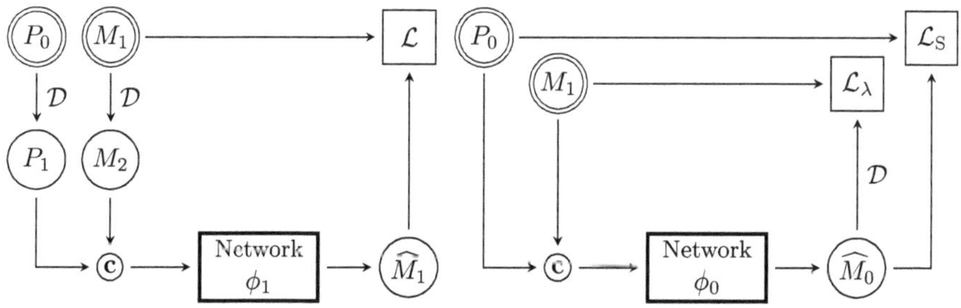

Fig. 7 Wald-like (left) and proposed (right) training frameworks. In the first case, training takes place in the reduced resolution domain, MS and PAN are immediately downgraded, and the latter is not used to compute the loss. In the proposed framework, only the original high-resolution PAN and MS are used for training, and they are both used to compute the loss

images are acquired at a fixed resolution. For example, all panchromatic images provided by the WorldView-3 sensor have a spatial resolution of 0.31 m. At this resolution, several small urban objects, like cars, traffic signs, and so on, are fully characterized with well-defined geometric shapes. With the help of low-resolution spectral information, they can be accurately recovered. However, with the resolution shift approach, the network sees only images of much lower resolution, 1.24 m (with 4.96 m multispectral) where these tiny objects lose completely their shape, reducing to very few pixels or even sub-pixels. Contrary to what happens in super-resolution, there is no 8 cm-resolution WorldView-3 image available to make up for this loss of information.

An additional problem is that resized images are much smaller than the original ones, providing much less data for training. Sticking to the WorldView-3 example, at low resolution there are 16 times fewer pixels than at full resolution. Considering the scarcity of training data, due to the restrictive policies of most data providers, this turns out to be a non-negligible drawback.

4.2 A New Loss Paradigm

These considerations, together with experimental results much below expectations, motivate our proposal of a full-resolution training framework. We will train pansharpening networks using the original data, thereby including full-resolution panchromatic images. Clearly, we must do without the ground truth images, which do not exist. Therefore, the cornerstone of our proposal is the definition of a new loss function that takes the role of the conventional full-reference loss.

Since we lack the full-resolution reference, M_0, we use the next most valuable pieces of information, that is, its projections on the low-resolution and panchromatic domains, M_1 and P_0. The network output $\widehat{M}_0$ is compared with these two references, in their respective domains, to ensure spectral and spatial consistency. Accordingly, the proposed loss becomes

$$\mathcal{L}(M_1, P_0; \widehat{M}_0) = \mathcal{L}_\lambda(M_1; \mathcal{D}(\widehat{M}_0)) + \beta \mathcal{L}_S(P_0; \mathcal{G}(\widehat{M}_0)) \qquad (6)$$

with β a suitable parameter that weighs the spectral and spatial loss terms.

Figure 8 illustrates our approach geometrically. The target image M_0 (red dot) is regarded as the combination of its M_1 and P_0 projections plus a third unknown component (call it U) which cannot be explained by either of the former two. By minimizing the loss of Eq. (6), we are pushing the estimate $\widehat{M}_0$ (black dot) towards the projections of M_0 on the (M_1, P_0) plane. The origin of the third component has been critically explored in [78], comparing alternative perspectives and assumptions. Our working hypothesis is that this unpredictable part is indeed small and, therefore, our final estimate will be very close to the actual image. It is left to the experimental results to say the final word in favour or against this hypothesis.

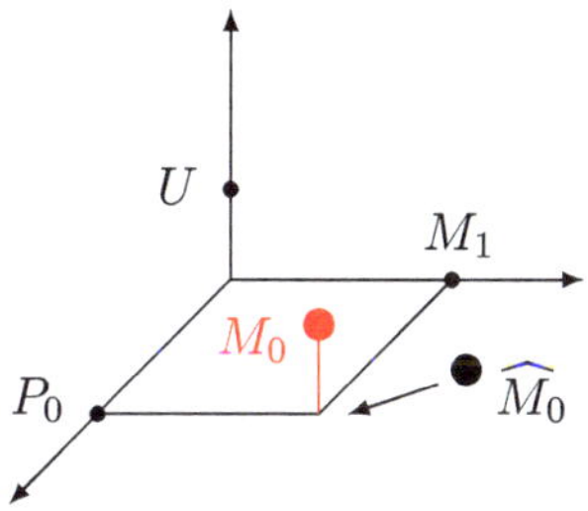

Fig. 8 Graphical sketch of the ideal proposed approach. Lacking a ground truth, the pansharpening process aims at generating an image, $\widehat{M}_0$, whose projections coincide with the known original data, P_0 and M_1. An unknown residual, U, orthogonal to this plane, remains unaccounted for. Our conjecture is that this latter component is small

At the very least, with our approach, we are not discarding any relevant data in the training process.

In the practical implementation, we depart slightly from the elegant symmetric formulation of Eq. (6). Indeed, while the $\mathcal{D}$ operator can be reasonably assumed to be known, such that $M_1 = \mathcal{D}(M_0)$, there is no consensus in the literature on the exact form and even on the conceptual correctness of the $\mathcal{G}$ operator. To circumvent this problem, this operator is bypassed, here, and the spatial loss term is computed as the sum of B individual contributions, one for each spectral band of $\widehat{M}_0$. Synthetically, the proposed loss reads as

$$\mathcal{L}(M_1, P_0; \widehat{M}_0) = \mathcal{L}_\lambda(M_1; \mathcal{D}(\widehat{M}_0)) + \beta \mathcal{L}_S(P_0; \widehat{M}_0) \tag{7}$$

A block diagram of the proposed training procedure is shown in Fig. 7, next to the Wald-like training scheme with resolution downgrading, for easy comparison. Visual inspection provides an immediate appreciation of the fundamental changes:

1. in the Wald-like framework, P_0 is immediately downgraded and never used further, therefore high-resolution information is lost forever;
2. in the proposed framework, instead, an additional spatial loss term $\mathcal{L}_S$ is introduced to take advantage of the information conveyed by the PAN;
3. in the proposed framework, the only resolution downgrade takes place *after* pansharpening, and only for the purpose of comparison with the original MS.

In the following two subsections, we describe in detail the spectral and spatial loss terms.

4.3 Spatial Loss

The role of the spatial loss is to inject in the pansharpened image the high-resolution structures observed in the PAN. Accordingly, the PAN can be used to perform a prediction, necessarily imperfect, of the output image bands and preferably a linear prediction, lacking any reasons to prefer more complex solutions. Following this point of view, here, we define the spatial loss term as a function of the correlation coefficient between the PAN and the spectral bands of the output image.

Let X and Y be two equal-size single-band images, and let σ_X^2, σ_Y^2 and σ_{XY} indicate their sample variances and covariance. Then, the correlation coefficient between X and Y is defined as

$$\rho_{XY} = \frac{\sigma_{XY}}{\sigma_X \sigma_Y}, \qquad -1 \leq \rho_{XY} \leq 1 \tag{8}$$

The correlation coefficient indicates to what extent one image can be linearly predicted from the other, with $|\rho| = 1$ implying perfect predictability and $\rho = 0$ total incorrelation.

Now, we expect to find in the pansharpened bands mostly the same spatial layout of the PAN and therefore a strong correlation with it. However, to preserve the spectral information, such a correlation cannot be unitary. Actually, it can be expected to vary spatially and from band to band, as a function of the observed scene. For example, in vegetated areas, we expect the PAN to have a strong correlation with the "green" band of the output and a weaker correlation with other bands, while the opposite will happen in other regions. In rare cases, even negative correlations are observed, due to local contrast inversions between the PAN and some MS bands [78]. This leads us to consider a three-dimensional spatial-spectral correlation field rather than a single coefficient. So, in Eq. (8), let X be a square patch of size $\sigma \times \sigma$ extracted from the PAN at spatial location (i, j) and let Y be the corresponding patch extracted from band b of $\widehat{M}_0$, then we obtain the three-dimensional correlation field

$$\rho^\sigma(i, j, b) \triangleq \rho^\sigma_{P,\widehat{M}_0(b)}(i, j) \tag{9}$$

which depends on spatial coordinates (i, j), spectral coordinate b, and the size parameter σ.

Now, we could think of defining a local spatial loss as

$$\ell^\sigma(i, j, b) = 1 - \rho^\sigma(i, j, b), \qquad 0 \leq \ell \leq 2 \tag{10}$$

and the global spatial loss term as its average. However, by doing so we would neglect the inherent spatial-spectral variability of the correlation mentioned above and push it uniformly towards 1. Therefore, to address this problem, we define an auxiliary reference correlation field, $\rho^{\sigma,\text{ref}}(i, j, b)$, computed between a low-pass filtered version of the PAN and an expanded version (plain interpolation) of the MS, and define the local loss as

$$\ell^\sigma(i, j, b) = \begin{cases} 1 - \rho^\sigma(i, j, b) & \rho^\sigma < \rho^{\sigma,\text{ref}} \\ 0 & \text{otherwise} \end{cases} \tag{11}$$

The reference correlation field can be computed exactly from the available data and provides a rough approximation of the target correlation field. A positive loss $\ell^\sigma = 1 - \rho^\sigma$ is incurred at site (i, j, b) whenever the local correlation is too small, forcing the output band to follow the spatial layout of the PAN. When ρ^σ exceeds the reference value $\rho^{\sigma,\mathrm{ref}}$, however, there is no further contribution to the global loss, and the network is free to optimize the output based on other inputs.

Although the use of correlation is certainly not new in pansharpening, we point out that our approach is very different from what is encountered in conventional methods. Component substitution, for example, relies on the strong assumption of a perfect *global* correlation between the pansharpened MS bands and the PAN [78]. When this assumption is violated, especially in the presence of occultation or contrast inversion phenomena, strong spectral aberrations are observed. In some traditional injection-based methods [66], instead, local correlation is used just to exert a consistency check. Injection of PAN details takes place only when the local correlation is high, switching to a plain upsampling of the MS bands otherwise. We assume a *generally* large local correlation between PAN and MS, but verify our hypothesis on the reference field, $\rho^{\sigma,\mathrm{ref}}$, and leverage deep learning with a suitable loss to exploit this dependence.

4.4 Spectral Loss

The spectral loss is computed in a straightforward manner by comparing the low-resolution projection of the pansharpened image, $\mathcal{D}(\widehat{M}_0)$, with its natural reference M_1

$$\mathcal{L}_\lambda = \| \mathcal{D}(\widehat{M}_0) - M_1 \|_1 \tag{12}$$

where $\| \cdot \|_1$ indicates the ℓ_1-norm.

As already said, the low-resolution projection operator $\mathcal{D}$ has been widely studied in the literature and can be assumed to be known. It consists of band-dependent low-pass filtering followed by spatial decimation at pace R

$$\mathcal{D}(M_n(\cdot, \cdot, b)) = [M_n(\cdot, \cdot, b) * h_b] \downarrow R \tag{13}$$

Under this assumption, $\mathcal{L}_\lambda$ can be expected to completely vanish in the presence of correct pansharpening, $\widehat{M}_0 = M_0$, a property not always satisfied by other quality indicators [70].

However, this is the case only if the original spectral bands are correctly aligned; otherwise a co-registration step is required. Indeed, in multi-resolution imagery, the MS bands are often misaligned. This is due to technological constraints of the sensing systems and may also depend on the specific product released. As a result, spectral aberrations appear in the image, easily spotted in false-colour representations as thin lines with weird colours near object boundaries. Therefore, it is good practice to co-register the MS spectral bands beforehand, a step often neglected by researchers and practitioners alike. Interestingly, in the proposed framework, bands are automatically co-registering. In fact, to maximize their spatial correlation with the PAN, they are eventually aligned with it and hence among themselves. This good thing, however, has a perverse effect. After decimation, the well-aligned low-resolution projection will be compared with a misaligned reference, generating a non-zero loss even in the presence of a perfect output. However, this problem is readily solved. The band-to-PAN shifts resulting after the fine-tuning are used in the decimation step to realign $\mathcal{D}(\widehat{M}_0)$ with M_1.

In the proposed loss function of Eq. (7), two critical hyper-parameters must be set: the patch size σ used in the spatial loss term, and the weight β which balances spatial and spectral losses. In Sect. 5.8 we describe and discuss the preliminary experiments carried out to select the values of σ and β used in our implementation.

4.5 *Target-Adaptive Operating Modality*

Remote sensing images present a large variability, due to the portrayed scene, the sensor characteristics, the acquisition conditions, etc. Even a large and well-designed dataset could hardly capture this wide variety, but the training sets used in practical applications consists often of just one or a few (large) images, often acquired by the same sensor. This is mostly due to the high cost of multi-resolution images and the scarcity of data freely available to the research community. Understandably, models trained in these conditions work poorly on new off-training images. To address this problem, target-adaptive pansharpening was proposed in [72]. This operating modality (see Fig. 9) consists of unfreezing the network weights, $\phi^{(0)}$, and performing a few cycles of fine-tuning to the target image, using

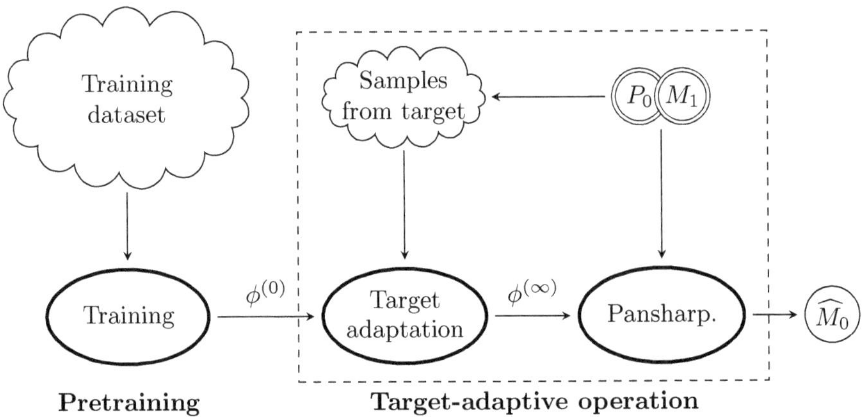

Fig. 9 High-level flowchart of target-adaptive pansharpening

some selected samples extracted by the target image itself. With a sensible choice of parameters, only a limited increase in complexity is incurred. On the positive side, the generalization ability improves sharply, with performance gains that may be also very significant, depending on the training-test mismatch. We, therefore, regard target adaptation as an essential ingredient for real-world pansharpening methods and an integral part of the proposed framework. At test time, the user is only asked to provide/select the pre-trained network, and then the algorithm runs a few iterations of fine-tuning to optimize the weights for the target image, before carrying out the actual pansharpening using the updated parameters, $\phi^{(\infty)}$. The default number of tuning iterations was set to 50 in [72]. Here, we raise it to 100, based on the experimental results discussed in Sect. 5.5.

5 Experimental Analysis

5.1 *Comparative Methods*

For all comparative analyses, we rely on the benchmark toolbox [86] which implements a large number of methods belonging to the four main categories recalled in the introduction: CS, MRA, VO, and Machine Learning (ML). All methods available in the toolbox are used in the experiments, except for a few VO solutions which suffer software compatibility issues. In addition, we consider two more state-of-the-art ML methods, PanNet [94] and DRPNN [93], re-trained on our datasets to ensure a fair comparison.

5.2 *Datasets*

Table 1 lists the datasets used for training, validation, and fine-tuning of the deep learning-based models and for testing of all methods. In some cases, we use baseline models pre-trained on other datasets detailed in the reference papers.

5.3 *Performance Measures*

Assessing the performance of pansharpening methods is an open issue, given the lack of full-resolution ground truths. A widespread approach is to measure performance objectively in a reduced resolution setting. Popular indexes used to this end are Spectral Angle Mapper (SAM) (Spectral Angle Mapper), *Erreur Relative Globale Adimensionnelle de Synthése* (ERGAS) (*Erreur Relative Globale Adimensionnelle de Synthése*), and $Q2^n$ (multiband extension of the Universal

Table 1 Datasets. GSD: PAN ground sample distance at nadir [m]. PAN/MS resolution ratio, $R = 4$. Adelaide and Washington, courtesy of DigitalGlobe©. Mexico City, Napoli, Waterford and Genova (DigitalGlobe©) provided by ESA

Sensor-site	# tiles	PAN size	GSD	Usage
WV3-Mexico City	1	2048×2048	0.31	Training
WV3-Mexico City	2	2048×2048	0.31	Validation
WV3-Adelaide	10	2048×2048	0.31	Testing
WV2-Napoli	1	2048×2048	0.46	Training
WV2-Napoli	2	2048×2048	0.46	Validation
WV2-Washington	13	2048×2048	0.46	Testing
GE1-Waterford	1	2048×2048	0.41	Training
GE1-Waterford	2	2048×2048	0.41	Validation
GE1-Genova	10	2048×2048	0.41	Testing

Image Quality Index, UIQI) [9, 35, 90], also provided in the benchmark toolbox [86]. However, this approach is at odds with our goals and is not followed here.

Instead, we consider full-resolution no-reference indexes, which assess separately spectral and spatial fidelity. Many such indexes have been proposed in recent years towards this end, for example [58, 65, 102]. For spectral fidelity, we consider here the spectral distortion index, D_λ, proposed by Khan [42], in the slightly modified implementation of the assessment toolbox [86], together with the reprojection indexes, R-SAM, R-ERGAS, and R-$Q2^n$, proposed in [70]. Note that R-$Q2^n$ equals 1-D_λ if the latter is implemented as originally proposed. For spatial fidelity, instead, we consider the spatial distortion index, D_S, proposed in [7], and the correlation distortion index, D_ρ, also proposed in [70]. Unlike the spectral case, these two indexes have a deeply different rationale and sometimes provide contrasting results. In particular, experiments carried out in [70] show that D_ρ correlates better than D_S with experts' visual assessment, especially for high-quality pansharpening.

5.4 Does Full-Resolution Training Improve Performance?

This subsection aims to prove that the proposed full-resolution training framework does indeed improve the performance of deep learning-based pansharpening, as measured by full-resolution quality indexes and especially visual inspection. Towards this end, we consider three state-of-the-art networks, PanNet [94], DRPNN [93], and A-PNN [72], a variant of PNN [55] with a skip connection for residual learning. For each network, we consider three versions. First of all, the basic *model*, as originally trained by the authors using losses based on L_1-norm (A-PNN) or L_2-norm (the others). By doing so, we have a solid starting point, the network optimized by the authors on their own data and available online. Then, we add two target-adaptive versions, with adaptation carried out at reduced resolution, with the

Table 2 For each *model* ∈ {A-PNN, PanNet, DRPNN}, we consider several versions, differing in pre-training and target adaptation Z-PNN is a proposed PNN variant later detailed (Sect. 5.5)

	Pre-training		Target-adaptation		
Full acronym	Dataset	Resolution	Applied	Resolution	# Iter.
model	authors'	reduced	no	–	–
*model**	ours	reduced	no	–	–
model-TA	authors'	reduced	yes	reduced	2000
model-TA-FR	authors'	reduced	yes	full	2000
Z-PNN (0 iter.)	ours	full	no	–	–
Z-PNN	ours	full	yes	full	100

Wald-like approach (*model*-TA), or, at full resolution, with the proposed framework (*model*-TA-FR). Unlike in normal operations, where only a few iterations are used to save time, we use a large number of iterations here, 2000, to ensure a very good adaptation to the target image. This allows the network to "forget" the initial parameters, removing possible biases due to the different pre-training conditions. At this point, differences in performance will depend only on the architecture and, for each architecture, on the use of the low- or high-resolution training framework. Table 2 summarizes these models and variants.[1]

Figures 10 and 11 report spectral and spatial quality indexes, respectively, obtained for the WorldView-3 Adelaide test image. Similar results are obtained with different test images.

The first observation concerns the significant performance gaps observed between different pre-trained models (light grey bins). For example, A-PNN has an R-SAM index about half that of DRPNN. However, such differences depend more on the limited generalization ability of the methods than on their intrinsic effectiveness. A-PNN was originally trained on data well-aligned with our WorldView-3 test image, something that probably did not happen with DRPNN. This interpretation is strongly supported by results obtained with the TA models (dark grey bins). In fact, with target adaptation, the performance improves almost always significantly, and the quality indexes become much more uniform across the three methods. Overall, target adaptation mitigates the mismatch between training and test set and the resulting indexes can be regarded as more reliable indicators of the actual potential of the various pansharpening tools.

We now turn to the real objective of our analysis, the performance obtained with target adaptation at high resolution (blue) is to be compared with that obtained at low resolution (dark grey). Regarding spectral quality, a significant gain is observed for all methods over all indexes (again, with minor exceptions), and the performance

[1] In the table we also include models used in subsequent experiments, that is, the versions re-trained on our datasets (marked by an asterisk), and Z-PNN, a PNN variant proposed here (Sect. 5.5). We warn the reader that the toolbox [86] uses a slightly different acronym (A-PNN-FT, Advanced PNN with Fine-Tuning) to indicate the reduced-resolution target-adaptive A-PNN [72] that we name here A-PNN-TA.

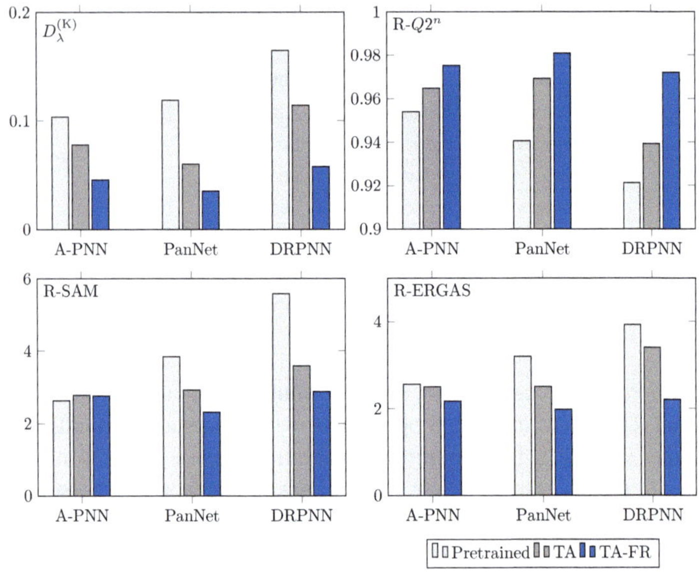

Fig. 10 Full-resolution *spectral* accuracy indexes for Adelaide

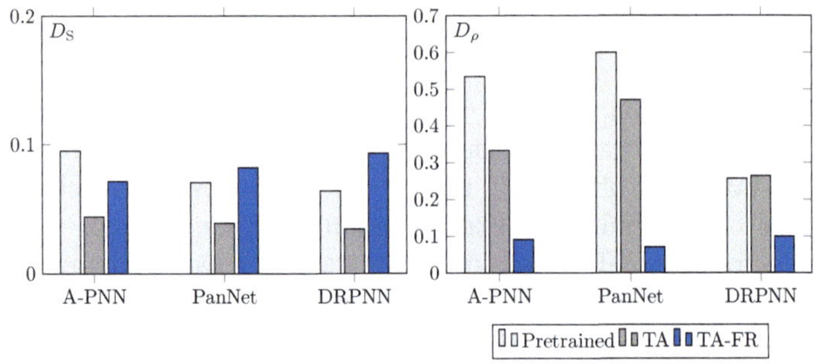

Fig. 11 Full-resolution *spatial* accuracy indexes for Adelaide

appears to be even more uniform than before. For spatial quality, instead, results are more controversial. While the D_ρ index drops, suggesting a large quality improvement, the D_S index grows again, indicating a spatial accuracy comparable to that of pre-trained models. Two facts motivate this strong mismatch. On one hand, we argue that D_S is not really a reliable indicator when quality is very high. Indeed, as also noted in [70], D_S does not really measure spatial quality, but rather a sort of cross-scale spatial quality consistency. So, it may be small even in the presence of strong spatial distortion, provided the same distortion occurs across the various scales of interest, and it may be large even with perfect pansharpening, $\widehat{M}_0 = M_0$. On the other hand, since the spatial loss used in our training framework follows closely the definition of D_ρ, this indicator may be biased in favour of TA-FR

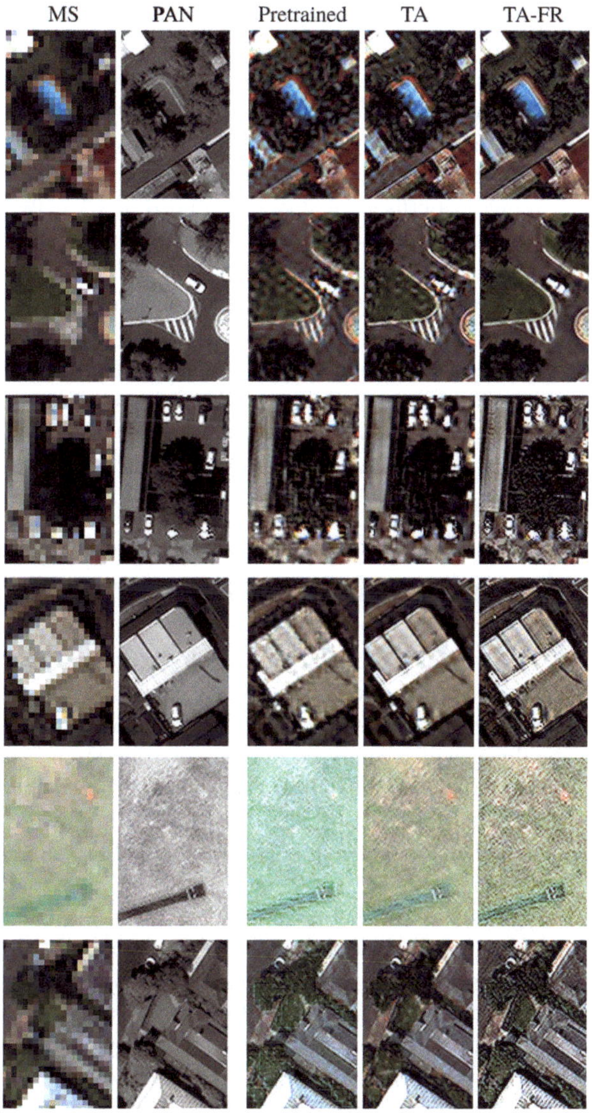

Fig. 12 Results on crops from the Adelaide image. From left to right: MS, PAN, pre-trained, TA, TA-FR. From top to bottom: A-PNN (rows 1-2), PanNet (rows 3-4), DRPNN (rows 5-6). Red, green, and blue bands are used for RGB composition

methods. Since such contradictions cannot be reconciled, we will keep using both indicators, leaving the final say to visual inspection.

In Fig. 12, for some crops of the Adelaide test image, we show the original MS and PAN data together with the output pansharpened images obtained with the pre-trained, TA, and TA-FR versions of the three CNN-based methods. Since we are interested in comparing training schemes against one another, not architectures, we show different crops for different architectures so as to offer a richer yet compact picture. First of all, visual inspection fully confirms the improvements in terms of spectral accuracy ensured by target adaptation. With respect to pre-trained models, colours are better preserved, and some evident errors are avoided. In addition, the TA-FR solutions seem to ensure clear improvements also in terms of spatial accuracy. Some strange patterns created by pre-trained or TA networks disappear. Small objects (e.g. the cars) are reconstructed with higher accuracy and, in general, all contours are sharper. High-frequency textures observed in the PAN are preserved (sometimes, even oversharpened). Overall, we see a huge improvement concerning the pre-trained models, as predicted by D_ρ, and also a consistent improvement concerning the TA versions. While further work is certainly necessary to obtain fully satisfactory pansharpening, we believe that these results represent convincing indications that high-resolution training is the right path to follow.

5.5 Z-PNN: A CNN-Based Pansharpening Method Pre-trained at Full Resolution

The analysis of the previous subsection sheds light on the potential of high-resolution training. However, it relies on intensive target adaptation, which has non-negligible costs in terms of both memory and time. Such costs are summarized in Table 3 for the three models analysed thus far, considering a 2048×2048-pixel multi-resolution image and an NVIDIA Quadro P6000 GPU. In practice, 2000 iterations require 1 hour of processing time or more. This was not the case with the target-adaptive method proposed in [72], as it worked on much smaller (16×) low-resolution images and used only 50 iterations.

To obtain *fast* high-quality pansharpening, we refined the original model weights through a further pre-training phase carried out at full resolution, using a dedicated training image for each sensor (see again Table 1). By doing so, we expect that much fewer iterations will be necessary for target adaptation. Since all three architectures appear to perform equally well, from now on we focus only on the simplest one, A-PNN. The resulting network will be referred to as Z-PNN, short for Zoom-PNN. We test the impact of this modification on an off-training WorldView-3 image. Figure 13 shows the progress of spectral and spatial loss terms, as adaptation proceeds, for

Table 3 Computation time (per iteration) and memory requirements to perform target adaptation on a 2048×2048 WV3 image

	Time [seconds]			GPU Memory [GB]		
	A-PNN	PanNet	DRPNN	A-PNN	PanNet	DRPNN
TA	0.862	0.346	0.615	0.38	0.74	2.03
TA-FR	1.885	1.885	11.172	8.96	12.84	24.09

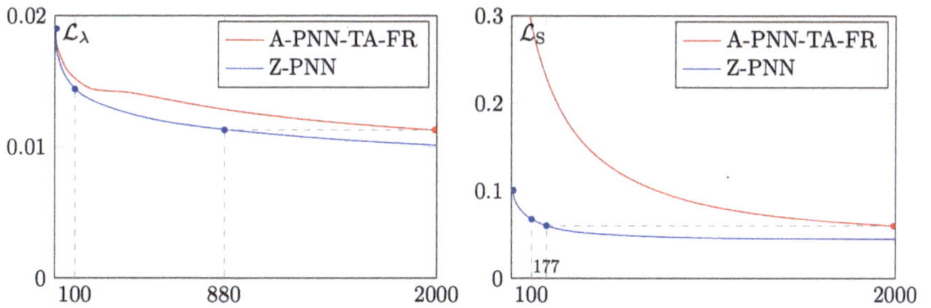

Fig. 13 Spectral (left) and spatial (right) losses vs. number of iterations for adapting A-PNN-TA-FR (red lines) and Z-PNN (blue lines) to the target image

the versions without (A-PNN-TA-FR) and with (Z-PNN) this further pre-training phase. The right part, concerning the spatial loss, is especially telling. While the A-PNN-TA-FR curve lowers very slowly, reaching eventually the value $\mathcal{L}_S \simeq 0.06$ after 2000 iterations, the Z-PNN curve reaches the same value after less than 200 iterations. Actually, the spatial loss is quite low from the beginning, $\mathcal{L}_S \simeq 0.10$, ensuring a good performance even in the absence of any adaptation. The left figure, instead, shows that the spectral loss benefits from fine-tuning also when starting from the Z-PNN weights. In any case, a small number of iterations seems to be sufficient to observe a significant improvement. Figure 14 shows, for two crops of the test image, the evolution of the pansharpened output as adaptation goes on. The images fully confirm all previous observations. In summary, it appears that Z-PNN could be safely used even without adaptation or with just a few iterations. In the following experiments, we set conservatively the number of iterations to 100. However, the user is free to change this value depending on both available resources and quality target.

5.6 *Comparative Analysis*

We can now move to a full-fledged comparative analysis. Experiments will be carried out on test images acquired by three different sensors (WV2, WV3, and GE1), listed in Table 1, and results will be compared with those of the reference methods summarized in Table 4.

Numerical results are shown in the bar graphs of Figs. 15, 16, and 17, for the WV3, WV2, and GE1 datasets, respectively. Each graph refers to a different full-resolution measure and each bar to a different method. The reference methods are those listed in Table 4, grouped according to their approach (CS, MRA, VO, ML), and shown with a different bar style for each group. Newly developed methods, Z-PNN with (100 iterations) and without (0) target adaptation, and A-PNN-TA-FR are shown in shades of blue at the end. Note that PanNet and DRPNN have been re-trained on our dataset to ensure a fairer comparison, an asterisk marks this version.

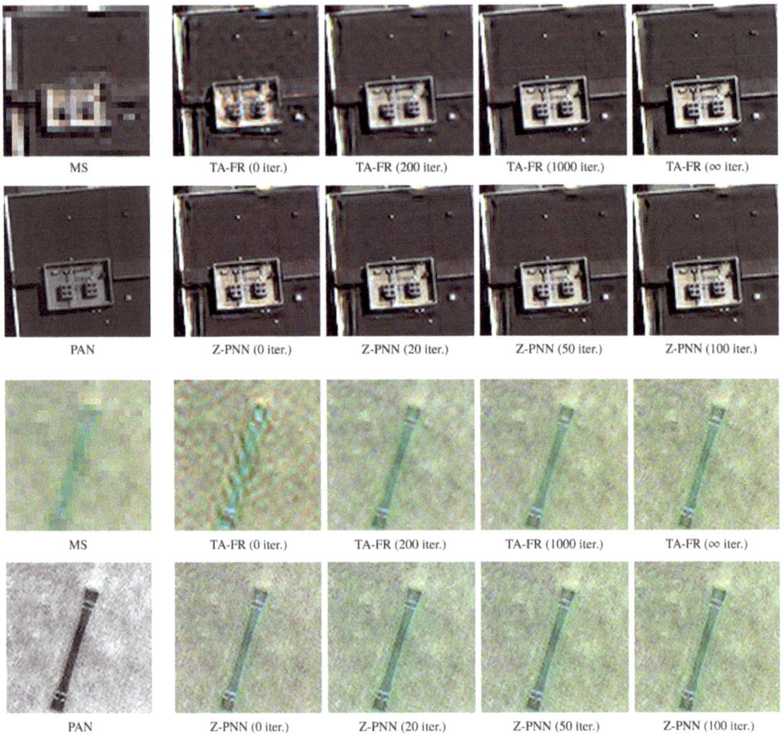

Fig. 14 Impact of target adaptation with increasing iterations on image quality for (A-PNN-)TA-FR (odd rows) and Z-PNN (even rows) for two WV3 crops. MS and PAN are shown on the left. Z-PNN reaches a satisfactory quality long before A-PNN-TA-FR

Table 4 Detailed list of all reference methods

Component substitution (CS)
BT-H [50], BDSD [36], C-BDSD [34], BDSD-PC [85], GS [45],
GSA [5], C-GSA [68], PRACS [19]
Multi-resolution analysis (MRA)
AWLP [10], MTF-GLP [10], MTF-GLP-FS [87], MTF-GLP-HPM [10],
MTF-GLP-HPM-H [50], MTF-GLP-HPM-R [88], MTF-GLP-CBD [11],
C-MTF-GLP-CBD [68], MF [69]
Variational optimization (VO)
[89], SR-D [80], TV [62]
Machine learning (ML)
PNN [55], PNN-IDX [55], A-PNN [72], A-PNN-TA [72], DRPNN* [93], PanNet* [94]

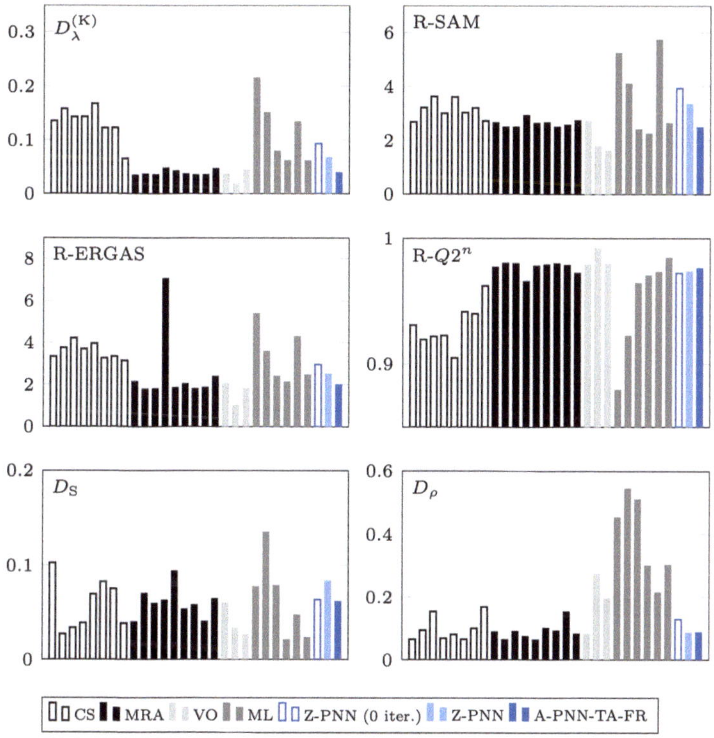

Fig. 15 Numerical results on the WV3-Adelaide dataset

To begin, let us focus on Fig. 15, also because similar considerations, with minor differences, hold for the other cases. The most notable outcome is that, contrary to the widespread belief, ML methods do not outperform conventional methods (smaller is better for all measures). As an example, the MRA methods (black) are generally[2] superior to ML methods (dark grey) in terms of both spectral quality and spatial quality indicators. This surprising result is due, in our opinion, to the low-resolution vs. high-resolution mismatch. ML methods are usually trained at low resolution, with the Wald-like protocol of Fig. 7(left), and then tested again with the Wald protocol. Therefore, it is not surprising that numerical results speak largely in their favour. Visual analyses on full-resolution data, however, have always cast some shadows on the superiority of ML methods. Such doubts are confirmed here, where results are computed only in terms of *high-resolution* indexes. These provide a more unbiased assessment of performance and are better predictors of the quality of pansharpened images the end users can expect.

[2] Note that individual methods show occasional failures on some images; we neglect these special cases in this high-level analysis.

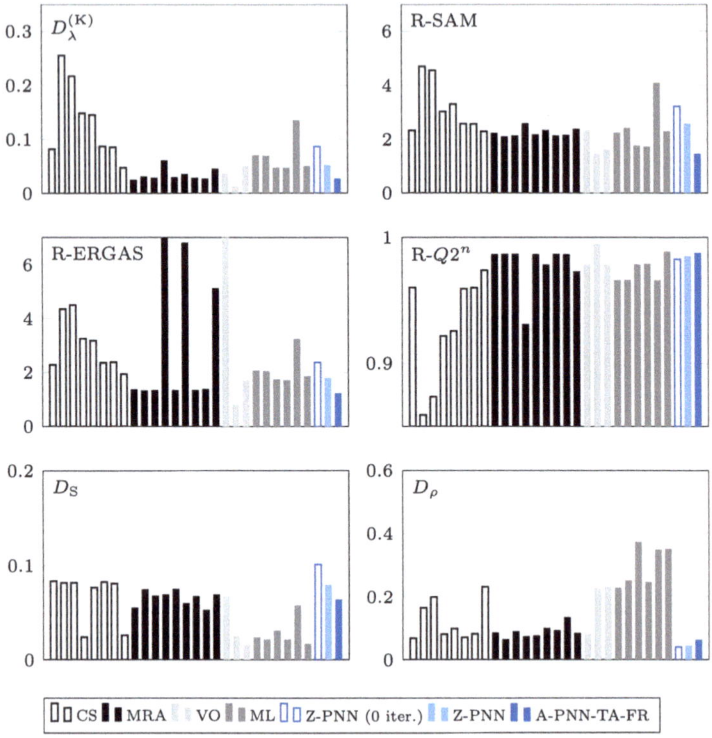

Fig. 16 Numerical results on the WV2-Washington dataset

The performance of ML methods improves significantly only when training takes place at high resolution, as proved by the last three (blue) bars. This behaviour is observed, more or less pronounced, with all sensors; see Figs. 16 and 17, and the improvement is especially significant in terms of spatial quality, according to the D_ρ indicator. In particular, the fully (2000 iterations) adapted method, A-PNN-TA-FR (last bar), has one of the smallest D_ρ values consistently on all datasets. Moreover, it has also very good spectral quality indicators, suggesting an excellent overall performance. On the other hand, it is fair to underline that D_S results depict a very different situation, almost opposite to D_ρ. Again, this calls for accurate visual inspection of pansharpened images, which is the next step of our analysis.

Figures 18, 19, and 20 show visual results for some crops acquired by the WorldView-3, WorldView-2, and GeoEye-1 sensors, respectively. For each crop, next to the original MS and PAN, we show the output of two methods trained at high resolution (A-PNN-TA-FR and Z-PNN), together with six reference methods. The latter are chosen as the best and second best-ranking methods in terms of D_ρ, D_S, and D_λ, respectively.

Let us consider Fig. 18, first, and let us compare the A-PNN-TA-FR with the original PAN-MS pair. By suitably enlarging the figure, one can fully appreciate the

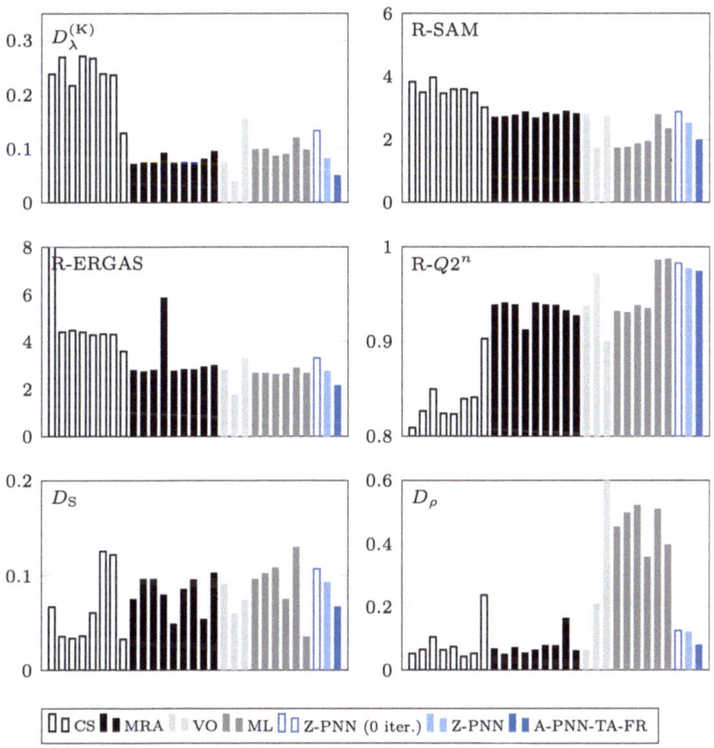

Fig. 17 Numerical results on the GE1-Genova dataset

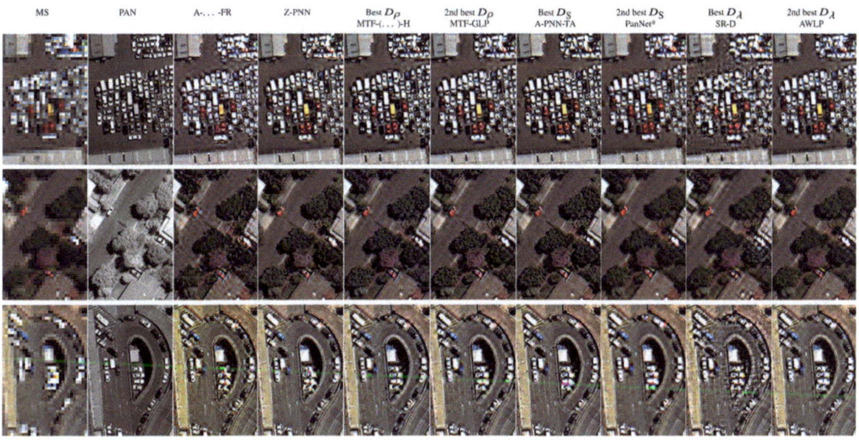

Fig. 18 Results for some WV3 crops. Left to right: MS, PAN, A-PNN-TA-FR, Z-PNN, best references for D_ρ, D_S, D_λ

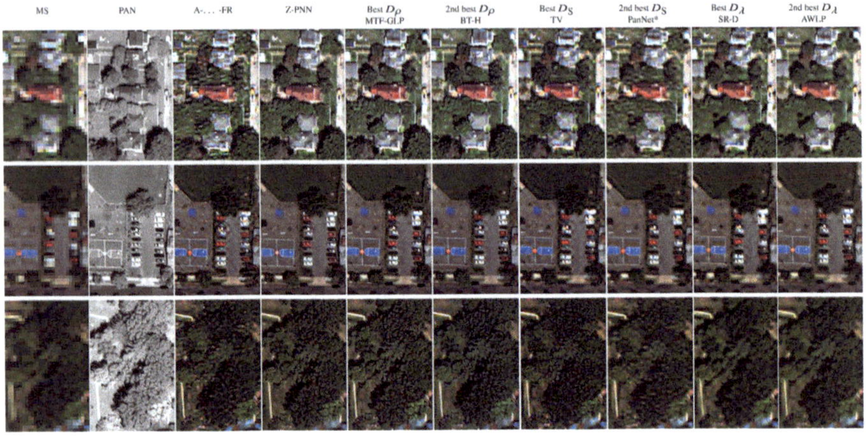

Fig. 19 Results for some WV2 crops. Left to right: MS, PAN, A-PNN-TA-FR, Z-PNN, best references for D_ρ, D_S, D_λ

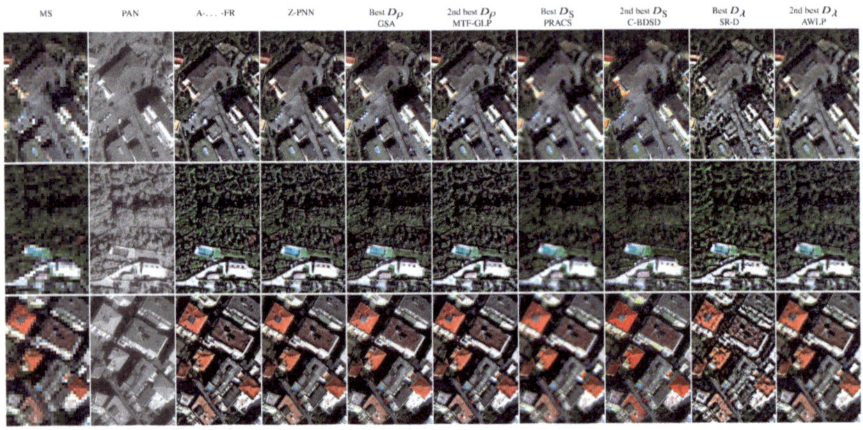

Fig. 20 Results for some GE1 crops. Left to right: MS, PAN, A-PNN-TA-FR, Z-PNN, best references for D_ρ, D_S, D_λ

impressive spatial quality of the result. All details are faithfully preserved with their original shapes and textures, and no alien pattern is introduced by the pansharpening process. Spectral quality is also very good, but this property is shared with several other methods. Z-PNN also provides very good results; we only observe a minor loss of spectral accuracy. Continuing along the row, MTF-GLP-HPM-H and MTF-GLP are the best reference methods in terms of D_ρ, and in fact, we observe a very good spatial fidelity also for them. Things are very different, instead, for A-PNN-TA and PanNet*, the best methods according to D_S. Besides a reduced precision on object shapes and some loss of resolution, especially for PanNet*, we observe annoying periodic patterns over the whole scene, confirming that D_S cannot be considered a fully reliable predictor of spatial fidelity. Finally, the best methods in terms of D_λ, SR-D, and AWLP ensure indeed a good spectral quality, although comparable to that of other methods, but exhibit some problems in terms of spatial fidelity.

Figures 19 and 20 show similar results for the WV2 and GE1 images. Beyond minor differences, the same phenomena described before are observed in all cases. A-PNN-TA-FR and Z-PNN keep providing very good results, especially in terms of spatial quality, only rarely matched by other methods, typically those performing best in terms of D_ρ.

5.7 The Challenge of Bands Weakly Correlated to the PAN

In an ideal scenario, the PAN and the MS bands should be highly correlated, meaning that the spectral information in the PAN band should be similar to that in the MS bands. However, for technological reasons, the PAN can acquire wavelengths that are not or only partially sampled by the MS bands and vice versa. This results in a possible weak correlation among the bands.

The lack of correlation may lead to challenges in achieving effective pansharpening results, as the algorithm relies on exploiting the relationships between the PAN and MS bands to enhance spatial resolution while preserving spectral information.

Indeed, this issue can produce:

(i) Loss of spectral fidelity, because the algorithm may struggle to accurately match the spectral characteristics between the PAN and MS bands;
(ii) Artefacts and distortions, because the algorithm may not effectively capture the relationship among different bands, resulting in unnatural colours, inconsistencies, and visual aberrations;
(iii) More in general, reduced performances, because the algorithm may not fully exploit the benefits of high spatial resolution provided by the PAN band.

The single bands acquired by WV3 satellites are shown, together with the auxiliary reference correlation field $\rho^{\sigma,\mathrm{ref}}$, computed between the low-pass filtered version of the PAN and the interpolated MS are presented in Fig. 21.

The $\rho^{\sigma,\mathrm{ref}}$ image provides a coarse indication of the correlation of a specific pixel of a band with the PAN. As already stated in Sect. 4.3, the correlation cannot be uniformly unitary. For example, for the vegetation areas, like those in Fig. 21, the index value is high for wavelengths describing green, yellow, and red edge and particularly low for some others, such as Coastal, Near-IR 1, and Near-IR 2.

Forcing the correlation, through the spatial loss function, to be unitary can produce the drawbacks effects just defined because it results in neglecting the inherent spatial-spectral variability of correlation. For this reason, a threshold calculated on $\rho^{\sigma,\mathrm{ref}}$ has been defined in the spatial loss function.

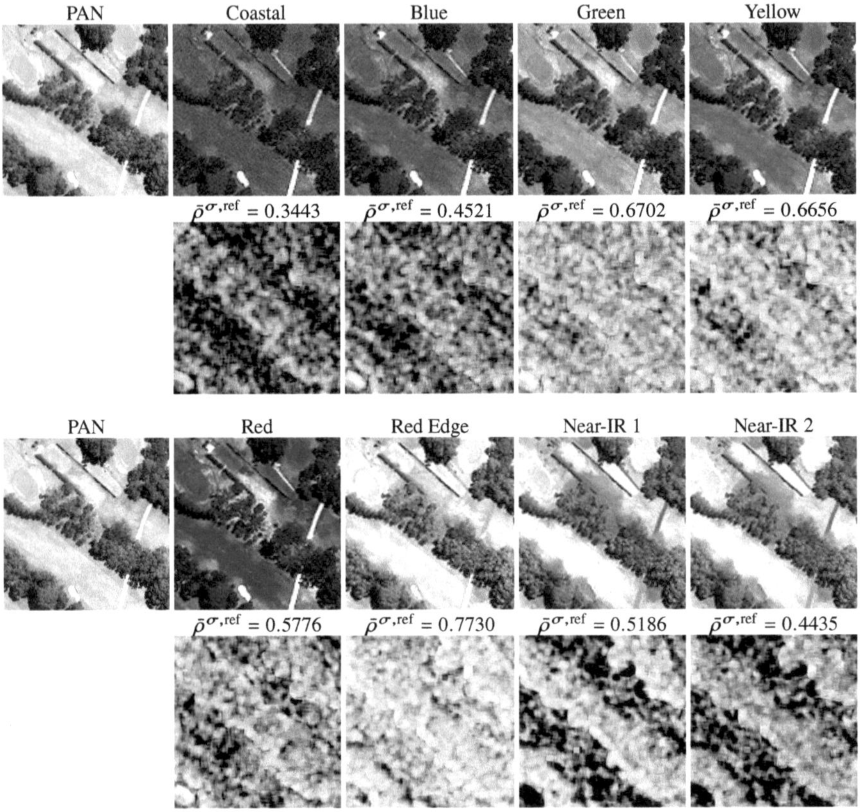

Fig. 21 Analysis of the local correlation coefficient computed for each band and a low-pass filtered version of the PAN. In odd rows are presented the PAN and the single MS bands. In even rows are shown the coarse local correlation maps $\left(\rho^{\sigma,\text{ref}}\right)$ for each MS band. Furthermore, for a better comprehension of the phenomenon is provided their average value $\left(\bar{\rho}^{\sigma,\text{ref}}\right)$

5.8 Setting Loss Hyper-parameters, Testing Alternative Losses

In all previous experiments, we used the loss of Eq. (7) with hyper-parameters σ and β optimized experimentally. Here, we discuss their impact on the performance and motivate experimentally the values selected in the implementation. In addition, we test an alternative loss function proposed in the literature for use in our framework.

5.8.1 Setting σ

The patch size σ is the only critical parameter of the proposed spatial loss. We already motivated the need to estimate the MS-PAN correlation on a *local* as opposed to a global scale (small σ), thereby limiting long-range spatial dependen-

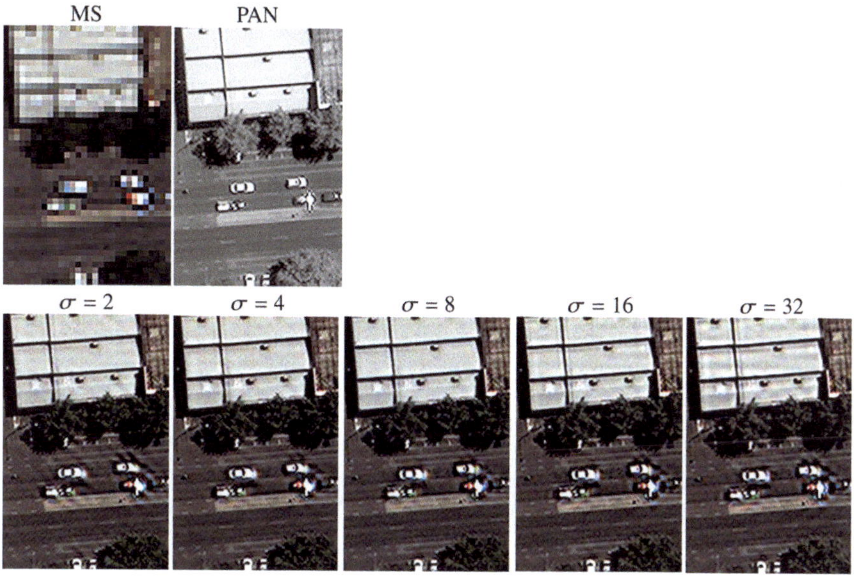

Fig. 22 Impact of patch size (σ) on pansharpening quality

cies and preserving spectral fidelity. On the other hand, with a very small value for σ, the correlation ends up being estimated on just a few points. Lacking any more precise theoretical guidance, we carried out several experiments on test images with σ doubling progressively from 2 to 32. A sample result is shown in Fig. 22. With $\sigma = 2$, obvious artefacts are visible on the roads, in the form of diagonal patterns. These disappear already with $\sigma = 4$ and then $\sigma = 8$; however, for larger values other spectral checkerboard aberrations appear on the building rooftops, like echoes of the existing black separation lines. We observed similar behaviour on many more test images, which suggests choosing small values for σ, between 4 and 8. Eventually, we set σ equal to the resolution ratio, R, which is always 4 for our images.

5.8.2 Setting β

The parameter β balances the relative importance of the spatial and spectral loss terms. When $\beta = 0$, only the spectral loss is taken into account, which negatively affects the spatial quality, and the opposite happens when $\beta \to \infty$. To quantify this behaviour, Fig. 23 reports the values of the spatial and spectral loss terms observed for A-PNN-TA-FR when β grows from 0.0001 to 10. There is a large range of values where both losses (solid lines) decrease concerning the case without adaptation (dashed lines). So, to gain a better insight, we resort again to visual inspection for a sample test image. In Fig. 24 we show the original MS (enlarged) and PAN, in

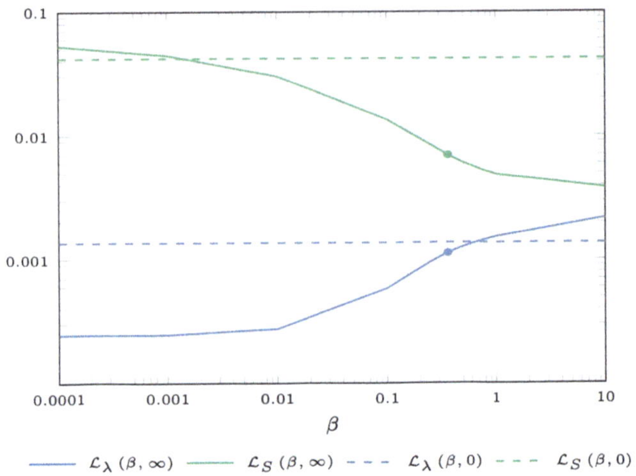

Fig. 23 Spatial and spectral losses as a function of β

the first row, the pansharpened outputs for various values of β, in the second row, and the difference between the former and an interpolated version of the MS, in the third row. For $\beta = 0.01$ and even 0.1, the output images appear blurred, with an insufficient spatial quality. For $\beta = 10$, instead, and to a lesser extent also for $\beta = 1$, there are colour distortions on the vegetation and other details, especially visible in the difference images. A good compromise is obtained with values between 0.1 and 1, and in fact we selected eventually $\beta = 0.25$ for GeoEye and $\beta = 0.36$ for WorldView.

5.8.3 Testing an Alternative Loss

Our system has been conceived based on a clear rationale, discussed in Sect. 4, and our training loss was designed to fulfil it. Nonetheless, one can legitimately wonder what happens if different losses are used in the same framework. So we fine-tuned the A-PNN-TA-FR model replacing our loss with a very different one, recently proposed [52] for high-resolution training.

This latter (call it $\mathcal{L}'$) comprises four terms

$$\mathcal{L}' = \mathcal{L}'_\lambda + \mathcal{L}'_S + \mathcal{L}'_{\text{QNR}} + \lambda \parallel \Theta \parallel_2^2 \quad (14)$$

The first two aim at improving spectral and spatial quality by minimizing combinations of MSE and structural similarity indexes in the upscaled multispectral and panchromatic domains. Instead, the third term directly targets the QNR [84] a well-known full-resolution quality measure, while the last one, the weighted norm of the parameters, serves only for regularization. The reader is referred to [52] for all details.

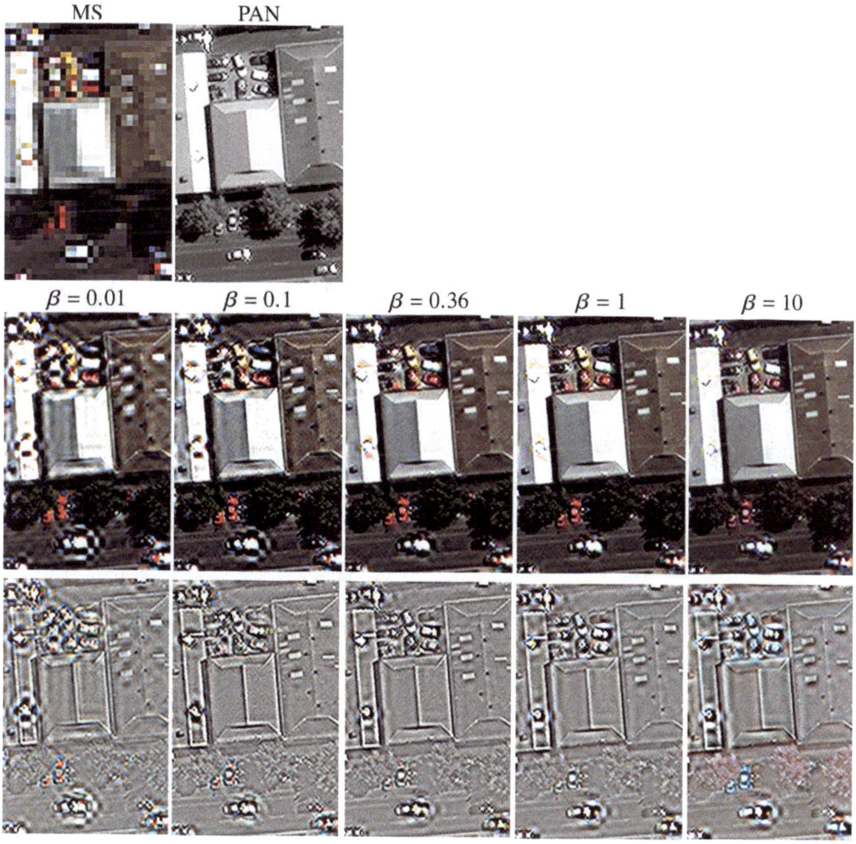

Fig. 24 Impact of loss balance (β) on pansharpening quality

Sample pansharpened crops are shown in Fig. 25, next to the original MS and PAN, for two samples for the WorldView3 Adelaide image. The spectral fidelity is quite good in both cases, although slightly better indexes are obtained with our loss, $D_\lambda = 0.03$ as opposed to 0.05. When considering spatial fidelity, however, an obvious performance gap appears. Images pansharpened with our loss are much sharper; fine textures and small details are much better preserved, as obvious from the comparison with the PAN; and no spurious pattern is generated in the process.

5.9 *Strengths and Weaknesses of the Proposed Framework*

Results of the previous subsections make clear what the main strengths of the proposed framework are. By using the original PAN-MS pairs to train a deep learning model, we make sure that the most informative data are taken into account

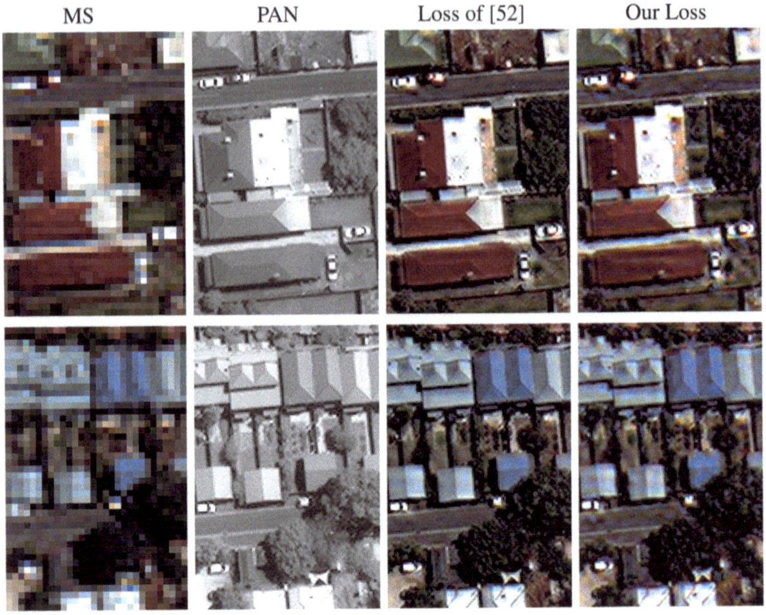

Fig. 25 Comparing the loss of [52] with the proposed loss

and lay the basis for obtaining high spectral and spatial fidelity in pansharpening. Results obtained with A-PNN, PanNet, and DRPNN are just examples of the potential of this approach. On the downside, working at high resolution incurs costs. Using the original data, without subsampling, causes pre-training to become much slower and memory intensive, a nuisance, but not a major problem, considering that pre-training takes place offline. On the other hand, target adaptation is important to ensure the best performance, and this process takes place online. For Z-PNN and 100 iterations, it requires about 3 minutes, as shown in Table 3. Depending on the application and mode of use, this may be overly annoying. With the following experiment, however, we show that this cost may be significantly reduced.

Figure 26 shows, on the left column, the original PAN-MS pair for a 128×128-pixel WV3 crop. In the middle column, we see the output of A-PNN-TA on the top and Z-PNN on the bottom, both adapted on the 2048×2048-pixel target image including our crop, displaying the by-now usual quality gap. Our focus, though, is on the right column. Here, adaptation is carried out *only* on the very same 128×128-pixel target crop, not the whole image, hence using much less data and computing time. While the quality of the A-PNN-TA image further degrades, likely due to the lack of sufficient data, this is not the case for the Z-PNN image, which is almost indistinguishable from the previous case. As this behaviour is observed consistently in our experiments, we conclude that Z-PNN can be safely fine-tuned on the very same scene of interest, even very small, providing stable and high-quality results. Needless to say, this comes at a fraction of the original computational cost, just

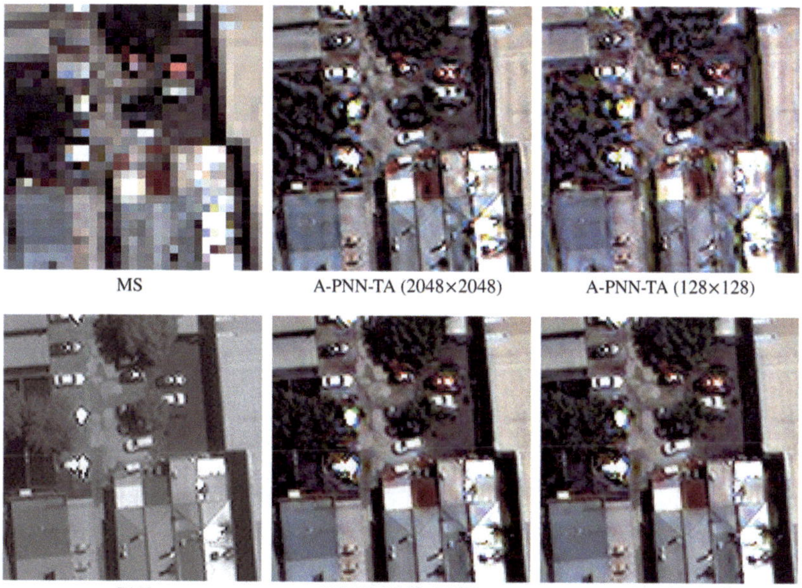

Fig. 26 For Z-PNN, the quality of fine-tuning does not appreciably depend on the size of the target scene, 2048×2048 (middle) or 128×128 (right). Therefore, it can be used to "zoom" in real time on any detail of interest

about 1 second in our example. Therefore, one can use Z-PNN in this modality to "zoom" on the details of interest, each time upgrading the original Z-PNN output (already good) in a matter of seconds.

Another critical point regards the different speeds of adaptation of the spectral and spatial loss terms (see Fig. 13). Since the latter improves much faster than the former, Z-PNN and A-PNN-TA-FR turn out to have a very similar spatial score (D_ρ) but, in some cases, a non-negligible gap in terms of spectral score (D_λ, R-SAM). This may call for a longer adaptation phase in the presence of very strict spectral accuracy requirements.

A valuable strength of the proposed framework is the automatic co-registration of pansharpened spectral bands. To better appreciate this feature, in Fig. 27 we show, for another WorldView-3 150×150-pixel crop, the input PAN-MS pair and the output images generated by Z-PNN and some reference methods where the co-registration problem is not addressed. This time, however, we use an unusual red-yellow-blue false-colour representation. While the red, green, and blue bands are usually well aligned, other bands, such as the yellow one, may be slightly shifted, due to the imaging system that acquires subsets of bands in slightly different time intervals. As expected, severe colour distortions are visible in all the output images except for Z-PNN, where spectral fidelity remains high also near sharp boundaries.

To conclude this Subsection, in Fig. 28 we show the output of *all* reference methods for a single small vegetation crop. Vegetation is extremely common in

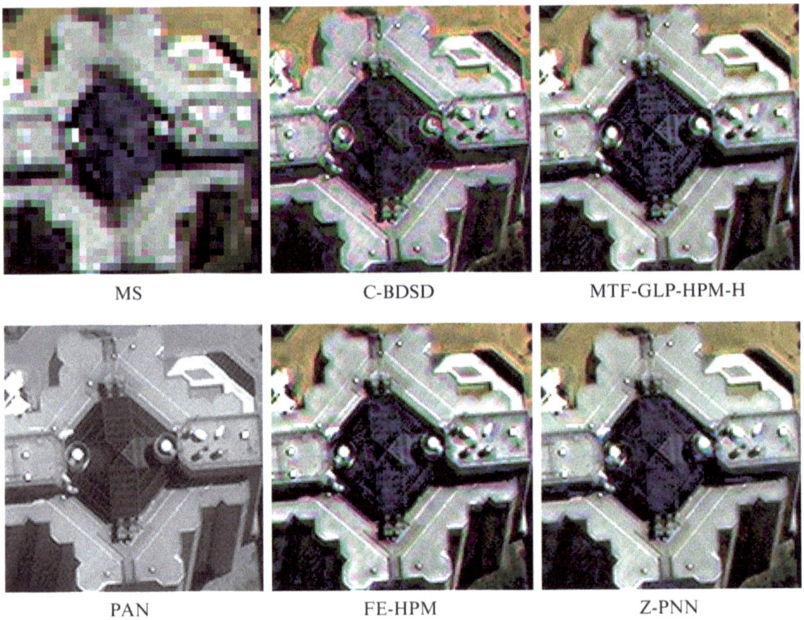

Fig. 27 In the red-yellow-blue display, the effects of MS bands misalignment is highlighted. Spurious green or magenta lines appear along object borders in all pansharpened images except Z-PNN's, where this issue is automatically addressed

multi-resolution imagery, but its correct pansharpening is often prohibitive due to the presence of fine textures at multiple scales. This is confirmed by the results in the figure. Apart from some methods that present a clear failure (e.g. DRPNN*), many more provide disappointing results, with large chromatic aberrations and/or a significant loss of detail. In general, MRA methods perform quite well on this image, much better than CS and VO. Also, ML methods trained at low resolution are among the worst in this task. Instead, thanks to the spatial loss based on local correlation, A-PNN-TA-FR and Z-PNN, trained with our high-resolution framework, show again a very good performance, preserving faithfully even the most subtle vegetation textures.

5.10 Implementation Details

All experiments were run on a server equipped with Nvidia Quadro P6000 GPU with 24GB memory, and all networks were implemented in PyTorch. Some of the tested CNN models, i.e. Z-PNN, PanNet*, and DRPNN*, needed a pre-training phase. For Z-PNN, as stated in Sect. 5.5, the model weights have been produced as a refinement of the original A-PNN model parameters [72], using a dedicated training image

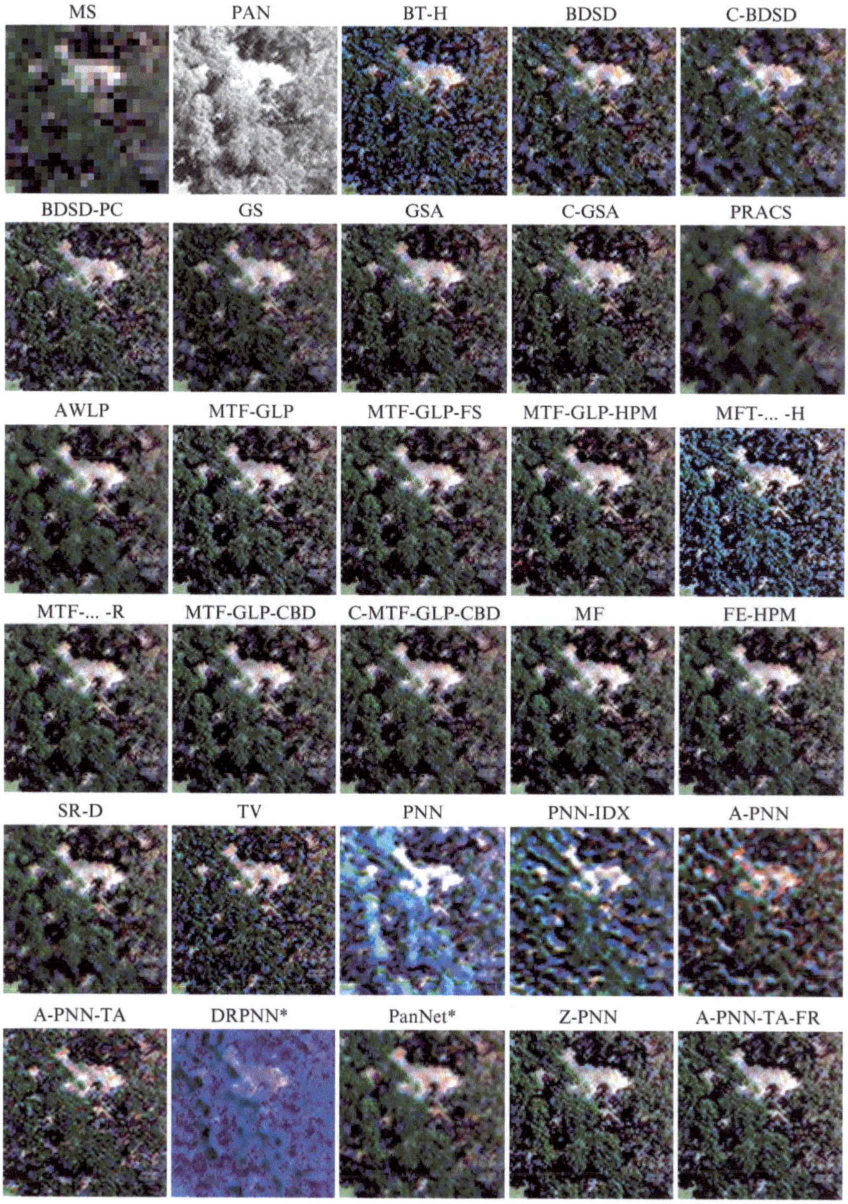

Fig. 28 Results of all methods on a small WV3 vegetation crop. Most methods, including ML methods trained at low resolution, show chromatic aberrations and resolution loss. ML methods trained at high resolution ensure high spatial and spectral fidelity

for each sensor, as indicated in Table 1. The whole image is used as a one-sample batch, running 2000 iterations that involve all layers, with a learning rate of 10^{-5} on WorldView-2/3 and of $5 \cdot 10^{-5}$ on GeoEye-1, and using the Adam optimizer with $\beta_1 = 0.9$ and $\beta_2 = 0.99$.

The models for PanNet* and DRPNN*, instead, have been reimplemented in PyTorch and trained from scratch on our training datasets for all three sensors, using the same hyper-parameters (learning rate, optimizer, loss, epochs, etc.) of the original versions. For these models, however, since the training occurs in the reduced resolution domain, we used 4×4 times larger tile, hence 8192×8192 pixels, to compensate for data volume reduction.

6 Conclusion

In this chapter, we have analysed the main drawbacks of the supervised training framework for deep learning-based pansharpening. More specifically, we have experimentally demonstrated that the hypothesis of scale-invariance, fundamental for the operational reliability of CNNs fitted with this training strategy, is not obvious. Moreover, we have highlighted how the endemic scarcity of data can emphasize the weakness of this hypothesis. For this reason, we have proposed a framework for full-resolution training of pansharpening models to exploit all the information carried by the original data, with no resolution downgrading. This has been achieved through the definition of a suitable compound loss, composed of two terms, which account separately for the spectral and spatial fidelity.

This training procedure can be easily applied to already existent or new neural networks, carrying out complete training or fine-tuning, on the fly, on the test image, greatly benefiting the overall quality of the outcomes.

To demonstrate the effectiveness of the procedure, we have applied this framework to train several state-of-the-art pansharpening models.

Experimental results are extremely encouraging. Besides numerical indicators, visual inspection confirms that the quality of the pansharpened images is largely improved thanks to high-resolution training. Beyond the framework itself and the trained pansharpening methods, though, the main contribution of this chapter is to prove the potential of this training approach, being it possibly a crucial element for unleashing the complete capabilities of deep learning in pansharpening. However, many improvements are certainly possible, such as:

(i) an algorithm capable of addressing the joint fine co-registration and pansharpening of multispectral images. A solution to this problem is presented in [20];
(ii) the optimization of target-adaptive phase, which pays even more computation overhead than the solution previously proposed in [72]. This problem has been addressed in different ways in [22] and [20];
(iii) the pansharpening of small moving objects.

With the publication of this chapter, we hope to stimulate the research on this new training paradigm, encouraging the definition of new and more accurate strategies and procedures for always better results and for solving the still open issues.

References

1. Aiazzi, B., Alparone, L., Baronti, S., Garzelli, A., Selva, M.: An MTF-based spectral distortion minimizing model for pan-sharpening of very high resolution multispectral images of urban areas. In: GRSS/ISPRS Joint Workshop on Remote Sensing and Data Fusion over Urban Areas, pp. 90–94 (2003)
2. Aiazzi, B., Alparone, L., Baronti, S., Garzelli, A., Selva, M.: Mtf-tailored multiscale fusion of high-resolution ms and pan imagery. Photogramm. Eng. Remote. Sens. **72**(5), 591–596 (2006)
3. Aiazzi, B., Alparone, L., Baronti, S., Garzelli, A., Selva, M.: Advantages of laplacian pyramids over"a trous"wavelet transforms for pansharpening of multispectral images. In: Image and Signal Processing for Remote Sensing XVIII, vol. 8537, pp. 12–21. SPIE, New York (2012)
4. Aiazzi, B., Alparone, L., Baronti, S., Garzelli, A., Selva, M., et al.: 25 years of pansharpening: a critical review and new developments. In: Signal and Image Processing for Remote Sensing, 2nd edn. pp. 533–548 (2012)
5. Aiazzi, B., Baronti, S., Selva, M.: Improving component substitution pansharpening through multivariate regression of ms + pan data. IEEE Trans. Geosci. Remote Sens. **45**(10), 3230–3239 (2007)
6. Alparone, L., Aiazzi, B., Baronti, S., Garzelli, A.: Remote sensing image fusion. Crc Press, New York (2015)
7. Alparone, L., Aiazzi, B., Baronti, S., Garzelli, A., Nencini, F., Selva, M.: Multispectral and panchromatic data fusion assessment without reference. Photogramm. Eng. Remote. Sens. **74**(2), 193–200 (2008)
8. Alparone, L., Baronti, S., Aiazzi, B., Garzelli, A.: Spatial methods for multispectral pansharpening: Multiresolution analysis demystified. IEEE Trans. Geosci. Remote Sens. **54**(5), 2563–2576 (2016)
9. Alparone, L., Baronti, S., Garzelli, A., Nencini, F.: A global quality measurement of pansharpened multispectral imagery. IEEE Geosci. Remote Sens. Lett. **1**(4), 313–317 (2004). https://doi.org/10.1109/LGRS.2004.836784
10. Alparone, L., Garzelli, A., Vivone, G.: Intersensor statistical matching for pansharpening: Theoretical issues and practical solutions. IEEE Trans. Geosci. Remote Sens. **55**(8), 4682–4695 (2017). https://doi.org/10.1109/TGRS.2017.2697943
11. Alparone, L., Wald, L., Chanussot, J., Thomas, C., Gamba, P., Bruce, L.: Comparison of pansharpening algorithms: Outcome of the 2006 GRS-S Data-Fusion Contest. IEEE Trans. Geosci. Remote Sens. **45**(10), 3012–3021 (2007)
12. Aly, H., Sharma, G.: A regularized model-based optimization framework for pan-sharpening. IEEE Trans. Image Process. **23**(6), 2596–2608 (2014)
13. Azarang, A., Ghassemian, H.: A new pansharpening method using multi resolution analysis framework and deep neural networks. In: 2017 3rd International Conference on Pattern Recognition and Image Analysis (IPRIA), pp. 1–6 (2017)
14. Ballester, C., Caselles, V., Igual, L., Verdera, J., Rougé, B.: A variational model for p+ xs image fusion. Int. J. Comput. Vis. **69**, 43–58 (2006)
15. Benedetti, P., Ienco, D., Gaetano, R., Ose, K., Pensa, R.G., Dupuy, S.: m^3Fusion: A deep learning architecture for multiscale multimodal multitemporal satellite data fusion. IEEE J. Sel. Top. Appl. Earth Obs. Remote Sens. **11**(12), 4939–4949 (2018)

16. Burt, P., Adelson, E.: The laplacian pyramid as a compact image code. In: Readings in Computer Vision, pp. 671–679. Elsevier, Amsterdam (1987)
17. Carper, W., Lillesand, T., Kiefer, R., et al.: The use of intensity-hue-saturation transformations for merging spot panchromatic and multispectral image data. Photogramm. Eng. Remote. Sens. **56**(4), 459–467 (1990)
18. Chavez, P., Sides, S., Anderson, J., et al.: Comparison of three different methods to merge multiresolution and multispectral data- landsat tm and spot panchromatic. Photogramm. Eng. Remote. Sens. **57**(3), 295–303 (1991)
19. Choi, J., Yu, K., Kim, Y.: A new adaptive component-substitution-based satellite image fusion by using partial replacement. IEEE Trans. Geosci. Remote Sens. **49**(1), 295–309 (2011). https://doi.org/10.1109/TGRS.2010.2051674
20. Ciotola, M., Poggi, G., Scarpa, G.: Unsupervised deep learning-based pansharpening with jointly enhanced spectral and spatial fidelity. IEEE Trans. Geosci. Remote Sens. **61**, 1–17 (2023). https://doi.org/10.1109/TGRS.2023.3299356
21. Ciotola, M., Ragosta, M., Poggi, G., Scarpa, G.: A full-resolution training framework for sentinel-2 image fusion. In: 2021 IEEE International Geoscience and Remote Sensing Symposium IGARSS, pp. 1260–1263 (2021). https://doi.org/10.1109/IGARSS47720.2021.9553199
22. Ciotola, M., Scarpa, G.: Fast full-resolution target-adaptive CNN-based pansharpening framework. Remote Sens. **15**(2) (2023). https://www.mdpi.com/2072-4292/15/2/319
23. Ciotola, M., Vitale, S., Mazza, A., Poggi, G., Scarpa, G.: Pansharpening by convolutional neural networks in the full resolution framework. IEEE Trans. Geosci. Remote Sens. **60**, 1–17 (2022). https://doi.org/10.1109/TGRS.2022.3163887
24. Cogliati, S., Sarti, F., Chiarantini, L., Cosi, M., Lorusso, R., Lopinto, E., Miglietta, F., Genesio, L., Guanter, L., Damm, A., et al.: The prisma imaging spectroscopy mission: overview and first performance analysis. Remote Sens. Environ. **262**, 112499 (2021)
25. Crespi, M., Colosimo, G., De Vendictis, L., Fratarcangeli, F., Pieralice, F.: Geoeye-1: analysis of radiometric and geometric capability. In: Personal Satellite Services: Second International ICST Confernce, PSATS 2010, Rome, Italy, February 2010 Revised Selected Papers 2, pp. 354–369. Springer, Berlin (2010)
26. Deng, L.J., Vivone, G., Paoletti, M., Scarpa, G., He, J., Zhang, Y., Chanussot, J., Plaza, A.: Machine learning in pansharpening: a benchmark, from shallow to deep networks. IEEE Geosci. Remote Sens. Mag. **10**(3), 279–315 (2022)
27. Do, M., Vetterli, M.: The contourlet transform: an efficient directional multiresolution image representation. IEEE Trans. Image Process. **14**(12), 2091–2106 (2005)
28. Dong, C., Loy, C., He, K., Tang, X.: Image super-resolution using deep convolutional networks. IEEE Trans. Pattern Anal. Mach. Intell. **38**(2), 295–307 (2016)
29. Dong, W., Hou, S., Xiao, S., Qu, J., Du, Q., Li, Y.: Generative dual-adversarial network with spectral fidelity and spatial enhancement for hyperspectral pansharpening. IEEE Trans. Neural Networks Learn. Syst. **33**(12), 7303–7317 (2021)
30. Dong, W., Zhang, T., Qu, J., Xiao, S., Liang, J., Li, Y.: Laplacian pyramid dense network for hyperspectral pansharpening. IEEE Trans. Geosci. Remote Sens. **60**, 1–13 (2021)
31. Fasbender, D., Radoux, J., Bogaert, P.: Bayesian data fusion for adaptable image pansharpening. IEEE Trans. Geosci. Remote Sens. **46**(6), 1847–1857 (2008)
32. Galar, M., Sesma, R., Ayala, C., Aranda, C.: Super-resolution for sentinel-2 images. Int. Arch. Photogramm. Remote. Sens. Spat. Inf. Sci. **42**, 95–102 (2019)
33. Gargiulo, M., Mazza, A., Gaetano, R., Ruello, G., Scarpa, G.: Fast super-resolution of 20 m sentinel-2 bands using convolutional neural networks. Remote Sens. **11**(22), 2635 (2019)
34. Garzelli, A.: Pansharpening of multispectral images based on nonlocal parameter optimization. IEEE Trans. Geosci. Remote Sens. **53**(4), 2096–2107 (2015). https://doi.org/10.1109/TGRS.2014.2354471
35. Garzelli, A., Nencini, F.: Hypercomplex quality assessment of multi/hyperspectral images. IEEE Geosci. Remote Sens. Lett. **6**(4), 662–665 (2009). https://doi.org/10.1109/LGRS.2009.2022650

36. Garzelli, A., Nencini, F., Capobianco, L.: Optimal MMSE pan sharpening of very high resolution multispectral images. IEEE Trans. Geosci. Remote Sens. **46**(1), 228–236 (2008)
37. Ghamisi, P., Rasti, B., Yokoya, N., Wang, Q., Hofle, B., Bruzzone, L., Bovolo, F., Chi, M., Anders, K., Gloaguen, R., et al.: Multisource and multitemporal data fusion in remote sensing: a comprehensive review of the state of the art. IEEE Geosci. Remote Sens. Mag. **7**(1), 6–39 (2019)
38. Ghassemian, H.: A review of remote sensing image fusion methods. Inf. Fusion **32**, 75–89 (2016)
39. Gleyzes, M., Perret, L., Kubik, P.: Pleiades system architecture and main performances. Int. Arch. Photogramm. Remote. Sens. Spat. Inf. Sci. **39**, 537–542 (2012)
40. He, K., Gkioxari, G., Dollár, P., Girshick, R.: Mask R-CNN. In: Proceedings of the IEEE International Conference on Computer Vision (ICCV), pp. 2980–2988 (2017)
41. He, X., Condat, L., Bioucas-Dias, J., Chanussot, J., Xia, J.: A new pansharpening method based on spatial and spectral sparsity priors. IEEE Trans. Image Process. **23**(9), 4160–4174 (2014)
42. Khan, M., Alparone, L., Chanussot, J.: Pansharpening quality assessment using the modulation transfer functions of instruments. IEEE Trans. Geosci. Remote Sens. **47**(11), 3880–3891 (2009)
43. Krizhevsky, A., Sutskever, I., Hinton, G.: Imagenet classification with deep convolutional neural networks. In: Advances in Neural Information Processing Systems, pp. 1106–1114 (2012)
44. Kwarteng, P., Chavez, A.: Extracting spectral contrast in landsat thematic mapper image data using selective principal component analysis. Photogramm. Eng. Remote. Sens. **55**(1), 339–348 (1989)
45. Laben., C., Brower, B.: Process for enhancing the spatial resolution of multispectral imagery using pan-sharpening. U.S. Patent 6011875, 2000 (2000)
46. Lateef, F., Ruichek, Y.: Survey on semantic segmentation using deep learning techniques. Neurocomputing **338**, 321–348 (2019)
47. Li, S., Kang, X., Fang, L., Hu, J., Yin, H.: Pixel-level image fusion: a survey of the state of the art. Inf. Fusion **33**, 100–112 (2017)
48. Li, S., Yang, B.: A new pan-sharpening method using a compressed sensing technique. IEEE Trans. Geosci. Remote Sens. **49**(2), 738–746 (2010)
49. Liu, X., Wang, Y., Liu, Q.: Psgan: A generative adversarial network for remote sensing image pan-sharpening. In: 2018 25th IEEE International Conference on Image Processing (ICIP), pp. 873–877 (2018)
50. Lolli, S., Alparone, L., Garzelli, A., Vivone, G.: Haze correction for contrast-based multispectral pansharpening. IEEE Geosci. Remote Sens. Lett. **14**(12), 2255–2259 (2017). https://doi.org/10.1109/LGRS.2017.2761021
51. Longbotham, N., Pacifici, F., Malitz, S., Baugh, W., Camps-Valls, G.: Measuring the spatial and spectral performance of worldview-3. In: Fourier Transform Spectroscopy and Hyperspectral Imaging and Sounding of the Environment, p. HW3B.2. Optica Publishing Group, Washington, D.C. (2015). https://doi.org/10.1364/HISE.2015.HW3B.2. https://opg.optica.org/abstract.cfm?URI=HISE-2015-HW3B.2
52. Luo, S., Zhou, S., Feng, Y., Xie, J.: Pansharpening via unsupervised convolutional neural networks. IEEE J. Sel. Top. Appl. Earth Obs. Remote Sens. **13**, 4295–4310 (2020). https://doi.org/10.1109/JSTARS.2020.3008047
53. Ma, J., Yu, W., Chen, C., Liang, P., Guo, X., Jiang, J.: Pan-GAN: an unsupervised pan-sharpening method for remote sensing image fusion. Inf. Fusion **62**, 110–120 (2020). https://doi.org/10.1016/j.inffus.2020.04.006. https://www.sciencedirect.com/science/article/pii/S1566253520302591
54. Mallat, S.: A theory for multiresolution signal decomposition: the wavelet representation. IEEE Trans. Pattern Anal. Mach. Intell. **11**(7), 674–693 (1989)

55. Masi, G., Cozzolino, D., Verdoliva, L., Scarpa, G.: Pansharpening by convolutional neural networks. Remote Sens. **8**(7), 594 (2016). https://doi.org/10.3390/rs8070594. http://www.mdpi.com/2072-4292/8/7/594
56. Masi, G., Cozzolino, D., Verdoliva, L., Scarpa, G.: CNN-based pansharpening of multi-resolution remote-sensing images. In: Joint Urban Remote Sensing Event 2017. Dubai (2017)
57. Mazza, A., Sica, F., Rizzoli, P., Scarpa, G.: Tandem-x forest mapping using convolutional neural networks. Remote Sens. **11**(24) (2019)
58. Meng, X., Bao, K., Shu, J., Zhou, B., Shao, F., Sun, W., Li, S.: A blind full-resolution quality evaluation method for pansharpening. IEEE Trans. Geosci. Remote Sens. **60**, 1–16 (2022). https://doi.org/10.1109/TGRS.2021.3087708
59. Möller, M., Wittman, T., Bertozzi, A., Burger, M.: A variational approach for sharpening high dimensional images. SIAM J. Imag. Sci. **5**(1), 150–178 (2012)
60. Nason, G., Silverman, B.: The stationary wavelet transform and some statistical applications. In: Wavelets and Statistics, pp. 281–299. Springer, Berlin (1995)
61. Padwick, C., Deskevich, M., Pacifici, F., Smallwood, S.: Worldview-2 pan-sharpening. In: Proceedings of the ASPRS 2010 Annual Conference, San Diego, CA, USA, vol. 2630, pp. 1–14 (2010)
62. Palsson, F., Sveinsson, J., Ulfarsson, M.: A new pansharpening algorithm based on total variation. IEEE Geosci. Remote Sens. Lett. **11**(1), 318–322 (2013)
63. Palsson, F., Ulfarsson, M., Sveinsson, J.: Model-based reduced-rank pansharpening. IEEE Geosci. Remote Sens. Lett. **17**(4), 656–660 (2019)
64. Pearlman, J., Carman, S., Segal, C., Jarecke, P., Clancy, P., Browne, W.: Overview of the hyperion imaging spectrometer for the NASA EO-1 mission. In: IGARSS 2001. Scanning the Present and Resolving the Future. Proceedings. IEEE 2001 International Geoscience and Remote Sensing Symposium (Cat. No. 01CH37217), vol. 7, pp. 3036–3038. IEEE, New York (2001)
65. Pradhan, P., King, R., Younan, N., Holcomb, D.: Estimation of the number of decomposition levels for a wavelet-based multiresolution multisensor image fusion. IEEE Trans. Geosci. Remote Sens. **44**(12), 3674–3686 (2006)
66. Ranchin, T., Aiazzi, B., Alparone, L., Baronti, S., Wald, L.: Image fusion–the arsis concept and some successful implementation schemes. ISPRS J. Photogramm. Remote Sens. **58**(1), 4–18 (2003). https://doi.org/10.1016/S0924-2716(03)00013-3. URL https://www.sciencedirect.com/science/article/pii/S0924271603000133. Algorithms and Techniques for Multi-Source Data Fusion in Urban Areas
67. Rao, Y., He, L., Zhu, J.: A residual convolutional neural network for pan-sharpening. In: 2017 International Workshop on Remote Sensing with Intelligent Processing (RSIP), pp. 1–4 (2017)
68. Restaino, R., Mura, M.D., Vivone, G., Chanussot, J.: Context-adaptive pansharpening based on image segmentation. IEEE Trans. Geosci. Remote Sens. **55**(2), 753–766 (2017)
69. Restaino, R., Vivone, G., Dalla Mura, M., Chanussot, J.: Fusion of multispectral and panchromatic images based on morphological operators. IEEE Trans. Image Process. **25**(6), 2882–2895 (2016). https://doi.org/10.1109/TIP.2016.2556944
70. Scarpa, G., Ciotola, M.: Full-resolution quality assessment for pansharpening. Remote Sens. **14**(8) (2022). https://doi.org/10.3390/rs14081808. https://www.mdpi.com/2072-4292/14/8/1808
71. Scarpa, G., Gargiulo, M., Mazza, A., Gaetano, R.: A CNN-based fusion method for feature extraction from sentinel data. Remote Sens. **10**(2), 236 (2018)
72. Scarpa, G., Vitale, S., Cozzolino, D.: Target-adaptive CNN-based pansharpening. IEEE Trans. Geosci. Remote Sens. **56**(9), 5443–5457 (2018). https://doi.org/10.1109/TGRS.2018.2817393
73. Seo, S., Choi, J.S., Lee, J., Kim, H.H., Seo, D., Jeong, J., Kim, M.: Upsnet: unsupervised pan-sharpening network with registration learning between panchromatic and multi-spectral images. IEEE Access **8**, 201199–201217 (2020). https://doi.org/10.1109/ACCESS.2020.3035802

74. Shao, Z., Cai, J.: Remote sensing image fusion with deep convolutional neural network. IEEE J. Sel. Top. Appl. Earth Obs. Remote Sens. **11**(5), 1656–1669 (2018)
75. Shensa, M., et al.: The discrete wavelet transform: wedding the a trous and mallat algorithms. IEEE Trans. Signal Process. **40**(10), 2464–2482 (1992)
76. Shettigara, V.: A generalized component substitution technique for spatial enhancement of multispectral images using a higher resolution data set. Photogramm. Eng. Remote. Sens. **58**(5), 561–567 (1992)
77. Starck, J.L., Candès, E., Donoho, D.: The curvelet transform for image denoising. IEEE Trans. Image Process. **11**(6), 670–684 (2002)
78. Thomas, C., Ranchin, T., Wald, L., Chanussot, J.: Synthesis of multispectral images to high spatial resolution: a critical review of fusion methods based on remote sensing physics. IEEE Trans. Geosci. Remote Sens. **46**(5), 1301–1312 (2008)
79. Tu, T.M., Huang, P., Hung, C.L., Chang, C.P.: A fast intensity-hue-saturation fusion technique with spectral adjustment for ikonos imagery. IEEE Geosci. Remote Sens. Lett. **1**(4), 309–312 (2004)
80. Vicinanza, M., Restaino, R., Vivone, G., Mura, M.D., Chanussot, J.: A pansharpening method based on the sparse representation of injected details. IEEE Trans. Geosci. Remote Sens. **12**(1), 180–184 (2015)
81. Vitale, S., Ferraioli, G., Scarpa, G.: A CNN-based model for pansharpening of worldview-3 images. In: IGARSS 2018—2018 IEEE International Geoscience and Remote Sensing Symposium, pp. 5108–5111 (2018). https://doi.org/10.1109/IGARSS.2018.8519202
82. Vitale, S., Scarpa, G.: A cross-scale loss for CNN-based pansharpening. In: IGARSS 2020—2020 IEEE International Geoscience and Remote Sensing Symposium, pp. 645–648 (2020). https://doi.org/10.1109/IGARSS39084.2020.9324219
83. Vitale, S., Scarpa, G.: A detail-preserving cross-scale learning strategy for CNN-based pansharpening. Remote Sens. **12**(3), 348 (2020)
84. Vivone, G., Alparone, L., Chanussot, J., Dalla Mura, M., Garzelli, A., Licciardi, G., Restaino, R., Wald, L.: A critical comparison among pansharpening algorithms. IEEE Trans. Geosci. Remote Sens. **53**(5), 2565–2586 (2014)
85. Vivone, G., Alparone, L., Garzelli, A., Lolli, S.: Fast reproducible pansharpening based on instrument and acquisition modeling: AWLP revisited. Remote Sens. **11**(19), 2315 (2019)
86. Vivone, G., Dalla Mura, M., Garzelli, A., Restaino, R., Scarpa, G., Ulfarsson, M., Alparone, L., Chanussot, J.: A new benchmark based on recent advances in multispectral pansharpening: revisiting pansharpening with classical and emerging pansharpening methods. IEEE Geosci. Remote Sens. Mag. **9**(1), 53–81 (2020)
87. Vivone, G., Restaino, R., Chanussot, J.: Full scale regression-based injection coefficients for panchromatic sharpening. IEEE Trans. Image Process. **27**(7), 3418–3431 (2018). https://doi.org/10.1109/TIP.2018.2819501
88. Vivone, G., Restaino, R., Chanussot, J.: A regression-based high-pass modulation pansharpening approach. IEEE Trans. Geosci. Remote Sens. **56**(2), 984–996 (2018). https://doi.org/10.1109/TGRS.2017.2757508
89. Vivone, G., Simões, M., Dalla Mura, M., Restaino, R., Bioucas-Dias, J., Licciardi, G., Chanussot, J.: Pansharpening based on semiblind deconvolution. IEEE Trans. Geosci. Remote Sens. **53**(4), 1997–2010 (2015). https://doi.org/10.1109/TGRS.2014.2351754
90. Wald, L.: Data fusion: definitions and architectures–fusion of images of different spatial resolutions. Les Presses de l'Ècole des Mines (2002)
91. Wald, L., Ranchin, T., Mangolini, M.: Fusion of satellite images of different spatial resolution: assessing the quality of resulting images. Photogramm. Eng. Remote. Sens. **63**(6), 691–699 (1997)
92. Wei, Y., Yuan, Q.: Deep residual learning for remote sensed imagery pansharpening. In: 2017 International Workshop on Remote Sensing with Intelligent Processing (RSIP), pp. 1–4 (2017)

93. Wei, Y., Yuan, Q., Shen, H., Zhang, L.: Boosting the accuracy of multispectral image pansharpening by learning a deep residual network. IEEE Geosci. Remote Sens. Lett. **14**(10), 1795–1799 (2017). https://doi.org/10.1109/LGRS.2017.2736020
94. Yang, J., Fu, X., Hu, Y., Huang, Y., Ding, X., Paisley, J.: Pannet: a deep network architecture for pan-sharpening. In: Proceedings of the IEEE International Conference on Computer Vision (ICCV) (2017). https://doi.org/10.1109/ICCV.2017.193
95. Yuan, Q., Wei, Y., Meng, X., Shen, H., Zhang, L.: A multiscale and multidepth convolutional neural network for remote sensing imagery pan-sharpening. IEEE J. Sel. Top. Appl. Earth Obs. Remote Sens. **11**, 978–989 (2018). https://doi.org/10.1109/JSTARS.2018.2794888
96. Zhang, G., Fang, F., Zhou, A., Li, F.: Pan-sharpening of multi-spectral images using a new variational model. Int. J. Remote Sens. **36**(5), 1484–1508 (2015)
97. Zhang, Y., De Backer, S., Scheunders, P.: Noise-resistant wavelet-based Bayesian fusion of multispectral and hyperspectral images. IEEE Trans. Geosci. Remote Sens. **47**(11), 3834–3843 (2009)
98. Zhang, Y., Liu, C., Sun, M., Ou, Y.: Pan-sharpening using an efficient bidirectional pyramid network. IEEE Trans. Geosci. Remote Sens. **57**(8), 5549–5563 (2019)
99. Zhao, Z., Zheng, P., Xu, S., Wu, X.: Object detection with deep learning: a review. IEEE Trans. Neural Networks Learn. Syst. **30**(11), 3212–3232 (2019)
100. Zhou, C., Zhang, J., Liu, J., Zhang, C., Fei, R., Xu, S.: Perceppan: towards unsupervised pan-sharpening based on perceptual loss. Remote Sens. **12**(14) (2020). https://doi.org/10.3390/rs12142318. https://www.mdpi.com/2072-4292/12/14/2318
101. Zhou, H., Liu, Q., Wang, Y.: PGMAN: an unsupervised generative multiadversarial network for pansharpening. IEEE J. Sel. Top. Appl. Earth Obs. Remote Sens. **14**, 6316–6327 (2021). https://doi.org/10.1109/JSTARS.2021.3090252
102. Zhou, J., Civco, D., Silander, J.: A wavelet transform method to merge landsat TM and SPOT panchromatic data. Int. J. Remote Sens. **19**(4), 743–757 (1998). https://doi.org/10.1080/014311698215973
103. Zhu, X., Grohnfeldt, C., Bamler, R.: Exploiting joint sparsity for pansharpening: the J-SparseFI algorithm. IEEE Trans. Geosci. Remote Sens. **54**(5), 2664–2681 (2015)

A Comprehensive Overview of Satellite Image Fusion: From Classical Model-Based to Cutting-Edge Deep Learning Approaches

Ivan Pereira-Sánchez, Eloi Sans, Julia Navarro, and Joan Duran

1 Introduction

At present, thanks to technological advances, there is a network of thousands of satellites orbiting the Earth that allows us to monitor any point on the planet in almost real-time. This capability opens up an endless number of applications and resources that are of vital importance in various fields, including environmental monitoring, meteorological observation, and cartography [73].

The knowledge derived from these satellites not only contributes to our understanding of Earth's intricate systems but also serves as the key to addressing broader challenges, from resource management to disaster preparedness. In an era characterized by ever-evolving global dynamics and with climate change accelerating, the advancement of tools to process the received data is essential [78].

Monitoring satellites can be divided into two groups: passive and active. On the one hand, active sensors capture the reflection of waves produced by the energy emitted by the sensor itself. Examples of active remote sensing technologies are RADAR (radio detection and ranging) and LiDAR (light detection and ranging). On the other hand, passive remote sensing sensors capture the reflection of waves produced by external sources. Examples of these sensors include photography, infrared cameras, charge-coupled devices, and radiometers [82, 96, 116].

In this chapter, we will focus on passive remote sensing satellites. Due to technical constraints related to on-board storage and bandwidth transmission,

The original version of the chapter has been revised. A correction to this chapter can be found at https://doi.org/10.1007/978-3-031-68106-6_10

I. Pereira-Sánchez (✉) · E. Sans · J. Navarro · J. Duran
Department of Mathematics and Computer Science and Institute of Applied Computing and Community Code (IAC3), Universitat de les Illes Balears (UIB), Palma, Islas Baleares, Spain
e-mail: i.pereira@uib.es; e.sans@uib.es; julia.navarro@uib.es; joan.duran@uib.es

M. Kawulok et al. (eds.), *Super-Resolution for Remote Sensing*, Unsupervised and Semi-Supervised Learning, https://doi.org/10.1007/978-3-031-68106-6_8

many of these satellites typically have sensors with complementary attributes. The panchromatic (PAN) sensor produces a high-resolution image that contains reflectance data representative of a wide range of wavelengths, thus accurately depicting the geometry of the scene. In contrast, each spectral sensor covers a narrower bandwidth, offering a detailed description of the chemical-physical composition of the objects. To compensate for the limited amount of energy available in a specific range of the spectrum, spectral sensors typically sample over a larger spatial extent, resulting in a low-resolution image. Our study focuses on two cases of satellite image fusion, pansharpening and hypersharpening, which mainly differ on the number of bands of the spectral image. In both cases, the goal is to combine the spatial and spectral information from the available data into a single high-resolution multispectral (MS) image and hyperspectral (HS) image, respectively.

In the following sections, we will review the state of the art of satellite image fusion, emphasizing the crucial role that modelling plays in performance and analysing how deep learning has changed the paradigm. We will propose a classification for different methods, encompassing both classic and cutting-edge techniques. Additionally, we will present a compendium of the most relevant satellites, along with the available data and their key features. Finally, we will evaluate the performance of fusion methods trained under the same conditions and using a proposed criterion for the training phase.

2 Overview of Satellite Image Fusion

Several book chapters and journal papers are dedicated to reviewing satellite image fusion techniques [10, 26, 31, 45, 62, 66, 94]. The most common approach is to classify the methods into three categories: component substitution, multi-resolution analysis, and variational optimization. More recent reviews have introduced a fourth category specifically for deep learning approaches.

Since, to a greater or lesser extent, most state-of-the-art fusion proposals are influenced by deep learning, we will roughly divide the methods into classic and deep learning categories. In turn, classic methods will be subclassified into component substitution (CS), multi-resolution analysis (MRA), and variational optimization (VO), while deep learning methods will be subclassified into pure deep learning and model-based deep learning.

In Tables 1 and 2, we introduce some abbreviations and notations that will be used throughout the chapter. Terms not listed will be defined according to the corresponding formula when necessary.

2.1 Classic Methods

In this chapter, classic methods are defined as those that do not use deep learning strategies and further classified into the subcategories CS, MRA, and VO. These techniques provide effective and understandable solutions for satellite image fusion.

Table 1 List of abbreviations to be used throughout the chapter. The abbreviations LRS and HRS indicate that they are valid for both MS and HS images

Abreviation	Concept
MS	Multispectral image
HS	Hyperspectral image
PAN	Panchromatic image
LR	Low-resolution
HR	High-resolution
LRS	LR spectral image
HRS	HR spectral image
CS	Component substitution
MRA	Multi-resolution analysis
VO	Variational optimization

Table 2 List of notations to be used throughout the chapter. When referring to low-resolution or upsampled spectral images without the number of channels being relevant, we will use F and $\widehat{F}$, respectively. We will employ F_m/F_h and $\widehat{F}_m/\widehat{F}_h$ only when the distinction between MS and HS is necessary

Symbol	Concept
Ω	HR domain
Ω_s	LR domain
H, W	Height and width in HR
h, w	Height and width in LR
s	sampling factor
C	Number of spectral channels
P	PAN
$\widehat{P}$	Low-pass version of PAN
U	Unknown HR MS/HS
$\widetilde{U}$	Estimated HR MS/HS
$\overline{U}$	Reference HR MS/HS
$F/F_m/F_h$	LR MS/HS
$\widehat{F}/\widehat{F}_m/\widehat{F}_h$	Upsampled MS/HR
η	Noise realization

2.1.1 Component Substitution

CS-based approaches rely on the use of colour decorrelation transforms to separate the spatial and spectral information of the upsampled MS or HS image. Fusion occurs by partially or totally substituting the component which is supposed to contain the geometry of the scene by the PAN image and applying the transformation back. The degree of correlation between the PAN image and the replaced component directly influences the performance of the fusion approach [92]. For this reason, histogram matching is usually applied before substitution to minimize distortions in the result.

The pioneering CS approaches were intensity-hue-saturation [18, 19], principal component analysis (PCA) [19, 51], and Brovey's method [36]. Subsequently, several more sophisticated transformations have been proposed, leveraging the pixel value relationship between PAN and MS/HS data. This is the case of Gram-Schmidt orthogonalization [5, 52], partial replacement adaptive CS [22], and nonlinear PCA

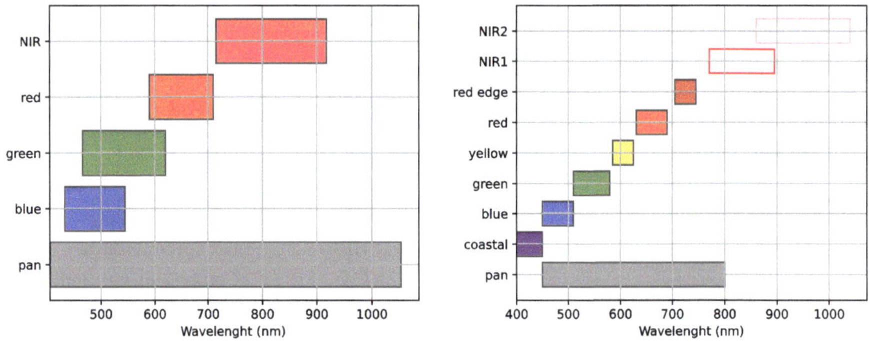

Fig. 1 Spectral response of the sensors of QuickBird (left) and WorldView2 (right) satellites to different wavelengths of light. In the case of QuickBird, the PAN sensor covers frequencies which are not covered by any of the others. On the contrary, for the WorldView2 system, there are frequencies covered for some spectral sensors that do not fall under the scope of the PAN sensitivity

[59]. Although CS methods were originally designed for pansharpening, many of them can be easily extended to HS data.

While CS methods usually perform well in rendering the spatial details of the scene, they often suffer from spectral distortions [5]. This is because, in general, the PAN sensor does not cover the same spectrum as the spectral sensors [3, 9, 92]. Figure 1 shows the spectral response of the sensors of WorldView2[3] and QuickBird[4] satellites to different wavelengths of light.

2.1.2 Multi-resolution Analysis

MRA methods inject the high frequencies of the PAN image into the upsampled spectral components through a multi-resolution decomposition. Accordingly, a spatial filter is applied to the PAN image to extract the spatial details. The fusion step can be summarized as [94]

$$U_k = \widehat{F}_k + \Theta_k \odot (P - \widehat{P}),$$

for $k = 1, ..., C$, where $\odot$ denotes pixel-wise product, $\widehat{P}$ is a low-pass version of P, and Θ_k encodes the coefficients of the gain factor.

The methods in this category differ on how the coefficients of G_k are defined and $\widehat{P}$ is generated. Laplacian pyramid [3, 4, 53], contourlet transform [88], curvelet transform [74], discrete wavelet transform [75, 76, 83, 109], and high-pass modulation [14, 19, 50] are mainly used. More sophisticated spatial filters have been proposed more recently in [47, 60, 84].

MRA techniques can better preserve spectral information, but they often experience spatial distortions such as ringing or staircasing effects [9, 92, 94]. However, when the frequency response of the low-pass filter used in the multiscale decomposition matches the modulation transfer function (MTF) of the spectral band, the performance in terms of spatial quality of MRA is comparable to that of CS [4].

Finally, hybrid methods that integrate both CS and MRA paradigms have also been explored [21, 57, 64].

2.1.3 Variational Optimization

Variational methods rely on observation models that describe how the acquired data is generated from the underlying reference images. Since inverting such a degradation process, which is usually guided by linear bounded operators, is an ill-posed inverse problem, prior knowledge on the solution one expects is required. The regularization theory assumes that the image which is to be reconstructed is sufficiently smooth. In the variational framework, this is formulated through the minimization of an energy functional that consists of data-fidelity terms, which impose constraints based on the observations, and regularization terms, which are related to hand-crafted priors and promote smooth solutions. In this setting, the fused image is obtained through optimization algorithms that minimize the proposed energy.

In general, variational methods offer excellent interpretability, robustness to distortions, and adaptability to diverse data and changes in the observation models. However, their performance is constrained by rigid priors that are not able to capture powerful representations of natural images.

We will now review the different observation models and regularizing priors that have been used in the literature for the variational formulation of the satellite image fusion problem.

2.1.4 Observation Models

The most common observation model for satellite image fusion was originally proposed by Ballester et al. [11, 77]. It assumes that the low-resolution MS/HS data is obtained from the underlying high-resolution MS/HS image by low-pass filtering followed by subsampling

$$F(x) = (DBU)(x) + \eta, \quad \forall x \in \Omega_s, \tag{1}$$

where B is the low-pass filter modelling the point spread function of the spectral sensors, D is the s-fold downsampling operator, $s \in \mathbb{Z}^+$ is the sampling factor, and η stands for the realization of noise. This is a widely assumed observation model in VO-based methods [11, 30, 42, 77]. Some other techniques tackle the inverse problem arising from (1) in a Bayesian framework [33, 99].

Ballester et al. [11] also assume that the PAN image results from a linear combination of the high-resolution spectral bands

$$P(x) = \sum_{k=1}^{C} \alpha_k U_k(x), \ \ \forall x \in \Omega, \tag{2}$$

where the coefficients satisfy $\sum_k \alpha_k = 1$. The constraint (2) is not only used in variational methods [11, 30, 77] but also in some deep learning-based techniques such as [56, 103]. However, as shown in Fig. 1, this is not, in general, a realistic assumption.

To somehow transfer the high frequencies of the PAN image to the fused product, several fidelity terms have been proposed. Fang et al. [32] replace the intensity profiles in (2) by their gradients, thus imposing the constraint

$$\nabla P(x) = \sum_{k=1}^{C} \alpha_k \nabla U_k(x), \ \ \forall x \in \Omega. \tag{3}$$

Zeng et al. [113] use in (3) a differential operator at four different directions instead of the gradient. On the contrary, some other energy fidelity terms are inspired by MRA mechanisms. In this setting, Moeller et al. [70] replaced (2) with the assumption that the wavelet coefficients of the sought solution at various resolution scales must match those of the low-resolution MS/HS or PAN image, depending on whether they correspond to low-level frequencies or provide information about the edges. Inspired by Brovey's method [36], Duran et al. [31] propose a fidelity term enforcing the preservation of the radiometric ratio between the PAN and each spectral component

$$U_k(x) - \widehat{F}_k(x) = \frac{\widehat{F}_k(x)}{\widehat{P}(x)} \left(P(x) - \widehat{P}(x)\right), \ \ \forall x \in \Omega, \tag{4}$$

for $k = 1, \ldots, C$, where $\frac{\widehat{F}_k(x)}{\widehat{P}(x)}$ are modulation coefficients that take the energy levels of the PAN and MS/HS data into account. Note that (4) can be rewritten as

$$\frac{U_k(x)}{\widehat{F}_k(x)} = \frac{P(x)}{\widehat{P}(x)}, \ \ \forall x \in \Omega.$$

Furthermore, the authors express P and $\widehat{P}$ in the reference of each spectral band, thus allowing the use of the constraint on misregistered data.

In order to preserve the chromaticity encoded in the spectral data, Moeller et al. [70] propose to preserve the spectral ratio as follows:

$$\frac{U_i(x)}{U_j(x)} = \frac{\widehat{F}_i(x)}{\widehat{F}_j(x)}, \ \ \forall x \in \Omega,$$

for any pair $i, j = 1, \ldots, C$ with $i \neq j$. The above constraint is equivalent to minimizing the spectral angle between each pixel frequency vector in the low-resolution MS/HS and fused images. This metric is widely used to measure spectral distortions [7].

Finally, some variational methods for satellite image fusion incorporate observation models from spectral unmixing, a procedure by which any reflected light in a pixel can be decomposed into a finite number of spectra, called endmembers, and a set of corresponding fractions or abundances that indicate the proportion of each endmember present in the pixel. Accordingly, the high-resolution MS/HS image can be obtained from a linear combination of the endmembers

$$U = EX + \eta, \tag{5}$$

where E is the matrix composed by the endmembers and X is the abundance matrix. One key aspect is determining the endmembers, which can be achieved, for instance, using dictionaries or Wavelet transforms. Non-negative matrix factorization [48, 110], low-rank sparse representation [85], and compressed sensing [35, 55, 121] are variational methods based, to a greater or lesser extent, on (5).

2.1.5 Regularization Terms

Relying solely on observational models can lead to ill-posed problems. To address this issue, regularization terms has been proposed, often involving some norm of the gradient in the unknown image.

Ballester et al. [11] regularize (1) by aligning all level lines of the PAN image and each high-resolution spectral band

$$\sum_{k=1}^{C} \int_{\Omega} \left(|\nabla U_k(x)| + \langle \operatorname{div}(\theta(x)), U_k(x) \rangle \right) dx, \tag{6}$$

where θ is the vector field that consists of all unit normal vectors of the level sets of the PAN data. Several other VO-based methods use also (6) as regularizing prior [8, 70].

The total variation, originally introduced by Rudin et al. [86] in the context of image denoising, assumes that images consist of regular regions (objects) surrounded by sharp contours and has inspired several regularization terms. In this regard, Passon et al. [77] proposed band-decoupled total variation for satellite image fusion:

$$\sum_{k=1}^{C} \int_{\Omega} |\nabla U_k(x)|^2 dx.$$

He et al. [42] incorporate the gradient of the PAN image into the total variation energy

$$\int_{\Omega} \sqrt{\sum_{k=1}^{C} |\nabla U_k(x)|^2 + \gamma^2 |\nabla P(x)|^2} dx,$$

where $\gamma > 0$ is a balancing parameter. Zeng et al. [113] combine anisotropic and isotropic total variation diffusion.

In order to exploit self-similarity within satellite images, in the sense that every small patch has many similar patches in the same image, Duran et al. [30, 31] propose the following nonlocal regularization term:

$$\sum_{k=1}^{C} \iint_{\Omega \times \Omega} (U_k(y) - U_k(x))^2 \, \omega_P(x, y) dy dx,$$

where $\omega_P(x, y)$ is the similarity distribution computed on the PAN image.

2.2 Deep Learning-Based Methods

While neural networks have been studied since the 1970s, it was not until the past two decades that they have become the most popular tools in computer vision and machine learning. Indeed, deep learning, that is, neural networks with multiple layers, has taken the lead in many challenging benchmarks and has become state of the art for a wide variety of problems, including satellite image fusion. This success can be mainly attributed to advances in computation and the availability of large datasets.

To the best of our knowledge, the pioneering work in satellite image fusion using neural networks was conducted by Huang et al. [43], who presented an autoencoder scheme for pansharpening inspired by the sparse denoising task [101]. Later on, based on a previous work for image super-resolution [27], Masi et al. [65] proposed the first architecture with convolutional neural networks for pansharpening.

In this chapter, we classify deep learning-based methods into pure deep learning, characterized by the designed architecture, and model-based deep unfolding, which unrolls the steps of an iterative optimization scheme into layers of neural networks.

2.2.1 Pure Deep Learning

Pure deep learning strategies are distinguished by their specific architecture, which plays a crucial role in the performance of the fusion method. Often, these techniques draw inspiration from layers proposed in other computer vision tasks such as image classification, image segmentation, or image super-resolution. In this setting, examples include residual networks [40, 112, 118], generative adversarial networks [37, 49], recurrent neural networks [20, 72, 87], attention modules [39, 61, 97], and transformer encoders [29, 105]. Although deep learning-based approaches are data-driven, the design of network architectures is usually inspired by MRA and CS mechanisms by considering, for instance, high-frequency extraction, multiscale procedures, or spectral decomposition.

To highlight the most influential methods in recent years and showcase cutting-edge approaches in this category, we have chosen the following methods: PanNet, [108], MSDCNN [111], DiCNN [41], FusionNet [25], SRPPNN [16], NLRNet [54], HSIT [12], and AWFLN [63]. With this selection, we aim to provide an overview of the architectures that have played a significant role in satellite image fusion.

Residual networks are formed by stacking residual blocks, which consist of a sequence of convolutional layers with a skip connection at the end. The skip connection, also called residual connection, provides the advantage of addressing the vanishing gradient problem to some extent. Specifically, PanNet [108] computes a high-pass filter for the PAN image and the upsampled spectral image and then concatenates the results. The concatenated data passes through various residual blocks, and, finally, it is added to a bicubic interpolation of the spectral image. The idea behind PanNet is to fuse the colour information encoded in the bicubic interpolated MS/HS image with the high frequencies estimated by the residual network. The architecture of PanNet is shown in Fig. 2.

There are several other methods that also use residual networks in their architectures. MSDCNN [111] (see Fig. 3) extracts multiscale features through the MSResB module to preserve the coarse structure as well as the texture details. DiCNN [41] (see Fig. 4) injects the details from a concatenation of PAN and upsampled MS/HS images using a block composed of convolutional layers and ReLU activation functions. FusionNet [25] (see Fig. 5) estimates the geometric details from the

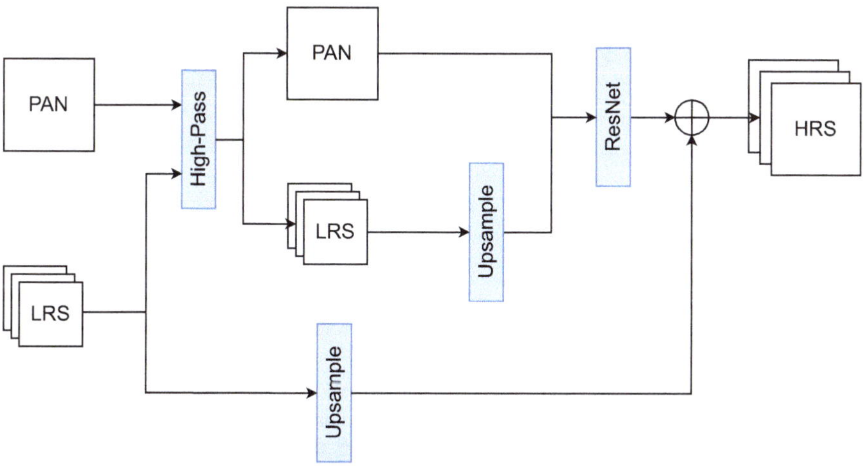

Fig. 2 Architecture of PanNet [108]. This method propagates the spectral information from the upsampled MS/HS image and preserve the spatial structure combining the PAN and MS/HS images via a ResNet [40]

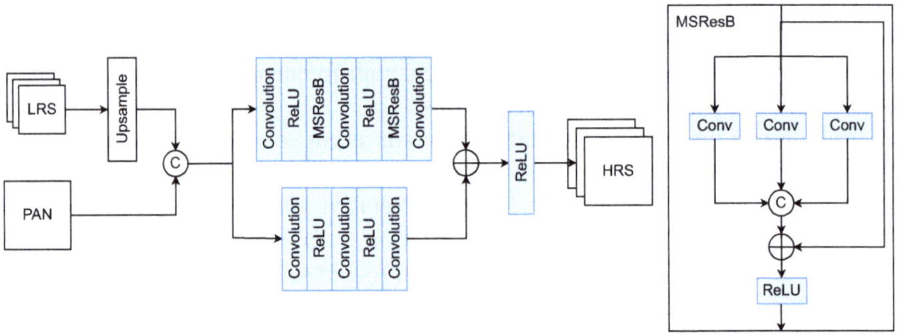

Fig. 3 Architecture of MSDCNN [111]. This model aims to extract multiscale features through the MSResB module to preserve the coarse structure as well as the texture details

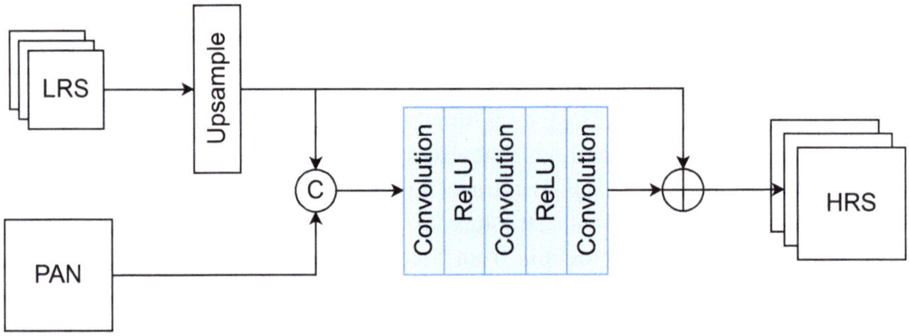

Fig. 4 Architecture of DiCNN [41]. This model aims to inject the details from a concatenation of PAN and upsampled MS/HS images using the block composed of convolutional layers and ReLU activation functions

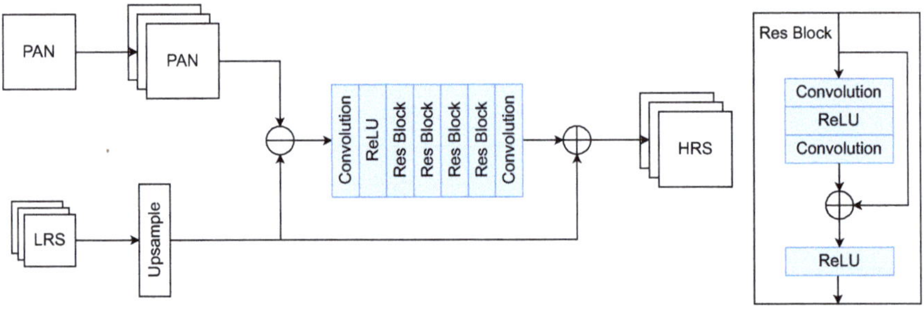

Fig. 5 Architecture of FusionNet [25]. This model estimates the geometric details from the difference between the PAN and the upsampled MS/HS data and then injects them into the upsampled MS/HS image

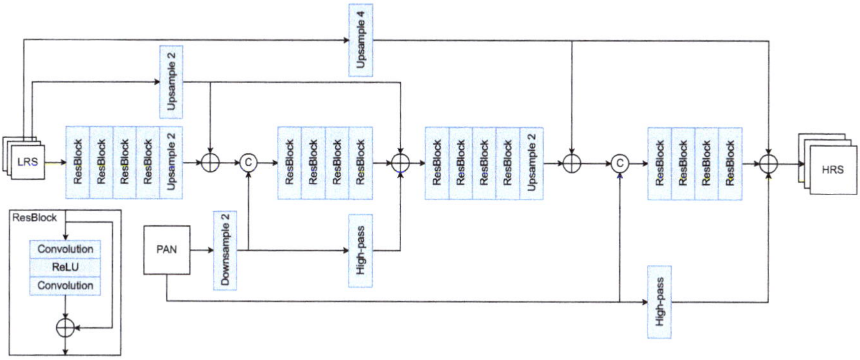

Fig. 6 Architecture of SRPPNN [16]. This model adapts a residual architecture for super-resolution to pansharpening by incorporating the PAN image at different scales and injecting their high frequencies to the processed upsampled MS/HS data

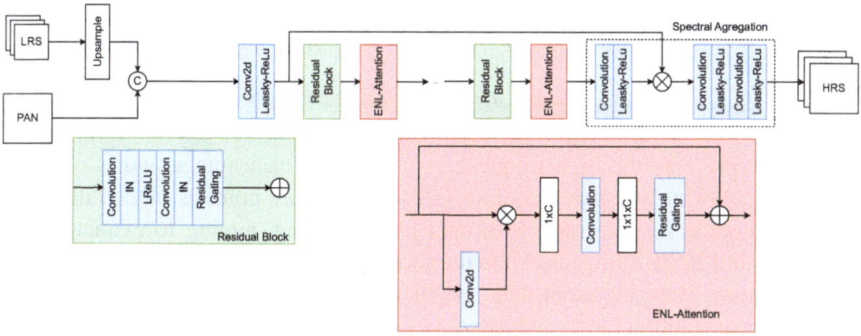

Fig. 7 Architecture of NLRNet [54]. This model combines residual blocks and efficient nonlocal attention module (ENL-Attention) to leverage image self-similarities in the fusion process. Then, a postprocessing is conducted by the spectral aggregation modules

difference between the PAN and the upsampled MS/HS data and then injects them into the upsampled MS/HS image. SRPPNN [16] (see Fig. 6) adapts a residual architecture for super-resolution to pansharpening by incorporating the PAN image at different scales and injecting their high frequencies to the processed upsampled MS/HS data.

Regarding attention modules, Wang et al. [97] propose an efficient way to compute a classic nonlocal filter [15] to leverage self-similarities in natural images. NLRNet [54] adapts the previous work and introduce a nonlocal attention module. The whole architecture, which is displayed in Fig. 7, is composed by a sequence of residual blocks followed by the efficient nonlocal channel attention module. Finally, the postprocessing is conducted by a spectral aggregation module that receives two feature maps and performs element-wise multiplication.

Transformer encoders have become very popular since they were proposed by Dosovitskiy et al. [29]. The foundation of these layers are the self-attention modules introduced by Wang et al. [97], but with a more sophisticated treatment. Specifically, the transformer layer combines the concatenation of various head-attentions, skip connections, and a multi-layer perceptron. Several works have incorporated transformer encoders for satellite image fusion [12, 67, 91, 114, 115, 120]. We will explain in detail the HSIT method proposed by Bandara et al. [12]. The method extracts features from the low-resolution MS/HS and PAN images to obtain queries and keys. The feature extraction process is carried out using a VGG-like architecture [89]. Subsequently, the attention module is applied, with the primary contribution being the Multi-Head Features Soft-Attention. Finally, a fusion module is added. The main backbone of the method employs the transformer mechanism at different scales and then fuses the outputs at each scale to obtain the estimated high-resolution MS/HS image. Figure 8 provides all the details of the architecture. This model needs to be pretrained due to the large number of parameters and its computational complexity.

To conclude this category, we present AWFLN [63], which does not fit completely into any of the previous network structures. The method estimates the fused image by adding missing spatial details to the upsampled MS/HS image

$$U = \widehat{F} + \widehat{D},$$

where $\widehat{D}$ represents the high frequencies of the reference image and is obtained through a neural network architecture. This architecture consists of two modules. The first module is concerned with data preprocessing, aiming to extract details from PAN and MS/HS images, while the second one focuses on extracting spectral-spatial features from the input data. In particular, this second module combines a multiscale convolution block with adaptive convolutional kernels based on spectral-spatial mixed attention. Although these kernels are computed using neural layers with learnable parameters, the resulting weights are adapted based on the data.

Figure 9 displays the complete architecture. Both attention modules operate on a local neighbourhood of the pixels, in contrast with the head-attention used by both NLRNet [54] and HSIT [12]. In fact, we can find several other works using attention mechanisms [38, 80, 119].

Except for HSIT, all deep learning methods referenced in this category have been proposed for pansharpening. However, their extension to hypersharpening is straightforward, since it only involves modifying the number of spectral bands of the input, of course, at some computational cost.

2.2.2 Model-Based Deep Unfolding

Although pure deep learning methods yield powerful representations, their training is often difficult. The architectures typically improve their performance as they become deeper, but this also increases the amount of necessary data, complexity, and computation time. They also require an expensive retraining whenever the parameters change. Furthermore, acquiring a sufficient amount of training data is

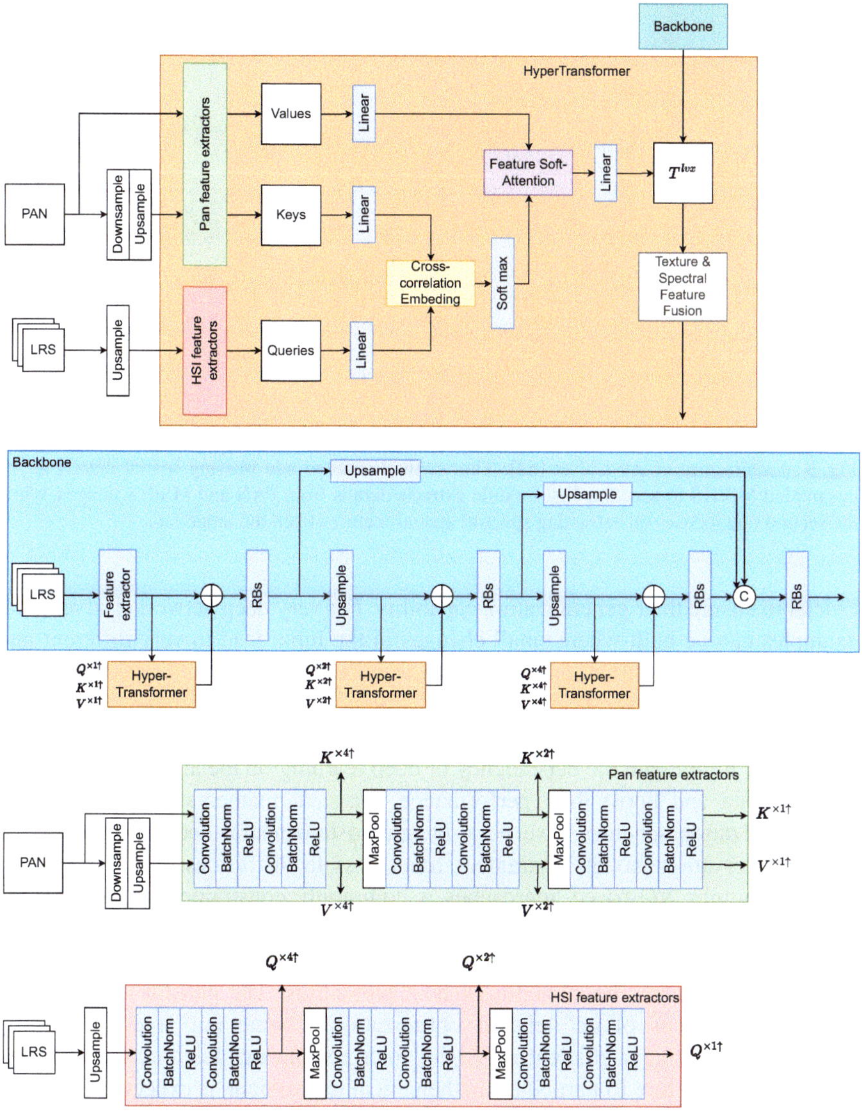

Fig. 8 Architecture of HSIT [12]. This model firstly extracts features at different scales and performs fusion through hyper-transformers modules and residual blocks (RBs). The architecture of the RBs is the same as that used in SRPPNN (refer to Fig. 6)

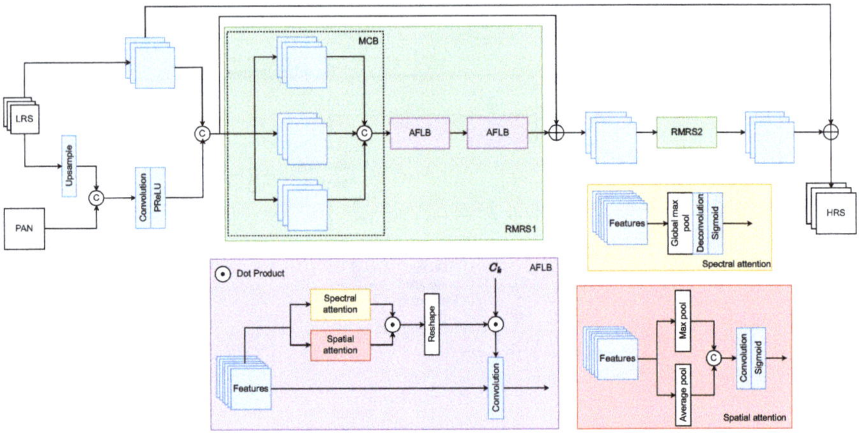

Fig. 9 Architecture of AWLFNet [63]. This method aims to add missing spatial details to the upsampled MS/HS image. The first module extracts details from PAN and MS/HS images, while the second one focuses on extracting spectral-spatial features from the input data

crucial to ensure their generalization capability. For very deep networks, adversarial examples can be built where small changes in the input lead to very different and nonsensical outputs, highlighting the poor stability of these systems. Even if much effort is dedicated to the design of explainable networks, they remain black boxes in many cases.

In order to mitigate the dependency of deep learning on the amount of available training data and boost their performance, classical mathematical formulations and optimization schemes have been integrated into neural networks. While the use of observation models guarantees robustness to distortions, the performance of, for example, VO-based approaches is commonly constrained by rigid priors. Conversely, pure data-driven learning techniques can adeptly learn natural priors but are less flexible and interpretable than model-based methods. Deep unfolding [1, 28, 71] combines the strengths of both paradigms. The general idea involves unrolling the steps of an iterative optimization scheme into a deep learning framework, resulting in efficient and highly interpretable architectures. Constraining the network structure to particular mathematical expressions is expected to reduce overfitting, improving generalization capabilities, and promote model interpretability. This is particularly crucial in scenarios where there might be a lack of labelled data, such as in satellite imaging.

In general, the backbone of deep unfolded methods is determined by the optimization algorithm that produces the fused product following some classical formulation. The operators involved are then substituted by neural architectures, which can be initialized with pretrained weights or scheduled for training.

Recently, deep unfolding has generated a considerable amount of research in satellite image fusion [17, 58, 79, 93, 100, 106]. We have selected GPPNN [103], MHFNet [102], MDCUN [107], MMNet [104], HyFPan [68], and S2DBPN [117] as the featured methods in this category. We intend to emphasize the crucial role that modelling plays in deep unfolding and its impact on the performance of the fusion method. In terms of notation, all norms used in this section refer to Frobenius norm.

To the best of our knowledge, the first unfolding method was MHFNet proposed by Xie et al. [102] for the fusion of a low-resolution HS and a high-resolution MS image. Adapting the architecture to pansharpening or hypersharpening only involves changes in the spectral dimensionality of the data. For this reason, we present MHFNet in the context of pansharpening/hypersharpening. The authors combine the data formation models (1) and (2) into a single one using unmixing theory (5) and then reformulate the associated inverse problem in the variational framework as follows:

$$\min_{Y} \|D(PA + YB) - F\|^2 + \lambda R(Y), \tag{7}$$

where R is a regularizing prior, $\lambda > 0$ a trade-off parameter, $D \in \mathbb{R}^{wh \times WH}$ the spatial response of the low-resolution spectral data (usually assumed to be the composition of a convolution and a downsampling operator), $A \in \mathbb{R}^{1 \times C}$ estimates an approximation of the low-resolution spectral data to the whole domain, $B \in \mathbb{R}^{(r-1) \times C}$ is the matrix corresponding to the endmembers coefficients, r is the rang of the endmembers matrix, and $Y \in \mathbb{R}^{HW \times (r-1)}$ is the matrix composed of the endmembers. The minimum of (7) is computed through the proximal gradient algorithm [13]:

$$Y^{k+1} = \text{prox}_{\lambda\sigma R}\left(Y^k - \sigma D^T\big(D\big(PA + Y^k B\big) - F\big)B^T\right).$$

The iterative scheme is unfolded by replacing the proximity operator with a residual network and the other involved operators with convolutional neural networks. Finally, the fused image is obtained as $U = PA + YB$. Figure 10 illustrates the proposed architecture.

Xu et al. [103] introduced GPPNN, which is also based on the observation models (1) and (2). Instead of defining an energy that combines both, the authors propose a two-level minimization problem

$$\begin{aligned} &\min_{U} \|DBU - F\|^2 + \lambda_1 R_1(U), \\ &\min_{U} \left\|\textstyle\sum_{k=1}^{C} \alpha_k U_k - P\right\|^2 + \lambda_2 R_2(U), \end{aligned} \tag{8}$$

where R_1 and R_2 are regularization terms and $\lambda_1, \lambda_2 > 0$ trade-off parameters. Proximal gradient is applied to solve each minimization problem in (8). The corresponding iterative schemes are computed one after another, using the output of the previous step as the input data

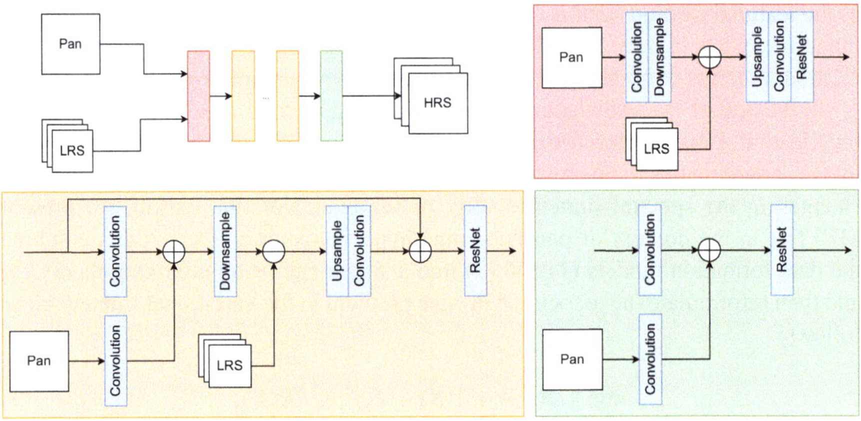

Fig. 10 Architecture of MHFNet [102]. The orange and red blocks correspond to one step of the unfolded scheme, with the difference that the red one assumes an initialized image with zeros. The green module corresponds to a postprocessing step. Refer to [40] for ResNet details

$$
\begin{aligned}
U^{n+1/2} &= \mathrm{prox}_{\lambda_1 \tau_1 R_1}\left(U^n - \tau \nabla f(U^n)\right), \\
U^{n+1} &= \mathrm{prox}_{\lambda_2 \tau_1 R_2}\left(U^{n+1/2} - \tau \nabla g(U^{n+1/2})\right),
\end{aligned}
\tag{9}
$$

where $f(U) = \|DBU - F\|^2$ and $g(U) = \left\|\sum_{k=1}^{C} \alpha_k U_k - P\right\|^2$. Finally, the proximal and other operators are replaced with convolutional layers. Figure 11 shows the detailed architecture.

MDCUN by Yang et al. [107] imposes (1) for the formation of the low-resolution spectral image. Furthermore, they introduce a nonlocal auto-regression prior which is capable of searching the similarities between long-range patches. More specifically, they assume that

$$
U = \mathrm{NL}(U) + \eta, \tag{10}
$$

where NL denotes a nonlocal operator. By introducing the above constraint, the observation model (1) is rewritten as $F = DB(\mathrm{NL}(U)) + \eta$. Therefore, the authors formulate the following energy:

$$
\begin{aligned}
\min_U \frac{1}{2}\|F - DBU\|^2 &+ \lambda R_1(U|P) \\
&+ \frac{\mu}{2}\|F - DB(\mathrm{NL}(U))\|^2 + \nu R_2(\mathrm{NL}(U)|P),
\end{aligned}
\tag{11}
$$

where R_1 and R_2 are regularizing priors guided by the PAN image, and $\lambda, \mu, \nu > 0$ are trade-off parameters. For minimization, they introduce two auxiliary variables

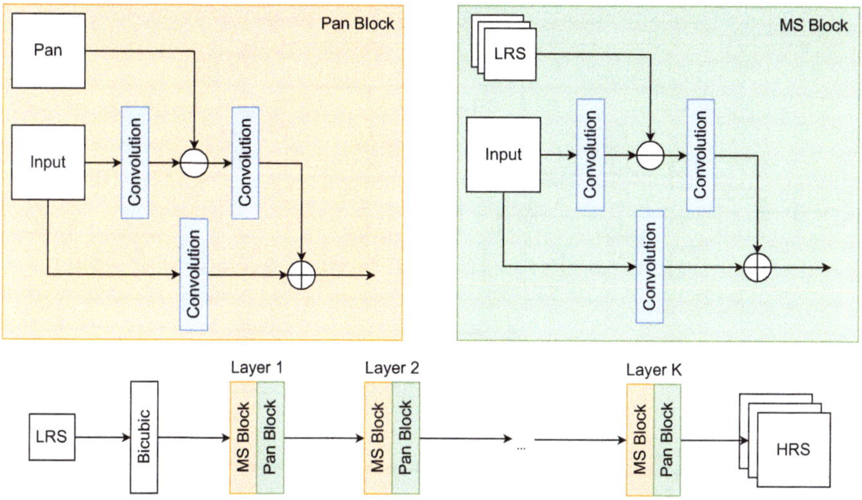

Fig. 11 Architecture of GPPNN [103]. The MS Block and PAN Block correspond to the unfolded steps for minimizing the first and second energies in (8)

and solve the resulting minimization problems with the proximal gradient algorithm. The regularization terms are replaced by convolutional layers that take advantage of the geometrical structures in the PAN image. The NL operator is replaced by a self-attention module [97] and the upsampling and downsampling operators by convolutional layers. Finally, memory modules are introduced to improve the transfer between stages of the method. Figure 12 displays the overall architecture as well as the details of the previously mentioned modules.

MMNet, as proposed by Yan et al. [104], exploits self-similarity in the regularization term instead of in the data-fidelity term, as in the case of MDCUN. The resulting energy is

$$\min_{U} \frac{1}{2} \|F - DBU\|^2 + \lambda_1 R(U|P) + \lambda_2 R_{\mathrm{NL}}(U|P), \tag{12}$$

where $\lambda_1, \lambda_2 > 0$ are trade-off parameters, R represents a local regularizing prior, and R_{NL} is a nonlocal regularization in the sense of (2.1.5). For optimization purposes, the authors introduce an auxiliary variable and solve the resulting minimization problems using the proximal gradient algorithm. In the unfolding step, the proximal operators are replaced with a layer that combines residual blocks with convolutions in an identical backbone, guiding the input variable with the PAN image. Importantly, the proximal operator associated to R_{NL} involves a nonlocal filter based on [97]. Finally, a memory mechanism that retains information from previous stages is introduced in each module to enhance the model capability. More details on the architecture are displayed in Fig. 13.

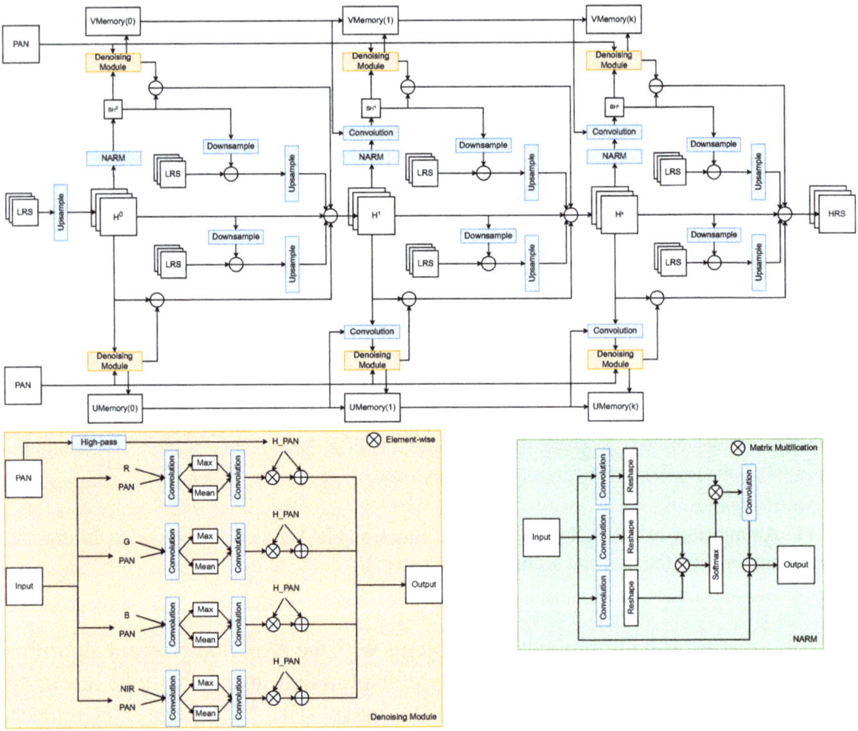

Fig. 12 Architecture of MDCUN [107]. The top diagram depicts the whole architecture, while the bottom diagrams displays the denoising and NARM modules, respectively

HyFPan by Mifdal et al. [68] presents a satellite image fusion method based on MHFNet [102] but also adopts the radiometric constraint (2.1.4). Accordingly, the proposed energy is

$$\min_{Y} \|D(PA + YB) - F\|^2 + \lambda\|\widehat{P}(PA + YB) - P\widehat{F}\|^2 + \beta R(Y),$$

where $\lambda, \beta > 0$ are trade-off parameters, R represents a general prior, and operators A and B are defined in (7). The authors minimize the above energy using proximal gradient and unfold the algorithm by replacing the involved operators with layers that combine both convolutional and residual blocks. Refer to Fig. 14 for an illustrative diagram of the architecture.

Finally, in contrast to all previous methods detailed in this category, the architecture of S2DBPN [117] is not based on an energy formulation but inspired by the back projection technique [23, 44]. This technique operates on the premise that the recovered image can be obtained by minimizing the reconstruction error between the target image and the output. The reconstruction error is computed under the

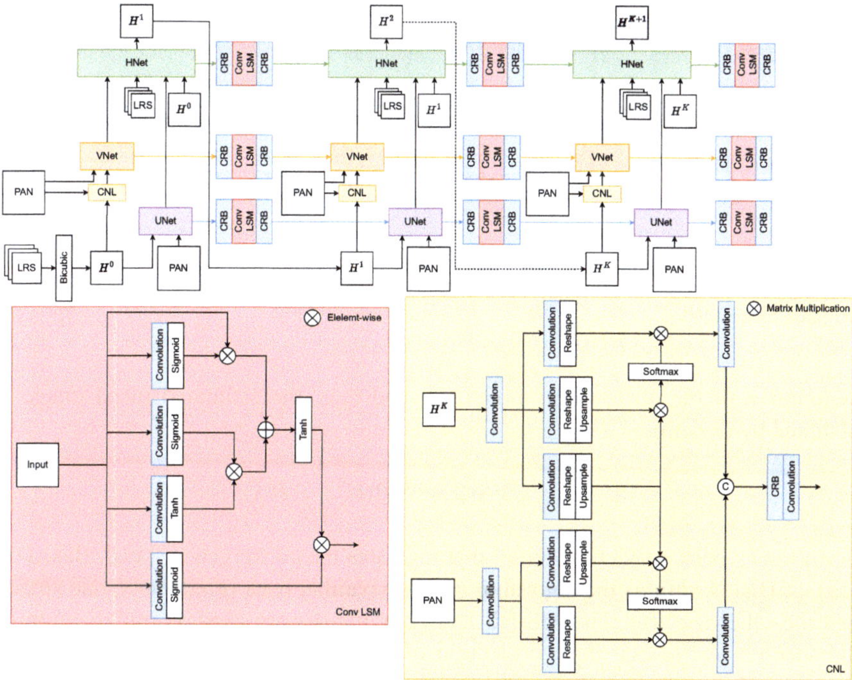

Fig. 13 Architecture of MMNet [104]. The top diagram shows the whole architecture, while the bottom diagrams displays the ConvLSM and CNL module, respectively. The UNet, VNet, and HNet represent the unfolded steps of the resulting iterative scheme

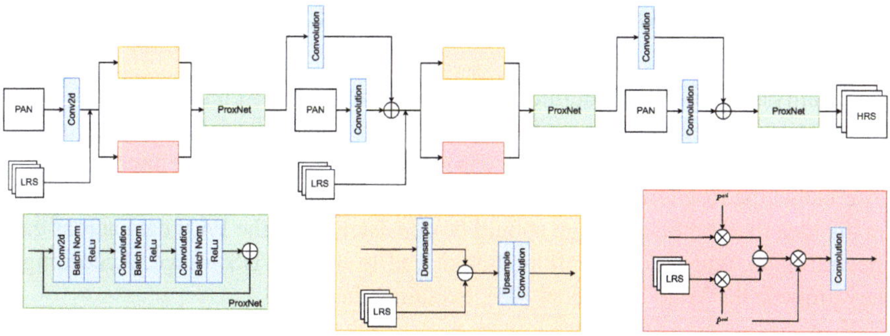

Fig. 14 Architecture of HyFPan [68]. This model is composed of an initialization step, one middle stage, and a final stage as postprocessing. The initial and middle steps correspond to the unfolding of the iterative scheme

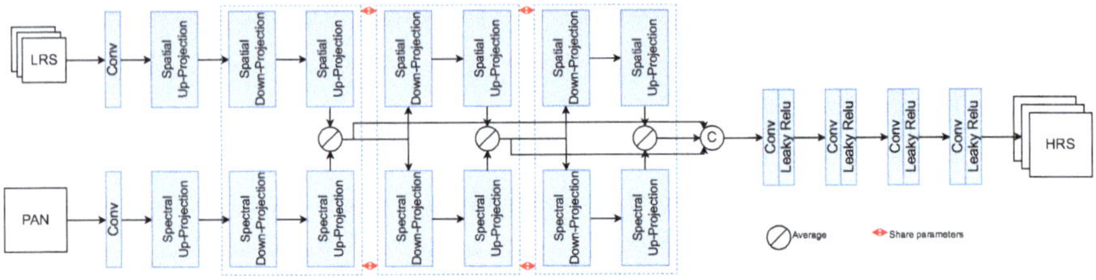

Fig. 15 Architecture of S2DBPN [117]. The spatial back projection is initialized with the MS/HS image, while the spectral one is initialized with the PAN image. The output at each stage is fused to obtain the results

assumption of the observation model (1), and thus, the estimated fused image is obtained by iterating

$$U^{k+1} = U^k + g * (DB(U^k) - HS), \tag{13}$$

where g is a back projection kernel that regulates the convergence speed. Based on (13), S2DBPN unfolds the algorithm into two parallel back projections, one spatial and the other spectral. The MS/HS and PAN images are used as initialization of the spectral and spatial back projection modules, respectively. At each iteration, the results of both back projections are averaged and used as the input for the following step. Finally, the outputs of all stages are fused to obtain the final result. The architecture of the method is given in Fig. 15.

3 Quality Assessment of Fused Products

The quality assessment of satellite image fusion methods is a challenging task due to the absence of reference images [19, 95]. When reference images are not available, a criterion to evaluate the performance is needed. Wald et al. [95] define two general criteria that the fused image must satisfy: consistency and synthesis. These are based on the idea that the fused image should be consistent with the PAN data across each spectral band, and its degradation should lead to the MS/HS image. Accordingly, they propose the Wald's protocol [95], explained in Sect. 3.1.

If one has access to reference images with which to compare the fused products, a quantitative measure is needed for performance evaluation. In this regard, the most commonly used metrics are described in Sect. 3.2. On the contrary, Sect. 3.3 lists non-reference metrics that allow evaluating the quality of the results by comparing with the spatial and spectral information encoded in the PAN and MH/HS images, respectively.

3.1 Wald's Protocol

In 1997, Wald et al. [95] proposed the so-called Wald's protocol for generating references from non-reference datasets. This protocol aims to provide a means to assess the performance of fusion methods based on the criteria defined in [95]. Since acquiring high-resolution spectral data may not be feasible, the low-resolution MS/HS images are considered as references. Then, both the MS/HS and PAN images are degraded by downsampling. Therefore, fusion methods can be evaluated using the degraded images as input, and their outputs can be assessed by comparing them with the original MS/HS image.

3.2 Reference Metrics

To evaluate the performance of the fusion methods, we have chosen some of the most widely used quality metrics: the global spatial measures *peak signal-to-noise ratio* (PSNR) [81] and *Erreur Relative Globale Adimensionnelle de Synthèse* (ERGAS) [83], the *spectral angle mapper* (SAM) [7], which measures the spectral quality, and $Q2^n$ [34], which is a generalization of the Q index [98] specifically designed to mimic human perception of image quality for images of any number of spectral bands.

In the following, we present the definitions of all the above metrics in terms of the estimated fused products $\widetilde{U} \in \mathbb{R}^{C\times W\times H}$ and the reference images $\overline{U} \in \mathbb{R}^{C\times W\times H}$. We use $\overline{U}_k$ and $\widetilde{U}_k$ to denote the kth band of $\widetilde{U}$ and $\overline{U}$, respectively. Furthermore, all norms used in this section refer to Frobenius norm.

PSNR is one of the most used metrics in image processing that accounts for spatial distortions. It is given in terms of the ratio between the maximum possible power of a signal and the power of corrupting noise that affects the fidelity of its representation

$$\mathrm{PSNR}(\overline{U}, \widetilde{U}) = 20\log_{10}\left(\frac{\mathrm{MAX}_I}{\mathrm{RMSE}(\overline{U}, \widetilde{U})}\right),$$

where MAX_I is the maximum value possible for a pixel and

$$\mathrm{RMSE}(\overline{U}, \widetilde{U}) = \frac{\|\overline{U} - \widetilde{U}\|}{\sqrt{C \cdot W \cdot H}}.$$

The optimal value of PSNR is ∞.

ERGAS is an indicator that provides an overall evaluation of the fused product. It is defined as

$$\mathrm{ERGAS}(\overline{U}, \widetilde{U}) = 100s\sqrt{\frac{1}{C}\sum_{k=1}^{C}\left(\frac{\mathrm{RMSE}(\overline{U}_k, \widetilde{U}_k)}{\mu_k}\right)^2},$$

where s is the sampling factor between high and low resolutions, μ_k is the mean value of the kth spectral band of the reference image, and

$$\mathrm{RMSE}(\overline{U}_k, \widetilde{U}_k) = \frac{\|\overline{U}_k - \widetilde{U}_k\|}{\sqrt{W \cdot H}}.$$

The optimal value of ERGAS is zero.

SAM is an index of spectral quality within a framework where each channel is treated as a separate coordinate axis. It is defined as

$$\mathrm{SAM}(\overline{U}, \widetilde{U}) = \frac{1}{W \cdot H}\sum_{x\in\Omega}\cos^{-1}\left(\frac{\langle \overline{U}(x), \widetilde{U}(x)\rangle}{\|\overline{U}(x)\|\|\widetilde{U}(x)\|}\right)$$

and its optimal value is zero.

Due to the limitations of error summation metrics such as RMSE or PSNR, some statistical indexes were introduced. In this context, the Q index or univariate UIQI index, proposed in [98], measures image distortion by combining correlation loss, contrast distortion, and luminance distortion. It was originally formulated for monochrome images as

$$\mathrm{Q}(\overline{U}_k, U_k) = \frac{\sigma_{\overline{U}_k, U_k}}{\sigma_{\overline{U}_k}\sigma_{U_k}}\frac{2\sigma_{\overline{U}_k}\sigma_{U_k}}{\sigma^2_{\overline{U}_k} + \sigma^2_{U_k}}\frac{2\mu_{\overline{U}_k}\mu_{U_k}}{\mu^2_{\overline{U}_k} + \mu^2_{U_k}}.$$

The statistics are computed on 8×8 patches, and the resulting values are then averaged over the whole image. The optimal value of Q index is 1.

To assess correlation loss, contrast distortion, and luminance distortion, the Q4 index [7] was introduced for multispectral images with four channels. Later, the $Q2^n$ index [34] was proposed to extend this idea for images with any number of spectral bands. These protocols were formulated by considering each pixel as a quaternion or hypercomplex number. Since a hypercomplex number has a dimension that is a power of 2, pixels with a number of channels not equal to a power of 2 are represented as hypercomplex numbers by adding zeros until they reach the next power of 2. The $Q2^n$ index can be easily applied to multiband images, allowing for the evaluation of both radiometric and spectral image distortions

$$\mathrm{Q2}^n(\overline{U}, U_k) = \frac{|\sigma_{\overline{Z}Z}|}{\sigma_{\overline{Z}}\sigma_Z}\frac{2\sigma_{\overline{Z}}\sigma_Z}{\sigma^2_{\overline{Z}} + \sigma^2_Z}\frac{2|\mu_{\overline{Z}}||\mu_Z|}{|\mu_{\overline{Z}}|^2 + |\mu_Z|^2}, \tag{14}$$

where $\overline{Z}$ and Z are the representation in hypercomplex numbers of $\overline{U}$ and U. The statistics are computed on 8×8 patches, and the resulting values are then averaged over the whole image. The optimal value of $Q2^n$ is 1.

3.3 Non-reference Metrics

For real satellite data, the absence of the ground truth makes more difficult to evaluate the performance of fusion methods. Non-reference metrics mitigate this problem since they measure the spectral similarity with the MS/HS image and the spatial similarity with the PAN image.

In this setting, the *quality with no reference* (QNR) protocol proposed in [6] uses the Q index to evaluate both spectral and spatial quality. On the one hand, D_S measures the spatial distortion by comparing the fused and the interpolated MS/HS images with the PAN image:

$$D_s = \frac{1}{C}\sum_{k=1}^{C} \left|\mathrm{Q}(\widehat{F}_k, P) - \mathrm{Q}(\widetilde{U}_k, P)\right|.$$

On the other hand, D_λ measures the spectral distortion by comparing the band correlation in the MS/HS image with that in the fused image:

$$D_\lambda = \frac{1}{C(C-1)}\sum_{k=1}^{C}\sum_{r=1,r\neq l}^{C} \left|\mathrm{Q}(F_k, F_r) - \mathrm{Q}(\widetilde{U}_k, \widetilde{U}_r)\right|.$$

From both indexes, QNR is finally computed as

$$\mathrm{QNR} = (1 - D_\lambda)^\alpha (1 - D_S)^\beta,$$

and its optimal value is 1. According to [2], we set $\alpha = 0.1$ and $\beta = 1$.

In [50], the FQNR protocol was introduced to improve QNR. The spectral distortion index is now computed as

$$D_\lambda^F = 1 - \mathrm{Q2}^n(\widetilde{U}_{L\downarrow}, F),$$

where the subscript $_{L\downarrow}$ indicates low-pass and decimation of the associated image and thc spatial distortion index as

$$D_s^F = \frac{1}{C}\sum_{k=1}^{C} \left|Q(\widehat{F}_{kH}, P_{L\downarrow H}) - Q(\widetilde{U}_{kH}, P_H)\right|,$$

where $\widehat{F}_{kH} = \widehat{F}_k - \widehat{F}_{kL}$, $\widetilde{U}_{kH} = \widetilde{U}_k - \widetilde{U}_{kL}$, $P_H = P - P_L$, and $P_{L\downarrow H} = P_{L\downarrow} - P_{L\downarrow L}$. Then, FQNR is defined as

$$\text{FQNR} = (1 - D_\lambda^F)^\alpha (1 - D_s^F)^\beta.$$

On the other hand, the HQNR protocol proposed in [2] combines the strengths of FQNR and QNR. It is defined as

$$\text{HQNR} = (1 - D_\lambda^F)^\alpha (1 - D_s)^\beta.$$

As suggested in [10], we set $\alpha = 1.08$ and $\beta = 1.1$ for FQNR and $\alpha = 0.98$ and $\beta = 0.89$ for HQNR.

4 Satellite Resources and Datasets

In recent years, numerous satellites with different sensor characteristics have been launched for Earth monitoring. Table 3 contains a list of those commonly used in image fusion.

To assess the performance of the fusion methods, we have selected Quickbird and WorldView2 datasets, which have MS and PAN sensors on board, and PRISMA dataset, which has HS and PAN sensors. Although most of the techniques were originally presented for either pansharpening or hypersharpening, we have adapted all of them to the three datasets since only a change in number of spectral channels is required. In Sect. 4.1, we provide extensive information on the satellites used in this review, and we explain how data has been generated in Sect. 4.2.

Researchers can download satellite data from various space agencies. The European Space Agency (ESA) allows downloads through their Earth Online Portal,[1] accessible after registering on the ESA User Services Portal.[2] This provides access to data from some WorldView,[3] QuickBird2,[4] GeoSat2,[5] and PRISMA,[6] among others.

---

1 https://earth.esa.int/eogateway/catalog/

2 https://esatellus.service-now.com

3 https://earth.esa.int/eogateway/catalog/worldview-2-full-archive-and-tasking

4 https://earth.esa.int/eogateway/catalog/quickbird-full-archive

5 https://earth.esa.int/eogateway/catalog/geosat-2-spain-coverage

6 https://www.asi.it/en/earth-science/prisma/

Table 3 Main satellites used in the state of the art for pansharpening/hypersharpening

Mission's name	Launch date	End of life date	Pan-resolution	Low-resolution	# Spectral bands
Ikonos-2	24 Sep 1999	31 Mar 2015	0.82 m	3.28 m	4
ALI (EO-1)	23 Nov 2000	30 Mar 2017	10 m	30 m	10
Hyperion (EO-1)	23 Nov 2000	30 Mar 2017	–	30 m	220
QuickBird-2	18 Oct 2001	17 Dec 2014	0.61 m	2.4 m	4
WorldView1	18 Sep 2007	Sep 2023	0.5 m	2 m	8
WorldView2	08 Oct 2009	–	0.5 m	2 m	8
Pleiades	17 Dec 2011	–	0.7 m	2.8 m	4
Gaofen-1	26 Apr 2013	2027	2 m	8 m	8
WorldView3	13 Aug 2014	–	0.31 m	1.24 m	8
WorldView4	11 Nov 2016	07 Jan 2019	0.31 m	1.24 m	8
VENuS	02 Aug 2017	–	–	5.3 m	12
PRISMA	22 Mar 2019	–	5 m	30 m	66

4.1 Selected Satellite Data

4.1.1 QuickBird

Launched in 2001, the QuickBird satellite was a groundbreaking commercial Earth observation satellite renowned for its exceptionally detailed imagery. It incorporated four MS bands (blue, green, red, and near-infrared) and a high-resolution PAN image. Offering a revisit time of 1–3.5 days, QuickBird proved valuable for various applications such as urban planning, agriculture, and environmental monitoring. Operating in a sun-synchronous orbit to ensure consistent lighting, the satellite boasted a wide swath width and made significant contributions to high-resolution satellite imagery until it concluded its operations in 2015. The MS data covers the nadir region from 450 to 890 nm with a 2.4 m resolution, while the PAN data has a spatial resolution of 60 cm, featuring a single band.

4.1.2 WorldView2

The information was obtained using the WorldView2 sensor, designed to operate within the visible and near-infrared spectrum. The MS sensor comprises eight spectral bands (coastal, blue, green, yellow, red, red edge, near-infrared region 1, and near-infrared region 2), with an additional PAN channel. The spatial sampling intervals are 1.84 m for MS and 0.46 m for PAN. The resolution ratio s is set at four, and the radiometric resolution is 11 bits.

4.1.3 PRISMA

In 2019, the Italian Space Agency initiated the PRISMA mission, dedicated to provide public data for a specific area, presenting significant potential for data fusion and resolution enhancement. The HS data encompasses the visible and near infrared (VNIR) region (400–1010 nm) and the short wavelength infrared (SWIR) range (20–2505 nm), providing 173 spectral bands with spatial resolution of 30 m. The PAN data consists of a single band with a spatial resolution of 5 m. PRISMA is in a sun-synchronous orbit, with an altitude of 614 km. The satellite primarily operated in a user-driven targeting mode with a re-targeting period of 7 days and a response time of 14 days.

4.2 *Data Processing*

As it is usual, the images obtained from the satellites have been cropped and processed into patches. For WorldView2 and QuickBird datasets, the Wald's protocol has been used, and the images have been cropped into non-overlapping patches of sizes 128×128 and 64×64, respectively. The original WorldView2 images were obtained through the portal, while the data used for QuickBird was provided by the public repository developed in [24]. The s-downsampling process, used during the Wald's protocol, has been conducted by averaging pixels within a square block of size s^2. This method has been selected to mimic the modulation transfer function [4].

On the other hand, for PRISMA dataset provided by Italian Space Agency, we cropped the original HS images into non-overlapping tiles of size 128×128. We specifically focused on the first 66 bands that cover the spectral response of the PAN sensor (the VNIR bands), excluding the SWIR bands. After checking all of the VNIR spectral bands, we noticed that three bands in most of the images are exactly zero. To eliminate unnecessary information, these bands have been removed. Then, we used the Wald's protocol, considering the HS image as the reference and generating the observed data by downsampling the original. However, we generated the PAN image using the spectral response sensors. This approach differs from applying downsampling directly to the PAN image. The main reason behind this strategy is that the sampling factor between the high- and low-resolution domains is 6, while most of the methods in the literature are designed for a sampling of 4, and their generalization requires non-trivial changes.

Finally, each dataset has been divided into training, validation, and testing sets using a proportion of 70/15/15.

5 Training Phase and Validation Process

In deep learning models, the training and validation phases play a fundamental role in the final performance of the model. In this context, the optimizer, the learning rate, the number of epochs, and the loss function are crucial to obtain the best results. In Table 4 we summarize the optimizer, the learning rate, the loss function, and the epochs used by the authors for training each compared model. Several methods use more sophisticated losses than the common MSE, L1, or SSIM. For a better understanding, in Sect. 5.1 we explain the different items labelled as *specific loss* in Table 4. Additionally, in Sect. 5.2, we will discuss the number of epochs needed to train the models and propose a criterion for early stopping. Finally, in Sect. 5.3, a comparison of the computational cost is conducted across all the datasets.

Table 4 Summary of the optimizer, learning rate, loss function, and number of epochs used by the authors for training each compared model. FIL stands for fixed initial learning rate

Model	Year	Epoch number	Pre-train epochs	Optimization algorithm	Initial learning rate	Learning rate tuning strategy	Loss function
PanNet	2017	450	No	Adam	1E-3	FIL	MSE
MSDCNN	2017	500	No	–	1E-1	×0.5 every 60 epochs	SSIM
MHFNet	2019	5000	No	Adam	–	FIL	Specific loss
DiCNN	2019	1000	No	Adam	2E-4	×0.5 every 200 epochs	MSE
SRPPNN	2020	–	No	Adam	4E-4	FIL	MSE
FusionNet	2021	4000	No	Adam	3E-4	FIL	MSE
GPPNN	2021	200	No	Adam	5E-4	FIL	L1
NLRNet	2022	–	No	Adam	1E-3	FIL	Specific loss
HSIT	2022	10k	10k	Adam	1E-4	×0.1 every 2k epochs	Specific loss
MMNet	2022	1000	No	Adam	8E-4	×0.5 at 200 epoch	L1
MDCUN	2022	1000	No	Adam	5E-4	×0.5 every 200 epochs	L1
HyPanF	2023	4000	No	Adam	1E-3	FIL	Specific loss
S2DBPN	2023	500	No	Adam	3E-5	FIL	MSE
AWFLN	2023	300	No	Adam	5E-4	×0.5 every 100 epochs	L1

5.1 Loss Functions

In this section we describe those loss functions labelled as *specific loss* in Table 4. That is, the training loss of the methods MHFNet [102], HyPanF [68], NLRNet [54], and HSIT[12].

The loss function used to train MHFNet [102] consists of the following three terms:

$$\mathcal{L}_{\text{MHFNet}}(\overline{U}, \widetilde{U}) = \|\overline{U} - \widetilde{U}\|^2 + \alpha \sum_{n=1}^{N} \|\overline{U} - \widetilde{U}^{(n)}\| + \beta \|\mathcal{D}own(\widetilde{U}) - F\|^2,$$

where $\overline{U}$ is the reference image, $\widetilde{U}^{(n)}$ is the output obtained at stage n, and $\mathcal{D}own$ is the unfolded downsampling operator. The first term is the common MSE loss function that forces the fused image to be similar to the reference one. The second one is added to ensure the convergence of the method with a low number stages. Finally, the third term forces to minimize the data fidelity term from the energy formulation.

Similarly to the previous case, the loss proposed in HyPanF [68] is

$$\begin{aligned}\mathcal{L}_{\text{HyPanF}}(\overline{U}, \widetilde{U}) =& \|\overline{U} - \widetilde{U}\|^2 + \alpha \sum_{n=1}^{N} \|\overline{U} - \widetilde{U}^{(n)}\| + \beta \|\mathcal{D}own(\widetilde{U}) - F\|^2 \\ &+ \gamma \|\widehat{P\overline{U}} - P\widehat{F}\|^2,\end{aligned}$$

where the authors propose to include a fourth term that promotes the minimization of the radiometric fidelity term, which is also included in their energy formulation.

The loss proposed to train NLRNet [54] is composed by a spatial term, a spectral one and a band-consistency term:

$$\begin{aligned}\mathcal{L}_{\text{NLRNet}}(\overline{U}, \widetilde{U}) =& \alpha \|\overline{U} - \widetilde{U}\|_1 + \beta \left(1 - \sum_{x \in \Omega} \frac{\langle \widetilde{U}(x), \overline{U}(x) \rangle}{\|\widetilde{U}(x)\| \|\overline{U}(x)\|}\right) \\ &+ \gamma \sum_{k=1}^{C-1} \left\|(\overline{U}_k - \overline{U}_{k+1}) - (\widetilde{U}_k - \widetilde{U}_{k+1})\right\|_1 \\ &+ \gamma \left\|(\overline{U}_1 - \overline{U}_C) - (\widetilde{U}_1 - \overline{U}_C)\right\|_1.\end{aligned}$$

The loss proposed for the model HSIT [12] also combines three terms as follows:

$$\begin{aligned}\mathcal{L}_{\text{HSIT}}(\widetilde{U}, \overline{U}) =& \alpha \|\widetilde{U} - \overline{U}\|_1 + \beta \|\phi_{\text{VGG}}(f_{\text{rgb}}(\widetilde{U})) - \phi_{\text{VGG}}(f_{\text{rgb}}(\overline{U}))\| \\ &+ \gamma \|f^s_{\text{fe}}(\widetilde{U}) - T^s\|,\end{aligned}$$

where ϕ_{VGG} is a perceptual loss based on the VGG network, which has demonstrated to be useful for RGB super-resolution tasks [46], and f_{rgb} is a function that synthesizes the HS image into an RGB image using a Gaussian-approximated spectral response curve. Finally, f_{fe}^{s} represents the feature map at the s-th spatial scale, and T^{s} denotes the transferred feature map at the s-th spatial scale.

5.2 *Convergence Criterion*

The training phase of neural networks models entail various dependencies and drawbacks. Neural networks designed for imaging purposes rely on factors such as channels, size, and patches within the dataset, along with occasional considerations for the dimensions of the images. On the one hand, channel dependencies determine the number of parameters in the model, thus influencing its complexity. On the other hand, the size of the dataset can impact the number of iterations or epochs required for the training process to converge. Additional attributes affecting the training process of neural networks include the choice of loss functions and performance metrics.

In the exploration and comparison of neural networks, various strategies exist in the literature. The first strategy involves fixing a dataset and training the models using specific parameters provided by the authors. The second strategy fixes the number of epochs and the dataset, training different models accordingly. Due to dataset dependencies, both strategies encounter similar issues; the convergence of training heavily relies on the dataset. Therefore, setting epochs or applying author-specified parameters may lead to models failing to converge.

In the literature, some works study early-stopping criteria. Miseta et al. [69] discuss a method to prevent overfitting based on Pearson correlation. Deng et al. [26] observed the curves of the training loss and validation loss to determine the convergence of the models.

Our contribution in this regard is to define a criterion for training based on the geometry of the learning curves. We propose a method that is independent of loss functions, the number of parameters, and the dataset itself. The method involves calculating the least-squares line of the last k epochs and stopping the training when the slope falls below a specified threshold, denoted as ϵ. With $f(x_i)$ being the loss function at epoch x_i and X_i^k the window of k epochs ending at epoch i, this line is given by

$$y = mx + c,$$

with

$$m = \frac{k \sum_{i=1}^{k} x_i f(x_i) - \sum_{i=1}^{k} x_i \sum_{i=1}^{k} f(x_i)}{k(\sum_{i=1}^{k} x_i^2) - (\sum_{i=1}^{k} x_i)^2}, \quad c = \frac{1}{k}\left(\sum_{i=1}^{k} f(x_i) - m \sum_{i=1}^{k} x_i\right).$$

Therefore, we propose to stop the training when the condition $m < \epsilon$ is reached.

The robustness of this method relies solely on the shape of the curve. In our implementation, we apply the criterion to the loss function, but this can be applied to any metric. The window size is a parameter vulnerable to the training volatility, so we choose a large window size to smooth the effect of outliers. In our tests, we selected a window size of $k = 100$ epochs.

This criterion is not enough due to the scale difference in loss functions. To solve this, we go further and introduce the term $\lambda(X_i^k) = 10^{\log_{10}(\mu(X_i^k))}$ where μ is the mean function. Then the stopping criterion would be $m < \lambda * \epsilon$. This way the stopping parameter ϵ is independent on the scale of the loss function. In Fig. 16 we can appreciate the scale independence. In Table 5 we can see a comparison of the best metrics after training each model with the PRISMA dataset with 1000 epochs, author's defined epoch number, and our criterion. It is important to remark that our criterion has not changed significantly the performance order of the models. Otherwise, our criterion helped those methods which the authors gave a lower maximum epoch number.

The entire strategy in the training process takes into account the update of the learning rate. We propose to train the models using a scheduler that reduces the learning rate by a factor of δ when a metric has ceased to improve for the last n epochs. With a selected $n = k/2$ and $\delta = \epsilon$. This approach establishes a connection between the updating strategy and the convergence criterion, allowing the sharing of parameters and logic.

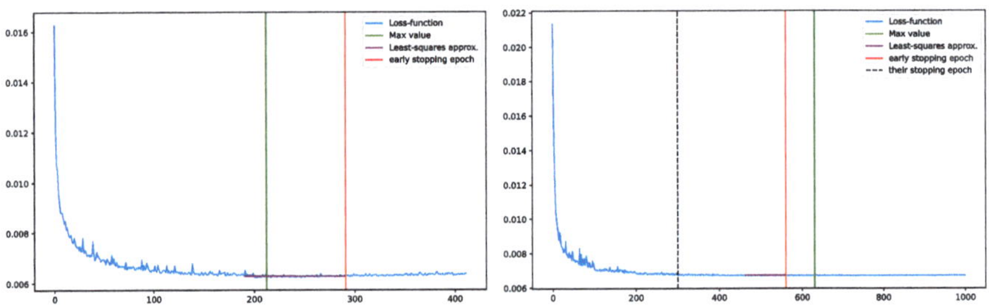

Fig. 16 Loss curves during training for MDCUN [107] (left) and AWFLN [63] (right). The left figure shows how the stopping would stop the training earlier but after reaching the minimum value of the loss. The right figure shows how the proposed epoch number in authors paper, 300, is not enough for this experiment. In both cases, we used a window size of $k = 100$ and $\epsilon = 1E - 5$

Table 5 Early-stopping comparison for PRISMA dataset. For each method, we show the loss (in terms of 1E-2) and PSNR values using 1000 epochs (first row), using author's defined epoch number (second row), and applying our proposed criterion (third row)

Model	Epochs	Loss	PSNR	Model	Epochs	Loss	PSRN
PanNet	1000	0.014	40.609	NLRNet	1000	65.92	41.239
	450	0.014	40.609		1000	65.92	41.239
	302	0.014	40.593		439	67.06	41.181
MSDCNN	1000	5.641	38.954	HSIT_pre/HSIT	10,000	6.375/90.61	27.144/40.674
	500	5.643	38.952		10,000	6.375/90.61	27.144/40.674
	556	5.641	38.954		3635/2968	6.375/90.61	27.144/40.674
MHFNet	1000	0.013	41.187	MMNet	1000	1.139	36.474
	1000	0.013	41.187		1000	1.139	36.474
	724	0.013	41.187		115	1.327	35.389
DiCNN	1000	0.016	40.094	MDCUN	412	0.624	42.149
	1000	0.016	40.094		1000	0.624	42.149
	500	0.016	39.965		291	0.624	42.149
SRPPNN	1000	0.011	41.756	HyPanF	1000	0.013	40.856
	1000	0.011	41.756		4000	0.013	40.856
	238	0.011	41.756		464	0.013	40.658
FusionNet	1000	0.013	40.968	S2DBPN	1000	0.769	40.514
	4000	0.013	40.968		500	0.824	39.747
	677	0.013	40.813		820	0.769	40.371
GPPNN	1000	0.721	40.922	AWFLN	1000	0.669	41.542
	200	0.724	40.922		300	0.675	41.470
	264	0.724	40.922		560	0.670	41.536

5.3 *Computational Cost Comparison*

Comparing the computational cost of each model necessitates considering various aspects. Firstly, the number of parameters involved in the model depends on the datasets it is trained on. Secondly, the floating-point operations per second in terms of gigaflops (GFlops) required to compute one image are also considered. This magnitude was calculated using the Python package *ptfops* [90]. These parameters are listed for each method and dataset in Table 6. The table reveals a clear correlation between the number of channels and the computational cost of the method. Additionally, it shows no apparent relationship between shallowness and unfolding. Importantly, MDCUN and HSIT are the most computationally expensive methods among those studied despite the difference on parameters. MDCUN is an unfolding model, while HSIT is a pure deep learning model. In addition we can see the large dependence of the channel number of MMNet and how it increases exponentially with the increment of channels. Finally, the table also reflects the trend of increasing computational cost over time. During the 6-year period analysed, the depth of neural networks has grown, leading to a subsequent rise in training effort.

Table 6 Number of parameters and floating-point operations per second in terms of gigaflops (GFlops) for each method and dataset

	QuickBird		WorldView2		PRISMA	
	Parameters	GFlops	Parameters	GFlops	Parameters	GFlops
PanNet	95,404	0.391	98,584	1.611	200,719	25.670
MSDCNN	189,852	0.780	228,556	3.756	706,736	112.300
MHFNet	137,235	0.564	181,323	2.909	2,919,498	376.161
DiCNN	42,180	0.174	46,792	0.771	110,207	16.291
SRPPNN	34,1840	0.553	351,072	2.288	478,012	30.100
FusionNet	76,324	0.618	78,632	2.510	110,367	27.271
GPPNN	119,808	0.370	238,592	2.862	1,871,872	196.521
NLRNet	66,704	2.665	675,808	10.870	883,488	123.783
HSIT	34,304,293	13.537	43,993,841	56.016	45,023,331	1,439.200
MMNet	51,822	0.762	204,498	12.136	12,546,300	6,685.260
MDCUN	98,347	15.156	140,907	121.209	797,497	8,615.124
HyPanF4	121,194	0.510	79,778	1.362	1,317,553	195.960
S2DBPN	16,187,812	10.715	16,192,552	42.867	1,6257,727	386.666
AWFLN	16,4195	0.656	181,391	2.685	1,352,396	35.949

6 Experimental Results

To conclude the comprehensive review and showcase the performance of each model, an assessment of the different methods is conducted in this section. The methods that will be compared are PanNet [108], MSDCNN [111], DiCNN [41], MHFNet [102], SRPPNN [16], FusionNet [25], GPPNN [103], NLRNet [54], HSIT [12], MMNet [104], MDCUN [107], HyFPan [68], AWFLN [63], and S2DBPN [117].

The comparison is carried out across the three described datasets from satellites with sensors of different characteristics. Firstly, the QuickBird satellite provides a MS image with four bands. Secondly, the WorldView2 satellite provides a MS image with eight bands. Finally, the PRISMA satellite provides a HS image with 63 bands. To train all the approaches in each dataset, we applied the proposed early stopping criterion presented in Sect. 5.2.

6.1 PRISMA Assessment

In this section, we compare quantitatively and qualitatively the performance of all methods on the PRISMA dataset. Quantitative results for both the validation and testing sets are displayed in Table 7 in terms of the reference metrics explained in Sect. 3.2. Since all experiments are conducted with a sampling factor of 4 and the factor between HS and PAN sensors is 6 in this dataset, the models cannot be

Table 7 Metrics provided by the results of each method on the PRISMA validation and test sets. The best metrics are highlighted with bold letters, with the second and third best metrics highlighted with double underline and single underline, respectively

	ERGAS↓		PSNR↑		SSIM↑		$Q2^n$ ↑		SAM↓	
	Validation	Test	Validation	Test	Validation	Test	Validation	Test	Validation	Test
PanNet	44.88	34.18	40.59	42.06	0.949	0.962	0.791	0.788	3.61	3.00
MSDCNN	52.69	39.59	38.95	39.65	0.944	0.958	0.761	0.759	4.41	3.96
MHFNet	46.81	33.16	41.19	42.67	0.954	0.965	0.804	0.798	3.43	2.80
DiCNN	48.15	37.58	39.97	41.34	0.940	0.953	0.770	0.755	3.83	3.26
SRPPNN	41.03	31.00	41.76	43.20	0.957	0.968	0.816	0.816	3.19	2.64
FusionNet	44.74	34.42	40.81	42.20	0.949	0.961	0.796	0.791	3.47	2.96
GPPNN	45.42	37.98	40.92	41.58	0.951	0.960	0.793	0.776	3.52	3.17
NLRNet	43.38	79.76	41.18	33.23	0.954	0.922	0.813	0.663	3.38	7.76
HSIT	46.40	33.89	40.67	42.07	0.950	0.962	0.770	0.781	3.65	3.01
MMNet	75.02	71.66	35.40	37.95	0.886	0.928	0.717	0.712	8.49	5.38
MDCUN	**38.86**	**30.30**	**42.15**	**43.34**	**0.959**	**0.968**	**0.840**	**0.820**	**3.04**	**2.60**
HyPanF	45.96	34.72	40.66	42.07	0.950	0.963	0.806	0.791	3.55	3.04
S2DBPN	46.17	35.67	40.37	41.63	0.949	0.963	0.792	0.786	3.72	3.16
AWFLN	42.01	32.02	41.54	42.85	0.955	0.965	0.816	0.803	3.26	2.75

evaluated with the raw data provided by the satellite. Consequently, the comparison without reference makes no sense in this case.

Table 7 reports that the best performance is achieved by MDCUN in both validation and testing sets. The next two best methods are SRPPNN and AWFLN. Surprisingly, although NLRNet is yielding good results in validation, its performance drops significantly when applied on the testing set. Actually, NLRNet is the method presenting the largest metrics differences between validation and testing sets, while the rest of the approaches yield near minimal disparities.

Additionally, qualitative results are displayed in Figs. 17 and 18, featuring images from the validation and testing sets, respectively. To visualize the results, we have generated the RGB image using channels 33, 45, and 55. The conclusions extracted by the quantitative evaluation are supported by the visual results. On the one hand, MDCUN, AWFL, and SRPPNN are the methods that better reconstruct the geometry while preserving color in both sets. On the other hand, MMNet, which yields the worst quantitative metrics, also produces poor visual results in the validation set when compared to the other methods. In addition, the visual quality of NLRNet results decreases between validation and test sets, which is in line with the quantitative evaluation. This comparison shows that NLRNet can correctly restore the geometry but fails processing the color in Fig. 18.

6.2 QuickBird Assessment

The comparison using the QuickBird dataset is also conducted both quantitatively and qualitatively. Table 8 shows the reference metrics for the validation and testing sets, while Table 9 shows the non-reference metrics with the raw data. The best results with reference images are NLRNet and MMNet in validation and test sets, respectively. In fact, the performance of MMNet is among the top 5 when evaluating all the metrics in the validation set; therefore, we can conclude that is the most favourable method in Table 8. Again, all methods except NLRNet yield metrics that are on pair between validation and test set, while NLRNet fails when applied on the testing set. Regarding Table 9 with non-reference metrics, the most promising methods are PanNet, NLRNet, and HSIT, which interestingly contrasts with the reference metrics.

Qualitative results on the testing set using the Wald's protocol and with the raw data are displayed in Figs. 19 and 20, respectively. To visualize the results, we have generated the RGB image using channels 2, 1, and 0, respectively. The qualitative analysis represented in Fig. 19 is in concordance with the quantitative evaluation. Overall, MMNet is the one that provides the best fused results, and NLRNet obtains the worst visual output. However, if we look at specific regions, other methods may perform best. For instance, if we focus on the bottom-right corner, we can appreciate that AWFLN achieves the best performance. In Fig. 20 we compare with the MS image. In contrast to the quantitative comparison with no reference, PanNet achieves

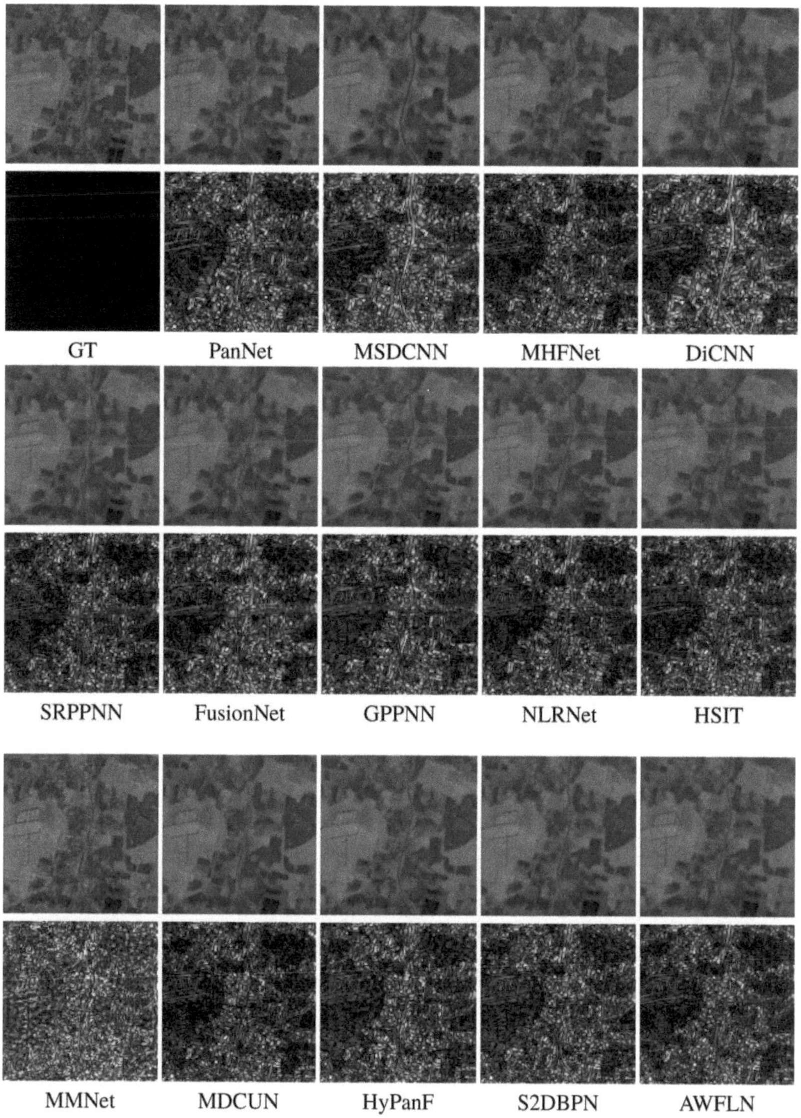

Fig. 17 Results on the PRISMA validation set. Output of each method (top) and error map comparing each method's output to the reference image (bottom). The error maps are computed as the mean of the absolute value difference across all channels. To enhance display quality, white balance is applied to the resulting image, along with gamma correction in each channel

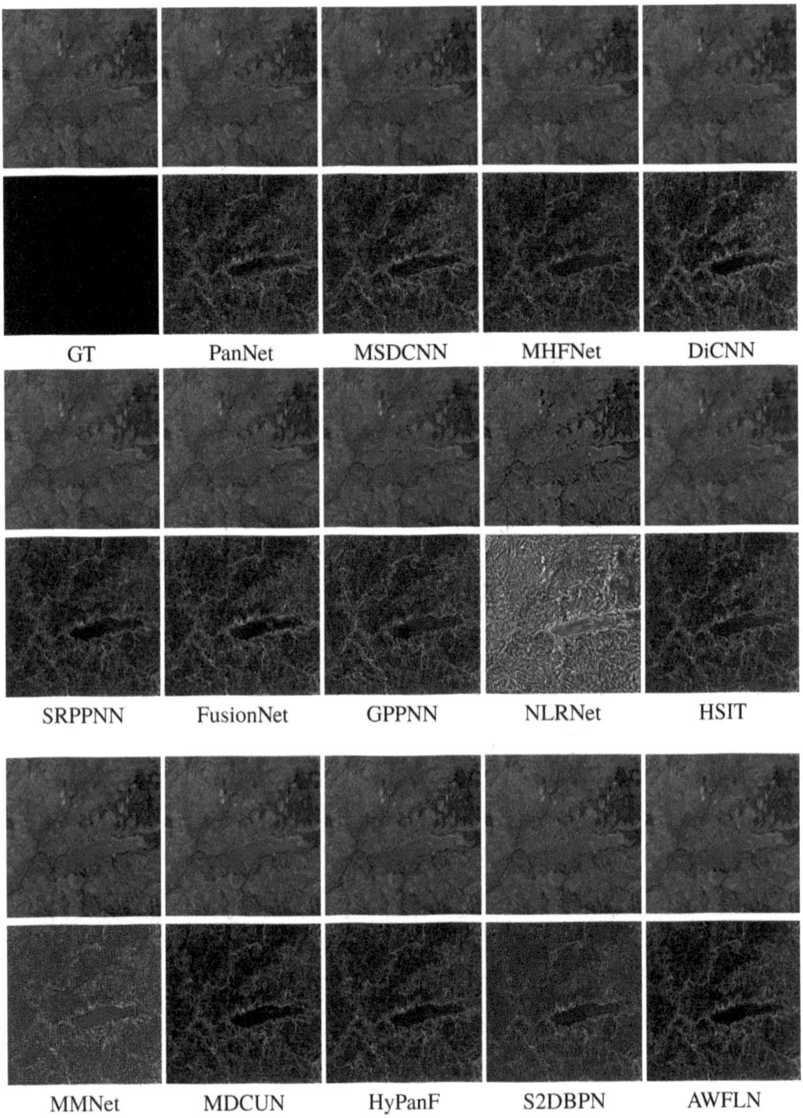

Fig. 18 Results on the PRISMA testing set. Output of each method (top) and error map comparing each method's output to the reference image (bottom). The error maps are computed as the mean of the absolute value difference across all channels. To enhance display quality, white balance is applied to the resulting image, along with gamma correction in each channel

Table 8 Metrics yielded by the results of each method on the QuickBird validation and test sets. The best metrics are highlighted with bold letters, with the second and third best metrics highlighted with double underline and single underline, respectively

	ERGAS↓		PSNR↑		SSIM↑		$Q2^n$ ↑		SAM↓	
	Validation	Test	Validation	Test	Validation	Test	Validation	Test	Validation	Test
PanNet	39.95	42.33	43.10	40.41	0.978	0.975	0.903	0.931	4.44	4.99
MSDCNN	39.20	41.81	43.27	40.52	0.979	0.976	0.903	0.932	4.36	4.93
MHFNet	39.97	41.02	43.16	40.70	0.979	0.976	0.903	**0.934**	4.40	4.83
DiCNN	40.79	43.88	42.93	40.10	0.977	0.973	0.901	0.928	4.56	5.17
SRPPNN	33.24	42.18	43.51	40.45	0.980	0.975	0.907	0.933	4.23	4.97
FusionNet	39.21	42.22	43.27	40.44	0.979	0.975	0.904	0.931	4.36	4.97
GPPNN	39.28	46.67	43.26	39.57	0.979	0.971	0.903	0.924	4.36	5.50
NLRNet	**36.00**	79.42	**43.97**	35.18	**0.982**	0.962	**0.911**	0.859	**4.01**	9.02
HSIT	40.82	47.22	42.88	39.45	0.978	0.971	0.903	0.926	4.55	5.55
MMNet	38.13	**40.75**	43.50	**40.74**	0.980	**0.977**	0.906	**0.934**	4.24	**4.80**
MDCUN	39.64	41.75	43.21	40.57	0.979	0.976	0.902	0.933	4.39	4.90
HyPanF	45.45	46.35	41.99	39.58	0.974	0.971	0.891	0.923	5.04	5.48
S2DBPN	36.83	45.20	43.87	39.84	0.981	0.972	0.910	0.929	4.05	5.34
AWFLN	38.88	41.28	43.34	40.65	0.979	0.976	0.904	0.933	4.32	4.85

Table 9 Average metrics on the testing set of QuickBird using the raw data. The best metrics are highlighted with bold letters, with the second and third best metrics highlighted with double underline and single underline, respectively

	QNR↑	FQNR↑	HQNR↑
PanNet	**0.7779**	**0.4936**	**0.7011**
MSDCNN	0.6739	0.2688	0.4939
MHFNet	0.6731	0.2692	0.4865
DiCNN	0.6639	0.2575	0.4847
SRPPNN	0.6728	0.2706	0.4900
FusionNet	0.6683	0.4841	0.2641
GPPNN	0.6776	0.2658	0.4896
NLRNet	0.7304	0.3273	0.5142
HSIT	0.6790	0.3144	0.5267
MMNet	0.6748	0.2726	0.4875
MDCUN	0.6705	0.2851	0.5146
HyPanF	0.6628	0.2530	0.4805
S2DBPN	0.6774	0.2786	0.4925
AWFLN	0.6709	0.2661	0.4899

the worst visual results since it provides a blurred image. While HSIT and MMNet produce halos in the edges, the other methods correctly fuse the image.

6.3 *WorldView Assessment*

In the WorldView2 dataset, the comparison was conducted similarly to the QuickBird dataset. The metrics with reference images in the validation and testing sets are shown in Table 10, and the metrics with the non-reference testing images can be seen in Table 11. According to Table 10, NLRNet achieves the best metrics in both validation and testing sets. With small differences, it is followed by MMNet and MDCUN, which are the second and third best approaches in both sets, respectively. In addition, we observe no significant disparities between validation and testing set metrics among all methods. For instance, the drop in performance observed for NLRNet with QuickBird does not occur in this case. Regarding the non-reference indexes displayed in Table 11, MHFNet is overall the method with the best metrics, closely followed by AWFLN.

Concerning the qualitative comparison, Fig. 21 shows the results of the different methods on one example in the testing set using the Wald's protocol, while Fig. 22 displays the results using the raw images. To visualize these results, we have generated the RGB image using channels 4, 2, and 1, respectively. Again, the qualitative comparison in Fig. 21 is in concordance with the quantitative one using reference metrics. On a global analysis, we can see how NLRNet, MMNet, and MDCUN are the ones yielding lower errors with respect to the ground truth image. However, in specific regions such as the bottom-left corner, HyPanF and MMNet clearly outperform the rest of the approaches.

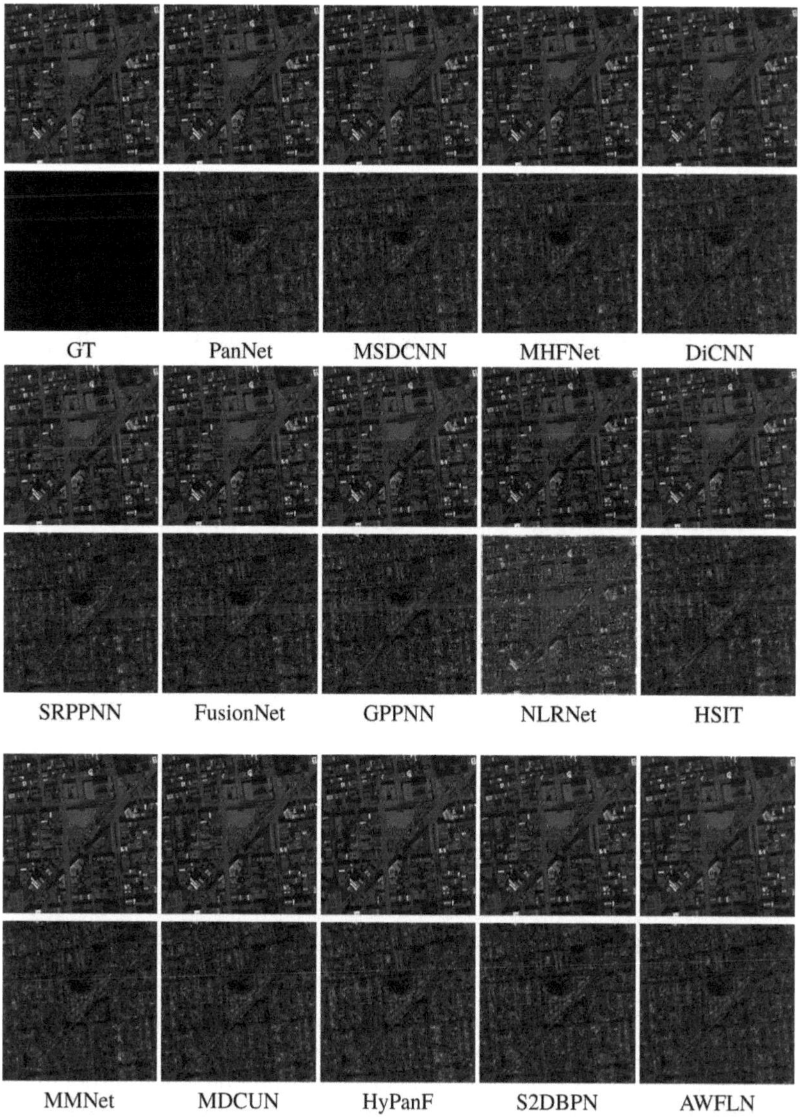

Fig. 19 Results of each method on one example of the QuickBird test set. For each method, we display the result (top) and the error map (bottom). The error maps are computed as the mean of the absolute value difference across all channels. To enhance display quality, white balance is applied to the resulting image, along with gamma correction in each channel

Finally, in Fig. 20 we compare with the MS image. In this case, the most favourable methods are MHFNet and AWFLN, as they provide a sharp result while correctly preserving color. Additionally, although in the comparison with the

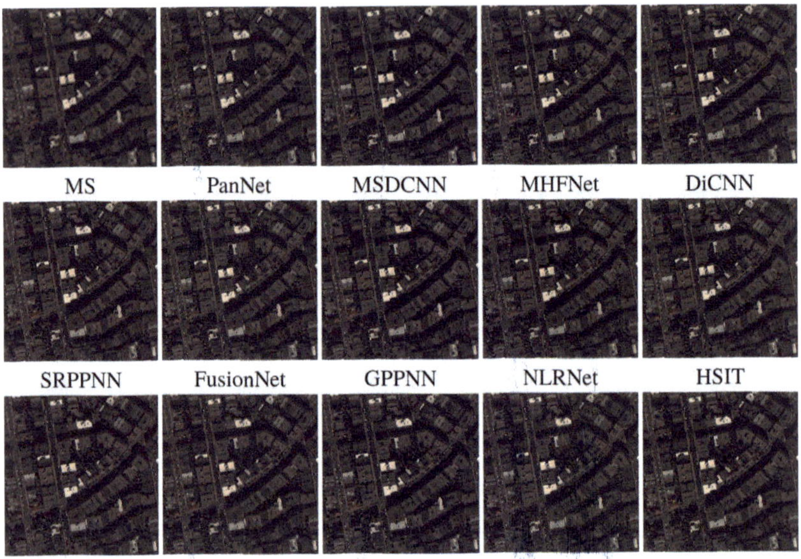

Fig. 20 Visual results on QuickBird raw data. To improve the display quality, white balance is applied to the resulting image, along with gamma correction in each channel

reference image NLRNet and MMNet are among the ones providing best results, with this data they fail when reconstructing color.

7 Conclusion

In this chapter, we have presented a detailed review of satellite image fusion, emphasizing the fundamental role that modelling plays in performance. We have analysed how classical approaches, based on component substitution, multi-resolution analysis, and variational optimization, have influenced cutting-edge deep learning methods, particularly, unfolding techniques. A quantitative and qualitative comparison of several state-of-the-art methods is conducted on Worldview2, Quickbird, and PRISMA datasets, which have distinct resolution and sensor characteristics.

To ensure a fair performance comparison, we have introduced a numerical method for early stopping based on the loss function proposed for the corresponding authors, moving beyond the conventional fixed maximum epoch approach. This method analyses the geometric features of the loss curve to determine the optimal stopping point. Our empirical evaluation has demonstrated that, while achieving

Table 10 Metrics of the results of each method on the WorldView2 validation and test sets. The best metrics are highlighted with bold letters, with the second and third best metrics highlighted with double underline and single underline, respectively

	ERGAS↓		PSNR↑		SSIM↑		$Q2^n$ ↑		SAM↓	
	Validation	Test	Validation	Test	Validation	Test	Validation	Test	Validation	Test
PanNet	18.50	18.98	40.42	40.12	0.970	0.968	0.928	0.931	2.65	2.72
MSDCNN	21.15	21.62	39.13	38.85	0.967	0.965	0.918	0.922	3.04	3.12
MHFNet	18.52	18.95	40.42	40.13	0.970	0.969	0.928	0.931	2.65	2.71
DiCNN	18.83	19.23	40.16	39.88	0.968	0.967	0.923	0.927	2.72	2.79
SRPPNN	17.45	17.87	40.89	40.60	0.973	0.971	0.932	0.936	2.51	2.58
FusionNet	18.31	18.92	40.51	40.17	0.971	0.969	0.928	0.931	2.63	2.72
GPPNN	18.03	18.55	40.58	40.27	0.972	0.970	0.930	0.932	2.61	2.69
NLRNet	**16.98**	**17.36**	**41.11**	**40.84**	**0.975**	**0.974**	**0.936**	**0.940**	**2.45**	**2.51**
HSIT	22.10	22.43	38.73	38.49	0.961	0.959	0.914	0.918	3.22	3.28
MMNet	17.51	17.89	40.94	40.67	0.973	0.972	0.931	0.936	2.50	2.56
MDCUN	17.45	17.86	40.90	40.62	0.973	0.971	0.933	0.937	2.51	2.57
HyPanF	18.29	18.67	40.58	40.30	0.972	0.970	0.927	0.931	2.59	2.65
S2DBPN	19.30	20.30	39.85	39.56	0.966	0.964	0.916	0.920	2.82	2.90
AWFLN	19.36	20.42	39.72	39.41	0.965	0.963	0.918	0.920	2.87	2.95

Table 11 Average metrics on the testing set of WorldView2 using the raw data. The best metrics are highlighted with bold letters, with the second and third best metrics highlighted with double underline and single underline, respectively

	QNR↑	FQNR↑	HQNR↑
PanNet	0.9435	0.7284	0.8971
MSDCNN	0.9284	0.6694	0.8685
MHFNet	**0.9469**	**0.7543**	0.8920
DiCNN	0.9318	0.6859	0.8801
SRPPNN	0.9322	0.6947	0.8833
FusionNet	0.9219	0.8722	0.6711
GPPNN	0.9131	0.6524	0.8583
NLRNet	0.9186	0.2652	0.3259
HSIT	0.9185	0.3581	0.4213
MMNet	0.9386	0.7245	0.8817
MDCUN	0.9249	0.6814	0.8779
HyFPan	0.9452	0.7475	0.8893
S2DBPN	0.9276	0.6834	0.8677
AWFLN	0.9452	0.7434	**0.8975**
MMNet	0.9386	0.7245	0.8817

comparable performance to the maximum epoch approach, our method leads to significant training time reductions. Notably, our approach is particularly beneficial in cases where the original maximum epoch is insufficient, enabling training convergence at a later stage.

With the experimental analysis, we conclude that the best-performing methods are MDCUN, AWFLN, and SRPPNN for PRISMA, NLRNet, and MMNet for QuickBird and MMNet, MDCUN, and NLRNet for WorldView. Since MDCUN has also shown to be competitive for QuickBird, it can be considered as the state of the art. All methods have shown to slightly differ when evaluated on the validation and testing sets. Actually, we have noticed that these discrepancies are more pronounced with pure deep learning methods, whereas model-based unfolding approaches demonstrate superior adaptability to different data. In addition, it is also interesting to note that the best performing methods, MDCUN, NLRNet, MMNet, and AWFLN, incorporate attention mechanisms into their architectures.

In conclusion, both pure deep learning and model-based unfolding methods exhibit good performance, with the latter providing more adaptability to different data. Moreover, attention mechanisms have demonstrated their effectiveness in improving the quality of the fused products. Looking ahead to future research, we consider that efforts should be directed towards developing model-based unfolding models that incorporate attention mechanisms and new architectures taken from pure deep learning approaches.

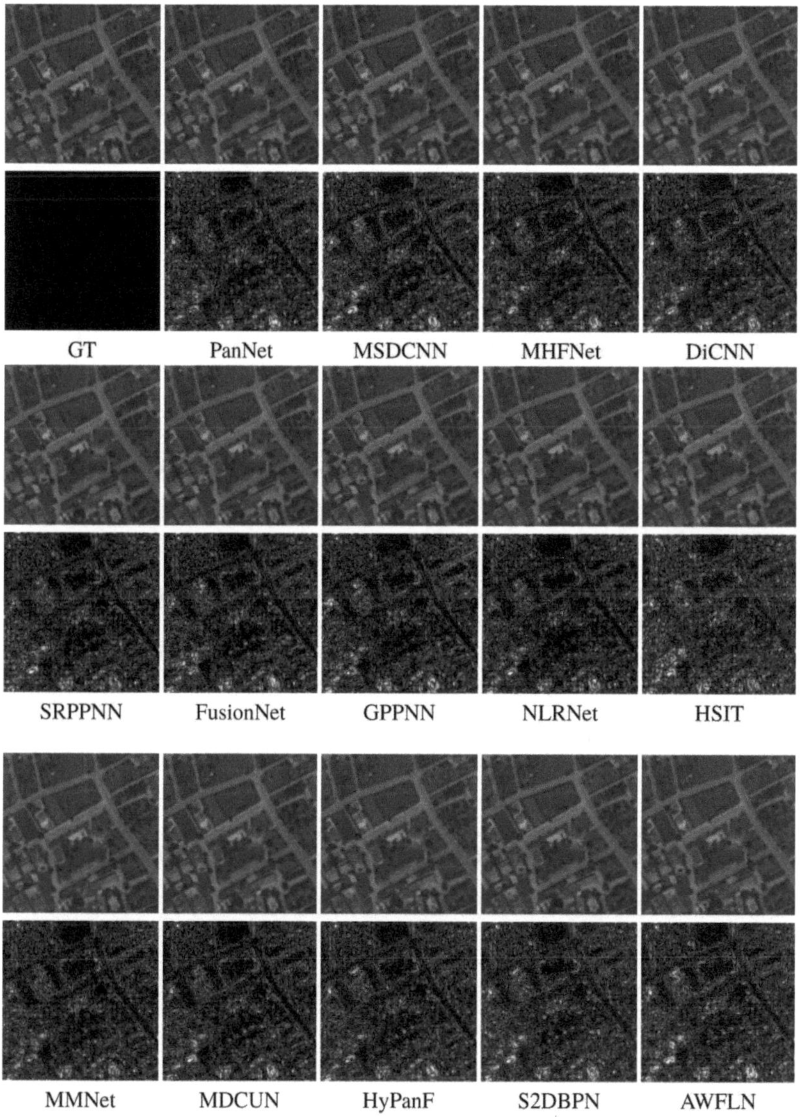

Fig. 21 Results of each method on one example of the WorldView2 test set. For each method, we display the result (top) and the error map (bottom). The error maps are computed as the mean of the absolute value difference across all channels. To improve the display quality, white balance is applied to the resulting image, along with gamma correction in each channel

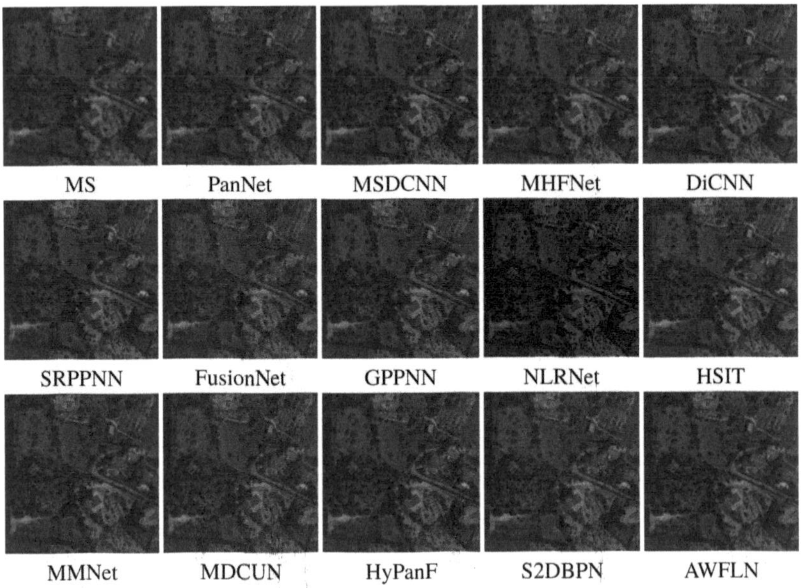

Fig. 22 Results of each method on WorldView2 raw data. To improve the display quality, white balance is applied to the resulting image, along with gamma correction in each channel

Acknowledgments The authors gratefully acknowledge the computer resources at Artemisa, funded by the EU ERDF and Comunitat Valenciana and the technical support provided by IFIC (CSIC-UV).

Competing Interests This study was funded by MCIN/AEI/10.13039/501100011033/ and by the European Union NextGenerationEU/PRTR via the MaLiSat project TED2021-132644B-I00; and also by MCIN/AEI/10.13039/501100011033 and by "ERDF A way of making Europe", European Union, via the MoMaLiP project Grant PID2021-125711OB-I00.

Ivan Pereira-Sánchez has received a predoctoral grant from Conselleria de Fons Europeus, Universitat i Cultura (GOIB) FPU2023-004-C.

The authors have no conflicts of interest to declare that are relevant to the content of this chapter.

References

1. Adler, J., Öktem, O.: Learned primal-dual reconstruction. IEEE Trans. Med. Imaging **37**(6), 1322–1332 (2018)
2. Aiazzi, B., Alparone, L., Baronti, S., Carlà, R., Garzelli, A., Santurri, L.: Full-scale assessment of pansharpening methods and data products. In: Image and Signal Processing for Remote Sensing XX, vol. 9244, p. 924402. SPIE (2014)
3. Aiazzi, B., Alparone, L., Baronti, S., Garzelli, A.: Context-driven fusion of high spatial and spectral resolution images based on oversampled multiresolution analysis. IEEE Trans. Geosci. Remote Sensing **40**(10), 2300–2312 (2002)
4. Aiazzi, B., Alparone, L., Baronti, S., Garzelli, A., Selva, M.: Mtf-tailored multiscale fusion of high-resolution ms and pan imagery. Photogramm. Eng. Remote Sensing **72**(5), 591–596 (2006)

5. Aiazzi, B., Baronti, S., Selva, M.: Improving component substitution pansharpening through multivariate regression of ms + pan data. IEEE Trans. Geosci. Remote Sensing **45**(10), 3230–3239 (2007)
6. Alparone, L., Aiazzi, B., Baronti, S., Garzelli, A., Nencini, F., Selva, M.: Multispectral and panchromatic data fusion assessment without reference. ASPRS J. Photogramm. Eng. Remote Sensing **74**, 193–200 (2008). https://doi.org/10.14358/PERS.74.2.193
7. Alparone, L., Baronti, S., Garzelli, A., Nencini, F.: A global quality measurement of pan-sharpened multispectral imagery. IEEE Geosci. Remote Sensing Lett. **1**(4), 313–317 (2004). https://doi.org/10.1109/LGRS.2004.836784
8. Aly, H.A., Sharma, G.: A regularized model-based optimization framework for pan-sharpening. IEEE Trans. Image Process. **23**(6), 2596–2608 (2014)
9. Amro, I., Mateos, J., Vega, M., Molina, R., Katsaggelos, A.K.: A survey of classical methods and new trends in pansharpening of multispectral images. EURASIP J. Adv. Signal Process. **2011**(1), 1–22 (2011)
10. Arienzo, A., Vivone, G., Garzelli, A., Alparone, L., Chanussot, J.: Full-resolution quality assessment of pansharpening: theoretical and hands-on approaches. IEEE Geosci. Remote Sensing Mag. **10**(3), 168–201 (2022)
11. Ballester, C., Caselles, V., Igual, L., Verdera, J., Rougé, B.: A variational model for p+ xs image fusion. Int. J. Comput. Vis. **69**, 43–58 (2006)
12. Bandara, W.G.C., Patel, V.M.: Hypertransformer: A textural and spectral feature fusion transformer for pansharpening. In: Proceedings of the IEEE/CVF Conference on Computer Vision and Pattern Recognition, pp. 1767–1777 (2022)
13. Beck, A., Teboulle, M.: A fast iterative shrinkage-thresholding algorithm for linear inverse problems. SIAM J. Imaging Sci. **2**(1), 183–202 (2009)
14. de Béthune, S., Muller, F., Donnay, J.P.: Fusion of multispectral and panchromatic images by local mean and variance matching filtering techniques. Fusion Earth Data **28**, 30 (1998)
15. Buades, A., Coll, B., Morel, J.M.: A non-local algorithm for image denoising. In: 2005 IEEE Computer Society Conference on Computer Vision and Pattern Recognition (CVPR'05), vol. 2, pp. 60–65. IEEE, Piscataway (2005)
16. Cai, J., Huang, B.: Super-resolution-guided progressive pansharpening based on a deep convolutional neural network. IEEE Trans. Geosci. Remote Sensing **59**(6), 5206–5220 (2020)
17. Cao, X., Fu, X., Hong, D., Xu, Z., Meng, D.: Pancsc-net: a model-driven deep unfolding method for pansharpening. IEEE Trans. Geosci. Remote Sensing **60**, 1–13 (2021)
18. Carper, W., Lillesand, T., Kiefer, R., et al.: The use of intensity-hue-saturation transformations for merging spot panchromatic and multispectral image data. Photogramm. Eng. Remote Sensing **56**(4), 459–467 (1990)
19. Chavez, P., Sides, S.C., Anderson, J.A., et al.: Comparison of three different methods to merge multiresolution and multispectral data- landsat tm and spot panchromatic. Photogramm. Eng. Remote Sensing **57**(3), 295–303 (1991)
20. Chen, L., Lai, Z., Vivone, G., Jeon, G., Chanussot, J., Yang, X.: Arbrpn: A bidirectional recurrent pansharpening network for multispectral images with arbitrary numbers of bands. IEEE Trans. Geosci. Remote Sensing **60**, 1–18 (2021)
21. Cheng, J., Liu, H., Liu, T., Wang, F., Li, H.: Remote sensing image fusion via wavelet transform and sparse representation. ISPRS J. Photogramm. Remote Sensing **104**, 158–173 (2015)
22. Choi, J., Yu, K., Kim, Y.: A new adaptive component-substitution-based satellite image fusion by using partial replacement. IEEE Trans. Geosci. Remote Sensing **49**(1), 295–309 (2010)
23. Dai, S., Han, M., Wu, Y., Gong, Y.: Bilateral back-projection for single image super resolution. In: 2007 IEEE International Conference on Multimedia and Expo, pp. 1039–1042. IEEE, Piscataway (2007)
24. Deng, L.J.: Pancollection (2023). https://github.com/liangjiandeng/PanCollection
25. Deng, L.J., Vivone, G., Jin, C., Chanussot, J.: Detail injection-based deep convolutional neural networks for pansharpening. IEEE Trans. Geosci. Remote Sensing **59**(8), 6995–7010 (2020)

26. Deng, L.J., Vivone, G., Paoletti, M.E., Scarpa, G., He, J., Zhang, Y., Chanussot, J., Plaza, A.: Machine learning in pansharpening: a benchmark, from shallow to deep networks. IEEE Geosci. Remote Sensing Mag. **10**(3), 279–315 (2022)
27. Dong, C., Loy, C.C., He, K., Tang, X.: Image super-resolution using deep convolutional networks. IEEE Trans. Pattern Anal. Mach. Intell. **38**(2), 295–307 (2015)
28. Dong, W., Wang, P., Yin, W., Shi, G., Wu, F., Lu, X.: Denoising prior driven deep neural network for image restoration. IEEE Trans. Pattern Anal. Mach. Intell. **41**(10), 2305–2318 (2018)
29. Dosovitskiy, A., Beyer, L., Kolesnikov, A., Weissenborn, D., Zhai, X., Unterthiner, T., Dehghani, M., Minderer, M., Heigold, G., Gelly, S., et al.: An image is worth 16x16 words: transformers for image recognition at scale (2020). arXiv preprint arXiv:2010.11929
30. Duran, J., Buades, A., Coll, B., Sbert, C.: A nonlocal variational model for pansharpening image fusion. SIAM J. Imaging Sci. **7**(2), 761–796 (2014)
31. Duran, J., Buades, A., Coll, B., Sbert, C., Blanchet, G.: A survey of pansharpening methods with a new band-decoupled variational model. ISPRS J. Photogramm. Remote Sensing **125**, 78–105 (2017)
32. Fang, F., Li, F., Shen, C., Zhang, G.: A variational approach for pan-sharpening. IEEE Trans. Image Process. **22**(7), 2822–2834 (2013)
33. Fasbender, D., Radoux, J., Bogaert, P.: Bayesian data fusion for adaptable image pansharpening. IEEE Trans. Geosci. Remote Sensing **46**(6), 1847–1857 (2008)
34. Garzelli, A., Nencini, F.: Hypercomplex quality assessment of multi/hyperspectral images. IEEE Geosci. Remote Sensing Lett. **6**, 662 – 665 (2009). https://doi.org/10.1109/LGRS.2009.2022650
35. Ghahremani, M., Ghassemian, H.: Remote sensing image fusion using ripplet transform and compressed sensing. IEEE Geosci. Remote Sensing Lett. **12**(3), 502–506 (2014)
36. Gillespie, A.R., Kahle, A.B., Walker, R.E.: Color enhancement of highly correlated images. ii. channel ratio and "chromaticity" transformation techniques. Remote Sensing Environ. **22**(3), 343–365 (1987)
37. Goodfellow, I., Pouget-Abadie, J., Mirza, M., Xu, B., Warde-Farley, D., Ozair, S., Courville, A., Bengio, Y.: Generative adversarial nets. In: Advances in Neural Information Processing Systems, vol. 27 (2014)
38. Guan, P., Lam, E.Y.: Multistage dual-attention guided fusion network for hyperspectral pansharpening. IEEE Trans. Geosci. Remote Sensing **60**, 1–14 (2021)
39. Guo, M.H., Xu, T.X., Liu, J.J., Liu, Z.N., Jiang, P.T., Mu, T.J., Zhang, S.H., Martin, R.R., Cheng, M.M., Hu, S.M.: Attention mechanisms in computer vision: a survey. Comput. Vis. Media **8**(3), 331–368 (2022)
40. He, K., Zhang, X., Ren, S., Sun, J.: Deep residual learning for image recognition. In: Proceedings of the IEEE Conference on Computer Vision and Pattern Recognition, pp. 770–778 (2016)
41. He, L., Rao, Y., Li, J., Chanussot, J., Plaza, A., Zhu, J., Li, B.: Pansharpening via detail injection based convolutional neural networks. IEEE J. Sel. Topics Appl. Earth Observ. Remote Sensing **12**(4), 1188–1204 (2019)
42. He, X., Condat, L., Chanussot, J., Xia, J.: Pansharpening using total variation regularization. In: 2012 IEEE International Geoscience and Remote Sensing Symposium, pp. 166–169. IEEE, Piscataway (2012)
43. Huang, W., Xiao, L., Wei, Z., Liu, H., Tang, S.: A new pan-sharpening method with deep neural networks. IEEE Geosci. Remote Sensing Lett. **12**(5), 1037–1041 (2015)
44. Irani, M., Peleg, S.: Motion analysis for image enhancement: resolution, occlusion, and transparency. J. Vis. Commun. Image Represent. **4**(4), 324–335 (1993)
45. Javan, F.D., Samadzadegan, F., Mehravar, S., Toosi, A., Khatami, R., Stein, A.: A review of image fusion techniques for pan-sharpening of high-resolution satellite imagery. ISPRS J. Photogramm. Remote Sensing **171**, 101–117 (2021)
46. Johnson, J., Alahi, A., Fei-Fei, L.: Perceptual losses for real-time style transfer and super-resolution. In: Computer Vision–ECCV 2016: 14th European Conference, Amsterdam, The

Netherlands, October 11-14, 2016, Proceedings, Part II 14, pp. 694–711. Springer, Berlin (2016)
47. Kaplan, N.H., Erer, I.: Bilateral filtering-based enhanced pansharpening of multispectral satellite images. IEEE Geosci. Remote Sensing Lett. **11**(11), 1941–1945 (2014)
48. Karoui, M.S., Djerriri, K., Boukerch, I.: Pansharpening multispectral remote sensing data by multiplicative joint nonnegative matrix factorization. Int. J. Remote Sensing **37**(4), 805–818 (2016)
49. Karras, T., Laine, S., Aila, T.: A style-based generator architecture for generative adversarial networks. In: Proceedings of the IEEE/CVF Conference on Computer Vision and Pattern Recognition, pp. 4401–4410 (2019)
50. Khan, M.M., Alparone, L., Chanussot, J.: Pansharpening quality assessment using the modulation transfer functions of instruments. IEEE Trans. Geosci. Remote Sensing **47**(11), 3880–3891 (2009). https://doi.org/10.1109/TGRS.2009.2029094
51. Kwarteng, P., Chavez, A.: Extracting spectral contrast in landsat thematic mapper image data using selective principal component analysis. Photogramm. Eng. Remote Sens **55**(1), 339–348 (1989)
52. Laben, C., Brower, B.: Process for enhancing the spatial resolution of multispectral imagery using pan–sharpening: US, us6011875. US US6011875 (2000)
53. Lee, J., Lee, C.: Fast and efficient panchromatic sharpening. IEEE Trans. Geosci. Remote Sensing **48**(1), 155–163 (2009)
54. Lei, D., Chen, H., Zhang, L., Li, W.: Nlrnet: an efficient nonlocal attention resnet for pansharpening. IEEE Trans. Geosci. Remote Sensing **60**, 1–13 (2022). https://doi.org/10.1109/TGRS.2021.3067097
55. Li, S., Yang, B.: A new pan-sharpening method using a compressed sensing technique. IEEE Trans. Geosci. Remote Sensing **49**(2), 738–746 (2010)
56. Li, X., Li, Y., Shi, G., Zhang, L., Li, W., Lei, D.: Pansharpening method based on deep non-local unfolding. IEEE Trans. Geosci. Remote Sensing **61**, 1–11 (2023)
57. Li, Z., Jing, Z., Yang, X., Sun, S.: Color transfer based remote sensing image fusion using non-separable wavelet frame transform. Pattern Recogn. Lett. **26**(13), 2006–2014 (2005)
58. Li, Z., Li, J., Zhang, F., Fan, L.: Cadui: cross attention-based depth unfolding iteration network for pan-sharpening remote sensing images. IEEE Trans. Geosci. Remote Sensing **61**, 1–20 (2023)
59. Licciardi, G.A., Khan, M.M., Chanussot, J., Montanvert, A., Condat, L., Jutten, C.: Fusion of hyperspectral and panchromatic images using multiresolution analysis and nonlinear pca band reduction. EURASIP J. Adv. Signal Process. **2012**(1), 1–17 (2012)
60. Liu, J., Liang, S.: Pan-sharpening using a guided filter. Int. J. Remote Sensing **37**(8), 1777–1800 (2016)
61. Liu, Q., Zhou, H., Xu, Q., Liu, X., Wang, Y.: Psgan: a generative adversarial network for remote sensing image pan-sharpening. IEEE Trans. Geosci. Remote Sensing **59**(12), 10227–10242 (2020)
62. Loncan, L., De Almeida, L.B., Bioucas-Dias, J.M., Briottet, X., Chanussot, J., Dobigeon, N., Fabre, S., Liao, W., Licciardi, G.A., Simoes, M., et al.: Hyperspectral pansharpening: a review. IEEE Geosci. Remote Sensing Mag. **3**(3), 27–46 (2015)
63. Lu, H., Yang, Y., Huang, S., Chen, X., Chi, B., Liu, A., Tu, W.: Awfln: An adaptive weighted feature learning network for pansharpening. IEEE Trans. Geosci. Remote Sensing **61**, 1–15 (2023)
64. Luo, Y., Liu, R., Zhu, Y.F.: Fusion of remote sensing image base on the pca+ atrous wavelet transform. Int. Arch. Photogramm. Remote Sensing Spatial Inform. Sci. **37**(Part B7), 1155–1158 (2008)
65. Masi, G., Cozzolino, D., Verdoliva, L., Scarpa, G.: Pansharpening by convolutional neural networks. Remote Sensing **8**(7), 594 (2016)
66. Meng, X., Shen, H., Li, H., Zhang, L., Fu, R.: Review of the pansharpening methods for remote sensing images based on the idea of meta-analysis: practical discussion and challenges. Inform. Fusion **46**, 102–113 (2019)

67. Meng, X., Wang, N., Shao, F., Li, S.: Vision transformer for pansharpening. IEEE Trans. Geosci. Remote Sensing **60**, 1–11 (2022)
68. Mifdal, J., Tomás-Cruz, M., Sebastianelli, A., Coll, B., Duran, J.: Deep unfolding for hyper sharpening using a high-frequency injection module. In: 2023 IEEE/CVF Conference on Computer Vision and Pattern Recognition Workshops (CVPRW), pp. 2106–2115 (2023). https://doi.org/10.1109/CVPRW59228.2023.00204
69. Miseta, T., Fodor, A., Vathy-Fogarassy, A.: Surpassing early stopping: A novel correlation-based stopping criterion for neural networks. Neurocomputing **567**, 127028 (2023). https://doi.org/10.1016/j.neucom.2023.127028
70. Möller, M., Wittman, T., Bertozzi, A.L., Burger, M.: A variational approach for sharpening high dimensional images. SIAM J. Imaging Sci. **5**(1), 150–178 (2012)
71. Monga, V., Li, Y., Eldar, Y.C.: Algorithm unrolling: interpretable, efficient deep learning for signal and image processing. IEEE Signal Process. Mag. **38**(2), 18–44 (2021)
72. Mou, L., Ghamisi, P., Zhu, X.X.: Deep recurrent neural networks for hyperspectral image classification. IEEE Trans. Geosci. Remote Sensing **55**(7), 3639–3655 (2017)
73. Navalgund, R.R., Jayaraman, V., Roy, P.: Remote sensing applications: an overview. In: Current Science, pp. 1747–1766 (2007)
74. Nencini, F., Garzelli, A., Baronti, S., Alparone, L.: Remote sensing image fusion using the curvelet transform. Inform. Fusion **8**(2), 143–156 (2007)
75. Nunez, J., Otazu, X., Fors, O., Prades, A., Pala, V., Arbiol, R.: Multiresolution-based image fusion with additive wavelet decomposition. IEEE Trans. Geosci. Remote Sensing **37**(3), 1204–1211 (1999)
76. Otazu, X., González-Audícana, M., Fors, O., Núñez, J.: Introduction of sensor spectral response into image fusion methods. application to wavelet-based methods. IEEE Trans. Geosci. Remote Sensing **43**(10), 2376–2385 (2005)
77. Palsson, F., Sveinsson, J.R., Ulfarsson, M.O., Benediktsson, J.A.: A new pansharpening method using an explicit image formation model regularized via total variation. In: 2012 IEEE International Geoscience and Remote Sensing Symposium, pp. 2288–2291. IEEE, Piscataway (2012)
78. Purkis, S.J., Klemas, V.V.: Remote Sensing and Global Environmental Change. John Wiley & Sons, New York (2011)
79. Qu, J., Dong, W., Li, Y., Hou, S., Du, Q.: An interpretable unsupervised unrolling network for hyperspectral pansharpening. IEEE Trans. Cybern. **53**(12), 7943–7956 (2023)
80. Qu, Y., Baghbaderani, R.K., Qi, H., Kwan, C.: Unsupervised pansharpening based on self-attention mechanism. IEEE Trans. Geosci. Remote Sensing **59**(4), 3192–3208 (2020)
81. Rabbani, M., Jones, P.W.: Digital Image Compression Techniques, vol. 7. SPIE Press (1991)
82. Raj, T., Hanim Hashim, F., Baseri Huddin, A., Ibrahim, M.F., Hussain, A.: A survey on lidar scanning mechanisms. Electronics **9**(5), 741 (2020)
83. Ranchin, T., Wald, L.: Fusion of high spatial and spectral resolution images: the arsis concept and its implementation. Photogramm. Eng. Remote Sensing **66**, 49–61 (2000)
84. Restaino, R., Vivone, G., Dalla Mura, M., Chanussot, J.: Fusion of multispectral and panchromatic images based on morphological operators. IEEE Trans. Image Process. **25**(6), 2882–2895 (2016)
85. Rong, K., Jiao, L., Wang, S., Liu, F.: Pansharpening based on low-rank and sparse decomposition. IEEE J. Sel. Topics Appl. Earth Observ. Remote Sensing **7**(12), 4793–4805 (2014)
86. Rudin, L.I., Osher, S., Fatemi, E.: Nonlinear total variation based noise removal algorithms. Phys. D: Nonlinear Phenom. **60**(1-4), 259–268 (1992)
87. Schuster, M., Paliwal, K.K.: Bidirectional recurrent neural networks. IEEE Trans. Signal Process. **45**(11), 2673–2681 (1997)
88. Shah, V.P., Younan, N.H., King, R.L.: An efficient pan-sharpening method via a combined adaptive pca approach and contourlets. IEEE Trans. Geosci. Remote Sensing **46**(5), 1323–1335 (2008)
89. Simonyan, K., Zisserman, A.: Very deep convolutional networks for large-scale image recognition (2014). arXiv 1409.1556

90. Sovrasov, V.: ptflops: a flops counting tool for neural networks in pytorch framework (2018–2023). https://github.com/sovrasov/flops-counter.pytorch
91. Su, X., Li, J., Hua, Z.: Transformer-based regression network for pansharpening remote sensing images. IEEE Trans. Geosci. Remote Sensing **60**, 1–23 (2022)
92. Thomas, C., Ranchin, T., Wald, L., Chanussot, J.: Synthesis of multispectral images to high spatial resolution: a critical review of fusion methods based on remote sensing physics. IEEE Trans. Geosci. Remote Sensing **46**(5), 1301–1312 (2008)
93. Tian, X., Li, K., Wang, Z., Ma, J.: Vp-net: an interpretable deep network for variational pansharpening. IEEE Trans. Geosci. Remote Sensing **60**, 1–16 (2021)
94. Vivone, G., Alparone, L., Chanussot, J., Dalla Mura, M., Garzelli, A., Licciardi, G., Restaino, R., Wald, L.: A critical comparison of pansharpening algorithms. In: 2014 IEEE Geoscience and Remote Sensing Symposium, pp. 191–194. IEEE, Piscataway (2014)
95. Wald, L., Ranchin, T., Mangolini, M.: Fusion of satellite images of different spatial resolutions: assessing the quality of resulting images. Photogramm. Eng. Remote Sensing **63**(6), 691–699 (1997)
96. Wang, J., Wang, L., Feng, S., Peng, B., Huang, L., Fatholahi, S.N., Tang, L., Li, J.: An overview of shoreline mapping by using airborne lidar. Remote Sensing **15**(1), 253 (2023)
97. Wang, X., Girshick, R., Gupta, A., He, K.: Non-local neural networks. In: Proceedings of the IEEE Conference on Computer Vision and Pattern Recognition, pp. 7794–7803 (2018)
98. Wang, Z., Bovik, A.: A universal image quality index. IEEE Signal Process. Lett. **9**(3), 81–84 (2002). https://doi.org/10.1109/97.995823
99. Wei, Q., Dobigeon, N., Tourneret, J.Y.: Bayesian fusion of multi-band images. IEEE J. Sel. Top. Signal Process. **9**(6), 1117–1127 (2015)
100. Xiang, Z., Xiao, L., Yang, J., Liao, W., Philips, W.: Detail-injection-model-inspired deep fusion network for pansharpening. IEEE Trans. Geosci. Remote Sensing **60**, 1–15 (2022)
101. Xie, J., Xu, L., Chen, E.: Image denoising and inpainting with deep neural networks. In: Advances in Neural Information Processing Systems, vol. 25 (2012)
102. Xie, Q., Zhou, M., Zhao, Q., Xu, Z., Meng, D.: Mhf-net: an interpretable deep network for multispectral and hyperspectral image fusion. IEEE Trans. Pattern Anal. Mach. Intell. **44**(3), 1457–1473 (2020)
103. Xu, S., Zhang, J., Zhao, Z., Sun, K., Liu, J., Zhang, C.: Deep gradient projection networks for pan-sharpening. In: Proceedings of the IEEE/CVF Conference on Computer Vision and Pattern Recognition, pp. 1366–1375 (2021)
104. Yan, K., Zhou, M., Zhang, L., Xie, C.: Memory-augmented model-driven network for pansharpening. In: European Conference on Computer Vision, pp. 306–322. Springer, Berlin (2022)
105. Yang, F., Yang, H., Fu, J., Lu, H., Guo, B.: Learning texture transformer network for image super-resolution. In: Proceedings of the IEEE/CVF Conference on Computer Vision and Pattern Recognition, pp. 5791–5800 (2020)
106. Yang, G., Cao, X., Xiao, W., Zhou, M., Liu, A., Chen, X., Meng, D.: Panflownet: a flow-based deep network for pan-sharpening. In: Proceedings of the IEEE/CVF International Conference on Computer Vision, pp. 16857–16867 (2023)
107. Yang, G., Zhou, M., Yan, K., Liu, A., Fu, X., Wang, F.: Memory-augmented deep conditional unfolding network for pan-sharpening. In: Proceedings of the IEEE/CVF Conference on Computer Vision and Pattern Recognition, pp. 1788–1797 (2022)
108. Yang, J., Fu, X., Hu, Y., Huang, Y., Ding, X., Paisley, J.: Pannet: A deep network architecture for pan-sharpening. In: Proceedings of the IEEE International Conference on Computer Vision (ICCV) (2017)
109. Yocky, D.A.: Image merging and data fusion by means of the discrete two-dimensional wavelet transform. JOSA A **12**(9), 1834–1841 (1995)
110. Yokoya, N., Yairi, T., Iwasaki, A.: Coupled nonnegative matrix factorization unmixing for hyperspectral and multispectral data fusion. IEEE Trans. Geosci. Remote Sensing **50**(2), 528–537 (2011)

111. Yuan, Q., Wei, Y., Meng, X., Shen, H., Zhang, L.: A multiscale and multidepth convolutional neural network for remote sensing imagery pan-sharpening. IEEE J. Sel. Top. Appl. Earth Observ. Remote Sensing **11**(3), 978–989 (2018)
112. Zagoruyko, S., Komodakis, N.: Wide residual networks (2016). arXiv preprint arXiv:1605.07146
113. Zeng, D., Hu, Y., Huang, Y., Xu, Z., Ding, X.: Pan-sharpening with structural consistency and $\ell 1/2$ gradient prior. Remote Sensing Lett. **7**(12), 1170–1179 (2016)
114. Zhang, F., Zhang, K., Sun, J.: Multiscale spatial–spectral interaction transformer for pan-sharpening. Remote Sensing **14**(7), 1736 (2022)
115. Zhang, F., Zhang, K., Sun, J., Wang, J., Bruzzone, L.: Drformer: Learning disentangled representation for pan-sharpening via mutual information-based transformer. IEEE Trans. Geosci. Remote Sensing **62**, 1–15 (2023)
116. Zhang, J.A., Liu, F., Masouros, C., Heath, R.W., Feng, Z., Zheng, L., Petropulu, A.: An overview of signal processing techniques for joint communication and radar sensing. IEEE J. Sel. Top. Signal Process. **15**(6), 1295–1315 (2021)
117. Zhang, K., Wang, A., Zhang, F., Wan, W., Sun, J., Bruzzone, L.: Spatial-spectral dual back-projection network for pansharpening. IEEE Trans. Geosci. Remote Sensing **61**, 1–16 (2023)
118. Zhang, K., Zuo, W., Chen, Y., Meng, D., Zhang, L.: Beyond a gaussian denoiser: residual learning of deep cnn for image denoising. IEEE Trans. Image Process. **26**(7), 3142–3155 (2017)
119. Zheng, Y., Li, J., Li, Y., Guo, J., Wu, X., Chanussot, J.: Hyperspectral pansharpening using deep prior and dual attention residual network. IEEE Trans. Geosci. Remote Sensing **58**(11), 8059–8076 (2020)
120. Zhou, H., Liu, Q., Wang, Y.: Panformer: a transformer based model for pan-sharpening. In: 2022 IEEE International Conference on Multimedia and Expo (ICME), pp. 1–6. IEEE, Piscataway (2022)
121. Zhu, X.X., Bamler, R.: A sparse image fusion algorithm with application to pan-sharpening. IEEE Trans. Geosci. Remote Sensing **51**(5), 2827–2836 (2012)

Super-Resolution for Spectral Image

Qiang Li and Qi Wang

1 Introduction

1.1 Background

With advancements in image sensor technology, spectral imaging system can capture target information across several, dozens, or even hundreds of continuous bands. It generates spectral images in the form of cubes, such as multispectral images and hyperspectral images. Spectral images possess distinct characteristics compared to natural images. They contain a significant amount of spectral dimension information and integrate spatial and spectral aspects. As a result, spectral images provide rich spatial-spectral information about the targets being observed. It reveals subtle spectral characteristics of the measured targets in detail, greatly enhancing the ability for target classification [46] and recognition [43]. Consequently, spectral images have been widely used in crop classification and assessment, medical diagnosis, pollution monitoring, soil mineral exploration, disaster monitoring, and other fields.

In a spectral imaging system, spatial and spectral resolution are essential parameters for evaluating image quality [28]. Spatial resolution refers to the ability of the sensor to distinguish the minimum distinguishable target size within the corresponding range of pixels of the observed image. It quantifies the level of detail in the spatial information of the image. Spectral resolution relates to the ability of the image to resolve spectral details of the target. In the same spectral range, a higher number of bands in the image leads to increased spectral resolution. However, spectral imaging systems face certain limitations due to their narrow spectral width.

Q. Li · Q. Wang (✉)
Name, School of Artificial Intelligence, Optics and Electronics (iOPEN), Northwestern Polytechnical University, Xi'an, Shaanxi, China
e-mail: qiangli@nwpu.edu.cn; crabwq@nwpu.edu.cn

M. Kawulok et al. (eds.), *Super-Resolution for Remote Sensing*, Unsupervised and Semi-Supervised Learning, https://doi.org/10.1007/978-3-031-68106-6_9

To maintain an acceptable signal-to-noise ratio, each pixel needs to cover a relatively large spatial area to collect an adequate light. This requirement limits the number of bands that can be accommodated within the restricted imaging spectrum range. Consequently, the spatial resolution of spectral images is generally lower than that of natural images [42]. This trade-off between spatial and spectral resolution is an inherent characteristic of spectral imaging systems. In addition, the imaging system is usually affected by many aliasing factors, such as imaging environment, relative motion, as well as transmission medium in the process of image acquisition. As a result, the obtained image is often degraded. This degradation leads to the loss of high-frequency detail from the real scene. Consequently, it becomes challenging to effectively discriminate targets, and it severely limits the perceptual and intelligent processing. Given these limitations, there is an urgent need to develop corresponding technologies that can enhance the quality of low-resolution spectral images. Significant benefits can be achieved in a variety of applications by improving the quality of these images. This improvement can enhance the ability to interpret and analyze subsequent tasks, leading to advancements in target identification, classification, and other relevant processes.

To enhance the resolution of spectral images, one approach is to improve the hardware equipment. However, existing technologies have limitations that prevent them from meeting all requirements. To overcome these constraints, researchers have turned to signal processing technology [2, 7, 31] to achieve spectral image super-resolution under restricted hardware conditions. This technique enables the enhancement of spatial resolution and leads to improved image content. Unlike natural images, practical applications often rely on the spectral reflection curve to determine the presence of specific targets in a scene. Therefore, it is crucial to preserve the spectral information during enhancing spatial resolution. This ensures that the super-resolution task effectively utilizes the spectral data. This chapter aims to propose methods that improve spatial resolution while maintaining spectral information. The approach enables the reconstructed image to recover lost texture details, resulting in a more realistic visual perception. This study is particularly valuable for addressing image degradation caused by satellite aerial photography and image transmission/storage/compression. Moreover, the study yields more accurate outcomes for interpretation and application in high-level tasks.

1.2 Difficulties and Challenges

Spectral image super-resolution has gained significant attention from researchers as an economical and effective method to enhance image quality. Over the years, the performance of related algorithms has seen continuous improvement. However, at the current stage, there are three main difficulties and challenges in the technical aspects of spectral image super-resolution. These challenges are as follows:

High-Dimensional Data Processing Spectral images usually have high-dimensional data, and each pixel contains information of multiple spectral bands. Such high-dimensional data requires efficient algorithms to analyze [52].

Spectral Information Preservation Spectral image super-resolution aims to improve the spatial resolution while maintaining the consistency of the original spectral information. This requires full consideration of spectral information retention [3, 30, 33, 34]. Accurate maintenance of spectral information is essential for many spectral applications. However, there are still difficulties.

Artifacts and Pseudo-Details Artifacts and pseudo-details may be created during image super-resolution, which refers to the unreal image architecture and details that appear in the results [57]. Therefore, improving the authenticity and naturalness of super-resolution results remains a challenge.

In short, the current spectral image super-resolution technology is faced with difficulties and challenges in the high-dimensional data processing, spectral information preservation, and image detail content. To solve these problems, it is necessary to conduct in-depth research and innovation. The algorithms are constantly improved by combining the knowledge of computer vision and spectral analysis, which promotes the development of spectral image super-resolution technology.

1.3 Dataset and Evaluation Metrics

1.3.1 Dataset

The spectral image super-resolution reconstruction model based on deep learning relies on extensive datasets during the training process. Compared with natural images, the number of available spectral images is relatively small. Fortunately, there is no need to annotate pixel categories when training the model for the image super-resolution task. Unlike constructing datasets for high-level tasks, image super-resolution randomly selects small patches from high-resolution images and degrades them to generate a large number of training sample pairs. This section introduces some commonly used datasets in the field of spectral image super-resolution. Figure 1 displays the corresponding RGB images of selected spectral images.

- CAVE: The CAVE dataset[1] is a spectral image dataset captured using the Apogee Alta U260 camera in 2008 [63]. It covers a wavelength range from 400 nm to 700 nm. The dataset comprises 32 different scenarios and is categorized into five parts: true and false, skin and hair, paint, food and beverages, and objects. Each spectral image in the CAVE dataset consists of 31 bands, with each band having a size of 512 × 512 pixels. The images are stored as 16-bit grayscale PNG images.

[1] http://www1.cs.columbia.edu/CAVE/databases/multispectral/.

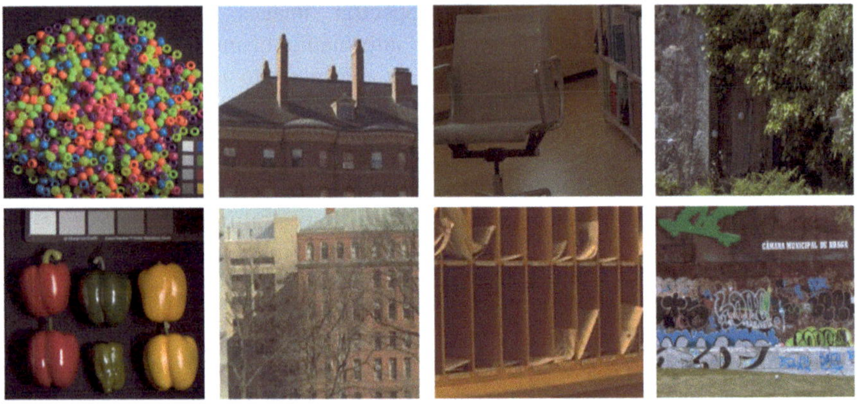

Fig. 1 Some spectral images correspond to RGB samples from CAVE and Harvard dataset

- Harvard: The Harvard dataset[2] was obtained using the Nuance FX CRI Inc. camera [4]. The captured images in this dataset have a wavelength range from 420 nm to 720 nm, with a step size of 10 nm. The dataset can be categorized into two main groups: 50 images captured in indoor and outdoor scenes under daylight conditions and 25 images captured under artificial and mixed lighting conditions. Each spectral image in the Harvard dataset consists of 31 bands, and the size of each band is 1040×1392 pixels. Unlike the CAVE dataset, the Harvard dataset is stored as a .mat file.
- Chikusei: The dataset[3] was acquired using the Headwall Hyperspec-VNIR-C imaging sensor over Chikusei, Ibaraki, Japan. Unlike above CAVE and Harvard datasets, the Chikusei dataset only has one image. The spectral image consists of 128 bands covering a spectral range from 363 nm to 1018 nm, where the resolution of each band is 2517×2335 pixels.

1.3.2 Evaluation Metrics

The primary objective of spectral image super-resolution is to enhance spatial resolution while preserving the spectral properties of the image. To assess the performance of reconstructed spectral images, two main evaluation approaches are commonly used: quantitative evaluation and qualitative evaluation. Quantitative evaluation involves comparing the reconstructed image with a high-resolution spectral reference image. Various mathematical formulas are applied to calculate the differences between the two images. This evaluation method focuses on objective measurements of image quality. On the other hand, qualitative evaluation relies on visual comparison between the reconstructed image and the reference image.

[2] http://vision.seas.harvard.edu/hyperspec/explore.html.

[3] http://naotoyokoya.com/Download.html.

Observers subjectively assess the spatial details and spectral fidelity of the image. In the field of image super-resolution, a combination of both evaluation approaches is often employed to assess the quality of reconstructed images. For quantitative evaluation, several indicators are usually used, including peak signal-to-noise ratio (PSNR), structural similarity (SSIM), and spectral angle mapper (SAM). These metrics provide quantitative measurements of image fidelity and similarity between the reconstructed image and the reference image. Let the reference image be $I_{HR} \in \mathbb{R}^{L \times W \times H}$ and the reconstructed image $I_{SR} \in \mathbb{R}^{L \times W \times H}$, where W and H represent the width and height of each band of the spectral image and L represents the total number of bands in the spectral image.

(1) Peak Signal-to-Noise Ratio, PSNR

$$PSNR = \frac{1}{L} \sum_{l=1}^{L} 10 log_{10} \left(\frac{MAX_l^2}{MSE_l} \right), \quad (1)$$

where MAX_l represents the maximum pixel value in the l band and MSE_l denotes the mean square error of the l band, which is defined as

$$MSE_l = \frac{1}{WH} \sum_{w=1}^{W} \sum_{h=1}^{H} \left(I_{SR}(l, w, h) - I_{HR}(l, w, h) \right)^2 . \quad (2)$$

(2) Structural Similarity, SSIM

$$SSIM = \frac{1}{L} \sum_{l=1}^{L} \frac{\left(2\mu_{I_{SR}}^l \mu_{I_{HR}}^l + c_1\right)\left(2\sigma_{I_{SR} I_{HR}}^l + c_2\right)}{m * n}, \quad (3)$$

where $\mu_{I_{SR}}^l$ and $\mu_{I_{HR}}^l$ denote the band mean in I_{SR} and I_{HR}, respectively. $\sigma_{I_{SR}}^l$ and $\sigma_{I_{HR}}^l$ denote the band variance in I_{SR} and I_{HR}, respectively. $\sigma_{I_{SR} I_{HR}}^l$ represents the band covariance between I_{SR} and I_{HR}. c_1 and c_2 are constants. Here, m and n are defined as

$$m = \left(\mu_{I_{SR}}^l\right)^2 + \left(\mu_{I_{HR}}^l\right)^2 + c_1, \quad (4)$$

$$n = \left(\sigma_{I_{SR}}^l\right)^2 + \left(\sigma_{I_{HR}}^l\right)^2 + c_2. \quad (5)$$

(3) Spectral Angle Mapper, SAM

$$SAM = arccos \left(\frac{< I_{SR}, I_{HR} >}{||I_{SR}||_2 ||I_{HR}||_2} \right), \quad (6)$$

where $arccos(\cdot)$ is the inverse cosine function, $< \cdot, \cdot >$ indicates dot product operation, and $|| \cdot ||_2$ represents L2 norm.

Among the three quality evaluation metrics, PSNR and SSIM are commonly employed as quality evaluation metrics to assess the spatial reconstruction performance of spectral images. Higher values of PSNR and SSIM indicate superior reconstruction performance. On the other hand, SAM serves as a measure to quantify the extent of spectral distortion. In this case, lower values of SAM indicate a better preservation of the original spectrum.

1.4 Main Research Contents

The trade-off between spectral and spatial resolution poses a significant challenge in spectral images. While high spectral resolution is achieved, spatial resolution is often compromised. In this study, we investigate the relationship between spectral and spatial information to enhance the quality of reconstructed images without sacrificing spectral resolution. The specific research contents and contributions are as follows:

(1) Spectral image super-resolution methods based on deep learning have achieved significant progress. However, due to the lack of exploration of larger spectral information in the recurrent model [52], the spatial-spectral information is inconsistent when the super-resolved images generated by different scale factors are transformed into images with the same scale factors. To address this issues, a spectral image super-resolution framework via two-stage learning is proposed. In the coarse stage, we enhance the exploration of spectral information by merging more adjacent bands based on the work [52] and group these bands together to facilitate the network to learn the information of the current band. Different from the work [52], we design intra-group/inter-group alternating fusion strategy to improve feature learning in group. However, we acknowledge that the coarse stage may still lack additional information along the spectral dimension. To solve this problem, we introduce a back-projection method in the fine stage. This method uses spectral angle constraints to study the spatial-spectral consistency in multi-scale images.

(2) Spectral image super-resolution methods using an auxiliary high-resolution RGB image have achieved great progress recently. However, the existing spectral image super-resolution methods fail to exploit the rich spectral characteristics of spectral images and the high spatial resolution of RGB images [13, 65]. In addition, the degradation process of the constructed low-resolution image follows a single mode. To solve these limitations, we propose a bidirectional feature propagation network. The model adopts bidirectional symmetrical architecture, which is convenient for network modulation, and effectively propagates and fuses information between spectral image and RGB image domain. Unlike previous studies [8, 65, 70], our goal is to improve the overall super-resolution performance by leveraging this bidirectional approach. To void the information ambiguity, considering the fixed conversion relationship between the super-resolved result and RGB image, we introduce a self-

supervised network at the end of the architecture. The network learns the global information, which ensures the consistency between the reconstructed spectral image and the RGB image. Besides, we design an image degradation strategy when constructing the dataset. It includes various blur kernels, downsampling, and random noise. This strategy achieves comprehensive training by simulating various degraded scenarios, resulting in a more robust model.

1.5 Section Organization

This chapter mainly focuses on the spectral image super-resolution reconstruction task for spectral image. First, we provide the introduction and related work for spectral image super-resolution. Then, inspired by the complementarity of adjacent bands and spatial-spectral combination, two solutions are proposed to explore the dependencies between band sequences. Finally, the research contents are summarized.

Section 1 introduces the research background and significance of spectral image super-resolution, research contents and contributions, datasets and evaluation metrics, etc. Section 2 describes the methods based on convolutional neural network (CNN), including single spectral image super-resolution method and spectral image super-resolution method based on auxiliary image fusion. Section 3 proposes a dual-stage learning method for spectral image super-resolution, including coarse stage and fine stage. Various experiments are conducted from two aspects of known degradation and unknown degradation. Section 4 presents a bidirectional symmetric architecture network for spectral image super-resolution and introduces the specific design details of the model in detail. Then, the performance of the model is discussed under various degradation mechanisms. Section 5 summarizes the work of the chapter and plans the next steps.

2 Related Work

Existing models in spectral image super-resolution can be broadly categorized into two main types based on the input data: single spectral image super-rcsolution method and spectral image super-resolution method via auxiliary image fusion. Single spectral image super-resolution method focuses on enhancing the spatial resolution of individual spectral images without any additional auxiliary information. These methods directly process the spectral image data and aim to reconstruct a high-resolution version of the same bands in spectral image. They employ various techniques to recover fine details and improve the overall quality of the image. Spectral image super-resolution method via auxiliary image fusion is constructed by incorporating additional information from auxiliary images, typically RGB images. The high spatial resolution of the RGB image is leveraged to enhance the super-resolution of the spectral image. These methods often involve the fusion of the spectral and RGB images at different stages to extract complementary information and improve the final reconstructed spectral image.

2.1 Single Spectral Image Super-Resolution

As for natural image super-resolution, early traditional algorithms [9, 44, 48] employ interpolation techniques such as nearest and bilinear interpolation. These methods aim to generate high-resolution pixels by averaging adjacent pixels in low-resolution images. Another approaches [18, 24] are based on image degradation model to construct model. These algorithms typically downsample the estimated image and calculate the difference between it and the actual low-resolution image to refine the super-resolved image. Moreover, some algorithms utilize the maximum a posteriori probability scheme to design the model. For example, Hardie et al. [14] present a maximum a posteriori probability estimation scheme to boost spatial resolution by leveraging high-resolution images generated by auxiliary sensor, e.g., Eismann et al. [10, 11]. However, traditional methods face two main challenges. On the one hand, spectral images often contain tens or even hundreds of spectral bands, and the super-resolution in band-by-band significantly increases the computational burden. On the other hand, these methods typically treat each band independently and do not effectively exploit the spectral information across different bands. These challenges have motivated the development of more advanced techniques in spectral image super-resolution to address these limitations and improve the overall performance.

Recently, the application of CNNs has revolutionized the natural image super-resolution task. CNN-based methods [21, 37, 39] have demonstrated remarkable improvements over traditional methods. These approaches utilize the powerful feature representation of CNNs to learn the mapping function from low-resolution to high-resolution images through supervised learning. Inspired by the approaches, the studies for spectral image super-resolution mainly explore 3D or 2D convolutions to extract relevant features. These convolutional operations are applied to the spectral dimension and spatial dimensions to capture and exploit the unique characteristics and dependencies within bands.

2.1.1 CNNs Using 2D Convolution

Spectral images consist of multiple bands. Motivated by current models, researchers regard each band as an independent image and improve the resolution of each individual band within a spectral image without relying on additional information. For instance, Li et al. [35] propose a spectral differential convolutional neural network. They incorporate a spatial constraint strategy to ensure that the reconstructed spectral image preserves the spectral curve, which is further refined through post-processing. Jia et al. [20] introduce a two-subnetwork framework. One subnetwork focuses on spatial feature extraction, while the other subnetwork specializes in spectral feature extraction. These subnetworks learn the mapping function and fine-tune the spectrum between low-resolution and high-resolution images, respectively.

Another model by Yuan et al. [64] leverages transfer learning for natural images to reconstruct spectral images. To ensure the preservation of spectral information, the model employs non-negative matrix decomposition to learn consistent spectral characteristics. All these methods typically involve a two-step process. Firstly, the spatial resolution is enhanced for the visual clarity of the image. Secondly, constraint criteria are applied to preserve the spatial information while avoiding spectral distortion. Unfortunately, the spatial content of the spectral image may change, when prioritizing spectral information preservation. Inspired by recurrent residual network [47], Li et al. [36] propose a grouped deep recurrent residual network for spectral image super-resolution. However, the approach exclusively relies on 2D convolution for feature extraction, which disregards the valuable spectral content. As a result, the reconstructed images lack clear edge details. Later, Jiang et al. [22] develop a progressive upsampling convolution scheme, which outperforms existing models using 2D convolution. Nevertheless, the progressive architecture is limited to generating results for larger-scale factors and cannot deal with smaller-scale factors, such as $\times 2$ and $\times 3$.

2.1.2 CNNs Using 3D Convolution

Spectral images possess not only abundant spatial information but also significant spectral information. To leverage this rich information, numerous spectral image super-resolution methods based on 3D convolution have been introduced in recent years. These methods specifically exploit the spectral characteristics of the data. As a result, they exhibit superior performance compared to methods relying solely on 2D convolution. For instance, Mei et al. [40] introduce a 3D full convolution neural network (3D-FCNN) to investigate the relationship between spatial information and neighboring pixels in spectral images. On this basis, Yang et al. [60] design a multi-scale wavelet 3D convolution neural network that generates reconstructed bands. In addition to above approaches, researchers explore the application of generative adversarial networks (GAN). Li et al. [27] propose a 3D-GAN super-resolution network. Similarly, Jiang et al. [23] present a GAN-based network which includes two parts: spectral feature extraction and spatial feature extraction. GAN-based networks are generally challenging to train, and they often struggle to extract both spatial and spectral features simultaneously. Unlike 2D convolution, 3D convolution convolves a 3D filter with a feature map, leading to a significant increase in network parameters. To address this drawback, researchers modify the filters by using $k \times 1 \times 1$ and $1 \times k \times k$ instead of $k \times k \times k$. Li et al. [25] develop double 1D-2D spatial spectral convolution neural networks, where 1D convolution extracts spectral features and 2D convolution extracts spatial features. A similar strategy is employed in SSRNet [51]. However, these methods often lack the differential treatment of multidomain content and do not widely mine spatial knowledge, resulting in poor model performance.

2.1.3 Representative Models

This section introduce several representative works by utilizing different types of convolution.

(a) **SSPSR:** There are many similarities between spectral image super-resolution and natural image super-resolution, and the main difference lies in the type of input image. Considering that there are no other obvious differences between the two technologies, Jiang et al. [22] utilize single RGB image super-resolution approach and propose a spatial-spectral prior network for spectral image super-resolution (SSPSR) using 2D convolution, as depicted in Fig. 2a. Obviously, only 2D convolution is not enough to acquire spectral knowledge.

(b) **3D-FCNN:** The regular 3D convolution described in Mei et al. [40] is originally employed to capture both spatial context and spectral correlation within neighboring bands, and 3D full convolutional neural network (3D-FCNN) is proposed. The simplified architecture of the method is shown in Fig. 2b. Although the regular 3D convolution can explore spectral information, it brings difficulties to effective network training. To address this limitation, the researchers introduce some enhancements, such as separable 3D convolution [23, 26], etc.

(c) **ERCSR:** Li et al. [32] claim that its main parallel module in MCNet [31] does not effectively combine 2D and 3D convolutions, resulting in structural redundancy. Considering this issue, Li et al. propose a novel alternate architecture that incorporates both 2D and 3D convolutions and explores the relationship between 2D/3D convolution for spectral image super-resolution (ERCSR), as shown in Fig. 2c. This architecture not only resolves module redundancy but also enhances representation learning in the spatial domain.

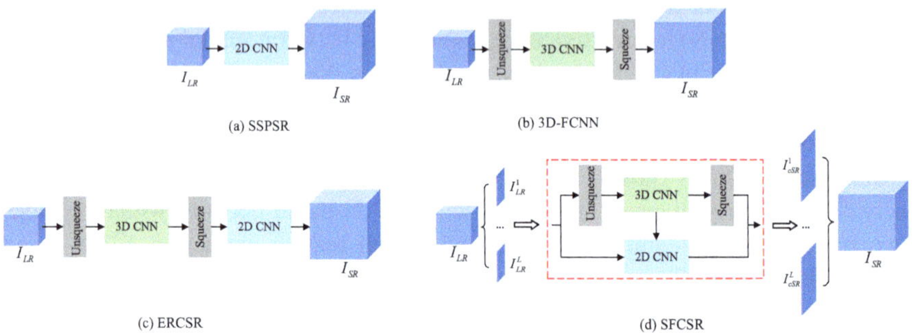

Fig. 2 Simplified architecture of several methods for spectral image super-resolution. I_{LR} refers to the low-resolution spectral image, while I_{SR} represents the super-resolved spectral image. Additionally, I^i_{LR} (where $1 \leq i \leq L$) denotes the i-th band of the low-resolution spectral image, where L corresponds to the total number of bands in the spectral image cube. Similarly, I^i_{cSR} (where $1 \leq i \leq L$) represents the i-th band of the coarse super-resolved spectral image. (**a**) SSPSR. (**b**) 3D-FCNN. (**c**) ERCSR. (**d**) SFCSR

(d) **SFCSR:** The approach proposed by Wang et al. [52] introduces a new input pattern where the information from a single band is analyzed together with its two adjacent bands and utilizes spectrum and feature context for spectral image super-resolution (SFCSR), as illustrated in Fig. 2d. This approach effectively reduces the memory footprint. However, its limitation is that it does not take into account more information of adjacent bands. Consequently, when employing a recurrent strategy for single band super-resolution, it can result in significant spectral distortion. Additionally, the recurrent mode used in this approach overlooks the overall learning and optimization of spectral images.

2.2 Image Super-Resolution Based on Auxiliary Image Fusion

It is common for spectral imagers to generate both RGB images and spectral images simultaneously. Due to the typically higher spatial resolution of RGB images, there are several available algorithms that leverage the combination of low-resolution spectral images and high-resolution RGB images for spectral image super-resolution [29, 38]. These algorithms include both traditional methods [1, 2, 7, 56] and deep learning-based approaches [17, 42, 55, 65]. The performance of algorithms for spectral image super-resolution has been continuously improving. However, existing traditional approaches rely on manual feature construction and the incorporation of prior knowledge in model design. Due to the limited feature representation ability of hand-crafted features, these methods struggle to recover fine details in real low-resolution images, which hampers their robustness. In contrast, deep learning-based methods have witnessed remarkable progress in spectral image super-resolution. Thanks to the powerful feature representation of CNNs [61, 62], researchers predominantly utilize deep learning techniques to build models for this task. Compared to traditional methods, deep learning-based approaches demonstrate superior performance in addressing the challenges presented by real low-resolution images. Based on the input mode, existing supervised methods for spectral image super-resolution can be roughly categorized into two groups: stack input mode and parallel input mode.

2.2.1 Stack Input Mode

In the stack input mode, the algorithms usually first upsample the low-resolution spectral image to the high-resolution RGB image with the same spatial resolution. These two images are then combined by stacking them as input to the model. For instance, Han et al. [12] propose a spatial and spectral fusion framework, where the high-resolution RGB image and upsampled low-resolution spectral image are concatenated into a cube. This input setting allows joint exploration of spectral attributes and rich spatial content. Similarly, Zhang et al. [67] adopt a similar approach when building input data. In contrast, Hu et al. [17] take a different strategy. They first downsample the high-resolution RGB image to the same spatial resolution as the low-resolution spectral image. The two images are then merged into a whole and upsampled using the PixelShuffle operation [45]. However, this method combines RGB images and spectral images prematurely without analyzing the differences between them. Currently, the stack input mode for building models in spectral image super-resolution is largely abandoned due to the above limitations and challenges.

2.2.2 Parallel Input Mode

At present, most existing spectral image super-resolution methods use parallel input, which inputs low-resolution spectral images and high-resolution RGB images into the model in parallel [42, 58, 70]. For example, Zhu et al. [70] propose a progressive zero-centric residual network (PZRes-Net) that incorporates low-resolution spectral images of different dimensions along the spectral direction. The aim is to learn a zero-centric residual content from both inputs. Wang et al. [53] develop a dense fusion framework, which consists of a measurement module and a fusion module. The measurement module learns the observation model, while the fusion module iteratively extracts spatial-spectral information. Similarly, MoG-DCN [8] follows a similar approach. It fuses the high-resolution RGB image with the image generated by the observation model. Additionally, the obtained observation image is merged with the low-resolution spectral image after various operations. These methods effectively explore the differences between individual images before fusion. However, they may not be able to learn the inherent properties of each image over a relatively long range. To address this challenge, Han et al. [13] present a multi-scale spatial-spectral fusion architecture with two pathways over a long range. Nevertheless, the fusion ambiguity is inevitably introduced into the model due to discrepancies in depth information during the fusion stage. In a different study, Zhang et al. [65] develop two networks to generates a general prior image and a specific super-resolution spectral image, respectively. Obviously, it faces two main challenges. On the one hand, it struggles to effectively distinguish and utilize the differences between spectral and RGB images. On the other hand, it fails to fully leverage the rich spectrum of spectral images and the high-resolution content of RGB images when constructing their models.

2.3 Conclusion

This section introduces reviews the current work on spectral super-resolution, including single spectral image super-resolution method and spectral image super-resolution method based on auxiliary image fusion. Additionally, the section highlights the existing issues and limitations within these methods, which can further improve the novelty of our design models.

3 Dual-Stage Learning for Single Spectral Image Super-Resolution

Spectral images contain valuable spectral information that can enhance the spatial reconstruction performance. However, some existing algorithms, such as those based on 3D convolution, allocate equal importance to the exploration in spatial and spectral domains when constructing models, which lacks more attention to spatial content [40]. Several spectral image super-resolution algorithms are developed based on the above motivation, e.g., [31, 32, 52]. The key concept behind these methods is to augment the model with additional 2D convolutions. The purpose of this manner is to enhance the feature learning of the model in the spatial domain, so as to improve the overall performance of spectral image super-resolution.

Spectral images typically consist of numerous bands, which poses a challenge for existing methods that require inputting all bands simultaneously into the model under a given batch size. This manner results in increased GPU memory usage. Interestingly, the issue is solved by developing a new super-resolution framework called SFCSR [52]. Inspired by the high similarity observed in adjacent bands [49, 50], SFCSR takes advantage of this similarity by integrating the information from a single low-resolution band and its two adjacent low-resolution bands for modeling. It achieves super-resolution for all bands in a recurrent manner. However, SFCSR for band reconstruction lacks the utilization of more spectral information and only relies on two adjacent bands. This limitation inevitably results in severe spectral distortion in the super-resolved spectral image. In fact, there are similarities observed between relatively distant bands, as shown in Fig. 3. To address this issue and compensate for the missing content during band reconstruction, it is crucial to effectively incorporate more adjacent bands. Additionally, the super-resolved images produced using different scale factors are transformed into images with the same scale factor. It leads to inconsistencies in spatial-spectral information. Therefore, the key challenge lies in finding a solution to effectively utilize relatively distant adjacent bands to guide the current band and achieve spatial-spectral consistency in super-resolved images with various scale factors.

To address the above challenges, this section presents a two-stage learning approach called DualSR for spectral image super-resolution, including coarse stage

Fig. 3 Illustration of the existence of high similarity in adjacent bands. We calculate the similarity between 17-th band and the other bands by the reciprocal of Euclidean distance

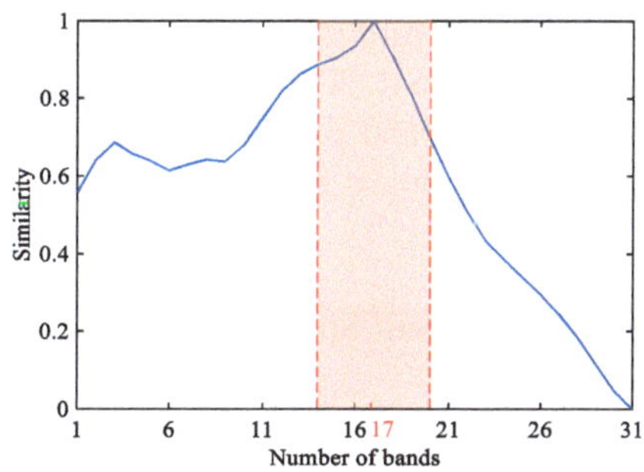

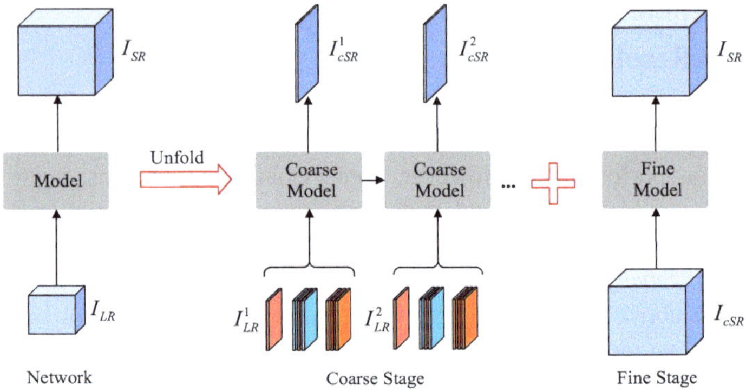

Fig. 4 Overview of the proposed DualSR for spectral image SR

and fine stage. The proposed method is illustrated in Fig. 4. In the coarse stage, a coarse stage model is constructed by partitioning the bands in spectral image into band groups. This model aims to generate initial super-resolved band for each group. It takes into consideration that even bands located farther can contribute valuable detail, which incorporates an adjacent spectral fusion mechanism to learn complementary content within and between groups. Then, the coarse super-resolved bands I_{cSR} are obtained in a recurrent manner. In the fine stage, an enhanced back-projection method is developed via spectral angle constraint to further guide the super-resolution process. To optimize the result I_{cSR}, the reconstruction error is introduced to refine the super-resolved image. Then, the spectral angle constraint ensures that the reconstructed spectral image has a similar spectral curve as the original low-resolution image. Finally, the reconstructed spectral image I_{SR} is obtained.

3.1 *Coarse Stage via Adjacent Spectral Partition*

Previous works for spectral image super-resolution typically process all bands synchronously. Inspired by the fact that the reconstruction of the current band is primarily influenced by adjacent bands, we follow the pipeline in [52] to design a novel architecture for spectral image super-resolution via adjacent spectral fusion, namely, CoarSR. The overview of coarse stage mainly covers three modules: neighboring band partition (NBP), adjacent spectral fusion mechanism (ASFM), and feature context fusion (FCF) (Fig. 5).

3.1.1 Adjacent Spectral Partition

In the SFCSR method, only two adjacent bands are considered for analysis, which neglects non-adjacent bands relative to the current band position. Figure 3 demon-

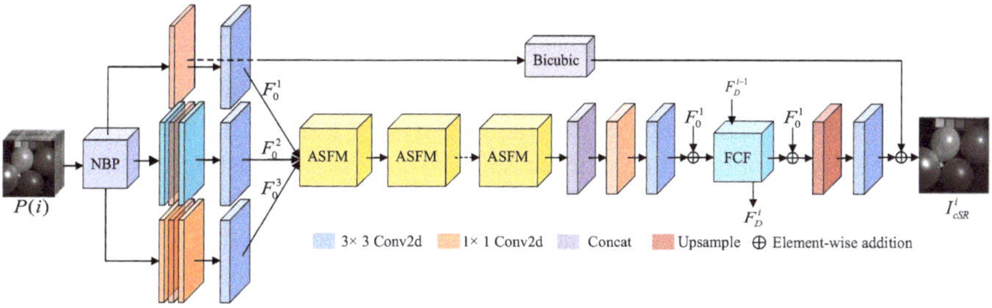

Fig. 5 Overview of the proposed CoarSR in coarse stage

strates that bands within a specific spectral range also exhibit significant similarity. This means that even relatively distant bands can provide explicit information for the reconstruction of the current band. In our proposed method, we take into account a single band, denoted as I_{LR}^i, and its four adjacent bands for analysis, i.e.,

$$P(i) = \begin{cases} \left[I_{LR}^1, I_{LR}^2, I_{LR}^3, I_{LR}^4, I_{LR}^5\right], & 1 \le i < 3 \\ \left[I_{LR}^{i-2}, I_{LR}^{i-1}, I_{LR}^i, I_{LR}^{i+1}, I_{LR}^{i+2}\right], & 3 \le i \le L-3, \\ \left[I_{LR}^{L-4}, I_{LR}^{L-3}, I_{LR}^{L-2}, I_{LR}^{L-1}, I_{LR}^L\right], & L-3 < i \le L \end{cases} \tag{7}$$

where L is total number of bands in the spectral image. To help the current bands by using adjacent bands, we propose a method called adjacent band partition (NBP), which divides the five bands into three different categories. This division is based on the similarity between these bands and the current band. As for the case $3 \le i \le L-3$, it can be expressed as

$$f_{BP}(P(i)) = \begin{cases} I_{LR}^i, \\ \left[I_{LR}^{i-1}, I_{LR}^i, I_{LR}^{i+1}\right], & 3 \le i \le L-3. \\ \left[I_{LR}^{i-2}, I_{LR}^i, I_{LR}^{i+2}\right], \end{cases} \tag{8}$$

When it comes to bands with larger or smaller indices, they are denoted as

$$f_{BP}(P(i)) = \begin{cases} I_{LR}^i, \\ \left[I_{LR}^1, I_{LR}^2, I_{LR}^3\right], & 1 \le i < 3, \\ \left[I_{LR}^4, I_{LR}^i, I_{LR}^5\right], \end{cases} \tag{9}$$

$$f_{BP}(P(i)) = \begin{cases} I_{LR}^i, \\ \left[I_{LR}^{L-2}, I_{LR}^{L-1}, I_{LR}^L\right], & L-3 < i \le L. \\ \left[I_{LR}^{L-3}, I_{LR}^i, I_{LR}^{L-4}\right], \end{cases} \tag{10}$$

3.1.2 Adjacent Spectral Fusion Mechanism

The adjacent bands not only share high similarity but also exhibit diverse textures in certain regions. This implies that the missing information in the current band can potentially be obtained from these neighboring bands. Existing works [17, 32] utilize small convolution kernels directly on the spectral image, which limits the extraction of features from non-adjacent bands. Consequently, the ability to borrow complementary content from relatively distant adjacent bands is hindered. To address this limitation, we propose an adjacent spectral fusion mechanism (ASFM). The ASFM modulates the network to enhance more complementary information from both intra-group and inter-group, which is depicted in Fig. 6.

Intra-Group Fusion After dividing these bands, the initial features F_0^1, F_0^2, F_0^3 are extracted from three groups using 2D convolution layers. To thoroughly explore the valuable information within each group, three parallel branches with same architecture are exploited independently. For the ease of explanation, we only describe the first branch in detail. Since the module mainly depends on spatial content, we deal with it in a way similar to the natural image super-resolution technology. To study the spatial features of these bands, we utilize the main residual module commonly used in the natural image super-resolution algorithm as a reference to construct the residual unit. Specifically, the residual unit within the intra-group consists of two identical blocks. Each block comprises two convolution layers with kernel size 3 × 3, followed by a ReLU activation function and a skip connection. Furthermore, to integrate complementary information from different groups, we concatenate the local features from the other two groups into the current group for fusion. Under the action of intra-group fusion, the features of current band I_{LR}^i in the spatial domain are fully developed.

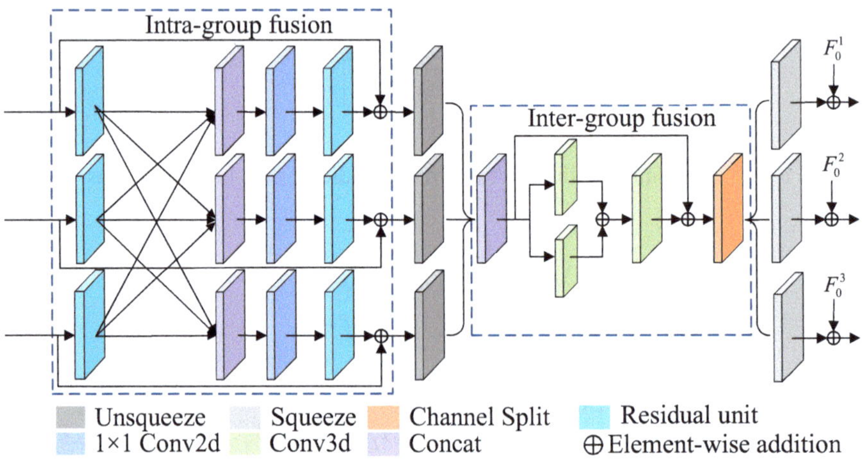

Fig. 6 Adjacent spectral fusion mechanism (ASFM)

Inter-Group Fusion Applying 2D convolution alone in the intra-group fusion limits the ability to capture content beyond the spatial dimension. To address this limitation and capture spectral features across different groups, we leverage 3D convolution in the inter-group fusion scheme. However, the use of regular 3D convolution directly results in a large number of network parameters, which increases the computational cost. To this issue, two convolution layers with kernels, i.e., $3 \times 1 \times 1$ and $1 \times 3 \times 3$, are attached with network to study spectral and spatial features, respectively. Then, the information between two domains is availably fused through an addition operation. As for spectral image super-resolution, its aim is to heighten spatial resolution. Meanwhile, it also keep high spectral fidelity. To achieve this end, the convolution layer with kernel $1 \times 3 \times 3$ is added after fusion to further enhance the learning ability of spatial domain. The inter-group fusion combines information from different groups, which allows the integration of knowledge in the spectral and spatial domain.

3.2 *Fine Stage via Spectral Angle Constraint*

Although our method currently employ five bands to restore a single band in the coarse stage, it fails to focus on the broader spectral perspective, resulting in spectral distortion. One way to solve this problem is to input all bands into the model and use supervision to optimize the results. However, this approach deviates from our original intention. To overcome this trouble, we explore an unsupervised approach to further refine the initial results. The back-projection algorithm, proposed by Irani et al. [19], is widely utilized to optimize the reconstruction error in an iterative manner. In this algorithm, multiple upsampling strategies are typically exploited to analyze the low-resolution image, and the reconstruction error is iteratively calculated. It has been successfully incorporated in various natural image super-resolution methods [15, 16, 59]. However, the optimal initialization for this algorithm remains unknown due to the predefined hyperparameters such as the number of iterations and the convolution kernel.

To enhance the algorithm, we make additional advancements in back-projection, which eliminates the need for predefined hyperparameters. This improved approach is illustrated in Fig. 7. According to scale factor s, the coarse result U is generated by the coarse model $\mathcal{F}$ for a low-resolution image I_{LR}. This process is denoted as

$$U = \mathcal{F}(s, I_{LR}). \tag{11}$$

Previous studies primarily resize the super-resolved image to match the size of the low-resolution image, which is followed by evaluating the reconstruction error between the two. However, when dealing with a large scale factor, this approach tends to discard a substantial amount of valuable information. It leads to negligible performance gains for the reconstruction task. To address this issue, we introduce

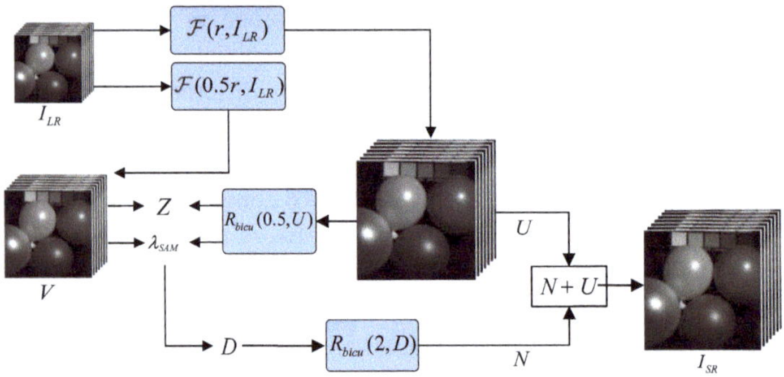

Fig. 7 Enhanced back-projection method via spectral angle constraint

coarse model again, which performs low-resolution image I_{LR} to obtain another initial SR result V for scale factor $s/2$, which is expressed as

$$V = \mathcal{F}\left(\frac{s}{2}, I_{LR}\right), \tag{12}$$

Subsequently, the coarse result U corresponding to the scale factor s is downsampled by bicubic interpolation. Once the low-resolution image M is generated, the reconstruction error is computed by means of the following calculation:

$$M = R_{bicu}(0.5, U), \tag{13}$$

$$Z = V - M, \tag{14}$$

To achieve high spectral fidelity, we embed the spectral angle mapper (SAM) as a means to constrain the error, i.e.,

$$\lambda_{SAM} = arccos\left(\frac{< M, V >}{||M||_2||V||_2}\right), \tag{15}$$

$$D = \begin{cases} \lambda_{SAM} \times Z, & \lambda_{SAM} < 1, \\ Z, & otherwise, \end{cases} \tag{16}$$

where $arccos(\cdot)$ represents the arccos function, $< \cdot, \cdot >$ denotes the dot product operation, and $|| \cdot ||_2$ denotes the L2 norm. The error is then upsampled using the Bicubic interpolation and added to the coarse result U. Eventually, the fine result I_{SR} is computed through

$$N = R_{bicu}(2, D), \tag{17}$$

$$I_{SR} = U + N. \tag{18}$$

Algorithm 1: Dual-stage spectral image super-resolution method (DualSR)

Input: Spectral image dataset with LR-HR pair, scale factor s
Output: Super-resolved spectral image I_{SR}

Randomly initialize coarse model parameters θ ;
while *not converged* **do**
  Select LR-HR batch;
  while $i \leq L$ **do**
    Divide bands in the spectral image into three groups by Eqs. (8)–(10);
    Update θ by conducting coarse model;
    $i \leftarrow i + 1$;
  end
end
Obtain coarse model parameters θ_c;
Generate coarse SR results U and V for scale factor s;
Calculate the reconstruction error by Eq. (16);
Acquire fine SR result I_{SR} by Eq. (18).

Our approach deviates from the convolution operation employed in the previous work [59]. Notably, its strategy eliminates the need for iterative procedures. By adopting an unsupervised refinement process, the method can be seamlessly integrated into any spectral image super-resolution algorithms, thereby optimizing the super-resolved image. This characteristic classifies our approach as a plug-and-play method. To outline the dual-stage super-resolution algorithm, we provide a summary in Algorithm 1.

3.3 Model Training

When generating label pairs for training, bicubic interpolation is commonly employed to create low-resolution images. However, the degradation process is more intricate in real-world scenarios due to factors such as noise and motion blur. Consequently, the performance of traditional super-resolution networks trained solely on bicubic kernels faces a significant decline when confronted with non-bicubic and real-world degradation kernels. To obtain the real low-resolution image, the high-resolution image undergoes degradation by anisotropic Gaussian. This process involves a kernel distribution p, where each kernel is determined by a covariance matrix α. Mathematically, it can be expressed as

$$\alpha = \begin{bmatrix} \cos(\theta) & -\sin(\theta) \\ \sin(\theta) & \cos(\theta) \end{bmatrix} \begin{bmatrix} \lambda_1 & 0 \\ 0 & \lambda_2 \end{bmatrix} \begin{bmatrix} \cos(\theta) & \sin(\theta) \\ -\sin(\theta) & \cos(\theta) \end{bmatrix}. \tag{19}$$

Here, the matrix has three parameters: an angle $\theta \sim U[0, \pi]$ and two eigenvalues $\lambda_1, \lambda_2 \sim U[0.2, 2r]$. Equation (19) is employed to blur the high-resolution image using the generated kernel. Subsequently, based on the scale factor r, the blurred image is downsampled by bicubic interpolation to generate a label pair $\{I_{LR}, I_{HR}\}$. These steps are repeated to construct the training dataset $\mathcal{D}$. The coarse network model is then optimized using the loss function $\mathcal{F}$, i.e.,

$$\mathcal{F}(t;\theta) = \min ||I^t_{cSR} - I^t_{HR}||_1, \tag{20}$$

where I^t_{cSR} represents the output of the network, which corresponds to the coarse estimation for t-th band. Similarly, I^t_{HR} represents the t-th band in high-resolution spectral image.

3.4 Experiment

3.4.1 Implementation Details

To provide the evaluation of the performance, a blur kernel with 21×21 is set to blur the high-resolution image according to Eq. (19). Subsequently, the low-resolution image is obtained by downsampling using bicubic interpolation. In our proposed network architecture, the convolution kernel with 1×1 is employed after concatenation to reduce the number of channels. The channel of convolution kernels with 3×3 within the network is set to 64. For model optimization, we utilize the ADAM optimizer with parameters $\beta_1 = 0.9$ and $\beta_2 = 0.999$. The batch size for input data is set to 64, and the entire network is trained for 200 epochs. The initial learning rate for all layers is set to 10^{-4}, and it is halved after every 30 epochs. The proposed model is implemented using the PyTorch framework, with version 1.8.1 and CUDA version 11.2.

3.4.2 Ablation Study

To assess the effectiveness of key modules, we conduct an ablation study on different versions of our model by removing each component, namely, inter-group fusion, intra-group fusion, and FCF. The results of the ablation study, evaluated using CoarSR, are presented in Table 1. As for intra-group fusion, the network directly utilizes 2D convolution to extract spatial information. Comparatively, when only

Table 1 Ablation study for scale factor ×4 on CAVE dataset. Bicubic super-resolution kernel is adopted

Module	Different combinations of modules			
Inter-group fusion	✓	×	✓	✓
Intra-group fusion	×	✓	✓	✓
FCF	×	×	×	✓
PSNR	38.720	39.521	39.635	39.665
SSIM	0.9295	0.9322	0.9327	0.9328
SAM	3.304	3.211	3.151	3.152

inter-group fusion is added, the network demonstrates improved performance. This suggests that the model should focus more on exploring spatial features to generate sharper results. When both inter-group and intra-group fusion are incorporated, the network yields superior results. It indicates that the combination effectively integrates complementary information between different groups. Furthermore, when all three modules (inter-group fusion, intra-group fusion, and FCF) are involved in feature extraction, the performance surpasses any combination of modules, except for SAM. This highlights the effectiveness and benefits of the proposed modules, as revealed through quantitative analysis. Overall, the ablation study provides clear evidence of the effectiveness of the key modules and their contributions to improving the performance of the model.

3.4.3 Study of Enhanced Back-Projection

To further refine the results obtained by CoarSR, we propose an enhanced back-projection method via spectral angle constraint. In this section, we investigate the effectiveness of the enhanced back-projection approach for bicubic kernel. Note that the proposed enhanced back-projection involves two different scale factors during the fine stage. As a result, the results are evaluated for scale factors $\times 4$ and $\times 8$ on two publicly available datasets.

Performance Comparison of Back-Projection The proposed enhanced back-projection method is a post-processing technique that is unsupervised and can be applied to any spectral image super-resolution algorithm to further improve its performance. To investigate the performance gain for different algorithms, we employ the enhanced back-projection as a post-processing and analyze the results. When the scale factor is set to $\times 4$, our designed back-projection method involves two different scale factors, namely, $\times 2$ and $\times 4$. As for SSPSR, the minimum scale factor is set to $\times 4$ due to progressive upsampling. As a result, we cannot provide the results for scale factor $\times 4$ when adding our designed back-projection. Table 2 presents the differences between the two scale factors (with or without post-processing) for various methods. It is evident that all methods exhibit superior performance when the enhanced back-projection is applied as post-processing, compared to without post-processing. Additionally, we also display the results obtained by using the back-projection method proposed by Yang et al. [59] as post-processing. Unlike our enhanced back-projection, the method from [59] only incorporates one scale factor. Therefore, we provide the results obtained by SSPSR for scale factor $\times 4$. As observed in Table 3, the improvement is not as significant as in our study while the method produces relatively good results. The comparison results show that our enhanced back-projection method is very effective in improving the performance of various super-resolution algorithms.

Effect of Spectral Angle Constraint In our work, we introduce the SAM as a means to constrain the reconstruction error and alleviate spectral distortion. To investigate the effectiveness of this constraint, we conduct experiments where we

Table 2 Performance gain by setting enhanced back-projection method. The bold indicates the best results

Post- processing	Scale	Metrics	CAVE				Harvard			
			SSPSR	SFCSR	ERCSR	CoarSR	SSPSR	SFCSR	ERCSR	CoarSR
×	×4	PSNR	38.366	39.192	39.224	**39.665**	39.293	40.077	40.211	**40.312**
		SSIM	0.9272	0.9321	0.9322	**0.9328**	0.9333	0.9373	0.9374	**0.9405**
		SAM	3.484	3.221	3.243	**3.152**	2.448	2.407	**2.384**	**2.384**
	×8	PSNR	34.151	35.294	35.102	**35.475**	34.435	35.097	35.252	**35.628**
		SSIM	0.8718	0.8828	0.8824	**0.8828**	0.8676	0.8737	0.8733	**0.8755**
		SAM	4.831	4.378	4.445	**4.269**	2.899	2.911	2.888	**2.794**
✓	×4	PSNR	–	39.372	39.334	**39.813**	–	40.257	40.308	**40.540**
		SSIM	–	0.9335	0.9334	**0.9339**	–	0.9394	0.9401	**0.9423**
		SAM	–	3.201	3.196	**3.130**	–	2.375	**2.367**	2.372
	×8	PSNR	34.453	35.341	35.479	**35.628**	34.855	35.414	35.564	**35.628**
		SSIM	0.8771	0.8846	0.8855	**0.8862**	0.8719	0.8775	0.8793	**0.8799**
		SAM	4.661	4.267	4.264	**4.141**	2.837	2.803	2.829	**2.774**

Table 3 Performance gain by setting [59] as post-processing. The bold indicates the best results

		CAVE				Harvard			
Scale	Metrics	SSPSR	SFCSR	ERCSR	CoarSR	SSPSR	SFCSR	ERCSR	CoarSR
×4	PSNR	38.426	39.225	39.244	**39.675**	39.350	40.089	40.224	**40.327**
	SSIM	0.9286	0.9326	0.9330	**0.9332**	0.9339	0.9380	0.9395	**0.9406**
	SAM	3.545	3.230	3.229	**3.165**	2.434	2.406	**2.382**	2.383
×8	PSNR	34.335	35.335	35.147	**35.497**	34.538	35.118	35.179	**35.270**
	SSIM	0.8730	0.8835	0.8830	**0.8832**	0.8680	0.8737	0.8735	**0.8756**
	SAM	4.922	4.398	4.467	**4.283**	2.905	2.925	2.858	**2.814**

Table 4 Performance gain without spectral angle constraint. The bold indicates the best results

		CAVE				Harvard			
Scale	Metrics	SSPSR	SFCSR	ERCSR	CoarSR	SSPSR	SFCSR	ERCSR	CoarSR
×4	PSNR	–	39.221	39.317	**39.774**	–	40.246	40.295	**40.541**
	SSIM	–	0.9330	0.9333	**0.9338**	–	0.9394	0.9396	**0.9422**
	SAM	–	3.205	3.209	**3.141**	–	2.381	**2.376**	2.385
×8	PSNR	34.453	35.341	35.479	**35.609**	34.928	35.417	35.574	**35.675**
	SSIM	0.8771	0.8846	0.8855	**0.8861**	0.8724	0.8774	0.8796	**0.8799**
	SAM	4.661	4.267	4.264	**4.145**	2.839	2.805	2.837	**2.786**

remove the SAM constraint and observe the changes in performance. Table 4 presents the results for scale factor ×4 on CAVE dataset. One can observe that some values remain unchanged when the calculated value for SAM is greater than 1. Consequently, the constraint does not have an effect through Eq. (16). In most cases, spectral angle constraint can improve spectral fidelity. In addition, the performance of other evaluation metrics is also improved to a certain extent. The results show that this method can effectively integrate the spatial information under constraint and make the spectral curves at different scales as consistent as possible.

3.5 Generalizability to Various Datasets

In this study, we evaluate the performance of our proposed DualSR by quantitative evaluation and qualitative comparison with several existing methods on CAVE and Harvard datasets. The comparisons include 3D-FCNN [40], EDSR [37], SSPSR [22], MCNet [31], SFCSR [52], and ERCSR [32].

3.5.1 Quantitative Evaluation

Tables 5 and 6 show the quantitative evaluation of different scale factors on two datasets by existing super-resolution methods. Our method shows excellent perfor-

Table 5 Quantitative evaluation on CAVE dataset. The bold and underline indicate the best and second best results

Scale	Metric	Bicubic	3D-FCNN	EDSR	SSPSR	MCNet	SFCSR	ERCSR	**CoarSR** (**Ours**)	**DualSR** (**Ours**)
$\times 4$	PSNR	35.755	37.626	38.587	38.366	39.026	39.192	39.224	39.665	**39.813**
	SSIM	0.9071	0.9195	0.9292	0.9272	0.9319	0.9321	0.9322	0.9328	**0.9339**
	SAM	3.944	3.360	3.804	3.484	3.292	3.221	3.243	3.152	**3.130**
$\times 8$	PSNR	31.805	32.956	31.554	34.151	35.320	35.294	35.102	35.475	**35.628**
	SSIM	0.8485	0.8600	0.8233	0.8718	0.8833	0.8828	0.8824	0.8828	**0.8862**
	SAM	5.291	4.384	10.670	4.831	4.423	4.378	4.445	**4.269**	4.4107

Table 6 Quantitative evaluation on Harvard dataset. The bold and underline indicate the best and second best results

Scale	Metric	Bicubic	3D-FCNN	EDSR	SSPSR	MCNet	SFCSR	ERCSR	**CoarSR** (**Ours**)	**DualSR** (**Ours**)
$\times 4$	PSNR	37.227	38.143	39.175	39.293	40.081	40.077	40.211	40.312	**40.540**
	SSIM	0.9122	0.9188	0.9324	0.9333	0.9367	0.9373	0.9374	0.9405	**0.9423**
	SAM	2.531	**2.363**	2.560	2.448	2.410	2.407	2.384	2.384	2.372
$\times 8$	PSNR	33.275	33.363	34.068	34.435	34.927	35.097	35.138	35.252	**35.628**
	SSIM	0.8518	0.8448	0.8598	0.8676	0.8732	0.8737	0.8733	0.8755	**0.8799**
	SAM	2.884	**2.726**	3.051	2.899	2.957	2.911	2.888	2.794	2.774

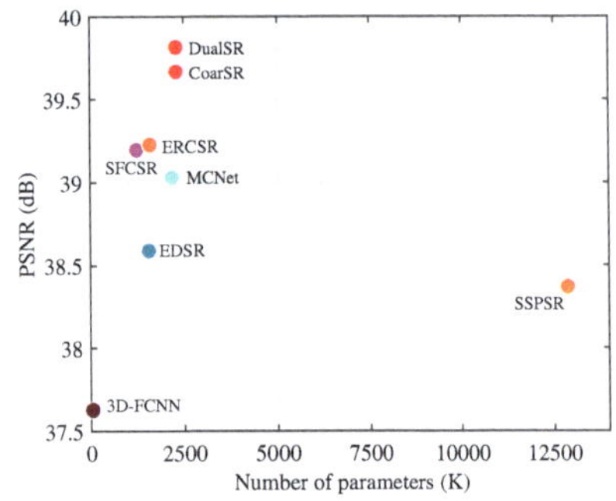

Fig. 8 PSNR performance versus number of parameters

mance with/without enhanced back-projection. Notably, among the approaches that solely employ 2D convolution, SSPSR achieves superior results across the datasets. This progressive upsampling approach helps to produce a clearer image. In contrast, the model using 3D convolution overall generally perform better performance than the model using 2D convolution, except for 3D-FCNN. The main reasons for 3D-FCNN are the shallow design and the lack of residual learning. With respect to recurrent networks such as SFCSR and CoarSR, they leverage the spectral architecture presented in multiple neighboring bands to capture complementary information. These networks prove beneficial in generating a single clear band. Although CoarSR has a higher number of parameters compared to SFCSR (as shown in Fig. 8), the results achieved by CoarSR significantly outperform SFCSR by +0.473 dB, 0.0007, and −0.069 for scale factor ×4. In our view, there are two main reasons for this improvement. Firstly, when spectral information is available, CoarSR incorporates additional representation learning in the spatial domain. Secondly, our method fully exploits the complementary information from multiple neighboring bands, effectively recovering missing details. The experiment reveals that our proposed DualSR method exhibits a noticeable increase in performance across all three metrics.

3.5.2 Qualitative Evaluation

We also conduct a qualitative evaluation. To show the visual results in the spatial domain, we provide two bands from the single super-resolved spectral image. Since the bands are grayscale, they may not effectively show the edge details. To clearly illustrate the differences between the super-resolved and high-resolution images, we give the absolute error map. Figures 9 and 10 present visual examples for scale factor ×4 on the two datasets. The results demonstrate that our method produces low absolute errors. Particularly, there are more distinct and shallow edges in certain regions. This shows that our proposed method can produce clearer edges and fine details, which is consistent with the analysis in Tables 5 and 6. It further emphasizes that our method can effectively learn spectral and spatial knowledge

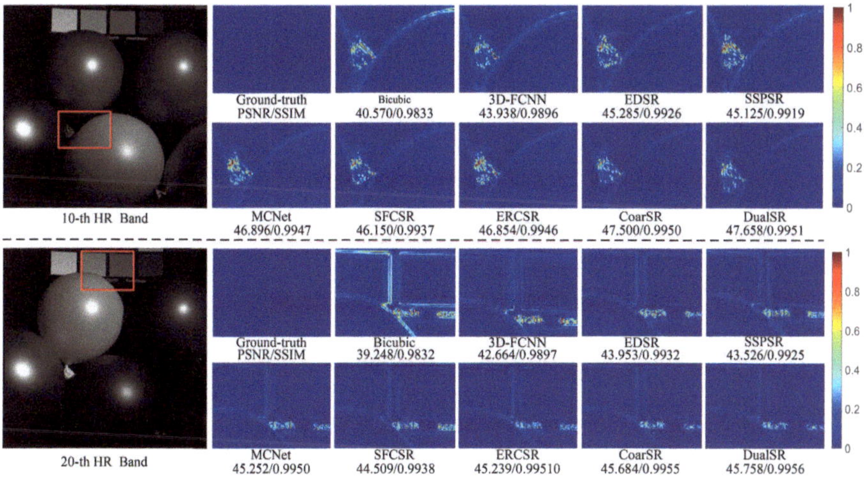

Fig. 9 Visual results in spatial domain on CAVE dataset. The first and second lines denote the results for 10-th and 20-th bands

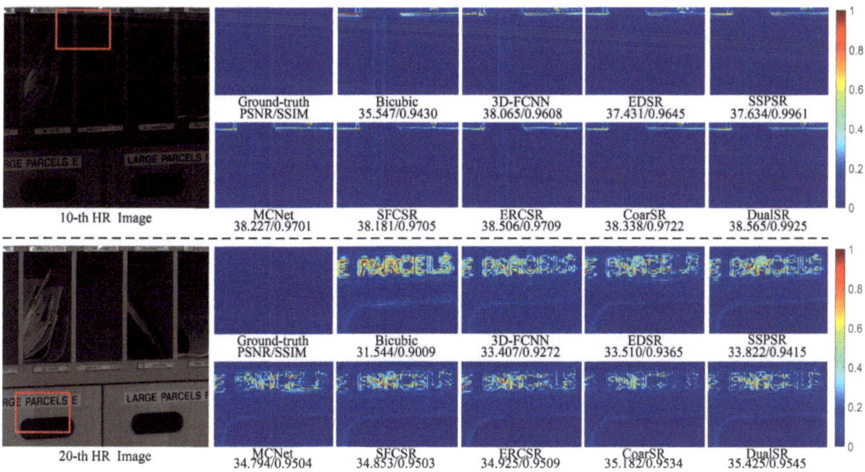

Fig. 10 Visual results in spatial domain on Harvard dataset. The first and second lines denote the results for 10-th and 20-th bands

while generating different textures. In addition to the spatial domain, we also present the visual results in the spectral domain. Specifically, we randomly select two pixels in the spatial domain and plot its spectral curve along the spectral dimension at the corresponding position. Figure 11 illustrates the spectral distortion of the super-resolved spectral image. We can observe that our DualSR maintains a similar spectral curve to the ground truth in most cases. This validates that our proposed method achieves higher spectral fidelity compared to other approaches.

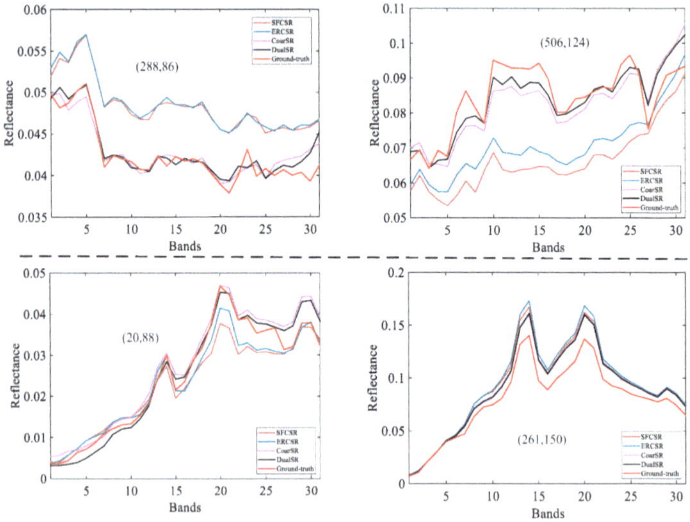

Fig. 11 Visual comparison in spectral domain. The first and second lines are the results corresponding to above images. Only two representative algorithms are compared with our method to avoid confusion

3.5.3 Generalizability to Various Blur Kernels

In this section, we present results on various blur kernels to demonstrate the generalizability of our method. To ensure a fair comparison, each image is downsampled using different downsampling kernels proposed in [41]. The low-resolution images are then input into model to obtain the corresponding restored images and the numerical results are obtained by calculating the average value. Here, the kernels used in our work include the *cubic*, *lanczos*, *box*, and *linear* to assess the robustness and effectiveness of these method across various scenarios.

Quantitative Evaluation Table 7 compares our performance with other methods on CAVE dataset. As can be seen, the performance of traditional super-resolution networks degrades very seriously due to the deviation between the fixed and unfixed kernel, which indicates that these methods cannot deal with other degradation kernels. On the contrary, our approach incorporates various kernels during the training phase. We observe that CoarSR can deal well with degraded images with unfixed kernel, and its performance exceeds that of SFCSR for scale factor $\times 4$ by a very large margin, i.e., +1.022 dB. From this point of view, CoarSR has better generalization than other methods. In other words, the proposed method is more robust in different scenarios.

Qualitative Evaluation We also adopt the same approach as described in Sect. 3.5.2 to present the visual results in both the spatial and spectral domains. To facilitate simple comparison, the results downsampled by linear kernel are

Table 7 Quantitative evaluation by various kernels on CAVE dataset. The bold and underline indicate the best and second best results

Scale	Metric	Bicubic	3D-FCNN	EDSR	SSPSR	MCNet	SFCSR	ERCSR	**CoarSR** (Ours)	**DualSR** (Ours)
×4	PSNR	35.652	37.263	38.204	37.932	38.317	38.410	38.390	<u>39.296</u>	**39.432**
	SSIM	0.9067	0.9175	0.9263	0.9253	0.9281	0.9281	0.9286	<u>0.9309</u>	**0.9312**
	SAM	3.871	<u>3.241</u>	3.601	3.578	3.410	3.390	3.403	3.242	**3.238**
×8	PSNR	31.735	32.813	34.250	33.920	34.601	34.635	34.503	<u>34.972</u>	**35.322**
	SSIM	0.8496	0.8584	0.8711	0.8699	0.8787	<u>0.8828</u>	0.8784	0.8802	**0.8842**
	SAM	5.134	<u>4.302</u>	4.855	4.893	4.638	4.893	4.727	4.462	**4.256**

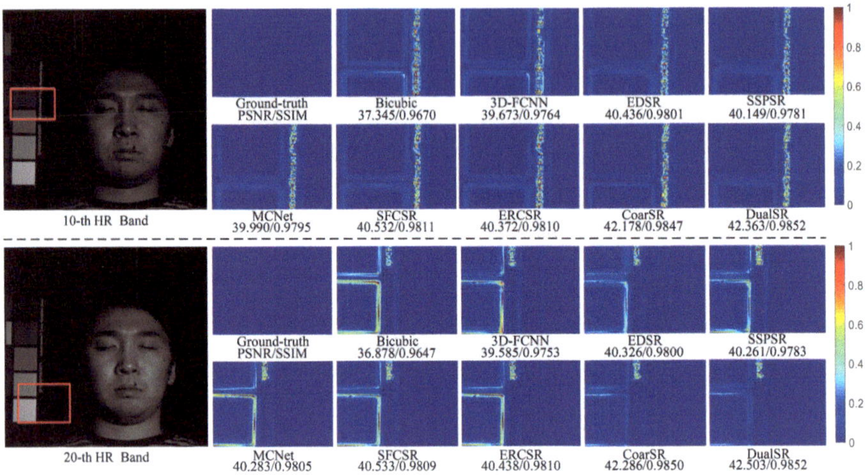

Fig. 12 Visual results downsampled by linear kernel in spatial domain. The first and second lines denote the results for 10-th and 20-th bands

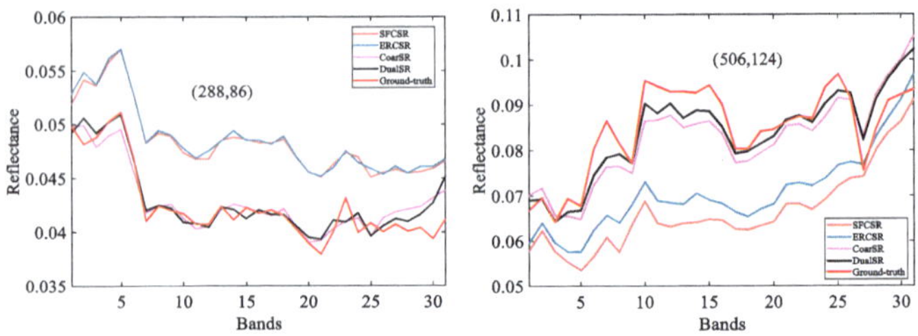

Fig. 13 Visual comparison in spectral domain. Only two representative algorithms are compared with our method to avoid confusion

exhibited, as shown in Figs. 12 and 13. As observed, our method still outperforms in two key aspects. Through the above analysis, it reveals the proposed model surpasses the networks over each dimension on various kernel conditions.

3.6 Conclusion

This section introduces a novel deep model for spectral image super-resolution that comprises two stages: coarse stage and fine stage. In the coarse stage, a band-by-band approach is employed to obtain a coarse super-resolved image. Unlike previous works, the proposed method incorporates additional adjacent bands to the model. It promotes the learning of potential information within the current band through grouping. Then, the intra-/inter-group fusion is utilized to extract complementary information from adjacent bands, resulting in enhanced details. Moreover, we design an enhanced back-projection strategy to ensure spectral-spatial consistency. This manner serves as a post-processing technique that significantly improves the performance of any spectral image super-resolution algorithm. Experimental results demonstrate that our method achieves state-of-the-art results.

4 Bidirectional Symmetric Network via Auxiliary Image for Spectral Image Super-Resolution

In practical optical imaging, a trade-off often exists between spatial resolution and spectral resolution. For instance, natural images tend to possess high spatial resolution but low spectral resolution, while spectral images exhibit high spectral resolution but low spatial resolution. In certain applications, it is necessary to acquire images with both high spatial and spectral resolution in order to achieve improved feature representation and learning. A common approach is to combine a spectral image with high spectral resolution and a natural image with high spatial resolution, so that the image has high resolution in both spatial and spectral domain.

In recent years, significant progress has made in super-resolution task by fusing low-resolution spectral images with high-resolution RGB images. This progress includes both traditional methods [5, 66] and deep learning methods [54, 65, 68]. Traditional methods typically rely on manual feature construction and incorporate prior knowledge from shallow images to design their models. However, the ability of shallow features to provide robust feature representation hinders the performance in challenging scenarios. In contrast, the methods based on deep learning show excellent performance. The recent studies [58, 70] take inspiration from algorithms developed for natural image super-resolution to solve this task. These approaches offer several advantages compared to traditional methods. They require less hypothetical prior knowledge. Although these methods obtain ideal results, there are still some obvious issues. Numerous techniques [12, 13, 17, 53, 67] initially combine or process the two images in parallel during the initial feature extraction stage. Indeed, there are distinct difference in edge textures between RGB images and spectral images, which presents challenges when integrating these two types early. It easily results in the performance degradation of the model. Ideally, the fusion of these attributes would facilitate a more comprehensive exploration of the spectral image information. Consequently, it is still an urgent problem to distinguish the difference between the two modalities and incorporate them into an effective super-resolution method.

Typically, low-resolution images suffer from various factors that degrade their quality, including noise, downsampling, blur, and more. Many existing methods [58, 65, 70] tackle this challenge by employing a single type of degradation in their dataset construction. When confronted with low-resolution images derived from real-world scenes, the performance of these methods faces a significant decline. This decline can be attributed to the fact that the actual degradation process of the actual low-resolution image is not consistent with the predetermined assumptions. To solve

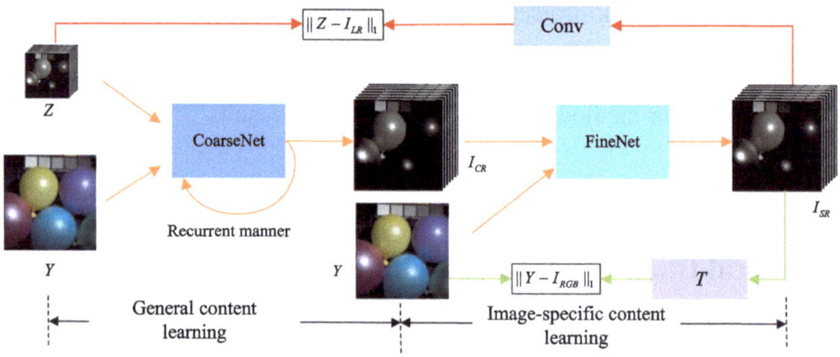

Fig. 14 The proposed architecture for spectral image super-resolution

these problems, this section proposes a bidirectional symmetric architecture network via auxiliary image for spectral image super-resolution, as depicted in Fig. 14. The architecture comprises two main components: CoarseNet and FineNet. CoarseNet considers the disparities between spectral image Z and RGB image Y and designs a bidirectional feature interaction model. This model effectively leverages the rich spectral inherent in spectral images and the high spatial resolution provided by RGB images. Through the bidirectional multistep propagation mechanism, the initial super-resolution results I_{CR} are recurrently generated. Given the fixed conversion relationship between the super-resolved result and RGB image, a self-supervised FineNet is appended to the end of CoarseNet. This FineNet learns global knowledge through cycle consistency loss, further enhancing the performance of the model to obtain super-resolved result I_{SR}. Moreover, to ensure the model effectively in dealing with real-world low-resolution images, this section randomly introduces two types of blur kernels, downsampling strategies, and Gaussian noise with diverse parameters to degrade high-resolution images and construct the dataset.

4.1 Symmetrical Feature Propagation Network via Auxiliary Image

The relationship between the low-resolution spectral image $Z \in \mathbb{R}^{w \times h \times L}$ and the high-resolution RGB image $Y \in \mathbb{R}^{W \times H \times 3}$ can be formulated as a super-resolution problem, where the goal is to estimate the high-resolution spectral image $X \in \mathbb{R}^{W \times H \times L}$ with higher spatial and spectral resolution. Among these images, the spectral response function $T \in \mathbb{R}^{L \times 3}$ is utilized to map the numerous spectra in X into three channels. The degradation function R is applied to degrade the spatial resolution from X to Z. The relationship is denoted as

$$Y = XT, \tag{21}$$

$$Z = RX. \tag{22}$$

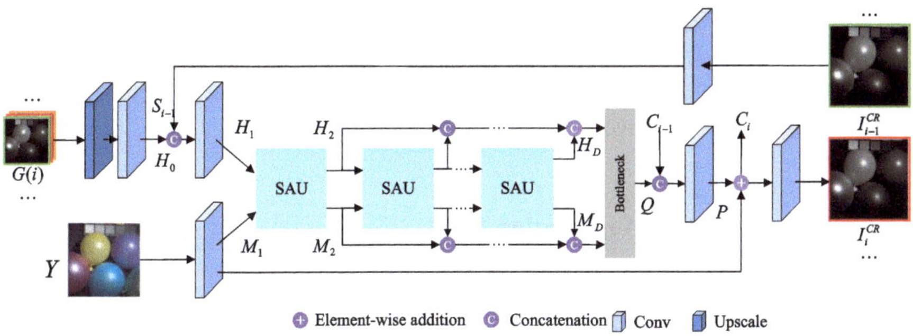

Fig. 15 The flowchart of proposed CoarseNet

Here, w and h correspond to the width and height of each band in the low-resolution spectral image, respectively. L represents the total number of bands. Similarly, W and H denote the width and height for RGB image. In this section, we describe a symmetrical feature propagation network via auxiliary image. The flowchart of our approach is depicted in Fig. 15. The architecture consists of three main components: feature extraction, separation-and-aggregation unit (SAU), and feature fusion reconstruction.

4.1.1 Feature Extraction

To restore the current band Z_i in the spectral image, we utilize three bands denoted as $G(i)$ from the spectral image Z and the corresponding RGB image Y. These three bands denoted as $G(i)$ are selected, i.e.,

$$G(i) = \begin{cases} [Z_1, Z_2, Z_3], & i = 1 \\ [Z_{i-1}, Z_i, Z_{i+1}], & 2 \leq i \leq L-1. \\ [Z_{L-2}, Z_{L-1}, Z_L], & i = L \end{cases} \tag{23}$$

To reduce the increase in parameters caused by multiple upsampling and downsampling operations, the bands $G(i)$ are first upsampled using Nearest interpolation to match the size of the RGB image Y. Subsequently, two branches are incorporated to analyze the respective features of $G(i)$ and Y using several convolutional layers. Given the high similarity between adjacent bands, the previously super-resolved band Z_{i-1} is introduced to assist in the reconstruction of the current band Z_i. Note that during the restoration of the first band Z_1, the feature context does not participate in the fusion with shallow features. As for the subsequent bands Z_i (where $i = 2, 3, \ldots, L$), the feature context is involved in the fusion process, which incorporates information from both shallow and previous super-resolved bands. The total process is defined as

$$H_1(i) = \begin{cases} H_0, & i = 1 \\ f_{3\times3}[I_{CR}^{i-1}, H_0], & 2 \le i \le L \end{cases}. \tag{24}$$

Here, $f_{3\times3}(\cdot)$ denotes convolution operation with kernel 3×3.

4.1.2 Separation-and-Aggregation Unit

Previous works often integrate the RGB image and low-resolution spectral image at the initial feature extraction. We believe that this approach cannot guarantee these two inherent forms throughout the super-resolution process. To address this, we introduce a separation-and-aggregation unit (SAU), as depicted in Fig. 16. It not only facilitates the feature fusion between the two images but also maintains the individuality particularity.

Separation Unit For simplicity, we focus on only one branch. In this case, we have shallow features denoted as H_1. These features are input into the SAU module, which is defined as

$$I_H = f_{3\times3}(ReLU(f_{3\times3}(H_1))), \tag{25}$$

where $ReLU(\cdot)$ is ReLU activation function. To leverage the complementary between the two branches, an effective approach is to combine their complementary features in a certain position based on their representation abilities. To accomplish this, we merge half of the feature maps from each branch $I_H, I_M \in \mathbb{R}^{C\times W\times H}$, resulting in new feature maps. i.e.,

$$I_H^c = \left[I_H^1, I_M^1\right], \tag{26}$$

$$I_M^c = \left[I_H^2, I_M^2\right], \tag{27}$$

where $I_H^1, I_H^2 \in \mathbb{R}^{(C/2)\times W\times H}$ and $I_M^1, I_M^2 \in \mathbb{R}^{(C/2)\times W\times H}$. C denotes the number of channels. This manner facilitates the bidirectional exchange of information between

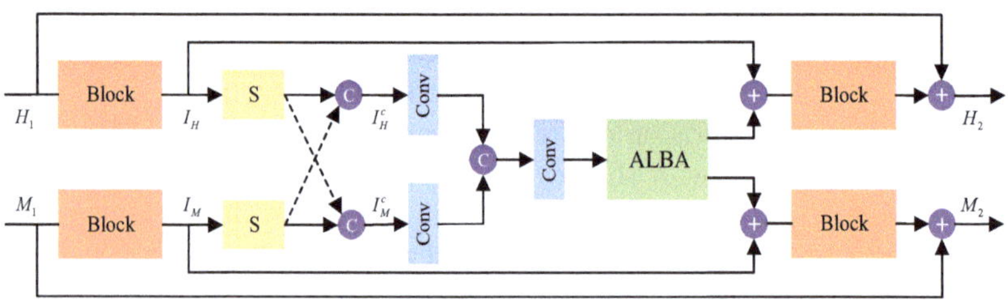

Fig. 16 Separation-and-aggregation unit (SAU)

data at the feature level. Then, each branch of the model heightens the integrated content through convolutions with kernel 3×3. The information from both branches is subsequently concatenated along the channel dimension.

Aggregation Unit The objects in the whole image sometimes occupy significant spatial coverage. Traditional small convolutions fail to adequately explore adjacent content. It hinders the ability of the model to make full use of similar information. While dilated convolutions can enhance feature learning within local context by increasing the receptive field, they cannot establish the corresponding relationships between similar content. To solve this problem, we propose an adaptive local block aggregation (ALBA) module. The main concept is to establish relationship between individual blocks and their local neighbors. The blocks with high similarity are aggregated to achieve effective context mining.

The relationship between blocks in an raw high-resolution RGB image can be accurately constructed without feature extraction, including position index I and similarity c. Specifically, the image Y is divided into small blocks with size $p \times p$. The similarity between the current block P_j and its adjacent blocks $P_{j,k}$ can be measured using cosine similarity. It is defined as

$$c_{j,k} = \frac{< P_j, P_{j,k} >}{\|P_j\|_2 \times \|P_{j,k}\|_2}, k = 1, 2, \ldots, 8, \tag{28}$$

Here, $< \cdot, \cdot >$ is dot product operation. $|| \cdot ||_2$ represents L2 norm to normalize the vectors involved in the computation. Since the current block exhibits similarity primarily with neighboring blocks, we sort the similarity values and select the top m blocks with the high similarity scores. These selected blocks contribute to generate the corresponding position index I. Figure 17 illustrates an example where three similar blocks are chosen. As for the blocks P_j located at the four corners of the image Y, only three adjacent blocks are considered in the calculation. Among other blocks, only five adjacent blocks are involved in the analysis.

To fuse the depth features, we utilize a block aggregation module. Similarly, the intermediate features $F_n \in \mathbb{R}^{W \times H}$ are divided into small blocks with size

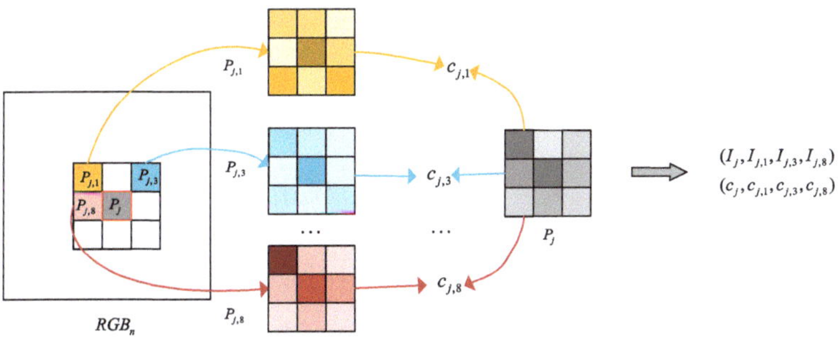

Fig. 17 Illustration of computing the position I and similarity c for block P_j

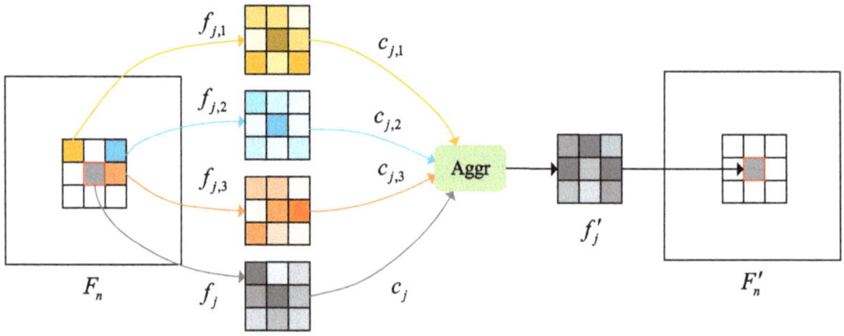

Fig. 18 Illustration of aggregating features among blocks in local

$p \times p$ using the same partitioning strategy. Here, n ranges from 1 to N. The blocks $f_{j,1}, f_{j,2}, \ldots, f_{j,m}$ that exhibit high similarity with the current block f_j are concatenated along the channel dimension. This concatenation is performed based on the constructed position index I and the corresponding similarity values c. As a result, a new block f_j' is formed through a convolutional layer with a 3×3 kernel. The overall process is represented as

$$f_j' = f_{3\times3}([f_j, f_{j,1} * c_{j,1}, \ldots, f_{j,m} * c_{j,m}]). \tag{29}$$

Figure 18 illustrates an example. After the fusion of information, the features from the two modalities are separated, and their respective branches are learned independently. This symmetrical architecture explicitly recalibrates the properties of each modality. The aggregated and separated results are then propagated to the subsequent unit in the encoder, forming a multistep propagation pattern. This mechanism encourages both streams to preserve their particular over a long range. It facilitates more accurate and effective coding.

4.1.3 Feature Fusion Reconstruction

Once the deep features are obtained at multiple levels, the features from the two modalities are gathered together at the bottleneck, i.e.,

$$U = f_{1\times1}([H_2, \ldots, H_D]), \tag{30}$$

$$V = f_{1\times1}([M_2, \ldots, M_D]), \tag{31}$$

$$Q = f_{3\times3}([U, V]), \tag{32}$$

where $f_{1\times1}(\cdot)$ represents convolution operation with kernel 1×1. We follow the study [52] and add the feature context. Concretely, we transfer the features C_{i-1}, which are obtained from the previous band super-resolution process, back into the network for the current band Z_i. This manner allows the network to leverage the contextual information from the previous band to enhance the performance of the current band. The process is denoted as

$$P = f_{1\times1}([Q, C_{i-1}]). \tag{33}$$

Finally, the model produces the estimated single band I_{CR}^i by applying a convolutional layer with kernel 3×3 followed by a residual connection.

Coarse Network Training To build real pairs, we generate five randomly kernels using both anisotropic and isotropic Gaussian. The kernel sizes include 7×7, 9×9, 13×13, 15×15, and 21×21. For anisotropic Gaussian kernels, the rotation angle is randomly selected from the range $[0, \pi]$, and the kernel width is chosen from the range $[0.2, s]$. For isotropic Gaussian kernels, the kernel width is selected from the range $[0.2, s]$. Using the obtained blur kernels, we employ two downsampling approaches to compute the downsampled spectral image. The one is that we directly convolve the image with the blur kernel. The other is that we first blur the image using the blur kernel and then downsample it using either bicubic or bilinear interpolation. Additionally, Gaussian noise is introduced to further degrade the downsampled spectral image. Its aim is to better represent the noise present in real scenes. According to the above procedures, we synthesize a large training data X, Z, where X represents the low-resolution spectral image and Z denotes the corresponding high-resolution spectral image. Subsequently, we train the coarse network by minimizing the loss function $\mathcal{L}_C$, which is expressed as

$$\mathcal{L}_C(\theta, i) = \min ||I_{CR}^i - X_i||_1, \tag{34}$$

$$I_{CR}^i = C(s, Y, G(i); \theta), i = 1, 2, ..., L, \tag{35}$$

where $I_{CR}^i \in \mathbb{R}^{W\times H}$ and $||\cdot||_1$ denotes L1 norm. To easily understand, we show the detailed training process in Algorithm 2.

4.2 *Refinement Network*

In the coarse stage, CoarseNet may not effectively capture image-specific details, especially for low-resolution spectral image with unknown degradation. To address this, the coarse super-resolved bands $I_{CR}^i \in \mathbb{R}^{W\times H}$, where $i = 1, 2, ..., L$, are combined to form $I_{CR} \in \mathbb{R}^{W\times H\times L}$. Then, a refinement network is built with the auxiliary RGB image Y and the initial results I_{CR}. This differs from the approach in [65], where only I_{CR} is used as the input. Specifically, a refinement network is embedded at the end of the coarse stage, as shown in Table 8. To reduce the potential mapping function, a cycle consistency loss is introduced using [69]. This

Algorithm 2: Coarse network training

Input: High-resolution dataset $\mathcal{D}$ and scale factor s
Output: Coarse model parameter θ_C
Initialize coarse model parameter θ;
while *not converged* **do**
  Sample image pair $\{X\}$ from dataset $\mathcal{D}$;
  Obtain images $\{X, Z\}$ by degradation, and generate high-resolution image Y by Eq. (21);
  Set $i = 1$;
  for $i \leq L$ **do**
    Compute $\mathcal{L}_C$ by Eq. (34);
    Update θ according to $\mathcal{L}_C$;
    $i \leftarrow i + 1$;
  end
end

Table 8 The architecture of refinement network

Type	Kernel size	Stride	Channel
Concat	–	–	34
Conv1	3×3	1	128
Conv2	3×3	1	128
Conv3	3×3	1	128
Conv4	3×3	1	128
Conv5	32×32	s	31

loss encourages the content between I and I^{SR} to be as consistent as possible, which can be expressed as

$$\mathcal{L}_{\mathcal{F}}(\sigma) = \min ||Y - I^{RGB}||_1 + \lambda||Z - I^{LR}||_1, \tag{36}$$

$$I^{RGB} = \mathcal{F}(s, Y, I_{CR}; \sigma)T, \tag{37}$$

$$I^{LR} = \mathcal{B}(s, I^{SR}; \sigma), \tag{38}$$

where λ denotes a balanced factor and it is fixed at 0.1. With respect to the test process, it is summarized in Algorithm 3.

4.3 Experiment

In this section, we analyze the effectiveness of the proposed method from the following aspects. First, we provide implementation details, as well as the comparison methods used for evaluation. Then, each module through model analysis is discussed. Finally, we compare our method with mainstream methods to evaluate its performance.

Algorithm 3: The proposed network test process

Input: High-resolution image I, coarse model parameter θ_C, scale factor s, and the number of iterations Q

Output: Super-resolved image I^{SR}

Generate RGB image Y using Eq. (21);
Downsample image I to obtain low-resolution image Z by different blur kernels, as display in Fig. 19. Then, add Gaussian noise;
Compute $I_{CR} = C(s, Y, G(i); \theta_c)$;
Initialize fine model parameter σ, and set $q = 1$;
for $q \leq Q$ **do**
    Calculate $\mathcal{L}_{\mathcal{F}}$ using Eq. (36);
    Update σ by $\mathcal{L}_{\mathcal{F}}$;
    $q \leftarrow q + 1$;
end
return $I^{SR}=\mathcal{F}(s, Y, I_{CR}; \sigma_{\mathcal{F}})$

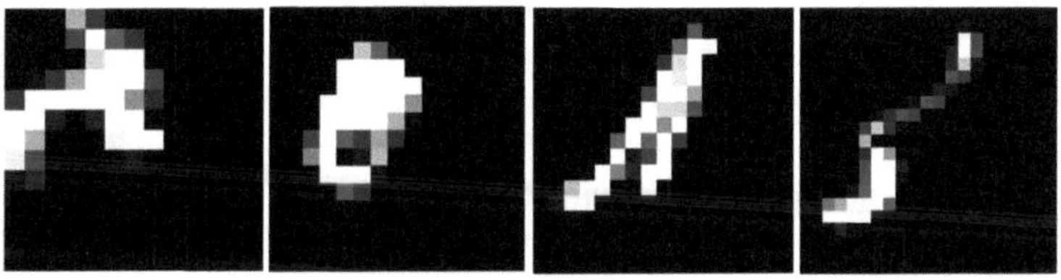

Fig. 19 Some visual results for different blur kernels

4.3.1 Implementation Details

With respect to the dataset, we randomly select 25 images from CAVE dataset for training, while the remaining images are reserved for testing. As for the Harvard dataset, we utilize only 50 images from real scenes for model learning. Among them, 40 images are allocated for training, while the remaining images are used for testing.

In the proposed model C, a 3×3 convolutional kernel is applied after the concatenation operation, with 64 kernels. The optimization of model C is performed using the ADAM optimizer, where the parameters are $\beta_1 = 0.9$ and $\beta_2 = 0.99$. The initial learning rate is set to 10^{-4}, and it is halved every 30 epoch. Similarly, the model $\mathcal{F}$ is optimized using the ADAM optimizer. The learning rate is set to 5^{-5} for the network $\mathcal{F}$.

In the experiment, we select five mainstream spectral super-resolution methods. These methods include LTTR [6], CMS [66], PZRes-Net [70], MHF-net [58], and UAL [65]. Based on their training modes, these methods can be categorized into supervised and unsupervised approaches. The unsupervised methods include LTTR and CMS, while the remaining methods are supervised.

Table 9 Ablation study for scale ×8 on CAVE dataset

Component	Different combinations of components			
ALBA module	×	√	×	√
Spectral context	×	×	√	√
PSNR	39.623	40.832	41.486	42.466
SSIM	0.9667	0.9713	0.9809	0.9855
SAM	6.267	5.667	4.981	4.412

4.4 Ablation Study

To enhance the feature representation of the proposed symmetrical feature propagation model, two key modules are introduced: adaptive local block aggregation (ALBA) module and spectral context module. In this section, the performance of the model is tested by selectively adding or removing these modules. In the experimental setup, a 13 × 13 blur kernel is employed to downsample the high-resolution spectral image through convolution, while Gaussian noise with zero mean and variance 0.001 is added to generate the low-resolution spectral image. Table 9 presents the performance of different module combinations. The ALBA module focuses on capturing local information, while the spectral context module considers global information. By combining these modules, the model enhances feature representation by learning similar content. The results demonstrate that this combined configuration outperforms other module combinations, reinforcing the effectiveness and superiority of the proposed modules.

4.5 Study of Adaptive Local Block Aggregation Module

To assess the effectiveness of individual components within the ALBA module, we conduct experiments for scale factor ×8 on CAVE dataset. The investigation focuses on three key aspects: block size, block number, and block fusion coefficient.

4.5.1 Block Size Analysis

To establish the block relation, the high-resolution RGB image is divided into non-overlapping blocks with equal size. In this study, we examine the impact of different block sizes on performance by considering four distinct sizes. The results of these experiments are summarized in Table 10. The results indicate that the choice of block size has a significant impact on the model performance. Notably, when the block size exceeds 4, a significant performance variation occurs. This phenomenon can be attributed to the inclusion of dissimilar content within larger blocks, which hampers effective aggregation learning. Conversely, smaller block sizes facilitate the exploration of more similar content, as evidenced by the data

Table 10 Performance of different block sizes

Block size	2 × 2	3 × 3	4 × 4	5 × 5
PSNR	42.490	42.466	42.873	42.023
SSIM	0.9854	0.9855	0.9831	0.9789
SAM	4.423	4.412	4.415	4.780

Table 11 Performance of different block numbers

Number of similar blocks	1	2	3
PSNR	41.796	42.381	42.466
SSIM	0.9803	0.9796	0.9855
SAM	4.795	4.707	4.412

presented in the table. Considering the trade-off between computational efficiency and performance, a block size with 3×3 is selected for subsequent experiments. This choice ensures an adequate capture of contextual information. Moreover, it minimizes the computational burden associated with position and similarity calculations.

4.5.2 Block Number Analysis

Within the SAU, the establishment between block relations involves the participation of adjacent blocks. To analyze the impact of the number of highly similar blocks selected for feature aggregation, this section explores three different numbers of adjacent blocks. The results are presented in Table 11. The table demonstrates that increasing the number of similar blocks involved in the establishment has a significant positive effect in the three metrics. However, it is important to consider the trade-off between performance enhancement and the increase in parameters. Taking this into account, we select three adjacent blocks for experiment.

4.5.3 Block Fusion Coefficient Analysis

When establishing the relationship between current block and its neighboring blocks, cosine similarity serves as the metric for calculating their similarity. Moreover, cosine similarity is utilized as the block fusion coefficient in the SAU to enable block fusion. To assess it, we conduct experiments where the block fusion coefficient is either fixed at 1 or allowed to be self-learning. Table 12 presents the results obtained with different block fusion coefficients. While the model produced satisfactory results when the fusion coefficient is self-learning or fixed at 1, it struggle to effectively handle new input low-resolution spectral images. In contrast, our proposed method dynamically adjusts the block fusion coefficient based on the cosine similarity and adapts to changes in the input image. This demonstrates the superiority of the cosine similarity coefficient in our approach.

Table 12 Performance of different block fusion coefficients

Coefficient	1	Self-learning	Cosine similarity
PSNR	42.383	42.457	42.466
SSIM	0.9849	0.9807	0.9855
SAM	4.307	4.564	4.412

Table 13 Performance comparison under blur kernels. The bold indicates the best results

Blur kernel	Metric	LTTR	CMS	PZRes-Net	MHF-net	UAL	CoarseNet
K_1	PSNR	31.309	31.905	34.213	36.000	40.549	**42.041**
	SSIM	0.6884	0.7632	0.8461	0.8719	0.9698	**0.9852**
	SAM	31.228	25.335	22.183	12.488	10.789	**4.461**
K_2	PSNR	31.331	31.766	34.161	35.956	40.534	**42.015**
	SSIM	0.6893	0.7609	0.8477	0.8716	0.9698	**0.9849**
	SAM	31.220	25.311	21.952	12.501	10.847	**4.485**
K_3	PSNR	31.346	32.133	34.292	36.082	40.572	**42.147**
	SSIM	0.6891	0.7649	0.8468	0.8724	0.9702	**0.9854**
	SAM	31.150	25.150	22.038	12.446	10.751	**4.437**
K_4	PSNR	31.035	31.857	34.203	35.975	40.414	**40.850**
	SSIM	0.6838	0.7600	0.8455	0.8698	0.9694	**0.9837**
	SAM	31.117	25.262	21.989	12.567	10.945	**4.731**

4.6 Generalizability to Various Degradations

To evaluate the generalization ability of the proposed method across diverse degradation conditions, we conduct experiments on CAVE dataset. These experiments involves different blur kernels, noise levels, and downsampling manner.

4.6.1 Study of Blur Kernels

In this section, we conduct experiments using four different blur kernels: 13×13, 15×15, 17×17, and 21×21. The corresponding visual results of these blur kernels are displayed in Fig. 19. For ease of description, we denote these kernels as K_1, K_2, K_3, and K_4 respectively. Following the strategy outlined in Sect. 4.4, low-resolution spectral images using these blur kernels are obtained. Table 13 displays the results yielded by various competitors under blur kernels. It is observed that none of the methods, including LTTR, CMS, PZRes-Net, and MHF-net, which employ predefined degenerate kernels, can produce satisfactory super-resolved results. Although UAL estimates the degradation kernel of the image, it struggles to accurately identify the differences between RGB image and low-resolution spectral image. Therefore, UAL produces relatively inferior results. In contrast, the proposed method adopts random blur kernels of various types and sizes. Additionally, it incorporates a symmetrical architecture that effectively exploits the content from both modalities. As a result, our approach outperforms all the competitors.

Table 14 Performance comparison under noise levels. The bold indicates the best results

Variance	Metric	LTTR	CMS	PZRes-Net	MHF-net	UAL	CoarseNet
0.0005	PSNR	34.123	34.207	36.870	38.102	42.515	**43.415**
	SSIM	0.7889	0.8403	0.8947	0.9136	0.9800	**0.9900**
	SAM	26.672	21.129	18.760	10.136	8.727	3.857
0.001	PSNR	31.309	31.905	34.213	36.000	40.549	**42.073**
	SSIM	0.6884	0.7632	0.8461	0.8719	0.9698	**0.9850**
	SAM	31.228	25.335	22.183	12.488	10.789	**4.491**
0.002	PSNR	28.651	29.572	31.565	33.838	38.540	**40.216**
	SSIM	0.5674	0.6600	0.7820	0.8122	0.9535	**0.9734**
	SAM	36.052	29.932	25.704	15.502	13.436	**5.485**

4.6.2 Study of Noise Levels

In this section, we investigate the performance of the comparison methods under different noise levels. Table 14 presents the comparison results of existing methods. Here, most methods assume simple degradation during the dataset construction, which makes them more sensitive to noisy images. As evident from the table, the performance of these methods is significantly affected by the various noise levels. Although UAL attempts to solve this issue through post-processing, its performance remains subpar. In contrast, the proposed method add random noise levels in the images, which attains the superior performance.

4.6.3 Study of Downsampling Approaches

This section analyzes the performance of existing methods under different downsampling strategies commonly used for high-resolution image, including convolution and interpolation. Specifically, we apply convolution, bicubic interpolation, and bilinear interpolation to downsample the high-resolution images. The results are summarized in Table 15. The table reveals that traditional methods exhibit unstable performance, while deep learning-based models demonstrate relatively more consistent results compared to the former. However, some deep learning models still show significant fluctuations across the evaluation metrics, which indicates limited robustness. As for our proposed method, it incorporates multiple downsampling ways and outperforms all other algorithms by a significant margin, achieving superior performance in terms of the evaluated metrics.

4.6.4 Generalizability to Different Datasets and Scales

To assess the generalizability of the proposed model, we compare it with five existing methods on three different datasets. In our experiments, we utilize a blur

Table 15 Performance comparison under downsampling approaches. The bold indicates the best results

Downsampling type	Metric	LTTR	CMS	PZRes-Net	MHF-net	UAL	CoarseNet
Convolution	PSNR	31.309	31.905	34.213	36.000	40.549	**42.018**
	SSIM	0.6884	0.7632	0.8461	0.8719	0.9698	**0.9852**
	SAM	31.228	25.335	22.183	12.488	10.789	**4.445**
Bicubic	PSNR	30.930	32.570	34.276	36.170	40.620	**42.466**
	SSIM	0.6840	0.7732	0.8514	0.8730	0.9700	**0.9855**
	SAM	31.261	24.888	21.607	12.474	10.799	**4.412**
Bilinear	PSNR	30.977	32.603	34.319	36.192	40.633	**42.191**
	SSIM	0.6856	0.7743	0.8517	0.8738	0.9703	**0.9854**
	SAM	31.202	24.838	21.612	12.439	10.772	**4.375**

kernel with size 13×13 to blur the high-resolution spectral image, followed by downsampling using bicubic interpolation. Then, Gaussian noise with mean 0 and variance 0.001 is added to generate the low-resolution image.

Table 16 presents the evaluation results on three datasets. Our proposed CoarseNet outperforms the other approaches in most cases. It delivers remarkable performance for small-scale factors and comparable performance for large-scale factors. After adding FineNet, it effectively alleviates this disadvantage. For example, compared to UAL with scale factor $\times 8$ on CAVE dataset, CoarseNet achieves the 1.8 dB improvement in PSNR. Interestingly, this improvement is further enhanced by at least 2.8 dB after incorporating FineNet to learn global information. Importantly, our proposed approach strikes a favorable balance between performance and model size, as illustrated in Fig. 20. Both the quantitative results and figure indicate that the proposed approach outperforms in terms of performance and parameter efficiency. Similarly, our method exhibits excellent performance over other competitors on Harvard dataset.

We also evaluate the models on Chikusei dataset. Note that this dataset lacks a corresponding spectral response function. Consequently, we exclude LTTR and CMS from the evaluation since they heavily rely on the spectral response function during super-resolution. Additionally, UAL utilizes an unsupervised adaptive network to optimize the model by minimizing the differences between the RGB image transformed by the spectral response function, the RGB image, the downsampled spectral image, and the input low-resolution spectral image. To ensure accurate results, we exclude FineNet during this test. Similar to the results obtained on the other two datasets, it is evident from the table that CoarseNet achieves desirable performance for small-scale factors, while it achieves comparable performance for large-scale factors.

Figures 21 and 22 provides some visual examples. To emphasize spatial details, a specific band is selected from the super-resolved spectral image, and the absolute difference between this band and the corresponding reference band is computed.

Table 16 Performance comparison of existing approaches on various datasets. The bold and underline indicates the best and second best results

Dataset	Scale factor	Metric	LTTR	CMS	PZRes-Net	MHF-net	UAL	**CoarseNet**	**CoarseNet +FineNet**
CAVE	×8	PSNR	30.930	32.570	34.276	36.170	40.620	42.466	**43.498**
		SSIM	0.6840	0.7732	0.8514	0.8730	0.9700	0.9855	**0.9863**
		SAM	31.261	24.888	21.607	12.474	10.799	4.412	**4.237**
	×32	PSNR	36.954	37.967	34.193	35.112	39.891	38.050	**40.134**
		SSIM	0.9204	0.9528	0.8656	0.9208	**0.9735**	0.9732	0.9700
		SAM	16.806	11.145	21.700	8.701	10.385	5.278	**5.240**
Harvard	×8	PSNR	30.032	32.492	33.404	34.212	39.235	41.896	**42.405**
		SSIM	0.6994	0.7917	0.8558	0.8470	0.9602	0.9737	**0.9748**
		SAM	23.542	18.864	17.133	13.869	7.787	**3.714**	3.758
	×32	PSNR	34.818	38.774	33.086	36.495	39.818	39.648	**40.684**
		SSIM	0.8952	0.9500	0.8806	0.9345	**0.9707**	0.9670	0.9695
		SAM	14.850	6.520	16.991	6.482	6.088	**4.135**	4.245
Chikusei	×8	PSNR	–	–	34.650	32.363	36.982	**38.005**	–
		SSIM	–	–	0.8784	0.8363	0.8966	**0.9551**	–
		SAM	–	–	13.228	13.282	**3.048**	8.452	–
	×32	PSNR	–	–	**36.428**	26.476	34.005	35.785	–
		SSIM	–	–	0.8954	0.8441	0.8411	**0.9403**	–
		SAM	–	–	12.675	12.587	**5.163**	11.174	–

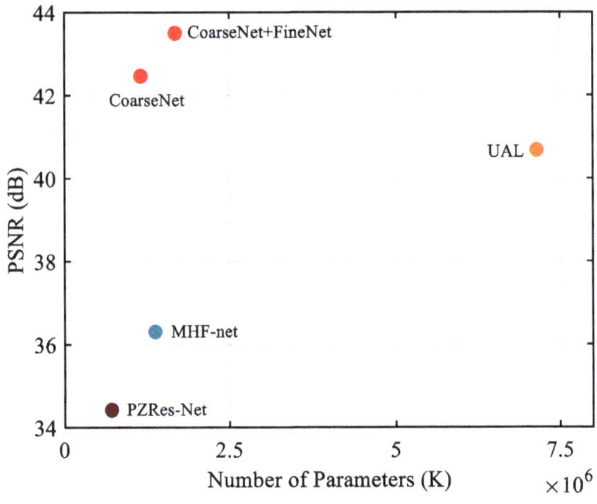

Fig. 20 PSNR performance versus number of parameter for scale factor ×32

As seen in this figure, our method effectively captures details within the magnified region. It indicates that the reconstructed image exhibits enhanced clarity and finer details. With respect to spectral fidelity, we randomly select a pixel in the spatial domain for each image in Fig. 21 and plot the corresponding spectral curve along the spectral dimension. Through the curve distribution, we observe that our method consistently aligns with the reference curve in most cases. This result demonstrates the robust generalization of our proposed method across diverse datasets and scale factors. It preserves spectral information and reconstructs the inherent spectral characteristics of the scenes.

4.7 Conclusion

This section develops a spectral image super-resolution architecture using bidirectional feature propagation. The proposed method focuses on promoting information propagation and fusion between low-resolution spectral images and RGB images through a multistep propagation strategy. This architecture is designed in a symmetric pattern, which enables both forms to retain their distinctive characteristics effectively. Additionally, an adaptive local block aggregation module is introduced to selectively aggregate blocks with high similarity. Furthermore, a refinement network is embedded to globally learn image-specific details. Experimental results demonstrate that the proposed approach outperforms existing methods in both quantitative and qualitative evaluations.

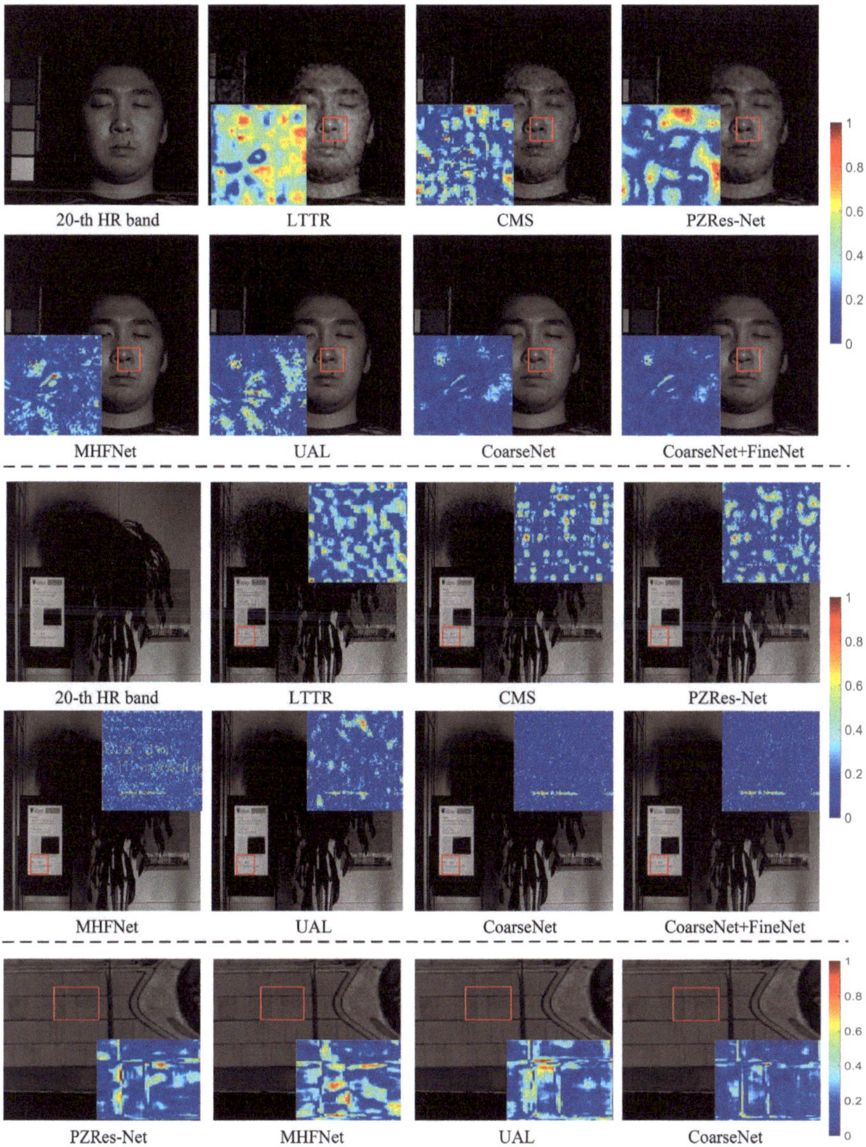

Fig. 21 Visual comparison in spatial domain. The first to third lines are the results corresponding to CAVE, Harvard, and Chikusei datasets. The first to third lines represent the visual results of the 20th band, 20th band, and 80th band, respectively

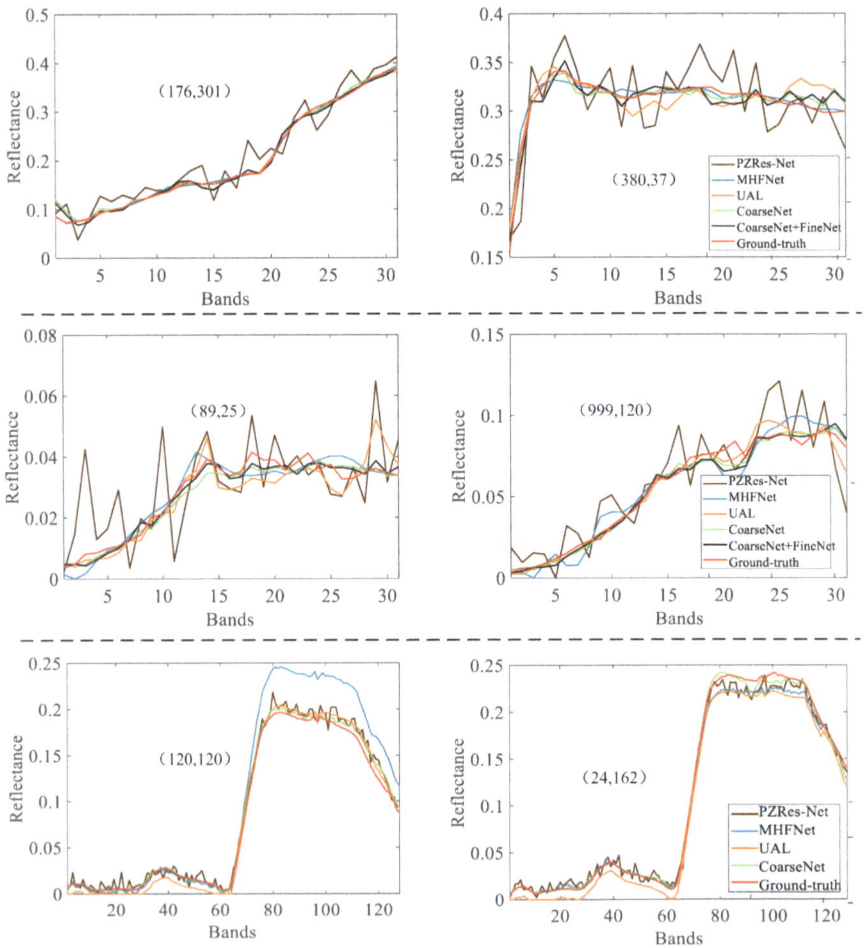

Fig. 22 Visual comparison in spectral domain. The first to third lines are the results corresponding to above images

5 Summary and Outlook

5.1 Summary

Spectral imaging often suffers from image degradation caused by factors such as atmospheric disturbance and limitations of satellite platforms. This degradation hampers the perception and analysis of spectral images, which makes it crucial to improve image quality through algorithms within existing hardware limitations. This chapter aims to improve the spatial resolution of spectral images while preserving spectral information and proposes two spectral image super-resolution methods. These methods exploit the internal relationships and essential attributes of the images to enhance both local and global details. The main research contents of this chapter include:

- A spectral image super-resolution framework via two stage learning is proposed. In the coarse stage, the framework selects the current band and its four most similar bands, forming a new groups. Within these groups, an adjacent spectral fusion strategy is employed to guide the model in capturing missing details during single band reconstruction. On this basis, a back-projection method via spectral angle constraint is introduced in the fine stage. This method enhances spatial and spectral consistency between multi-scale images and optimizes the results obtained in the coarse stage. Experimental results demonstrate that the proposed framework achieves excellent performance and obtains a low number of parameters.
- A blind spectral image super-resolution method via bidirectional feature propagation is proposed. The method explores the contents of spectral image and RGB image through a symmetrical architecture. A bidirectional multistep propagation manner is adopted to encourage information exchange while preserving the particularity of each modality. Additionally, a refinement network is embedded at the end of the architecture to explore spatial content across all bands. Experimental results show the effectiveness of the proposed method in handling real low-resolution images.

5.2 Outlook

Although the spectral image super-resolution technology has made great progress, the existing methods are still limited in some aspects. To further explore these issues, the future research work is conducted from the following aspects:

- Rich training data. The development of supervised methods for spectral image super-resolution models is hindered due to a scarcity of training images. To overcome this limitation, there are two potential approaches. One can focus on building a larger dataset that contains multiple scenes. This expanded dataset would provide more diverse and abundant data, which ultimately enhances the training process. Alternatively, transfer learning can be employed by utilizing rich natural images to build model.
- General super-resolution model construction. When training the model, the existing super-resolution methods train the model separately according to the dataset, so that the trained model only has the ideal result on the dataset, while the performance of the model is degraded on other datasets. Therefore, a key issue is to design a general super-resolution framework that enables the model to perform effectively across various datasets.
- Lightweight network design. To address the trouble posed by existing methods that employ 3D convolutions, which contribute to increased model parameters

and slower runtime, there is a need to design lightweight networks. The objective is to reduce the computational requirements of the model. It becomes possible to strike a balance between model performance and computational efficiency.

Acknowledgments This work was supported in part by the National Natural Science Foundation of China under Grant 62301385 and in part by the Key Research and Development Program of Shaanxi (Program No.2024GX-YBXM-130).

References

1. Akhtar, N., Shafait, F., Mian, A.: Sparse spatio-spectral representation for hyperspectral image super-resolution. In: Proceedings of the IEEE Conference on Computer Vision and Pattern Recognition, pp. 63–78 (2014)
2. Akhtar, N., Shafait, F., Mian, S.: Bayesian sparse representation for hyperspectral image super resolution. In: Proceedings of the IEEE Conference on Computer Vision and Pattern Recognition, pp. 3631–3640 (2015)
3. Bu, L., Dai, D., Zhang, Z., et al.: Hyperspectral super-resolution reconstruction network based on hybrid convolution and spectral symmetry preservation. Remote Sens. **15**(13), 3225 (2023)
4. Chakrabarti, A., Zickler, T.: Statistics of real-world hyperspectral images. In: Proceedings of the IEEE Conference on Computer Vision and Pattern Recognition, pp. 193–200 (2011)
5. Dian, R., Li, S.: Hyperspectral image super-resolution via subspace-based low tensor multi-rank regularization. IEEE Trans. Image Process. **28**(10), 5135–5146 (2019)
6. Dian, R., Li, S., Fang, F., et al.: Learning a low tensor-train rank representation for hyper-spectral image super-resolution. IEEE Trans. Neural Networks Learn. Syst. **30**(9), 2672–2683 (2019)
7. Dong, W., Fu, F., Shi, S., et al.: Hyperspectral image super-resolution via non-negative structured sparse representation. IEEE Trans. Image Process. **25**(5), 2337–2352 (2016)
8. Dong, W., Zhou, C., Wu, J., et al.: Model-guided deep hyperspectral image super-resolution. IEEE Trans. Image Process. **30**, 5754–5768 (2021)
9. Duchon, C.E.: Lanczos filtering in one and two dimensions. J. Appl. Meteorol. Climatol. **18**(8), 1016–1022 (1979)
10. Eismann, M.T., Hardie, R.C.: Application of the stochastic mixing model to hyperspectral resolution enhancement. IEEE Trans. Geosci. Remote Sens. **42**(9), 1924–1933 (2004)
11. Eismann, M.T., Hardie, R.C.: Hyperspectral resolution enhancement using high-resolution multispectral imagery with arbitrary response functions. IEEE Trans. Geosci. Remote Sens. **43**(3), 455–465 (2005)
12. Han, X.H., Shi, B., Zheng, Y., et al.: SSF-CNN: spatial and spectral fusion with CNN for hyperspectral image super-resolution. In: Proceedings of the IEEE International Conference on Image Processing, vol. 31, pp. 2506–2510 (2018)
13. Han, X.H., Zeng, Y., Chen, Y.W., et al.: Multi-level and multi-scale spatial and spectral fusion cnn for hyperspectral image super-resolution. In: Proceedings of the IEEE Conference on Computer Vision and Pattern Recognition Workshops, pp. 4330–4339 (2019)
14. Hardie, R.C., Eismann, M.T., Wilson, G.L.: Map estimation for hyperspectral image resolution enhancement using an auxiliary sensor. IEEE Trans. Image Process. **13**(9), 1174–1184 (2004)
15. Haris, M., Shakhnarovich, G., Ukita, N.: Deep back-projection networks for super-resolution. In: Proceedings of the IEEE Conference on Computer Vision and Pattern Recognition Workshops, pp. 1664–1673 (2018)
16. Haris, M., Shakhnarovich, G., Ukita, N.: Recurrent back-projection network for video super-resolution. In: Proceedings of the IEEE Conference on Computer Vision and Pattern Recognition Workshops, pp. 3897–3906 (2019)

17. Hu, J.F., Huang, T.Z., Deng, L.J., et al.: Hyperspectral image super-resolution via deep spatiospectral attention convolutional neural networks. IEEE Trans. Neural Networks Learn. Syst. **33**(12), 7251–7265 (2021)
18. Irani, M., Peleg, S.: Improving resolution by image registration. CVGIP: Graphical Models and Image Processing **53**(3), 231–239 (1991)
19. Irani, M., Peleg, S., Ukita, N.: Improving resolution by image registration. CVGIP: Graphical Models and Image Processing **53**(3), 231–239 (1991)
20. Jia, J., Ji, L., Zhao, Y., et al.: Hyperspectral image super-resolution with spectral–spatial network. Int. J. Remote Sens. **39**(21–22), 7806–7829 (2018)
21. Jia, W., Zhao, Y., Wang, R., et al.: Are recent sisr techniques suitable for industrial applications at low magnification? IEEE Trans. Ind. Electron. **66**(12), 9828–9836 (2019)
22. Jiang, J., Sun, H., Liu, X., et al.: Learning spatial-spectral prior for super-resolution of hyperspectral imagery. IEEE Trans. Comput. Imaging **6**, 1082–1096 (2020)
23. Jiang, R., Li, X., Gao, A.: Learning spectral and spatial features based on generative adversarial network for hyperspectral image super-resolution. In: Proceedings of the IEEE International Conference on Acoustics, Speech, and Signal Processing, pp. 3161–3164 (2019)
24. Kim, S., Bose, K.N., Valenzuela, H.M.: Recursive reconstruction of high resolution image from noisy undersampled multiframes. IEEE Trans. Acoust. Speech Signal Process. **38**(6), 1013–1027 (1990)
25. Li, J., Cui, R., Li, B., et al.: Dual 1D-2D spatial-spectral cnn for hyperspectral image super-resolution. In: Proceedings of the IEEE International Geoscience and Remote Sensing Symposium, pp. 3113–3116 (2019)
26. Li, J., Cui, R., Li, B., et al.: Hyperspectral image super-resolution with 1d–2d attentional convolutional neural network. IEEE Trans. Geosci. Remote Sens. **11**(23), 2859 (2019)
27. Li, J., Cui, R., Li, Y.: Multitemporal hyperspectral image super-resolution through 3D generative adversarial network. In: Proceedings of the IEEE International Workshop on the Analysis of Multitemporal Remote Sensing Images, pp. 1–4 (2019)
28. Li, J., Yuan, Q., Shen, H., et al.: Hyperspectral image super-resolution by spectral mixture analysis and spatial-spectral group sparsity. IEEE Geosci. Remote Sens. Lett. **13**(9), 1250–1254 (2016)
29. Li, Q., Gong, M., Yuan, Y., et al.: Symmetrical feature propagation network for hyperspectral image super-resolution. IEEE Trans. Geosci. Remote Sens. **60**, 1–12 (2022)
30. Li, Q., Gong, M., Yuan, Y., et al.: RGB-induced feature modulation network for hyperspectral image super-resolution. IEEE Trans. Geosci. Remote Sens. **61**, 1–11 (2023)
31. Li, Q., Wang, Q., Li, X.: Mixed 2D/3D convolutional network for hyperspectral image super-resolution. Remote Sens. **12**(10), 1660 (2020)
32. Li, Q., Wang, Q., Li, X.: Exploring the relationship between 2D/3D convolution for hyperspectral image super-resolution. IEEE Trans. Geosci. Remote Sens. **59**(10), 8693–8703 (2021)
33. Li, Q., Yuan, Y., Jia, X., et al.: Dual-stage approach toward hyperspectral image super-resolution. IEEE Trans. Image Process. **31**, 7252–7263 (2022)
34. Li, Q., Yuan, Y., Wang, Q.: Multiscale factor joint learning hyperspectral image super-resolution. IEEE Trans. Geosci. Remote Sens. **61**, 1–10 (2023)
35. Li, R., Hu, J., Zhao, X., et al.: Hyperspectral image super-resolution using deep convolutional neural network. Neurocomputing **266**, 29–41 (2017)
36. Li, Y., Zhang, L., Ding, C., et al.: Single hyperspectral image super-resolution with grouped deep recursive residual network. In: Proceedings of the IEEE International Conference on Multimedia Big Data, pp. 1–4 (2018)
37. Lim, B., Son, S., Kim, H., et al.: Enhanced deep residual networks for single image super-resolution. In: Proceedings of the IEEE Conference on Computer Vision and Pattern Recognition, pp. 1132–1140 (2017)
38. Liu, Y., Hu, J., Kang, X., et al.: Interactformer: interactive transformer and CNN for hyperspectral image super-resolution. IEEE Trans. Geosci. Remote Sens. **60**, 1–15 (2022)
39. Ma, J., Wang, X., Jiang, J., et al.: Image super-resolution via dense discriminative network. IEEE Trans. Ind. Electron. **67**(7), 5687–5695 (2020)

40. Mei, S., Yuan, X., Ji, J., et al.: Hyperspectral image spatial super-resolution via 3D full convolutional neural network. Remote Sens. **9**(11), 1139 (2017)
41. Michaeli, T., Irani, M.: Nonparametric blind super-resolution. In: Proceedings of the IEEE Conference on Computer Vision and Pattern Recognition, pp. 945–952 (2013)
42. Qu, Y., Qi, H., Kwan, C.: Unsupervised sparse dirichlet-net for hyperspectral image super-resolution. In: Proceedings of the IEEE Conference on Computer Vision and Pattern Recognition, pp. 2511–2520 (2018)
43. Ren, H., Chang, C.I.: Automatic spectral target recognition in hyperspectral imagery. IEEE Trans. Aerosp. Electron. Syst. **39**(4), 1232–1249 (2003)
44. Robert, K.: Cubic convolution interpolation for digital image processing. IEEE Trans. Acoust. Speech Signal Process. **29**(6), 1153–1160 (1981)
45. Shi, W., Caballero, J., Huszár, F., et al.: Real-time single image and video super-resolution using an efficient sub-pixel convolutional neural network. In: Proceedings of the IEEE Conference on Computer Vision and Pattern Recognition, pp. 1874–1883 (2016)
46. Sun, Y., Liu, B., Yu, X., et al.: Perceiving spectral variation: unsupervised spectrum motion feature learning for hyperspectral image classification. IEEE Trans. Geosci. Remote Sens. **60**, 1–17 (2022)
47. Tai, Y., Yang, J., Liu, X.: Image super-resolution via deep recursive residual network. In: Proceedings of the IEEE Conference on Computer Vision and Pattern Recognition, pp. 3147–3155 (2017)
48. Takeda, H., Farsiu, S., Milanfar, P.: Kernel regression for image processing and reconstruction. IEEE Trans. Image Process. **16**(2), 349–366 (2007)
49. Wang, Q., Li, Q., Li, X.: Hyperspectral band selection via adaptive subspace partition strategy. IEEE J. Sel. Top. Appl. Earth Obs. Remote Sens. **12**(12), 4940–4950 (2019)
50. Wang, Q., Li, Q., Li, X.: A fast neighborhood grouping method for hyperspectral band selection. IEEE Trans. Geosci. Remote Sens. **59**(6), 5028–5039 (2020)
51. Wang, Q., Li, Q., Li, X.: Spatial-spectral residual network for hyperspectral image super-resolution. arXiv (2020). ArXiv:2001.04609
52. Wang, Q., Li, Q., Li, X.: Hyperspectral image superresolution using spectrum and feature context. IEEE Trans. Ind. Electron. **68**(11), 11276–11285 (2021)
53. Wang, W., Zeng, W., Huang, Y., et al.: Deep blind hyperspectral image fusion. In: Proceedings of the IEEE Conference on Computer Vision and Pattern Recognition, pp. 4150–4159 (2019)
54. Wang, X., Chen, J., Wei, Q., et al.: Hyperspectral image super-resolution via deep prior regularization with parameter estimation. IEEE Trans. Circuits Syst. Video Technol. **32**(4), 1708–1723 (2021)
55. Wang, Z., Chen, B., Lu, R., et al.: Fusionnet: an unsupervised convolutional variational network for hyperspectral and multispectral image fusion. IEEE Trans. Image Process. **29**, 7565–7577 (2020)
56. Wei, Q., Bioucas-Dias, J., Dobigeon, N., et al.: Hyperspectral and multispectral image fusion based on a sparse representation. IEEE Trans. Geosci. Remote Sens. **53**(7), 3658–3668 (2015)
57. Xiao, Y., Yuan, Q., Jiang, K., et al.: EDiffSR: an efficient diffusion probabilistic model for remote sensing image super-resolution. In: IEEE Transactions on Geoscience and Remote Sensing (2023)
58. Xie, Q., Zhou, M., Zhao, Q., et al.: Multispectral and hyperspectral image fusion by MS/HS fusion net. In: Proceedings of the IEEE Conference on Computer Vision and Pattern Recognition, pp. 1585–1594 (2019)
59. Yang, J., Wright, J., Huang, T.S., et al.: Super-resolution via sparse representation. IEEE Trans. Image Process. **19**(11), 1423–1438 (2010)
60. Yang, J., Zhao, Y., Chan, J.C., et al.: A multi-scale wavelet 3D-CNN for hyperspectral image super-resolution. Remote Sens. **11**(13), 1557 (2019)
61. Yang, M., Jiao, L., Liu, F., et al.: Coarse-to-fine contrastive self-supervised feature learning for land-cover classification in sar images with limited labeled data. IEEE Trans. Image Process. **31**, 6502–6516 (2022)

62. Yang, M., Jiao, L., Liu, F., et al.: DPFL-Nets: deep pyramid feature learning networks for multiscale change detection. IEEE Trans. Neural Networks Learn. Syst. **33**(11), 6402–6416 (2022)
63. Yasuma, F., Mitsunaga, T., Iso, D., et al.: Generalized assorted pixel camera: postcapture control of resolution, dynamic range, and spectrum. IEEE Trans. Image Process. **19**(9), 325–329 (2010)
64. Yuan, Y., Zheng, X., Lu, X.: Hyperspectral image superresolution by transfer learning. IEEE J. Sel. Top. Appl. Earth Obs. Remote Sens. **10**(5), 1963–1974 (2017)
65. Zhang, L., Nie, J., Wei, W., et al.: Unsupervised adaptation learning for hyperspectral imagery super-resolution. In: Proceedings of the IEEE Conference on Computer Vision and Pattern Recognition, pp. 3073–3082 (2020)
66. Zhang, L., Wei, W., Bai, C., et al.: Exploiting clustering manifold structure for hyperspectral imagery super-resolution. IEEE Trans. Image Process. **27**(12), 5969–5982 (2018)
67. Zhang, X., Huang, W., Wang, Q., et al.: Ssr-net: spatial-spectral reconstruction network for hyperspectral and multispectral image fusion. IEEE Trans. Geosci. Remote Sens. **59**(7), 5953–5965 (2021)
68. Zheng, Y., Li, J., Guo, J., et al.: Edge-conditioned feature transform network for hyperspectral and multispectral image fusion. IEEE Trans. Geosci. Remote Sens. **60**, 5953–5965 (2021)
69. Zhu, J.Y., Park, T., Isola, P., et al.: Unpaired image-to-image translation using cycle-consistent adversarial networks. In: Proceedings of the IEEE Conference on Computer Vision and Pattern Recognition, pp. 2242–2251 (2017)
70. Zhu, Z., Hou, J., Chen, J., et al.: Hyperspectral image super-resolution via deep progressive zero-centric residual learning. IEEE Trans. Image Process. **30**, 1423–1438 (2021)

Correction to: A Comprehensive Overview of Satellite Image Fusion: From Classical Model-Based to Cutting-Edge Deep Learning Approaches

Ivan Pereira-Sánchez, Eloi Sans, Julia Navarro, and Joan Duran

Correction to:
Chapter 8 in: M. Kawulok et al. (eds.), *Super-Resolution for Remote Sensing*, Unsupervised and Semi-Supervised Learning, https://doi.org/10.1007/978-3-031-68106-6_8

The original version of chapter **A Comprehensive Overview of Satellite Image Fusion: From Classical Model-Based to Cutting-Edge Deep Learning Approaches** was inadvertently published with incomplete Competing Interests text as shown below. The chapter has been updated.

Competing Interests This study was funded by MCIN/AEI/10.13039/50110001 1033/ and by the European Union NextGenerationEU/PRTR via MaLiSat project TED2021-132644B-I00.

And has now been corrected as follows:

Competing Interests This study was funded by MCIN/AEI/10.13039/50110001 1033/ and by the European Union NextGenerationEU/PRTR via the MaLiSat project TED2021-132644B-I00; and also by MCIN/AEI/10.13039/501100011033 and by "ERDF A way of making Europe", European Union, via the MoMaLiP project Grant PID2021-125711OB-I00.

The updated version of this chapter can be found at https://doi.org/10.1007/978-3-031-68106-6_8

M. Kawulok et al. (eds.), *Super-Resolution for Remote Sensing*, Unsupervised and Semi-Supervised Learning, https://doi.org/10.1007/978-3-031-68106-6_10

Index

M. Kawulok et al. (eds.), *Super-Resolution for Remote Sensing*, Unsupervised and Semi-Supervised Learning, https://doi.org/10.1007/978-3-031-68106-6

MIX
Papier aus verantwortungsvollen Quellen
Paper from responsible sources
FSC® C105338

If you have any concerns about our products,
you can contact us on
ProductSafety@springernature.com

In case Publisher is established outside the EU,
the EU authorized representative is:
Springer Nature Customer Service Center GmbH
Europaplatz 3, 69115 Heidelberg, Germany

Printed by Libri Plureos GmbH
in Hamburg, Germany